Advanced Electrical and Electronic Systems

Advanced Electrical and Electronic Systems

Editor: Norman Schultz

NY RESEARCH
P R E S S

New York

Published by NY Research Press
118-35 Queens Blvd., Suite 400,
Forest Hills, NY 11375, USA
www.nyresearchpress.com

Advanced Electrical and Electronic Systems
Edited by Norman Schultz

International Standard Book Number: 978-1-63238-625-0 (Hardback)

Cataloging-in-Publication Data

Advanced electrical and electronic systems / edited by Norman Schultz.
    p. cm.
Includes bibliographical references and index.
ISBN 978-1-63238-625-0
1. Electrical engineering. 2. Electronic systems. 3. Electronics. I. Schultz, Norman.
TK145 .A38 2019
621.3--dc23

Contents

Preface

This book was inspired by the evolution of our times; to answer the curiosity of inquisitive minds. Many developments have occurred across the globe in the recent past which has transformed the progress in the field.

Electrical and Electronic Systems fall under the branch of engineering that deals with the design and integration of active and passive electrical components to create innovative power systems. The field has applications in photonics, renewable energy generation, automation, robotics, telecommunications, electric and hybrid vehicle technologies among others. This book strives to present the latest theoretical and empirical studies to broaden the understanding of electrical and electronic systems. The topics covered in this extensive book deal with the core aspects of this area of study. Engineers, physicists and students actively engaged in this field will find this book full of crucial and unexplored concepts.

This book was developed from a mere concept to drafts to chapters and finally compiled together as a complete text to benefit the readers across all nations. To ensure the quality of the content we instilled two significant steps in our procedure. The first was to appoint an editorial team that would verify the data and statistics provided in the book and also select the most appropriate and valuable contributions from the plentiful contributions we received from authors worldwide. The next step was to appoint an expert of the topic as the Editor-in-Chief, who would head the project and finally make the necessary amendments and modifications to make the text reader-friendly. I was then commissioned to examine all the material to present the topics in the most comprehensible and productive format.

I would like to take this opportunity to thank all the contributing authors who were supportive enough to contribute their time and knowledge to this project. I also wish to convey my regards to my family who have been extremely supportive during the entire project.

Editor

Soft Computing Techniques Applications and their Comparisons with Traditional pq Theory Based Control Schemes for Filter in Aircraft System

Saifullah Khalid*

Department of Electrical Engineering, IET Lucknow, India

Abstract

Constant Instantaneous Power Control Technique for extracting reference currents for shunt APF (active power filters) have been modified using Artificial Neural Network and Fuzzy logic control and their performances have been compared. The acute analysis of different comparisons of the compensation capability supported total harmonic distortion and speed is going to be done, and suggestions are going to be given for the selection of technique to be used. The simulated results using MATLAB model are shown, and that they can without doubt prove the importance of the projected control technique of aircraft shunt APF.

Keywords: Aircraft electrical system; Shunt active power filter; Constant instantaneous power control strategy; ANN; THD

Introduction

Non-linear loads cause the unbalancing, harmonics, distortion etc into the power arrangement and these unwanted problems turn out profusely of problems within the system. Whenever application of such loads can increase, source gets distorted and unbalanced. These currents foul the supply purpose of the utility. Therefore, it's necessary to compensate unbalance, a harmonic and reactive element of the load currents. Whereas once source is unbalanced and distorted, these problems worsen the system [1-3].

Today, the soft computing techniques like Fuzzy algorithms, ATS algorithms, Genetic Algorithm [4-12], particle swarm optimization [13], ANN [14-18] applied in every machinery and filter devices for optimization of the system applied or in the various control system.

In this paper, ANN based mostly and fuzzy logic (FLC) controller are wont to mend the whole performance of active filter for reduction of harmonics and other related drawbacks generated into the balanced, unbalanced and distorted system because of the nonlinear loads [1]. The simulation results clearly show their effectiveness. The simulation results non-heritable with the new model are abundant improved than those of traditional methodology.

The paper has been described during a successive manner. The APF define and also the load underneath contemplation is mentioned in Section II. The control algorithmic rule for APF converses in Section III. MATLAB/Simulink based mostly simulation results are conferred in Section IV, and eventually Section V concludes the paper.

System Depiction

The aircraft power system is a three-phase power system with the source frequency of 400 Hertz. As depicted in Figure 1, Shunt Active Power Filter gets better the power quality and balances the harmonic currents in the source supply system [19-23]. The APF is realized by using VSI connected at the PCC to a common Direct Current (DC) link voltage [24-27].

One three-phase rectifier in parallel with an inductive load and an unbalanced load connected in a phase with the midpoint, one three phase rectifier connects a pure resistance directly, and one three phase rectifier connects a pure resistance directly. These loads have been considered as load 1, 2 and 3 respectively. All three loads are connected

to both of the supply at such interval or together such that ability of APF can be evaluated and it has been tested for 15 cycles. The circuit parameters are given in Appendix.

Control Theory

The proposed control of APF depends on constant instantaneous power control (CIPC) strategy, and it has been optimized for artificial Neural Network and Fuzzy Logic Control.Constant instantaneousPower Control strategy has been discussed in brief in this section. The following section also deals with the primary application of ANN and FLC in the control schemes [24,28,29].

Constant Instantaneous Power Control Strategy (C.I.P.C.)

The control illustration of APF using CIPC approach is exposed in Figure 2. Four low pass filters have been out in the open in the control block. Three are with cut off frequency of 6.4 Kilo Hertz. They are useful to filter the voltages. Even as the other one is used for the power p_0. The phase voltages cannot be unswervingly used in the control block. The

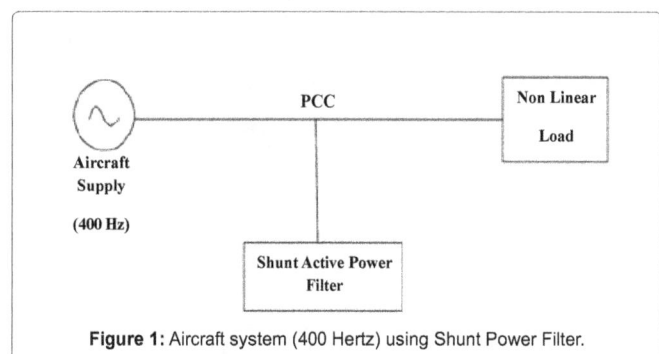

Figure 1: Aircraft system (400 Hertz) using Shunt Power Filter.

---

*Corresponding author: Saifullah Khalid, Department of Electrical Engineering, IET Lucknow, India, E-mail: saifullahkhalid@Outlook.com

grounds are the instability matter. The voltage harmonics beyond the 6.4 kilo hertz in the array of resonance frequency are blocked using the low pass filters.

P, q, p_0, v_α and v_β are completed following the computation from α-β-0 conversion and send to the α-β current reference block. α-β Current reference block computes $I'_{c\alpha}$ and $i'c_\beta$. To finish, α-β-0 inverse conversion block computes the current references. Thereafter, it is applied to the Pulse Width Modulation (PWM) current control.

Application of ANN based control

In this paper, CIPC approach has been represented by an ANN fabricated of two hidden layers among 10 neurons every one and one output layer with three neurons. The logarithmic commencement function is the support of the two hidden layers neurons, and linear commencement function intended for the output layer neurons.

As shown in Figure 3, the ANN has seven inputs (v_a, v_b, v_c, dc voltage error, i_a, i_b, i_c) and three outputs (i_{ra}, i_{rb}, i_{rc}) as observed in the different strategies. The reference current generation unit and dc voltage controller unit has been modeled and their individual and simultaneous effect has been observed.

Application of fuzzy logic control

The fuzzy logic control has been employed in the dc voltage control round of the APF. In fuzzy, the plan employs centrifugal defuzzification

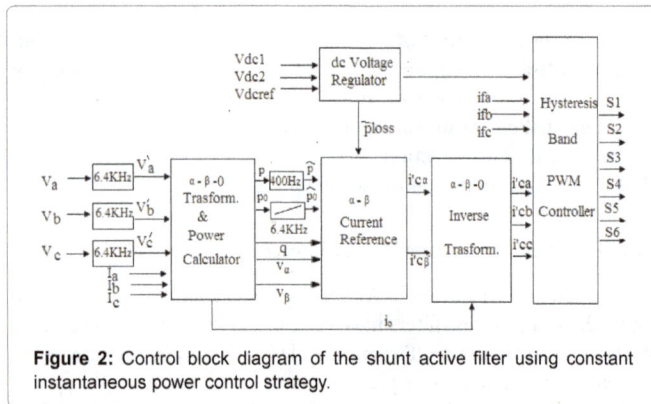

Figure 2: Control block diagram of the shunt active filter using constant instantaneous power control strategy.

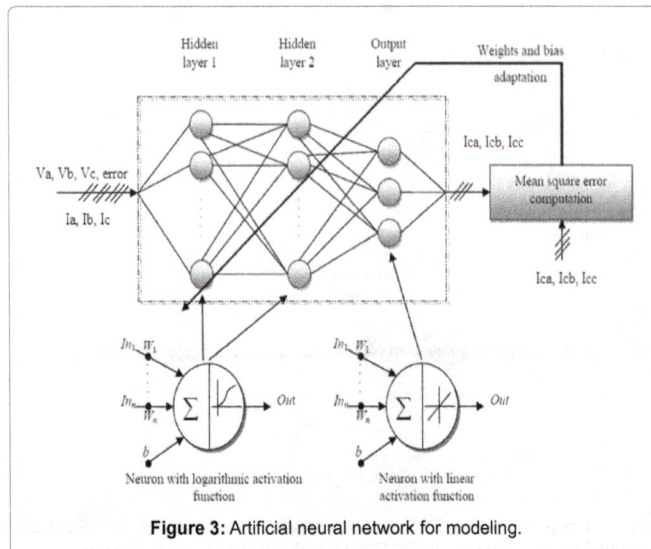

Figure 3: Artificial neural network for modeling.

manner. There are 2 inputs; error and its derivative and 1 output, that is the command signal. The two inputs uses Gaussian membership functions while the output uses triangle membership function. Table 1 explains the fuzzy control rule and Figure 4 illustrates the membership functions used.

Simulation Results and Discussions

The proposed scheme of APF is simulated in MATLAB environment to estimate its performance. Three loads have been applied together at a different time interval to check the affectivity of the control schemes for the reduction of harmonics. A small amount of inductance is also connected to the terminals of the load to get the most effective compensation. The simulation results clearly reveal that the scheme can successfully reduce the significant amount of THD in source current and voltage within limits.

Uncompensated system

When all three loads are connected as per the configuration discussed in the previous section, Total harmonic distortion (THD) of source (supply) current has been observed as nine point five percent and THD of source (supply) Voltage were one point five five percent. In this duration, the APF is not connected. The results are shown in Figure 5. By observant this information, we are going to merely understand that they are not within the limit of the international standard.

Compensated System

The performance of APF under different loads connected, when utilizing ANN Control has been discussed below for the control strategy given below.

For constant instantaneous power control strategy

From Figure 6 it has been empiric that that the THDs of source current and source voltage were 3.01% and 1.88%respectively. The compensation time was 0.0147 sec. At t=0.0147 sec, it is apparent that the waveforms for source voltage and source current have become sinusoidal. Figure 6 shows the waveforms of compensation current, DC capacitor voltage, and load current.

de/dt \ Error	Negative	Zero	Positive
Negative	Big Negative	Positive	Big Positive
Zero	Big Negative	Zero	Big Positive
Positive	Big Negative	Negative	Big Positive

Table 1: Fuzzy logic rule.

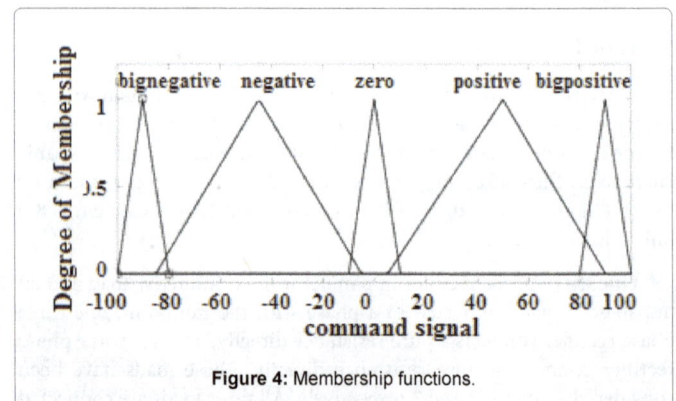

Figure 4: Membership functions.

The aberration in dc voltage can be acutely apparent in the waveforms. As per claim for accretion the compensation current for accomplishing the load current demand, it releases the energy, and after that it accuses and tries to achieve its set value. If we carefully observe, we can acquisition out that the compensation current is, in fact, accomplishing the appeal of load current, and afterward the active filtering the source current and voltage is affected to be sinusoidal.

For constant instantaneous power control strategy using ANN

THDs of source current and source voltage have been found 2.84% and 1.78% respectively after making observations from the simulation results shown in Figure 7. The waveforms for source voltage and source current have become sinusoidal at t=0.0066 sec. Compensation time

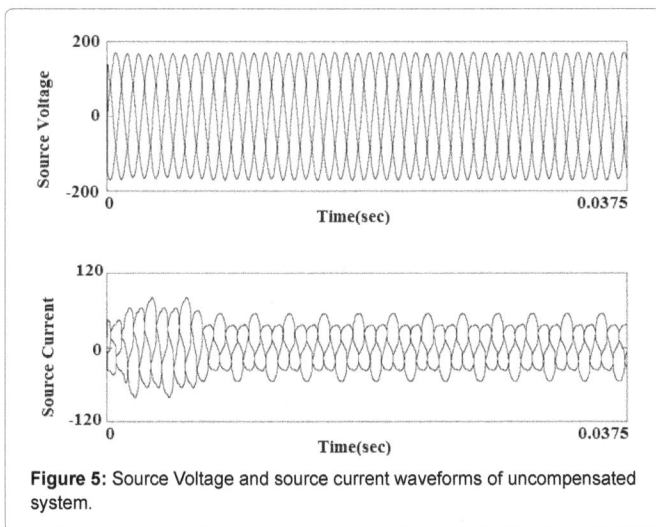

Figure 5: Source Voltage and source current waveforms of uncompensated system.

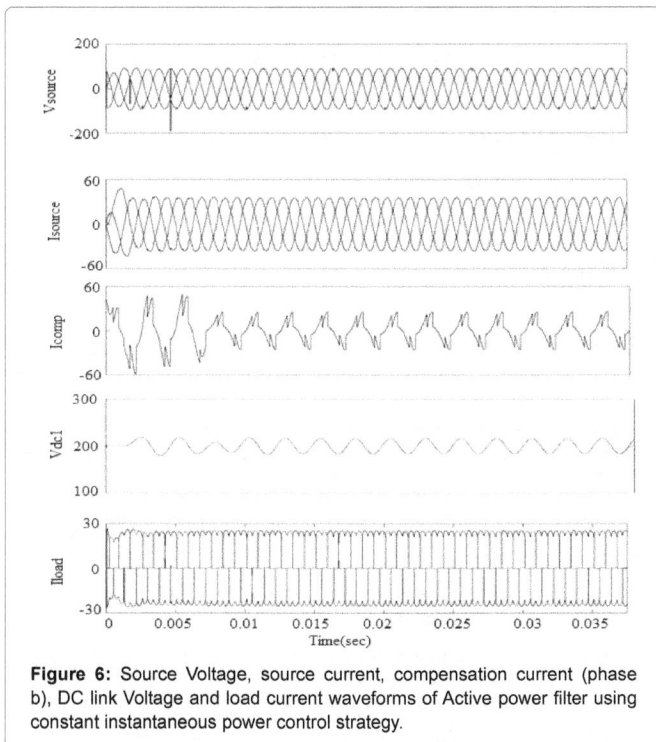

Figure 6: Source Voltage, source current, compensation current (phase b), DC link Voltage and load current waveforms of Active power filter using constant instanteous power control strategy.

is 0.0066 sec. The waveforms of compensation current, dc capacitor Voltage, and load current have been shown in Figures 6 and 7. Waveforms show the variations in dc capacitor voltage. Whenever the demand for high load current comes, it releases the energy that in turn increases the compensation current. Later on, it charges and tries to regain its previous set value. By making a simple observation, we can say that compensation current is fulfilling the demand of load current. After the active filtering, the source current and voltage is forced to be sinusoidal.

For constant instantaneous power control strategy using fuzzy logic control

THDs of source current and source voltage have been found 2.33% and 1.03% respectively after making observations from the simulation results shown in Figure 8. The waveforms for source voltage and source current have become sinusoidal at t=0.0044 sec. Compensation time is 0.0044 sec. The waveforms of compensation current, dc capacitor Voltage, and load current have been shown in Figure 8. Waveforms show the variations in dc capacitor voltage. Whenever the demand for high load current comes, it releases the energy that in turn increases the compensation current. Later on, it charges and tries to regain its previous set value. By making a simple observation, we can say that compensation current is fulfilling the demand of load current. After the active filtering, the source current and voltage is forced to be sinusoidal.

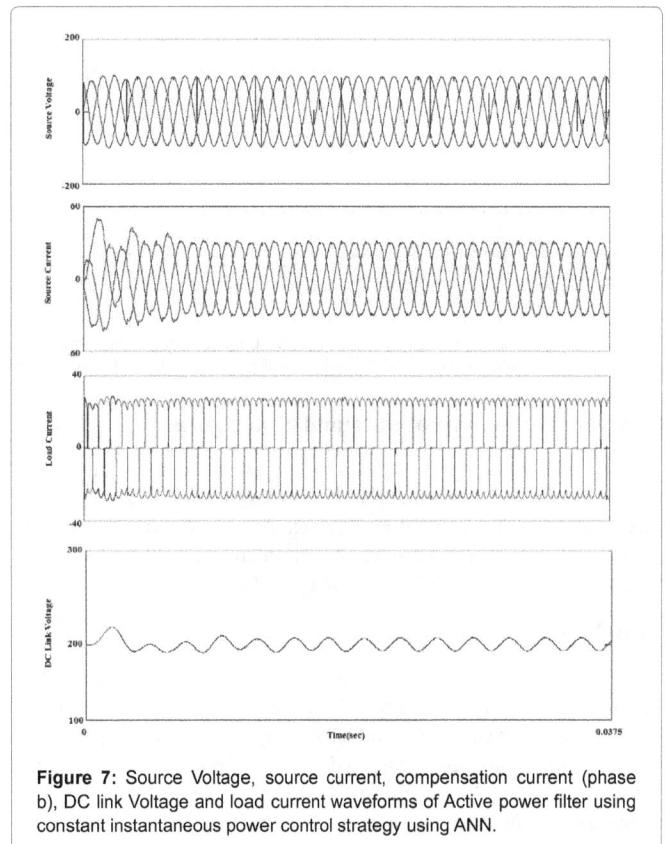

Figure 7: Source Voltage, source current, compensation current (phase b), DC link Voltage and load current waveforms of Active power filter using constant instantaneous power control strategy using ANN.

Strategy	THD-I (%)	THD-V (%)	Compensation Time (sec)
CIPC	3.01	1.88	0.0147
CIPC-ANN	2.84	1.78	0.0066
CIPC-FLC	2.33	1.03	0.0044

Table 2: Summary of simulation results.

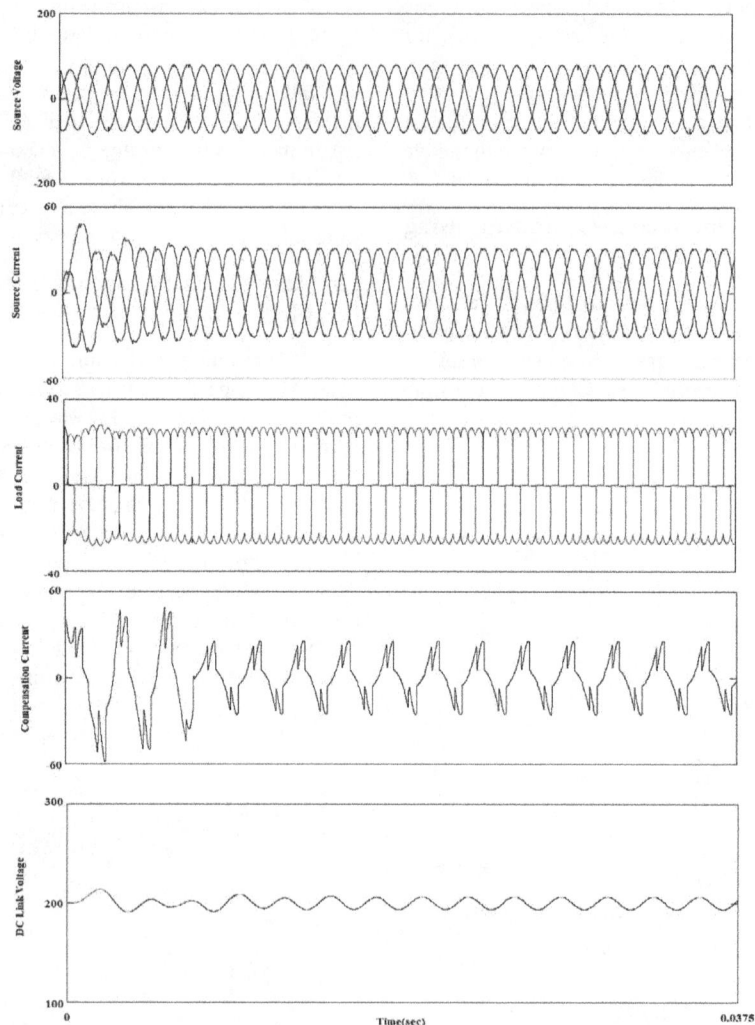

Figure 8: Source Voltage, Source Current, Load Current, Compensation Current (Phase b) and DC Link Voltage Waveforms of Active Power Filter using Constant Instantaneous Power Control Strategy using Fuzzy Logic Control with All Three Loads Connected for Aircraft System.

Comparative analysis of the simulation results

From the Table 2, we can easily say that Constant Instantaneous Power Control Strategy using Fuzzy Logic Controller (CIPC-FLC) has been found best for current and voltage harmonic reduction. When these results have been compared based on compensation time, it has been also found that CIPC-FLC strategy is also fastest one.

Conclusion

This paper has done an acute analysis of traditional, ANN and Fuzzy Logic control for shunt APF in aircraft power utility of 400 HZ. Optimum selection of control strategy based on compensation time and THD has been suggested. Overall Constant Instantaneous Power Control Strategy using FLC (CIPC-FLC) has been observed as an optimum choice. Constant Instantaneous Power Control Strategy's performance has been improved, which itself an achievement for the case of optimization in traditional strategies.

Parameters

The aircraft system parameters are [1]:

Three-phase source voltage: 115V/400 Hz.

Filter capacitor: 5 µF,

Filter inductor=0.25 mH.

Dc capacitor: 4700 µF.

Dc voltage reference: 400 V.

References

1. Donghua C, Guo T, Xie S, Zhou B (2005) Shunt Active Power Filters Applied in the Aircraft Power Utility. 36th Power Electronics Specialists Conference, PESC 5: 59-63.

2. Saifullah K, Bharti D (2014) Comparative Evaluation of Various Control Strategies for Shunt Active Power Filters in Aircraft Power Utility of 400 Hz. Majlesi Journal of Mechatronic Systems 3: 1-5.

3. Saifullah K, Bharti D (2013) Application of AI techniques in implementing Shunt APF in Aircraft Supply System. Proceeding of SPRINGER- SOCROPROS Conference, IIT-Roorkee 1: 333-341.

4. Guillermin P (1996) Fuzzy logic Applied to Motor Control. IEEE Transactions on Industrial Application 32: 51-56.

5. Hew Wooi AHAP, Hamzah A, Mowed HAF (2002) Fuzzy Logic Control of a three phase Induction Motor using Field Oriented Control Method. Society of Instrument and Control Engineers, SICE Annual Conference 264-267.

6. Jain SK, Agrawal P, Gupta H (2002) Fuzzy logic controlled shunt active power

filter for power quality improvement. IEE Proceedings of the Electric Power Applications 149: 317-328.

7. Norman M, Samsul B, Mohd N, Jasronita J, Omar SB (2004) A Fuzzy logic Controller for an Indirect vector Controlled Three Phase Induction Motor. Proceedings Analog And Digital Techniques In Electrical Engineering, TENCON 2004, Chiang Mai, Thailand 4: 1-4.

8. Afonso JL, Fonseca J, Martins JS, Couto CA (1997) Fuzzy Logic Techniques Applied to the Control of a Three-Phase Induction Motor. Proceedings of the UK Mechatronics Forum International Conference, Portugal. pp. 142-146.

9. Chiewchitboon P, Tipsuwanpom P, Soonthomphisaj N, Piyarat W (2003) Speed Control of Three-phase Induction Motor Online Tuning by Genetic Algorithm. Fifth International Conference on Power Electronics and Drive Systems, PEDS 1: 184-188.

10. Kumar P, Mahajan A (2009) Soft Computing Techniques for the Control of an Active Power Filter. IEEE Transactions on Power Delivery 24: 452-461.

11. Ismail KB, Abdeldjebar H, Abdelkrim B, Mazari B, Rahli M (2008) Optimal Fuzzy Self-Tuning of PI Controller Using Genetic Algorithm for Induction Motor Speed Control. Int. J. of Automation Technology 2: 85-95.

12. Guicheng W, Min Z, Xu X, Changhong J (2006) Optimization of Controller Parameters based on the Improved Genetic Algorithms. IEEE Proceedings of the 6th World Congress on Intelligent Control and Automation, Dalian, China.

13. Radha T, Chelliah TR, Pant M, Ajit A, Grosan C (2010) Optimal gain tuning of PI speed controller in induction motor drives using particle swarm optimization. Logic Journal of IGPL Advance Access.

14. Joao OP, Bimal BK, Eduardo BSL (2001) A Stator-Flux-Oriented Vector-Controlled Induction Motor Drive with Space-Vector PWM and Flux-Vector Synthesis by Neural Networks. IEEE Transaction on Industry Applications 37: 1308-1318.

15. Rajasekaran S, Vijayalakshmi PGA (2005) Neural Networks, Fuzzy Logic and Genetic Algorithm: Synthesis and Applications. Prentice Hall of India, New Delhi, fifth printing.

16. Rojas R (1996) Neural Network- A Systematic Introduction. Spriger-Verlag, Berlin.

17. Zerikat M, Chekroun S (2008) Adaptation Learning Speed Control for a High-Performance Induction Motor using Neural Networks. Proceedings of World Academy of Science, Engineering and Technology 35: 294-299.

18. Seong-Hwan K, Tae-Sik P, YooJi-Yoon, Gwi-Tae P (2001) Speed-Sensorless Vector Control of an Induction Motor Using Neural Network Speed Estimation. IEEE Transaction on Industrial Electronics 48: 609-614.

19. Saifullah K, Bharti D (2011) Power Quality Issues, Problems, Standards & their Effects in Industry with Corrective Means. International Journal of Advances in Engineering & Technology (IJAET) 1: 1-11.

20. Dugan RC, McGranaghan MF, Beaty HW (1996) Electrical Power Systems Quality. New York: McGraw-Hill.

21. Saifullah K, Bharti D (2010) Power Quality: An Important Aspect. International Journal of Engineering, Science and Technology 2: 6485-6490.

22. Ghosh A, Ledwich G (2002) Power Quality Enhancement Using Custom Power Devices. Boston, MA: Kluwer.

23. Khalid S, Vyas N (2009) Application of Power Electronics to Power System. University Science Press, INDIA.

24. Aredes M, Hafner J, Heumann K (1997) Three-Phase Four-Wire Shunt Active Filter Control Strategies. IEEE Transactions on Power Electronics 12: 311-318.

25. Saifullah K, Bharti D (2013) Power quality improvement of constant frequency aircraft electric power system using Fuzzy Logic, Genetic Algorithm and Neural network control based control scheme. International Electrical Engineering Journal (IEEJ) 4: 1098-1104.

26. IEEE Recommended Practices and Requirements for Harmonic Control in Electrical Power Systems. IEEE Standard 519-1992.

27. Saifullah K, Bharti D, Agrawal N, Kumar N (2007) A Review of State of Art Techniques in Active Power Filters and Reactive Power Compensation. National Journal of Technology 3: 10-18.

28. Saifullah K, Bharti D (2013) Comparison of Control Strategies for Shunt Active Power Filter under balanced, unbalanced and distorted supply conditions. Proceedings of IEEE Sponsored National Conference on Advances in Electrical Power and Energy Systems (AEPES-2013).

29. Saifullah K, Bharti D (2013) Comparative Critical Analysis of SAF using Soft Computing and Conventional Control Techniques for High Frequency (400 Hz) Aircraft System. Proceeding of IEEE- CATCON Conference.

2

Space Vector Pulse Width Modulation Based Indirect Vector Control of Induction Motor Drive

Ranjit M*

Department of Electrical and Electronics Engineering, VNR Vignana Jyothi Institute of Engineering and Technology, Hyderabad, India

Abstract

This paper presents the implementation of a different SPACE VECTOR PWM techniques applied to the indirect vector controlled induction motor (IM) drive involves decoupling of the stator current into torque and flux producing components of an induction motor. The drive control generally involves a fixed gain proportional-integral controller. Space vector pulse width modulation technique is widely used in inverter and rectifier controls. Compared to the sinusoidal PWM (SPWM), SVPWM is more suitable for digital implementation and can increase the obtainable maximum output voltage with maximum line voltage approaching 70.7% of the DC link voltage (compared to SPWM's 61.2%) in the linear modulation range. This paper presents the indirect vector controlled induction motor drive using different Space Vector PWM techniques are implemented in MATLAB-SIMULINK. The corresponding harmonic spectrum is calculated for various PWM techniques and the results are compared.

Keywords: Indirect vector control; Space vector pulse width modulation; DPWM; Induction motor

Introduction

The electrical machine that converts electrical energy into mechanical energy and vice versa, is the workhorse in a drive system. Induction motors have been used for over a century because of their simplicity, ruggedness and efficiency [1]. The asynchronous or induction motor is the most widely used electrical drive. Separately excited dc drives are simpler in control because independent control of flux and torque can be brought about [2]. In contrast, induction motors involve a coordinated control of stator current magnitude and the phase, making it a complex control. The stator flux linkages can be resolved along any frame of reference. This requires the position of the flux linkages at every instant. Then the control of the ac machine is very similar to that of separately excited dc motor. Since this control involves field coordinates it is also called field oriented control (Vector Control) [1,2]. Depending on the method of measurement, the vector control is divided into two subcategories: direct and indirect vector control. In direct vector control, the flux measurement is done by using the flux sensing coils or the Hall devices. The most common method is indirect vector control. In this method, the flux angle is not measured directly, but is estimated from the equivalent circuit model and from measurements of the rotor speed, the stator current and the voltage [2].

The Main purpose of two level inverter topologies is to provide a three phase voltage source, where the amplitude, phase, and frequency of the voltages should always be controllable [3]. PWM methods, the carrier-based PWM is very popular due to its simplicity of implementation, known harmonic waveform characteristics, and low harmonic distortion [3]. Space vector pulse width modulation (SVPWM) technique is widely used in inverter and rectifier controls [4,5]. In a space-vector PWM inverter, which is widely used, the voltage utilization factor can be increased to 0.906, normalized to that of the six step operation [6]. In the conventional Space Vector PWM technique complexity is involved due to sector identification and angle calculation. To reduce this complexity various discontinuous algorithms are proposed [5,6].

This paper presents the different Space Vector PWM techniques, which can be applied to the three phase VSI fed indirect vector controlled

induction motor (IM) drive. The performance of the Induction Motor (IM) is analyzed in steady state and transient conditions.

Induction Motor Modelling

The steady-state model and equivalent circuit are useful for studying the performance of machine in steady state. This implies that all electrical transients are neglected during load changes and stator frequency variations. The dynamic model of IM is derived by using a two-phase motor in direct and quadrature axes, where $ds-qs$ correspond to stator direct and quadrature axes, and dr −qr correspond to rotor direct and quadrature axes [3].

The dynamic analysis and description of revolving field machines is supported by well established theories. An Induction Motor of uniform air gap, with sinusoidal distribution of mmf is considered. The saturation effect and parameter changes are neglected [7].

The stator and rotor voltage equations in synchronous reference frame as:

$$V_{qs} = R_s i_{qs} + \frac{d}{dt}\psi_{qs} + \omega_e \psi_{ds} \tag{1}$$

$$V_{ds} = R_s i_{ds} + \frac{d}{dt}\psi_{ds} - \omega_e \psi_{qs} \tag{2}$$

$$V_{qr} = R_r i_{qr} + \frac{d}{dt}\psi_{qr} + (\omega_e - \omega_r)\psi_{dr} \tag{3}$$

$$V_{dr} = R_r i_{dr} + \frac{d}{dt}\psi_{dr} - (\omega_e - \omega_r)\psi_{qr} \tag{4}$$

*****Corresponding author:** Ranjit M, Assistant Professor, Department of Electrical and Electronics Engineering, VNR Vignana Jyothi Institute of Engineering and Technology, Hyderabad, India, E-mail: ranjit221@gmail.com

Where vas, vbs and vcs are the three phase voltages and v_{qs}^s v_{qs}^s v_{ds}^s, are the q-d axes voltages. These equations are also applicable to the current and flux linkage transformation.

The electromagnetic torque obtained from machine flux linkages and currents is as:

$$T_e = \frac{3}{2}\frac{P}{2}(\psi_{dr}i_{qr} - \psi_{qr}i_{dr}) \tag{5}$$

Where Te, P, Ψdr, Ψqr are the electromagnetic torque, number of poles, rotor d-q axes fluxes respectively. The electromagnetic dynamic equation describing the mechanical model of the induction motor is given by

$$T_e = J\frac{d\omega_m}{dt} + T_L + B\omega_m \tag{6}$$

$$\omega_m = \int(T_e - T_L)\frac{P}{2J}dt \tag{7}$$

Where J, T_L, B, ω_m are the moment of inertia of motor and the load torque, the friction coefficient and the mechanical speed. The equations (1) to (7) form the mathematical model equations of a three phase induction motor.

Indirect Vector Control

In the indirect vector control the unit vector signals (Cos θ_e and Sin θ_e) are generated in feed forward manner, indirect vector control is very popular in industrial application. The $d^s - q^s$ axes are fixed on the stator, and $d^r - q^r$ axes are fixed on the rotor moves at speed $\grave{u}_r$ as shown in Figure 1.

Synchronously rotating axes $d^e - q^e$ is rotating ahead of the $d^r - q^r$ axes by the positive slip angle θ_{sl} corresponding to slip frequency ω_{sl}. Since the rotor pole is directed on the d_e axes and $\omega_e = \omega_r + \omega_{sl}$ we can write

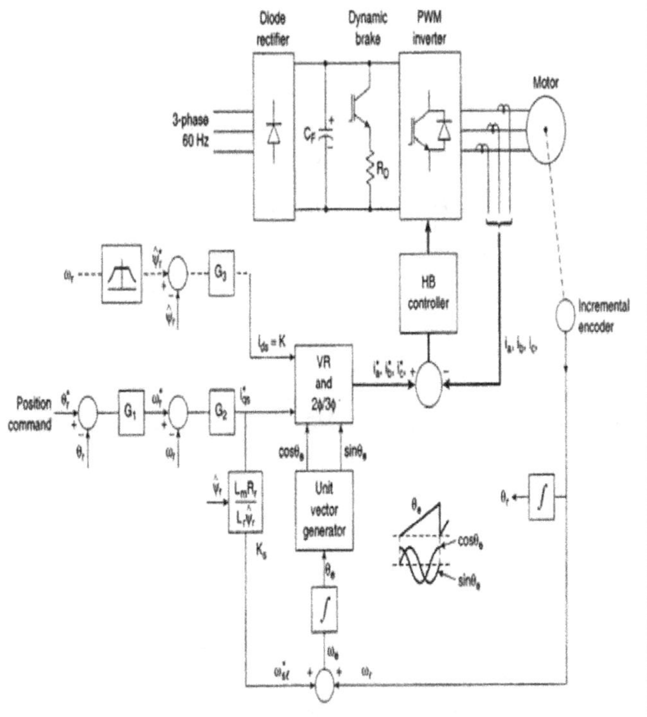

Figure 1: Block diagram of Indirect Vector controlled IM drive.

$$\theta_e = \int \omega_e dt = \int(\omega_r + \omega_{s1})dt = \theta_r + \theta_{s1} \tag{8}$$

The field component of the stator current

$$\hat{\psi}_r = \frac{L_m i_{qs}^*}{(1+\tau_r s)} \quad i_{ds}^* = \frac{\hat{\psi}_r}{L_m} \tag{9}$$

The torque component of the stator current

$$i_{qs}^* = \left(\frac{2}{3}\right)\left(\frac{2}{P}\right)\left(\frac{L_r}{L_m}\right)\left(\frac{T_e^*}{\hat{\psi}_r}\right) \tag{10}$$

Where, $\hat{\psi}_r = \frac{L_m i_{qs}^*}{(1+\tau_r s)}$

Therefore the slip speed

$$\omega_{sl}^* = \left(\frac{L_m i_{qs}^*}{\tau_r \hat{\psi}_r}\right) \tag{11}$$

Space Vector Pulse Width Modulation

The SVPWM technique can increase the fundamental component by up to 27.39% that of SPWM. The fundamental

Voltage can be increased up to a square wave mode where a modulation index of unity is reached. SVPWM is a form of PWM proposed in the mid-1980s that is more efficient compared to natural and regularly sampled PWM three-phase mathematical system can be represented by a space space vector. For example, given a set of three-phase voltages, space vector can defined by

$$V(t) = \frac{3}{2}\left[V_a(t)e^{j0} + V_b(t)e^{j\frac{2}{3}} + V_c(t)e^{j\frac{4}{3}}\right] \tag{12}$$

Where $V_a(t), V_b(t)$ and $V_c(t)$ are three sinusoidal voltages of the same amplitude and frequency but with +1200 phase shifts.

In the space vector modulation, a three phase two level inverter can be driven to eight switching states where the inverter has six active states (1-6) and two zero states (0 and 7).

The basic principle of SVPWM is based on the eight switch combinations of a three phase inverter. Each switching circuit generates three independent pole voltage V_{a0}, V_{b0}, and V_{c0}, which are the inverter output voltages with respect to the mid-terminal of the DC source marked as 'O' on the same Figure 2. These voltages are also called pole voltages. The pole voltages that can be produced are either $\frac{V_{dc}}{2}$ or $\frac{-V_{dc}}{2}$. For example, when switches S_1, S_6, S_2 are closed, corresponding pole voltages are $V_{a0} = \frac{V_d}{2}, V_{b0} = -\frac{V_d}{2}$, and $V_{c0} = -\frac{V_d}{2}$. This state is denoted as (1,0,0) and, according to equation (12), may be depicted as the space vector $V(t) = \frac{3}{2}[V_{dc}e^{j0}]$. Repeating the same procedure, we can find the remaining active non-active states.

The three-phase inverter is therefore controlled by six switches and eight inverter configurations. The eight inverter states can transformed into eight corresponding space vectors. In each configuration, the vector identification uses a 'O' to represent the negative phase voltage level and a '1' to represent the positive phase voltage level.

The relationship between the space vector and the corresponding switches states is given in Table 1 and Figure 3. In addition, the switches in one inverter branch are in controlled in a complementary fashion (1

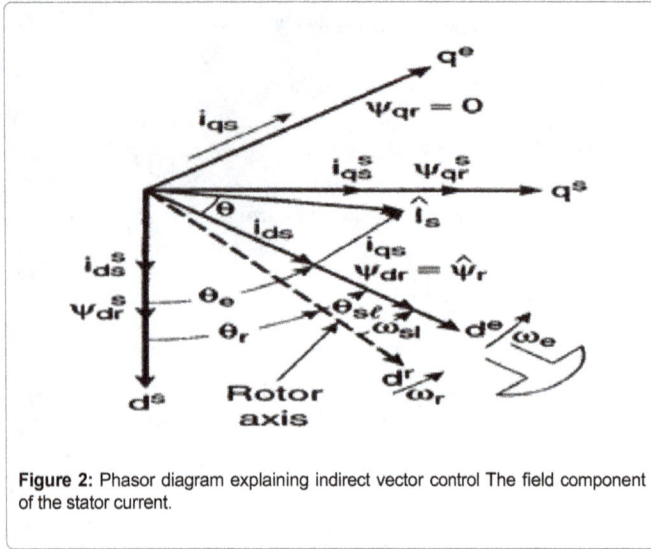

Figure 2: Phasor diagram explaining indirect vector control The field component of the stator current.

Sector	Upper Switches: S1,S3,S5	Lower Switches S4,S6,S2
1	$S_1 = T_a + T_b + \frac{T_0}{2}$ $S_3 = T_b + \frac{T_0}{2}$ $S_5 = \frac{T_0}{2}$	$S_4 = \frac{T_0}{2}$ $S_6 = T_a + \frac{T_0}{2}$ $S_2 = T_a + T_b + \frac{T_0}{2}$
2	$S_1 = T_a + \frac{T_0}{2}$ $S_3 = T_a + T_b + \frac{T_0}{2}$ $S_5 = \frac{T_0}{2}$	$S_4 = T_b + \frac{T_0}{2}$ $S_6 = \frac{T_0}{2}$ $S_2 = T_a + T_b + \frac{T_0}{2}$
3	$S_1 = \frac{T_0}{2}$ $S_3 = T_a + T_b + \frac{T_0}{2}$ $S_5 = T_b + \frac{T_0}{2}$	$S_4 = T_a + T_b + \frac{T_0}{2}$ $S_6 = \frac{T_0}{2}$ $S_2 = T_a + \frac{T_0}{2}$
4	$S_1 = \frac{T_0}{2}$ $S_3 = T_a + \frac{T_0}{2}$ $S_5 = T_a + T_b + \frac{T_0}{2}$	$S_4 = T_a + T_b + \frac{T_0}{2}$ $S_6 = T_b + \frac{T_0}{2}$ $S_2 = \frac{T_0}{2}$
5	$S_1 = T_b + \frac{T_0}{2}$ $S_3 = \frac{T_0}{2}$ $S_5 = T_a + T_b + \frac{T_0}{2}$	$S_4 = T_a + \frac{T_0}{2}$ $S_6 = T_a + T_b + \frac{T_0}{2}$ $S_2 = \frac{T_0}{2}$
6	$S_1 = T_a + T_b + \frac{T_0}{2}$ $S_3 = \frac{T_0}{2}$ $S_5 = T_a + \frac{T_0}{2}$	$S_4 = \frac{T_0}{2}$ $S_6 = T_a + T_b + \frac{T_0}{2}$ $S_2 = T_b + \frac{T_0}{2}$

Table 1: Comparision of %THD of different PWM sequences.

if the switch is on and 0 if it is off). Therefore,

$$S_1 + S_4 = 1$$
$$S_3 + S_6 = 1$$
$$S_5 + S_2 = 1$$

We use orthogonal coordinates to represent the three-phase two-level inverter in the phase diagram. There are eight possible inverter states that can generate eight space vectors. These are given by the complex vector expressions as

$$V_k = \{2/3V_{dr}e(k-1)\frac{\pi}{3} \qquad \text{if } k=1,2,3,4$$
$$= 0 \qquad \qquad \text{if } k=0,7$$

Output patterns for each sector are based on a symmetrical sequence. There are different schemes in space vector PWM and they are based on their repeating duty distribution. Based on the equations for T_a, T_b, T_0, T_7, and according to the principle of symmetrical PWM, the switching sequence in Table 1 is shown for the upper and lower switches.

Proposed Discontinuous SVPWM Technique

Continuous PWM (CPWM) suffers from the drawbacks like computational burden and inferior performance at high modulation indices. Moreover continuous PWM (CPWM) method the switching losses of the inverter are also high. Hence to reduce the switching losses and to improve the performance in high modulation region several discontinuous PWM (DPWM) methods have been reported. It uses the concept of imaginary switching times. The imaginary switching time periods are proportional to the instantaneous values of the reference phase voltages are defined as

$$T_a = \left(\frac{T_s}{V_d}\right)V_a \tag{14}$$

$$\tag{15}$$

$$T_b = \left(\frac{T_s}{V_d}\right)V_b \tag{16}$$

$$T_a = \left(\frac{T_s}{V_d}\right)V_c$$

Where T_S is the sampling period and V_{dc} is the dc link voltage. V_a, V_b, V_c are the phase voltages. The active vector switching times T_1 and T_2 may be expressed as

$$T_2 = T_x - T_{min} \tag{17}$$

$$T_2 = T_x - T_{min} \tag{18}$$

Where $T_x \in (T_{as}, T_{bs}, T_{cs})$ and is neither maximum nor minimum switching time. The effective time is the duration in which the reference voltage vector lies in the corresponding active states, and is the difference between the maximum and minimum switching times as given

$$T_{eff} = T_1 + T_2 \tag{19}$$

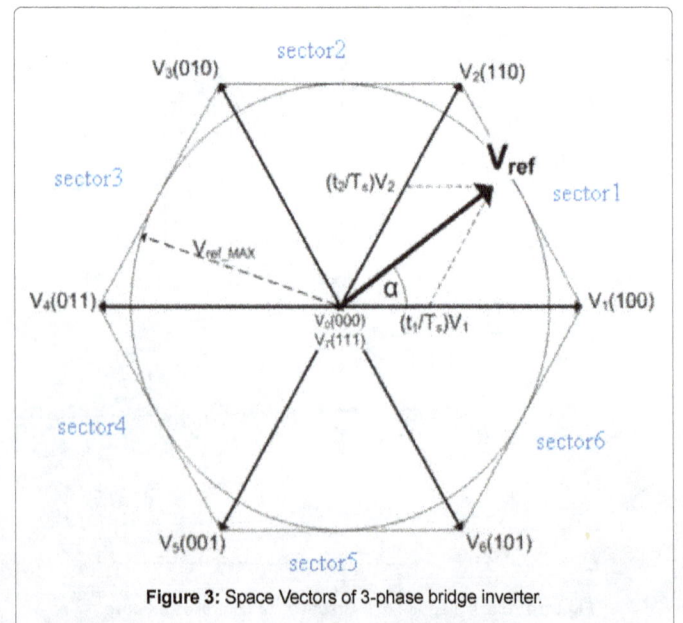

Figure 3: Space Vectors of 3-phase bridge inverter.

$$T_0 = T_s - T_{eff} \qquad (20)$$

In the proposed method the zero state time will be divided between two zero states as $T_0\mu$ for V_0 and $T_0 (1-\mu)$ for V_7 respectively, where μ lies between 0 and 1. The μ can be defined as $\mu = 1-0.5(1 + \text{sg n} (\cos3(\omega t + \delta))$ where ω is the angular frequency of the reference voltage, sgn(y) is the sign function, sgn(y) is 1, 0.0 and -1 when y is positive, zero and negative respectively. The modulation phase angle is represented by δ. When $\mu = 1$ any one of the phases is clamped to the positive bus for 120 degrees and when and $\mu = 0$ any one of the phases is clamped to the

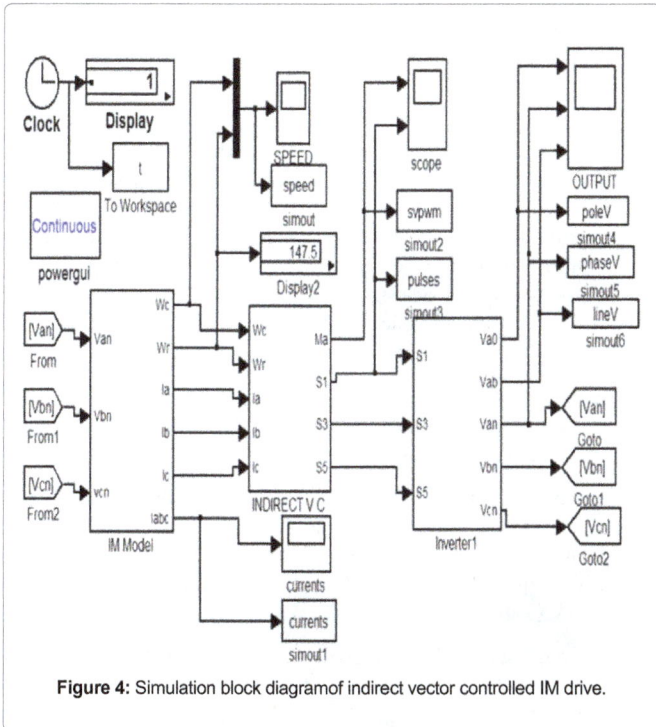

Figure 4: Simulation block diagramof indirect vector controlled IM drive.

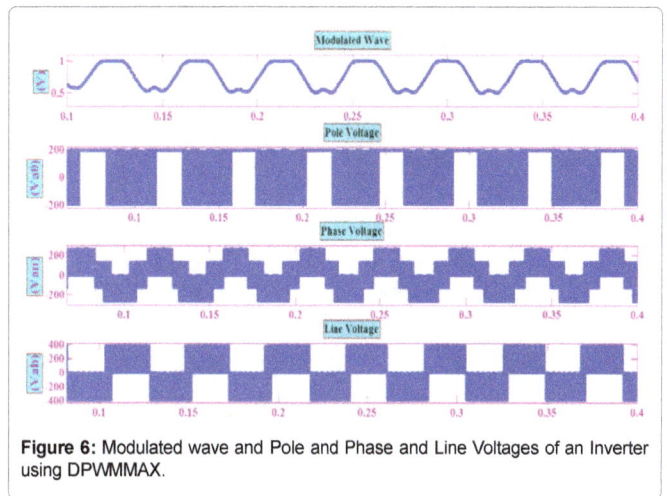

Figure 6: Modulated wave and Pole and Phase and Line Voltages of an Inverter using DPWMMAX.

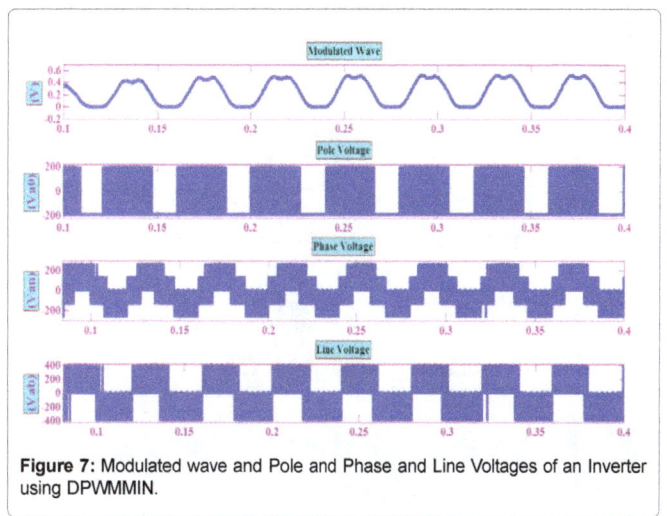

Figure 7: Modulated wave and Pole and Phase and Line Voltages of an Inverter using DPWMMIN.

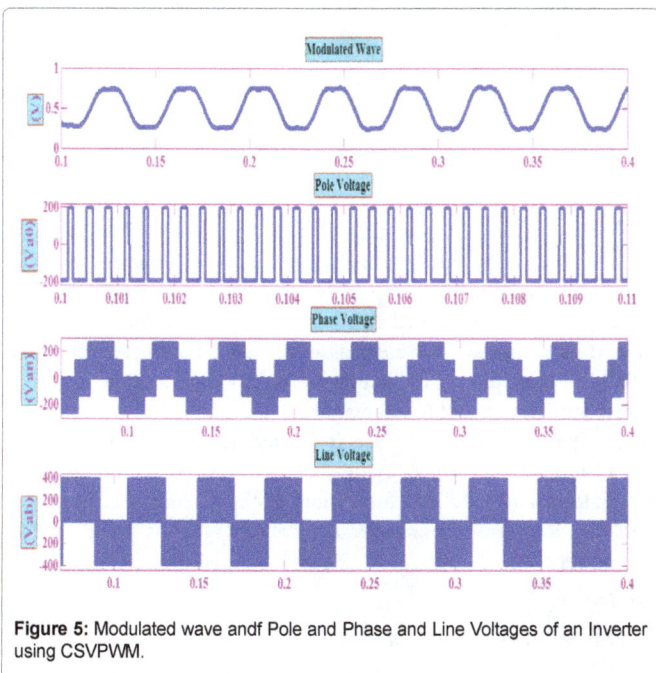

Figure 5: Modulated wave andf Pole and Phase and Line Voltages of an Inverter using CSVPWM.

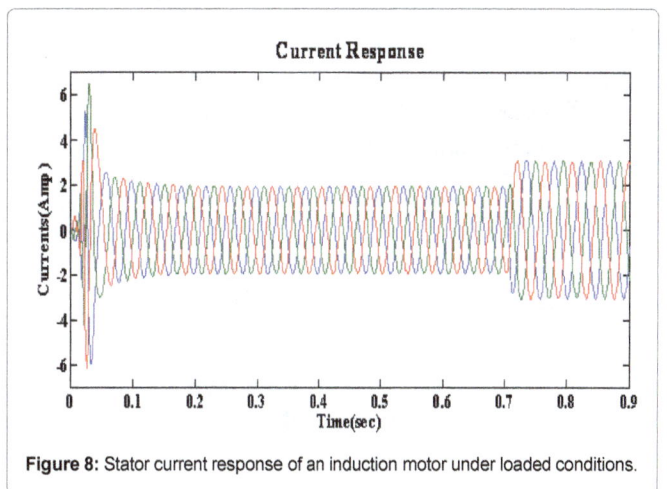

Figure 8: Stator current response of an induction motor under loaded conditions.

negative bus for 120 degrees. When $\mu=0$ and $\mu=1$, DPWMMAX and DPWMMIN are obtained respectively and $\mu=0.5$ results CPWM.

Simulation Results

To validate Space Vector Based indirect vector controlled induction motor drive. Indirect Vector controlled IM drive is implemented in

MATLAB SIMULINK. It consists of Induction motor, Indirect vector Control and SVPWM blocks. Induction motor is supplied from the Variable Voltage and Variable Frequency 3-phase Voltage Source

SVPWM Technique	Fundamental Voltage (V)	Current THD's (%)	Voltage THD's (%)
Continuous	191	19.15	34.92
Discontinuous (max clamping)	176.33	13.62	28.93
Discontinuous (min clamping)	175.2	13.83	27.90

Table 2: Switching sequence.

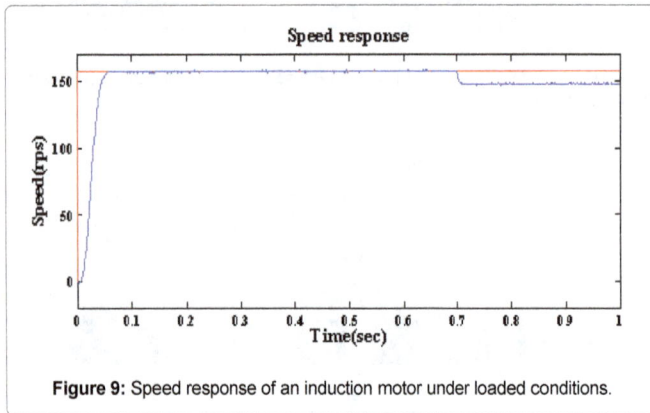

Figure 9: Speed response of an induction motor under loaded conditions.

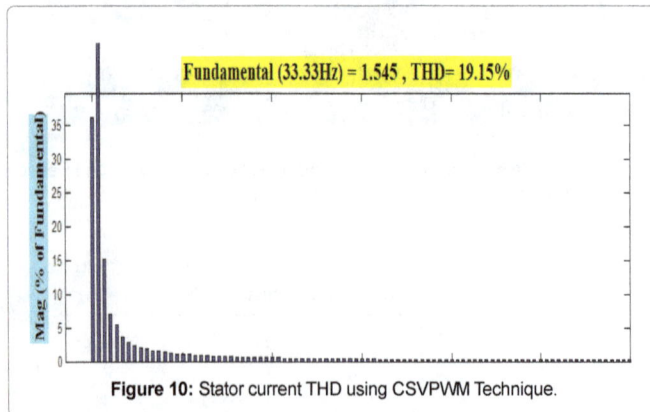

Figure 10: Stator current THD using CSVPWM Technique.

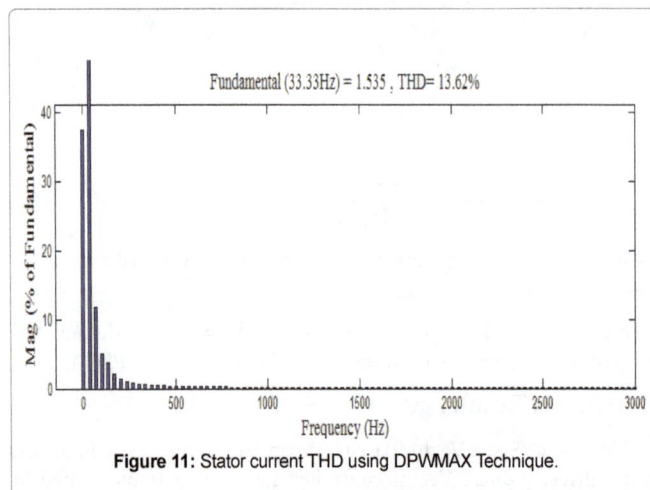

Figure 11: Stator current THD using DPWMAX Technique.

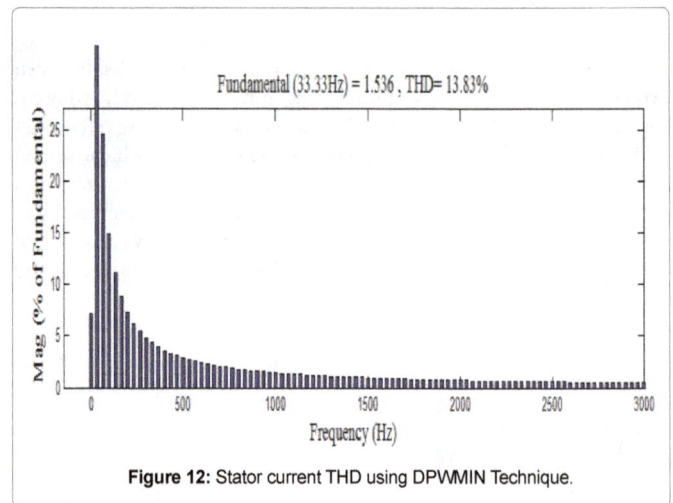

Figure 12: Stator current THD using DPWMIN Technique.

Inverter. Inverter switching Pattern is done with the help Space Vector Pulse Width Modulation Technique. Induction Motor Output Rotor Speed is taken as a Feedback signal and given to indirect vector control. Indirect Vector Control generates the 3-phase voltages to the Space Vector PWM. SVPWM generates Pulses to the Inverter.

The induction motor used in this case study is a 4 KW, 1440 rpm, 4-pole, 3-phase and the corresponding modula-ting, pole voltage, phase and line voltages are shown in Figures 4-8.

Inverter DC input Voltage Vd is 390 V, pulses to the 2-level inverter are applied using continuous SVPWM technique, so that the pole voltage (Va0) is (Vd/2)=195 V; Phase Voltages (Van) are (Vd/3) and (2Vd/3) the values are 130 V and 260 V; Line Voltage (Vab) is (Vd)=390 V (Table 2).

The figure shows that the stator current response of an IM, it reaches the Steady state reaches at 0.07 sec, currents are 2 Amp and the load torque of 2.5 N-M applied at 0.7 sec.

From the speed response the IM attains steady state at 0.07 sec and the speed is nearly at its rated speed i.e. 150 rad/sec.

Until load is applied at 0.7 sec (Figures 9-12).

Conclusion

In This work, indirect vector controlled induction motor drive fed from two-level 3-phase Space Vector PWM inverter is implemented. SVPWM uses the dc bus voltage than SPWM. In Continuous PWM (CPWM) technique the switching losses of the inverter are high. To reduce the switching loss and complexity in CPWM technique DPWM techniques are implemented. The modulating waveforms of different DPWM sequences are obtained based on their clamping sequences. In DPWMMAX method, the clamping of 120° takes place at the middle of 0°-180° for every 360° of fundamental voltage. In DPWMMAX method the clamping of 120° takes place at the middle of 180°–360° for every 360° of fundamental voltage. The corresponding harmonic spectrum is calculated and tabulated for various PWM techniques. Therefore, switching losses of the inverter are reduced greatly using DPWM techniques over CPWM.

References

1. Kumar R, Gupta RA, Bhangale SV (2007) "Indirect Vector Controlled Induction Motor Drive with Fuzzy Logic Based Intelligent Controller". ICTES 68-373.

2. Blaschke F (1972) "The principle of field orientation as applied to the new trans vector closed-loop control system for rotating field Machines", Siemens Review 34: 217-220.

3. Holtz J (1992) Pulse width modulation-A Survey. IEEE Trans. Ind. Electron 39: 410-420.

4. Narayanan G (2008) "Space vector based hybrid PWM Techniques for reduced current ripple," IEEE, Trans 55.

5. Reddy NR, Reddy TB "Simplified Space Vector Based Hybrid PWM Algorithm for Reduced Current Ripple", Int J Recent Trends in Engineering 2.

6. Narayanan G, Ranganathan VT (2005) "Analytical evaluation of harmonic distortion in PWM AC drives using the notion of stator flux ripples," IEEE Trans. Power Electronics 20: 466-474.

7. Hava AM, Lipo TA (1999) "Simple Analytical and Graphical Methods for Carrier-based PWM-VSI Drives", IEEE Trans on power electronics 14.

3

Fault Location in Underground Cables using ANFIS Nets and Discrete Wavelet Transform

Barakat S[1]*, Eteiba MB[2] and Wahba WI[2]

[1]*Electrical Engineering Department, Beni Suief University, Beni Suief, Egypt*
[2]*Electrical Engineering Department, Fayoum University, Fayoum, Egypt*

Abstract

This paper presents an accurate algorithm for locating faults in a medium voltage underground power cable using Adaptive Network-Based Fuzzy Inference System (ANFIS). The proposed method uses five ANFIS networks and consists of 2 stages, including fault type classification and exact fault location. In the first part, an ANFIS is used to determine the fault type, applying four inputs, i.e., the maximum detailed energy of three phase and zero sequence currents. Other four ANFIS networks are utilized to pinpoint the faults (one for each fault type). Four inputs, i.e., the maximum detailed energy of three phase and zero sequence currents, are used to train the Neuro-fuzzy inference systems in order to accurately locate the faults on the cable. The proposed method is evaluated under different fault conditions such as different fault locations, different fault inception angles and different fault resistances.

Keywords: ANFIS; Fault location; Underground cable; Wavelet Transform

Introduction

Underground cables have been widely implemented due to their reliability and limited environmental concerns. To improve the reliability of a distribution system, accurate identification of a faulted segment is required in order to reduce the interruption time during fault. Therefore, a rapid and accurate fault detection method is required to accelerate system restoration, reduce outage time, minimize financial losses and significantly improve the system reliability.

Various fault location algorithms for underground cables have been developed so far. For example, Ningkang and Yuan introduced a mathematical model that is based on calculating the impedance across a tested transmission line to localize all fault locations [1].

Although their model was satisfactory, they only used the post-fault phase magnitude current to identify fault location; however, their method is not applicable to the distribution system due to asymmetrical network. An alternative approach to identify fault location for a radial cable employed wavelet transform to extract valuable information from transient signals and eventually localize faults through a fuzzy logic system is presented in [2]. Javad implemented another approach locate faults in a combined overhead transmission line with underground power cable using ANFIS [3]. Wavelet transform is used to obtain the current patterns in [4]; the proposed methodology consists on training the ANFIS system with a fault database registers obtained from the power distribution system. The performance of the ANFIS nets was good and the 99.14% of the current patterns were correctly classified.

Here, we build upon previously presented methods and describe a fast and accurate method that to detect fault location in underground cables. The proposed method uses a novel wavelet-ANFIS combined approach. The ANFIS is used to extract information from the available Discrete Wavelet Transform coefficients to obtain coherent conclusions regarding fault location. Similar to any rule-based system, the rules are gathered through a Fuzzy Inference System (FIS) [5]. The efficacy of the proposed model was validated under different fault conditions.

This paper is organized as follows. Section 2, provides a brief introduction to wavelet transform. The theoretical background on ANFIS is presented in section 3, while the methodology that was used to locate faults is presented in section 4. The training and testing results are presented in section 5 followed by conclusions in section 6.

Wavelet Transform Analysis

Recently, Wavelet Transform (WT) techniques have been effectively used for multi-scale representation and analysis of many signals [6]. A wavelet transform decomposes transients into a series of wavelet components, each of which corresponds to a time-domain signal that covers a specific frequency band that contains more detailed information. Wavelets localize the information in the time–frequency plane; in particular, by exchanging one type of resolution by another. This makes wavelet transform techniques more suited for analyzing non-stationary signals. Wavelet analysis is also appropriate for rapid changes in transient signal analysis. Its major strength is its ability to demonstrate the local features of a particular area of a large signal. WT divides up data, operates for different frequency components, and then studies each component with a resolution that is matched to its scale [7].

Theory of discrete wavelet transform

In a Discrete Wavelet Transform (DWT), a time-scale representation of a digital signal is obtained using digital filtering techniques. In this case, filters of different cut-off frequencies are used to analyze the signal at different scales [8]. The signal then passes through a series of high pass filters to analyze the high frequencies, and through a series of low pass filters to analyze its low frequency components. Hence the signal (S) will be decomposed into two types of components, namely an approximation (termed A) and a detail (D). The approximation is the high scale, low-frequency component of the signal while the detail is the low-scale, high-frequency components. This decomposition process can be iterated, using successive approximations which can be further decomposed such that any given signal is broken down into many lower resolution components. This approach is called the wavelet decomposition tree which is shown in Figure 1.

---

*Corresponding author: Barakat S, Electrical Engineering Department, Beni Suief University, Beni Suief, Egypt, E-mail: shimaabara@yahoo.com

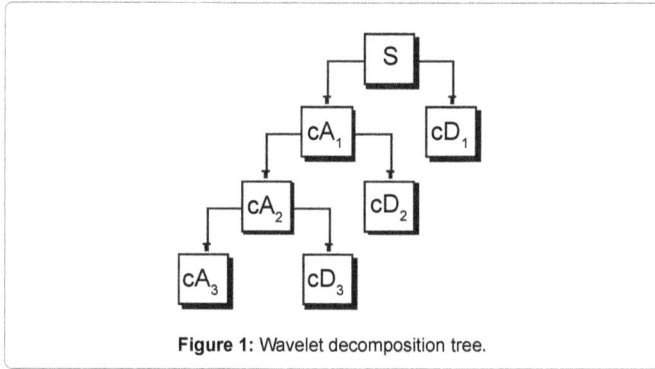

Figure 1: Wavelet decomposition tree.

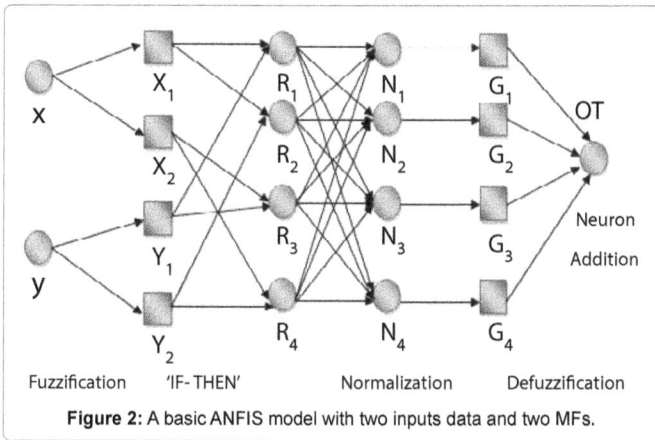

Figure 2: A basic ANFIS model with two inputs data and two MFs.

The ANFIS Concept

ANFIS is a simple data learning technique that adopts a fuzzy inference system model to transform a given input into a target output. This transformation involves membership functions, fuzzy logic operators and conditional rules. ANFIS is a Sugeno model from a development of Fuzzy Inference System (FIS) [9,10]. Sugeno-type ANFIS implements a first-order polynomial to the output system in order to replace zero-order in Sugeno FIS model. There are five main processing stages in a typical ANFIS operation, which include input fuzzification, application of fuzzy operators, application method, and output aggregation and defuzzification [2]. These stages are demonstrated in the ANFIS structure as shown in Figure 2. Each ANFIS layer has specific functions for calculating input and output parameter sets as described below [9].

Layer 1: Fuzzification of input data is performed in this layer. The following equations are used for this process:

$$X_i(x) = \frac{1}{1 + \left(\frac{x - c_i}{a_i}\right)^{2b_i}} \tag{1}$$

$$OT = \sum G_i \, i = 1, 2, 3, 4 \tag{2}$$

Where, $X_i(x)$ and $Y_i(y)$ are fuzzied values, and a_i, b_i and c_i are the parameters sets that are calculated by Gaussian input membership function.

Layer 2: This layer involves fuzzy operators and it uses the product (AND) operator to fuzzify the inputs. The following fuzzy relationship represents the products of fuzzy operators:

$$R_i = X_i(x) \times Y_i(y) \, i = 1, 2, 3, 4 \tag{3}$$

Layer 3 (Normalization): In this stage, every gained signal is divided to the total of gained signal by the following equation,

$$N_i = \frac{R_i}{R_1 + R_2 + R_3 + R_4} \, i = 1, 2, 3, 4 \tag{4}$$

Layer 4 (Defuzzification): In this stage, a normalized signal is gained again through a linear equation that is formed from the membership function of the output signal as shown in the following equation,

$$G_i = N_i (p_i x + q_i y + r_i) \, i = 1, 2, 3, 4 \tag{5}$$

Where, p_i, q_i and r_i being the membership function parameters for the linear signal.

Layer 5: The last process in the ANFIS operation is called neuron addition in which all defuzzification signals, Gi are added together as shown below:

$$OT = \sum G_i \, i = 1, 2, 3, 4 \tag{6}$$

OT is a predicted value.

Proposed Fault Location Method

The proposed method consists of two main stages namely, fault type classification and exact fault location. The presented algorithm contains five ANFISs. The first network is for fault type classification, while the remaining four networks are for accurate fault location (one for each fault type).

Features extraction using DWT

Here, we used the line current signals as the input to the DWT. Daubechies DB4 wavelet, is employed since it has been demonstrated to high performance. The fault transients of the study cases are analyzed through discrete wavelet transform at the level five. Both approximation and details information related fault current are extracted from the original signal with the multi-resolution analysis.

When a fault occurs in the cable, its effect can be observed as variations within the decomposition coefficient of the current signals that contain useful fault signatures. Figure 3 shows the DWT detailed coefficients at level 1 to level 5 for a particular type of fault studied in the work. The nature of the plot of detailed coefficients at level 1 reveals a sharp spike which corresponds to the fault initiation process. According to DWT theory, this spike represents the highest frequency within the fault signal. It is however, not practical to identify a fault based on this spike only since such spikes will occur every time there is a sudden change in the cable current signal. This will thus not be able to clearly differentiate between faults of different types and at different locations.

The nature of level 5 detailed coefficients (Figures 4 and 5) shows that along with the high spike, there is a certain side band containing some smaller spikes. The nature of this side band along with the dominant spike has been observed to change appreciably with variations in fault types and locations. Detailed coefficients at still higher levels, however, have been found to contain much wider side bands that complicated correlation with possible fault types and locations.

Therefore, we decided to start with, extracting some meaningful features from the level 5 detailed coefficients that can be correlated to possible fault types and locations. With this in mind, the maximum detailed energy of three phase and zero sequence currents have been used as features of the fault classification and location scheme. The proposed methodology consists of training the ANFIS system with a fault database obtained from the simulation of the cable.

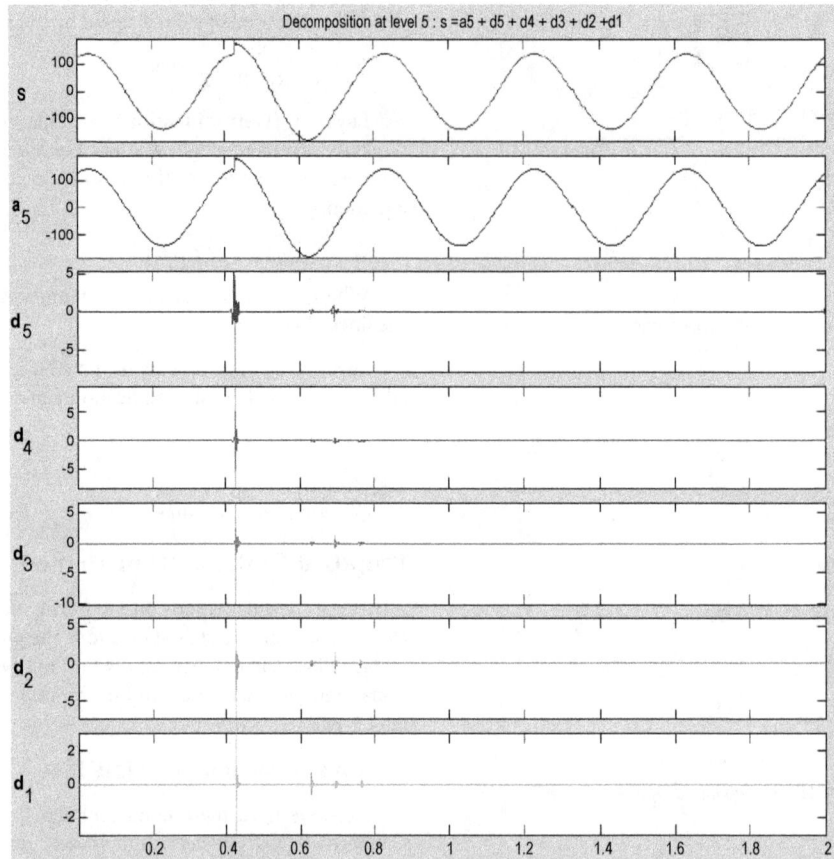

Figure 3: The DWT detailed coefficients at level 1 to level 5 for a three phase to ground fault (phase A).

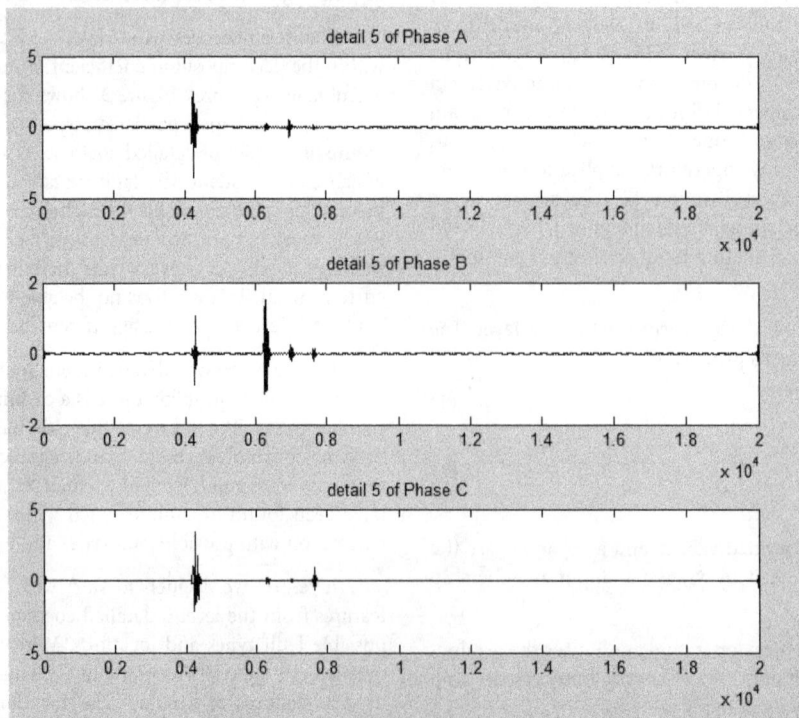

Figure 4: Level 5 detailed coefficients of Three Phase to Ground Fault case.

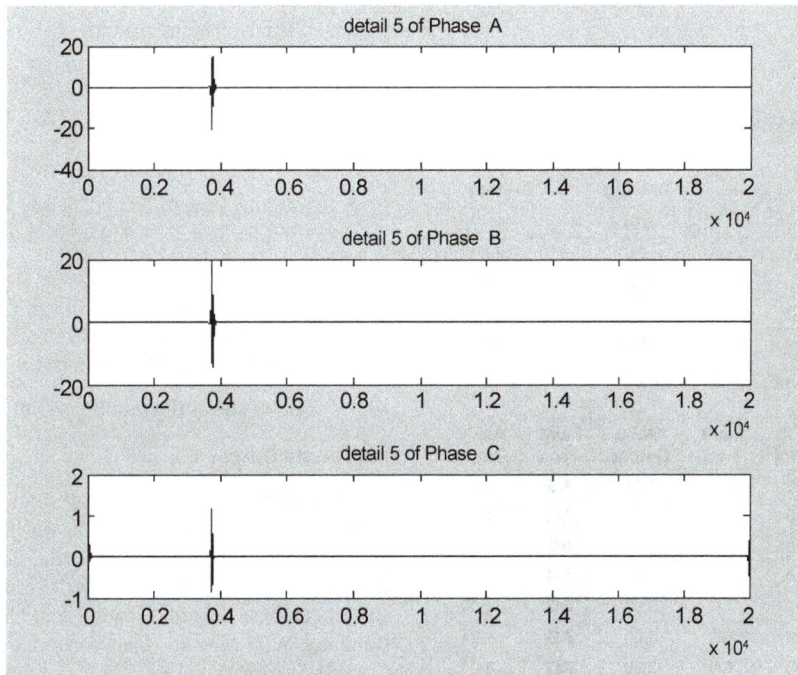

Figure 5: Detailed coefficients of Line to Line Fault case.

Fault classification scheme

In order to design an ANFIS, it is crucial to train it efficiently and correctly. The training set must be carefully chosen such that it can include a diversity of fault conditions such as different fault inception angles, different fault resistances and different fault locations are considered. The performance of the ANFIS is then tested using both patterns within and outside of the training set.

An acceptable and simple criterion that we used here is that the ANFIS input should provide more information for fault location than those not selected. Therefore, for fault type classification (ANFIS1), the maximum detailed energy of three phases and zero sequence currents are selected as inputs and the desired output is the fault type as set in Table1.

Fault location scheme

In this stage, four different ANFISs are trained for fault location based on the knowledge of the fault type. Once the fault is classified, the relevant ANFIS for fault location is activated. The inputs for these networks are the same as those for the inputs of ANFIS1. The output, however, is the distance of the fault point from the sending end of the cable in Km.

Tests and Results

Test system

The Alternative Transient Program (ATP) is used to simulate medium voltage underground cable model [11] with a sampling frequency of 200 KHz. The single line diagram is shown in Figure 6. The components that we used are the three phase voltage source, a tested cable and a fixed load. The specifications of cable material for 11 Kv are presented in Table 2.

Three phase voltage source: V=11KV with f=50 Hz

Fault Type	ANFIS Target
Three Phase to ground (ABCG)	1
Double line to ground (ABG)	2
line to line (AB)	3
Single line to ground (AG)	4

Table 1: Training target for ANFIS1 fault type classification.

Specification of MV underground cable material (XLPE Stranded Copper Conductor - 6 Km - Bergeron model)	
Radius (mm)	r1=6.75, r2=10.15, r3=12.05 r4=12.2, r5=13.8
Core conductor	ρ=1.7 E-8 Ω.m , μ=1.0
Insulation	μ=1.0 , ε=2.7
Sheath	ρ=2.5 E-8 Ω.m , μ=1.0

ρ : Resistivity of the conductor material.
μ: Relative permeability of the conductor material.
μ (ins.): Relative permeability of the insulating material outside the conductor.
ε (ins.): Relative permittivity of the insulating material outside the conductor.

Table 2: Specification of Cable material for 11 Kv

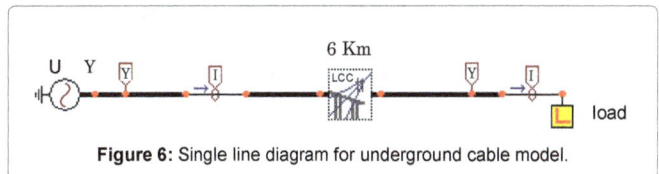

Figure 6: Single line diagram for underground cable model.

Load: Three-Phase 2 MVA Grounded-Wye load with parallel R, L elements (power factor=0.85, R=71.157 Ω, L=365.475mH).

Simulation of MV underground cable faults depends on four main fault parameters (fault type, fault distance, fault resistance, inception angle).

Training scenarios for the simulation

Fault type

- Single line to ground (AG)

- Double line to ground (ABG)

	Training	Testing
Three phase to ground	630	120
Single line to ground (AG)	630	180
Line to line (AB)	630	120
Double line to ground (ABG)	630	120
Total	2520	540

Table 3: Number of simulations.

Fault Type	ANFIS Output	Fault Type	ANFIS Output	Fault Type	ANFIS Output	Fault Type	ANFIS Output
ABCG	1.00	ABG	1.88	AB	3.01	AG	4.01
ABCG	1.00	ABG	1.88	AB	3.01	AG	4.01
ABCG	1.00	ABG	1.98	AB	3.01	AG	4.01
ABCG	0.94	ABG	2.00	AB	3.00	AG	4.00
ABCG	0.99	ABG	2.00	AB	3.00	AG	4.00
ABCG	1.00	ABG	2.00	AB	3.00	AG	4.01
ABCG	0.99	ABG	1.99	AB	3.00	AG	4.01
ABCG	1.00	ABG	2.01	AB	3.00	AG	4.03
ABCG	0.99	ABG	1.99	AB	3.00	AG	4.01
ABCG	1.00	ABG	2.01	AB	3.00	AG	4.03
ABCG	1.01	ABG	2.01	AB	3.00	AG	4.00
ABCG	1.02	ABG	2.00	AB	2.99	AG	4.00
ABCG	1.02	ABG	2.00	AB	3.01	AG	4.01
ABCG	1.02	ABG	2.00	AB	3.00	AG	4.01
ABCG	1.02	ABG	2.01	AB	2.99	AG	4.00
ABCG	1.16	ABG	2.03	AB	2.99	AG	4.01
ABCG	1.01	ABG	2.00	AB	3.02	AG	4.01
ABCG	1.01	ABG	2.03	AB	3.00	AG	4.00
ABCG	1.00	ABG	1.99	AB	3.00	AG	4.03
ABCG	1.00	ABG	1.97	AB	3.01	AG	4.00

Table 4: The fault classification training output results for all types of fault.

- Line to line (AB)

- Three phase to ground

Fault resistance: {0, 10, 30, 50, 100, 200 Ω}

Inception angle: {0°, 45°, 90°, 135°, 180°}

Fault distance:

- Training: [0.5, 0.75, 1, 1.25, 1.5, 1.75, 2, 2.25, 2.5, 2.75, 3, 3.25, 3.5, 3.75, 4, 4.25, 4.5, 4.75, 5, 5.25, 5.5] Km from the sending end.

- Testing: [0.625, 0.875, 1.125, 1.375, 1.625, 1.875, 2.125, 2.375, 2.625, 2.875, 3.125, 3.375, 3.625, 3.875, 4.125, 4.375, 4.625, 4.875, 5.125, 5.375] Km from the sending end.

- Table 3 shows the number of simulations used in this work.

Fault data base

To obtain a fault data-base, the four fault types were simulated in the MV underground cable shown in Figure 5: Single phase-fault to ground (Fault phase A to ground, Line to line fault (Fault phase A and B), Double line to ground fault (Fault phase A and B to ground) and three-phase to ground fault (fault phases A, B, C to ground). The simulated ATP files are then converted to a Microsoft Office Excel Comma Separated Values File (*.csv file) and imported to Matlab to perform wavelet analysis and the extracted features are stored in a database file. As a result 3060 simulations have been obtained. These were used to train and validate the ANFIS system.

ANFIS fault type classification results

Table 4 lists part of the fault classification training output results for all types of fault. As shown, the obtained predicted values are quite similar to that of the training target values. This demonstrates that ANFIS is able to recognize and classify fault correctly.

Fault location system tool

Development of the fault location system tool (FL) is based on Matlab GUI as shown in Figure 7. The purposes of building the GUI tool are to:

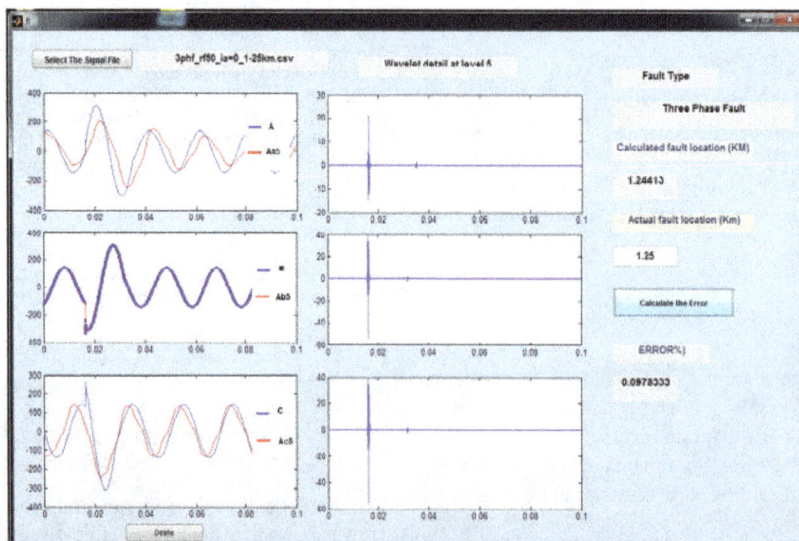

Figure 7: The application window of fault location system tool.

- Test the data using ANFIS models,

- Display the original signal, the wavelet approximation and detail coefficients at level 5,

- Produces fault types,

- Calculate fault location and

- Calculate the percentage error.

The FL needs only a standard data format of three-phase current as an input to pin point the fault.

ANFIS fault location estimation training results

In ANFIS Fault Location Estimation, all the four types of fault (ABCG, ABG, AB and AG) were trained separately. As a result, a total of four networks were used to estimate the fault distance. The results for each network are presented below.

The location error is defined as [12]:

$$\%Error = \left| \frac{ExactDistance - ANFISoutput}{TotalcableLength} \right| \times 100\%$$

where ,Total cable Length=6 Km

Three phase to ground ANFIS fault location model: It can be noticed from Figure 8 that the Percentage error of three phases to ground fault is less than 1.65%.

Double line to ground ANFIS fault location model: It can be noticed from Figure 9 that the Percentage error of double Line to ground fault is less than 1%.

Line to line ANFIS fault location model: It can be noticed from

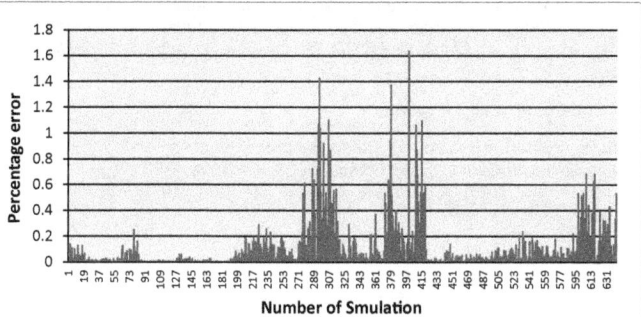

Figure 8: Percentage error of three phase to ground fault (training).

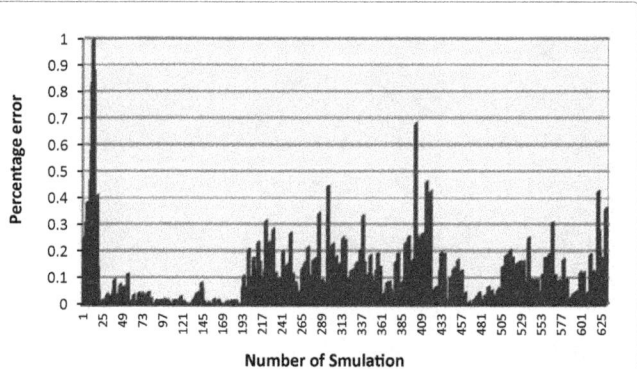

Figure 9: Percentage error of double Line to ground fault (training).

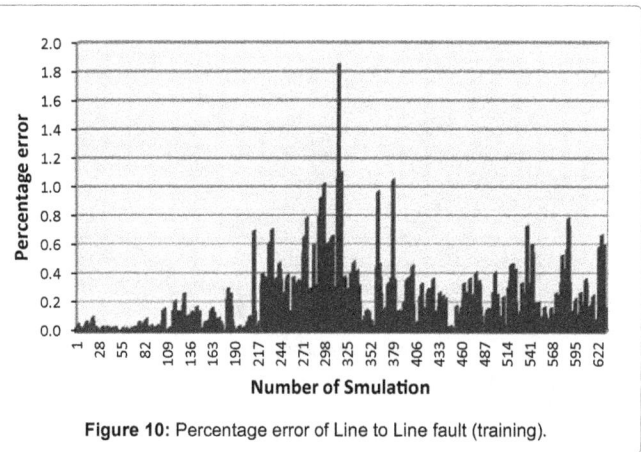

Figure 10: Percentage error of Line to Line fault (training).

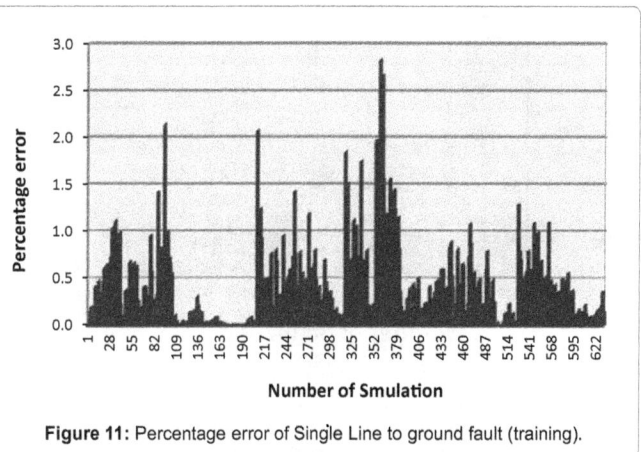

Figure 11: Percentage error of Single Line to ground fault (training).

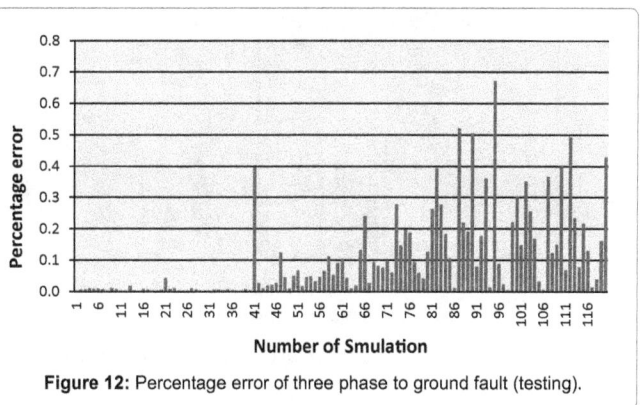

Figure 12: Percentage error of three phase to ground fault (testing).

Figure 10 that the Percentage error of Line to Line fault is less than 2%.

Single line to ground ANFIS fault location model: It can be noticed from Figure 11 that the Percentage error of Single Line to ground fault is less than 3%.

ANFIS fault location estimation testing results

Three phase to ground fault: It can be noticed from Figure 12 that the Percentage error of three phases to ground fault is less than 0.7%.

Double line to ground fault: It can be noticed from Figure 13 that the Percentage error of double Line to ground fault is less than 1.7%.

Line to line fault: It can be noticed from Figure 14 that the

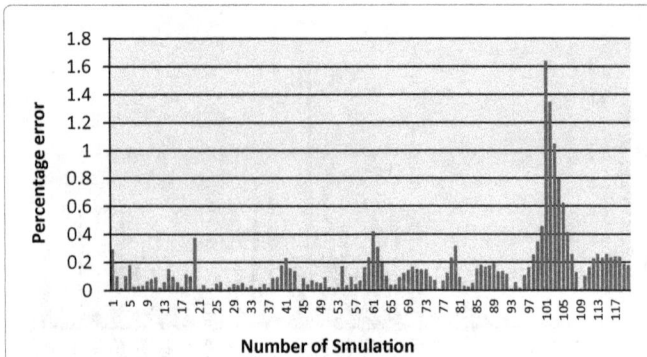

Figure 13: Percentage error of double Line to ground fault (testing).

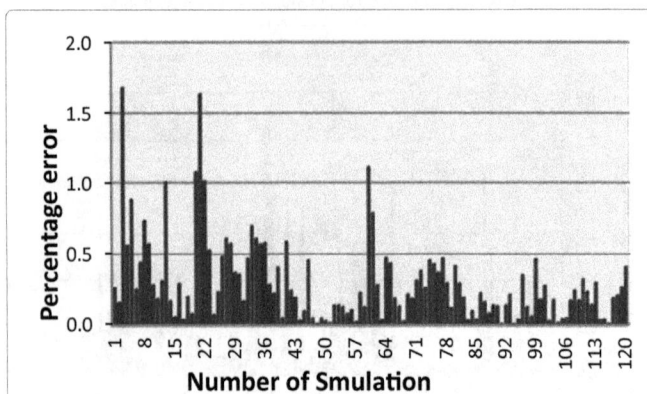

Figure 14: Percentage error of Line to Line fault (testing).

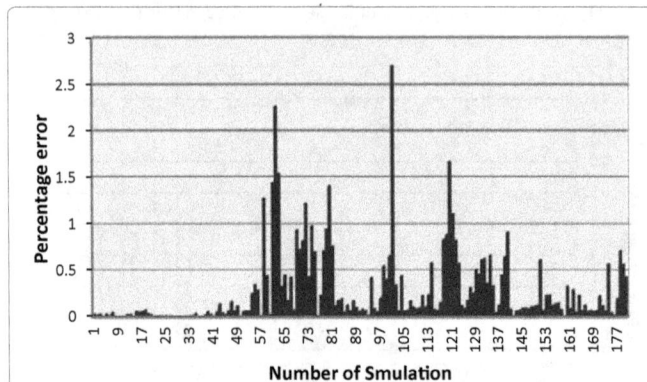

Figure 15: Percentage error of Single Line to ground fault (testing).

Percentage error of Line to Line fault is less than 1.8%.

Single Line to ground fault: It can be noticed from Figure 15 that the Percentage error of Single Line to ground fault is less than 3%.

Conclusion

This paper presented an application of fault location method to localize faults in a medium voltage underground cable based on the theory of Wavelet and Adaptive Network Fuzzy Inference System (ANFIS). The Proposed ANFIS uses only post-fault three-phase currents as inputs. It predicts the distance of the fault from the sending end point. The results show that the approach can accurately identify the faults types and locate the faults.

References

1. Ningkang, Yuan Liao (2010) Fault Location Estimation Using Current Magnitude Measurements. Proceedings of the IEEE Southest Conference 214-217.

2. Moshtagh J, Aggarwal RK (2004) A new approach to fault location in a single core underground cable system using combined fuzzy logic & wavelet analysis. The Eight IEE International Conference on Developments in Power System Protection 228-231.

3. Sadeh J, Afradi H (2009) A new and accurate fault location algorithm for combined transmission lines using Adaptive Network-Based Fuzzy Inference System. Electric Power Systems Research, 79: 1538–1545.

4. Mora JJ, Carrillo G,Perez L (2006) Fault Location in Power Distribution Systems using ANFIS Nets and Current Patterns, IEEE PES Transmission and Distribution Conference and Exposition Latin America, Venezuela 1-6.

5. Rasli, Hussain, Fauzi (2012) Fault Diagnosis in Power Distribution Network Using Adaptive Neuro-Fuzzy Inference System (ANFIS), Fuzzy Inference System - Theory and Applications.

6. Robertson DC, Camps OI, Mayer JS, Gish WB (1996) Wavelet and electromagnetic power system transients. IEEE Trans Power Deliver, 11: 1050-1058.

7. Bhowmik PS, Purkait P, Bhattacharya K (2009)   A novel wavelet transform aided neural network based transmission line fault analysis method. Int J Elec Power and Energy Systems 31: 213–219.

8. Misiti M, Misiti Y, Oppenheim G, Poggi JM (2001) Wavelet toolbox user's guide. The Math Work, Inc.

9. Jang JSR (1993) ANFIS: Adaptive-Network-Based Fuzzy Inference System. IEEE Transaction on System, Man, and Cybernetics society 23: 665 – 685.

10. Mora, JJ, Carrillo, G and Perez, L 2006. Fault Location in Power Distribution System Using ANFIS nets and Current Patterns. International Conference on Transmission and Distribution (TDC -06), Latin America, IEEE Power & Energy Society

11. International Cables Co. SAE

12. IEEE (2005) IEEE Guide for Determining Fault Location on AC Transmission and Distribution Lines, IEEE Standard C37.114–2004, 1-36.

4

Three Phase Shield Wire Schemes Unbalance and Voltage Fluctuation Reduction Using H-Bridge Cascaded Static Synchronous Compensator (STATCOM)

Alidou Koutou* and Mamadou Lamine Doumbia

Department of Electrical Engineering, Université du Québec à Trois-Rivières, Trois-Rivières, Québec, G9A 5H7, Canada

Abstract

Unbalance and voltage fluctuation are considered as a major problem within the 3-phase Shield Wire Schemes networks (SWS). This paper deals with the methods to reduce the unbalance and voltage fluctuation by using the H-bridge cascaded Static Synchronous Compensator (STATCOM). In unconventional networks without STATCOM, the unbalance factor is generally more than 27%. In this study, we proved that after voltage compensation by the H-bridge cascaded STATCOM, unbalance factor is reduce to 0.15% for ground impedance variation. The ground resistance has been varied between 33 and 8000 Ω, and the ground inductance between 10 and 110 H. Moreover, the voltage fluctuation is reduced. The obtained results correspond to those recommended in power distribution for the proper operation of electrical equipment.

Keywords: Cable shielding; Power cable insulation; Distribution feeders; Earth return path, Shield Wire scheme; Voltage unbalance; Voltage fluctuation, H-bridge cascaded STATCOM

Introduction

The Shield Wire Scheme (SWS) is a technical method to supply power to the villages and communities located along the High Voltage (HV) lines, up to 100 km distant from the HV transforming stations.

The SWS consists of insulating the shield wires (SWs) from the towers of the HV lines and energizing the SWs with Medium Voltage (MV) (20-34.5 kV) from the HV/MV transformer station at one end of the HV line [1] (Figure 1). This technique reduces the cost of investments while adapting to the low power requirements of these communities [2]. Supply of the villages along the HV line through fused single phase or three-phase transformers, depends on the scheme chosen and on the nature of the load to be supplied [3]. The overruling criterion has been to propose solutions that require only conventional distribution equipment, without power electronic devices, to provide a reliable service with simple and ordinary operational methods.

In fact, the three-Phase SWS is an unsymmetrical system with voltage fluctuation; because the three-Phase SWS supplying conventional MV/LV distribution transformers operated with one MV terminal permanently grounded [4]. Moreover, it has been reported that the drying up of land leads to an uncontrolled increase in grounding impedance [5]. However, the limit values for this increase are still unknown. Increasing the value of the impedance results in a high voltage unbalance which engenders a degradation of the unbalance factor of up to 8.18% for resistance variation and 27.31% for the inductance variation [6]. The value of unbalance factor must be contained within 1-2% for a correct operation of equipment [7].

The balancing of voltage can be obtained by applying different methods. But all, use only passive components, using a series resistor-reactor in the earth path and capacitors connected between the two SWs and between each SW and ground [8]. While these techniques allow the reduction of voltage unbalance in three-phase Shield wire, these types of passive compensators do not fit when a voltage variation occurs. They are functional only for predefined unbalance conditions.

To address this problem, this paper presents the use of H-bridge STATCOM to reduce unbalance and voltage fluctuation on the three-phase shield wire.

Matlab/Simulink software has been used to simulate the three-phase Shield wire of 34.5 KV and, a significant improvement of the unbalance and voltage fluctuation rate is obtained.

Investigated three-phase shield wire system

Three phase shield wire scheme description: Figure 1 presents a general representation of three-phase Shield Wire Scheme. A 34.5 kV shield wires medium voltage is used to provide 145/240 V. A medium-voltage transformer terminals MV/LV cables are connected to two cables and the third cable is grounded by an electrode that contains a resistance and an inductance (Figure 2).

The three-Phase SWS is planned to be implemented in Western Africa. It has a HV 225kV-50Hz-338 km long transmission line with an intermediate substation at 134 km from the sending end. The 225kV line is equipped with AAAC conductors with Φ=31.04 mm, R20°C=0.0583 Ω/km, whereas the SWs are ACSR conductors with diameter=10.02 mm, R20°C=0.58 Ω/km [9].

Cascade H-Bridge Topology and Control: STATCOM H-bridge can use in the SWS network to increase the system performance (unbalance and voltage fluctuation). The interest for multilevel CHB converters comes from their capacity to provide higher power, to generate good quality of waveforms, to operate at low switching frequency with low loss of energy and the low effort on statics devices [10]. The cascade H-bridge (CHB) STATCOM topology considers the

*Corresponding author: Alidou Koutou, Department of Electrical Engineering, Universite du Quebec a Trois-Rivieres, Québec Canada
E-mail: alidou.koutou@uqtr.ca

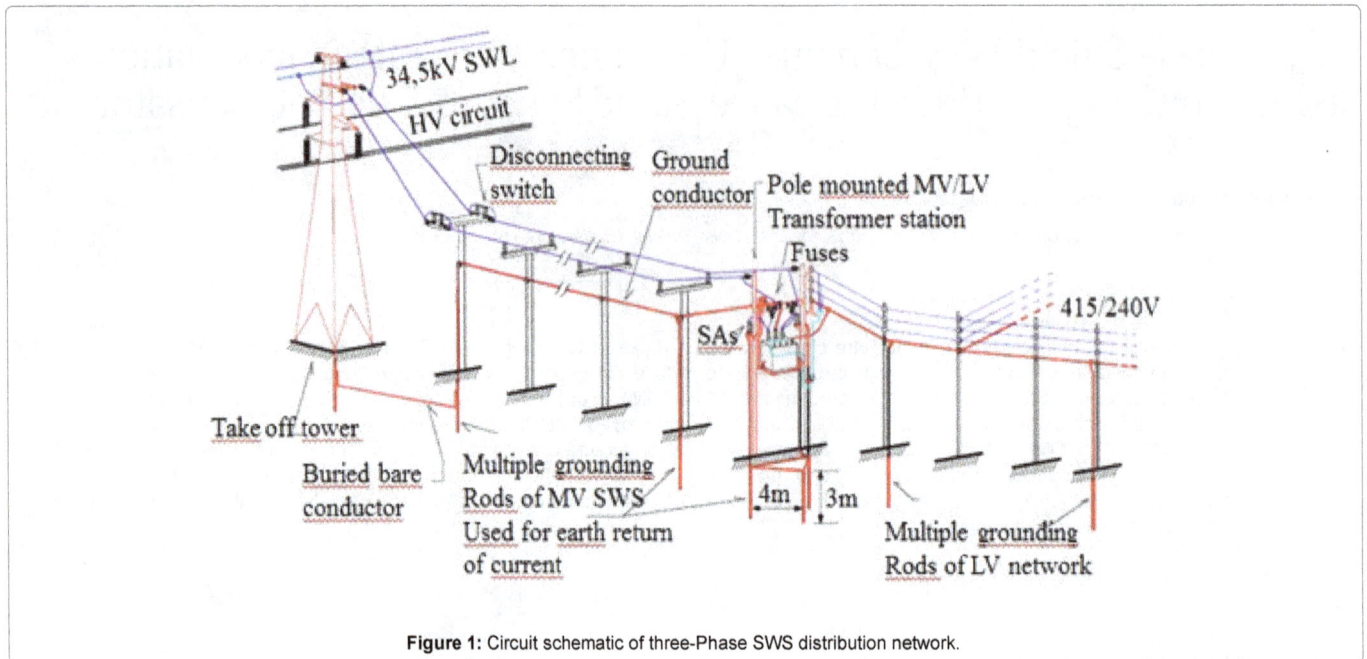

Figure 1: Circuit schematic of three-Phase SWS distribution network.

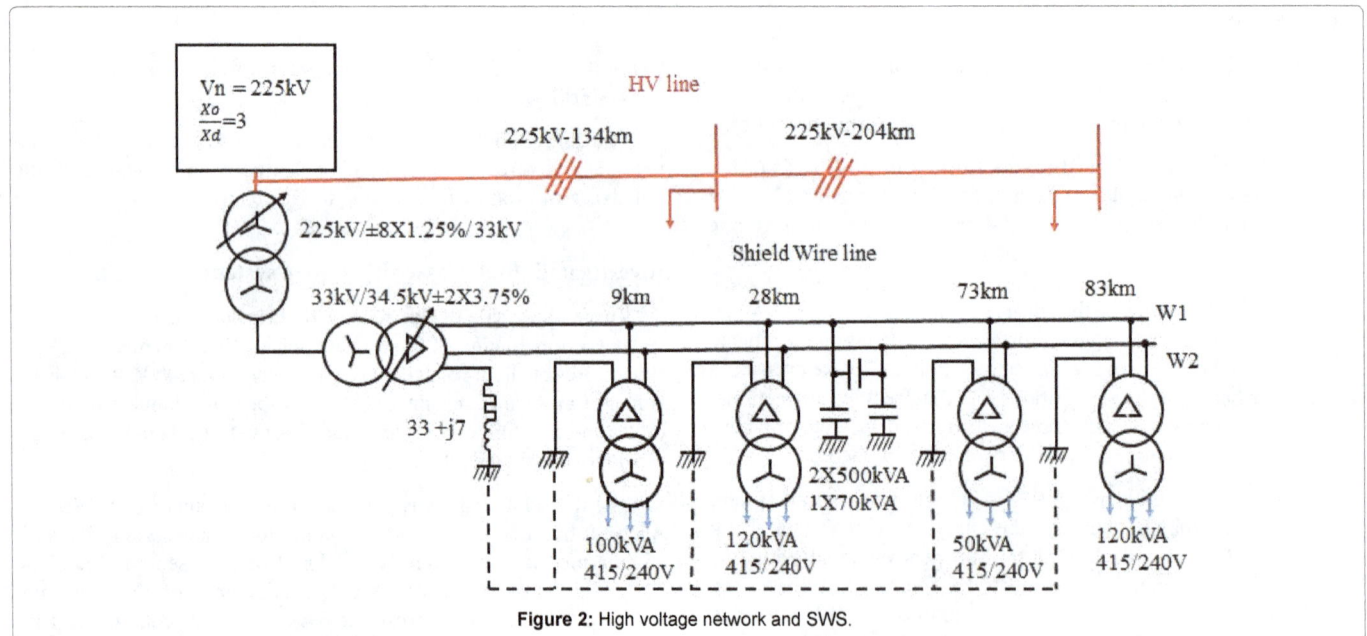

Figure 2: High voltage network and SWS.

series connection of several H-Bridges modules in order to share the total load voltage among all the modules. This allows a low voltage rating of the semiconductor devices (Figure 3).

The Sinusoidal PWM (SPWM) control method is used for this application. It is one of the most widely used control method in VSCs owing to its simplified mathematical requirements and easy implementation even with basic microcontrollers. The SPWM method is based on comparing a sinusoidal modulating signal with a triangular carrier waveform that is arranged according to required switching outputs [11,12] (Figures 4 and 5).

Methodology

A mathematical analysis and simulation were used. In mathematical analysis the technical characteristics of the network and the existing theories have served to determine the essential parameters for the functioning of the integrated network STATCOM. Elements of Matlab Simulink library Simpower Systems were used for the simulation. For the simulation we have given different values for the impedance and raise the voltage on the SWSs line. The unbalance factor was calculated from the values of voltage composed by the method proposed in the literature [13].

Passive filter's design

As any other voltage controlled source (VCS) based topology, this configuration needs a passive filter in order to absorb the instantaneous voltage differences between the converter and the grid, and to reduce the harmonic content of the injected AC current.

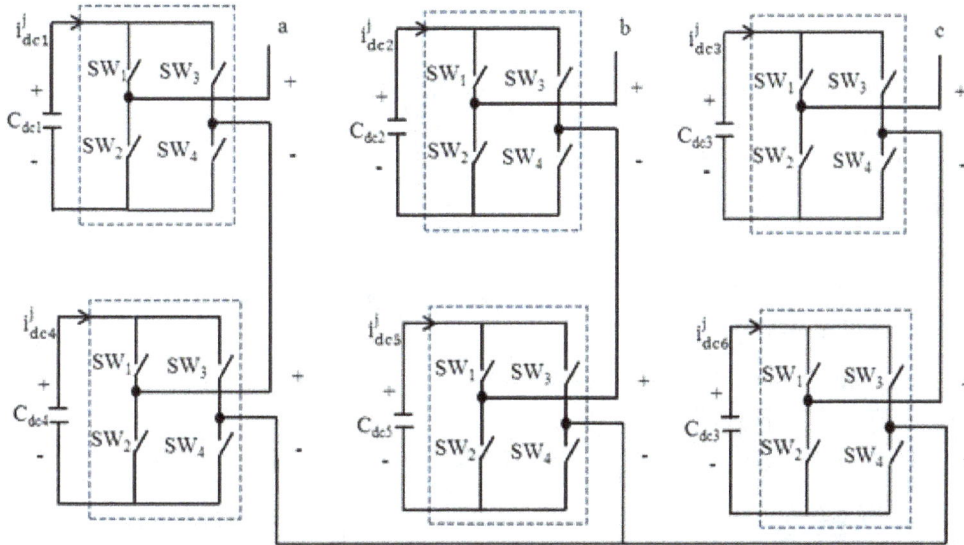

Figure 3: Topology for five level Cascaded H-Bridge Inverter.

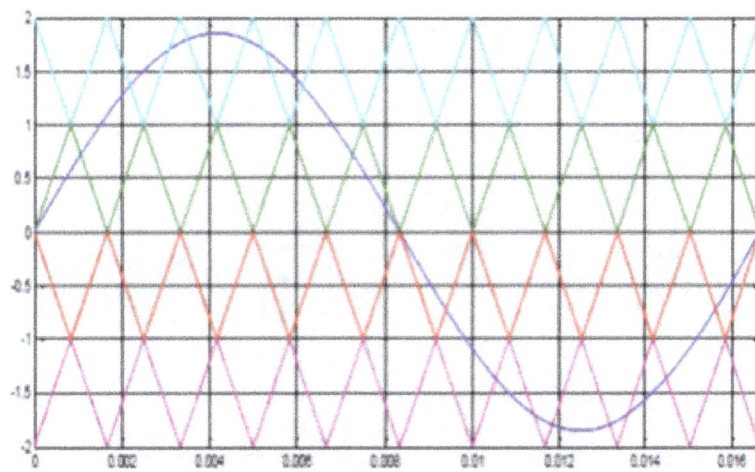

Figure 4: Control SPWM five level Cascaded H-Bridge Inverter.

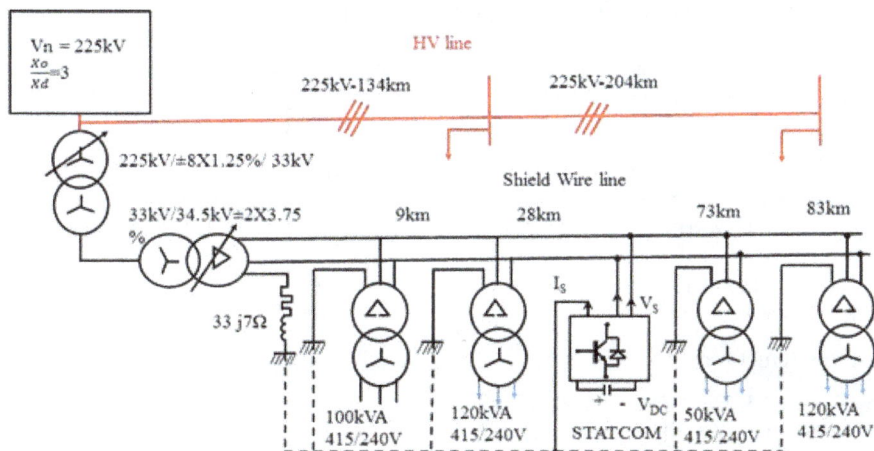

Figure 5: Three-Phase SWSs and STATCOM H-bridge.

Applying the voltage law in the AC side for the phase a of the STATCOM, the passive filter equation is obtained [14].

$$L_c \frac{d}{dt}i_c^a + R_c i_c^a + n_c s_c^a v_{dc}^a = v_s^a \qquad (1)$$

Where, s_c^a is the commutation function of modules of phase a and the resistor Rc is the parasitic (series) resistance of the inductor Lc.

Input voltage calculation

In order to evaluate the stationary behavior of the CHB STATCOM, the state variable equations must be evaluated in steady state [15]

Results and analysis of simulation

The Figure 5 has been implemented in Matlab/Simulink/SimPowerSystems. The nominal voltage V_n at the bus bar is 34.5 kV. The impact of variation of grounded resistor and inductance are evaluated [13].

$$\begin{cases} 0 = -\frac{R_c}{L_c}I_c^d + \omega_c I_c^q - \frac{n_c G_{ac} M_c^d V_{dc}}{L_c} + \frac{V_s^d}{L_c} \\ 0 = -\omega_s I_c^d - \frac{R_c}{L_c}I_c^d - \frac{n_c G_{ac} M_c^d V_{dc}}{L_c} + \frac{V_s^d}{L_c} \\ 0 = \frac{G_{ac}(M_c^d I_c^d + M_c^q I_c^q)}{2C_c} - \frac{V_{dc}}{R_{dc}C_{dc}} \end{cases} \qquad (2)$$

Where the variables in capital letters denote the constant value of each variable in an arbitrary operating point. The voltage should ensure that the summation of the DC voltages of all the power cells must be higher than the amplitude |Vs| of the source voltage. This constraint rises because VCSs operate with a DC voltage Vdc that must be higher than the peak value of the voltage in the AC side. A suitable value for Vdc should consider at least a 10% security margin. Then V_{dc} is:

$$V_{dc} = \frac{k_{dc}}{n_c}dc\sqrt{(V_s^d)^2 + (V_s^q)^2} \qquad (3)$$

Where k_{dc} kdc = 1.1 to ensure the 10% margin.

Case study 1: Variation of resistor

The grounded resistance is varied from 33 to 8 000 Ω (Figure 6). The system's three phase-to-phase voltages with passive components (capacitor bank). The rated voltage V_n is 34.5 kV. Grounded resistance variation affects the phase-to-phase voltage V_{ab}, V_{bc} and V_{ca}. The voltage rise or reduction varies +1.3% (for V_{ab}) to -6.6% (for V_{bc}). The most voltage reduction on V_{bc} and V_{ca} is due to the connection of phase C to the ground. The voltage unbalance factor varies from 2.8% to 8.18% (Figure 7).

In this case, the voltage amplitude variations varies between +0.37% (for V_{ab}) and -7.33% (for V_{ca}) (Figure 8). The use of the cascade H-bridge STATCOM makes an improvement in unbalance factor which varies between 0.15% and 3.24% (Figure 9).

Case study 2: Variation of the inductance

The grounded inductance is varied from 10 to 110 H (Figure 10). The rated voltage V_n is 34.5 kV. Grounded inductance variation affects the phase-to-phase voltage V_{ab}, V_{bc} and V_{ca}. The voltage rise or reduction varies +1.3% (for V_{ab}) to -23.3% (for V_{ca}). The most voltage

reduction on V_{ca} and V_{bc} is due to the connection of phase C to the ground. The voltage unbalance factor varies from 3.56% to 27.31% which is well above standard values (Figure 11).

In this case, the voltage amplitude variations varies between -5.31% (for V_{ca}) and +31.59% (for V_{bc}) (Figure 12). The use of the cascade H-bridge STATCOM makes an improvement in unbalance factor which varies between 0.2% and 2.31% (Figure 13).

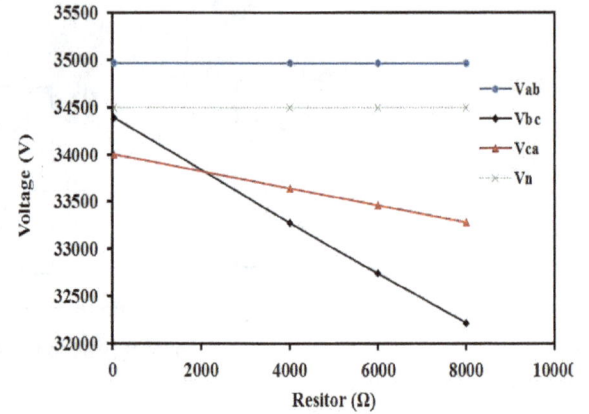

Figure 6: Voltage amplitude variation with passive components.

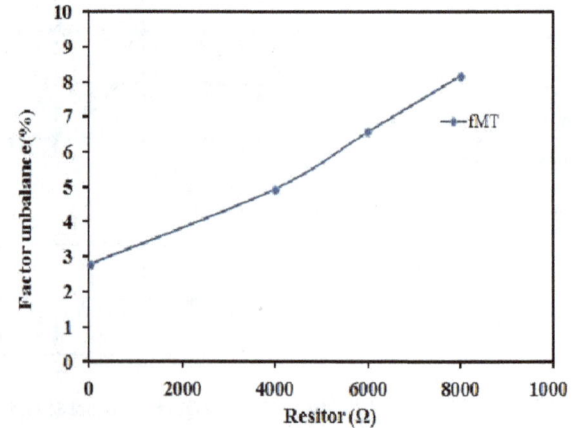

Figure 7: Factor unbalance with passive components.

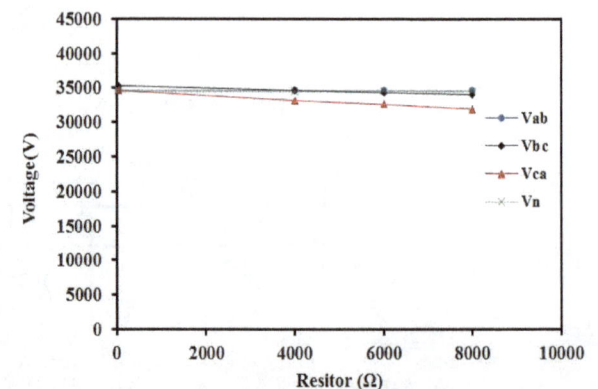

Figure 8: Voltage amplitude with cascaded H-bridge STATCOM: The system's three phase-to-phase voltages with cascaded H-bridge STATCOM.

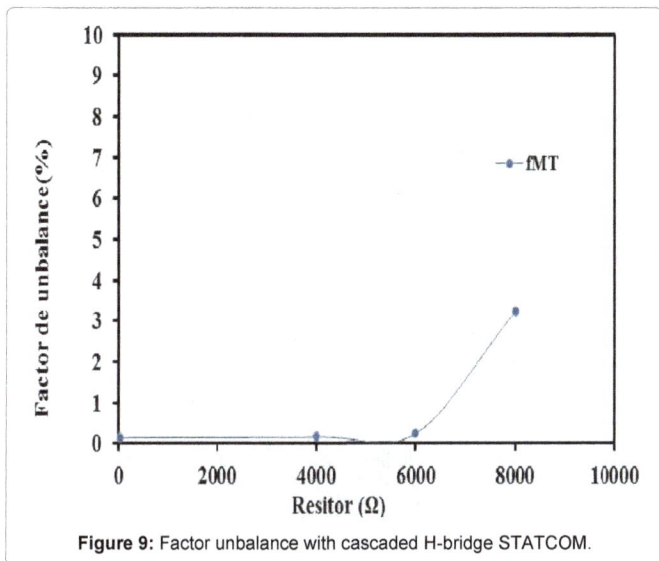

Figure 9: Factor unbalance with cascaded H-bridge STATCOM.

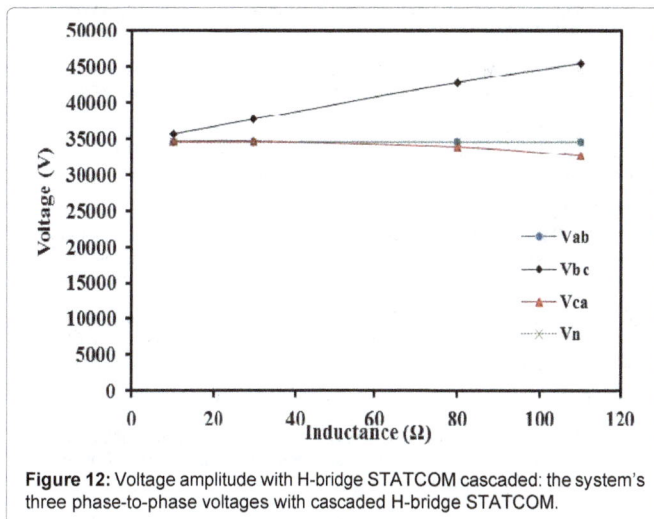

Figure 10: Voltage amplitude with passive components.

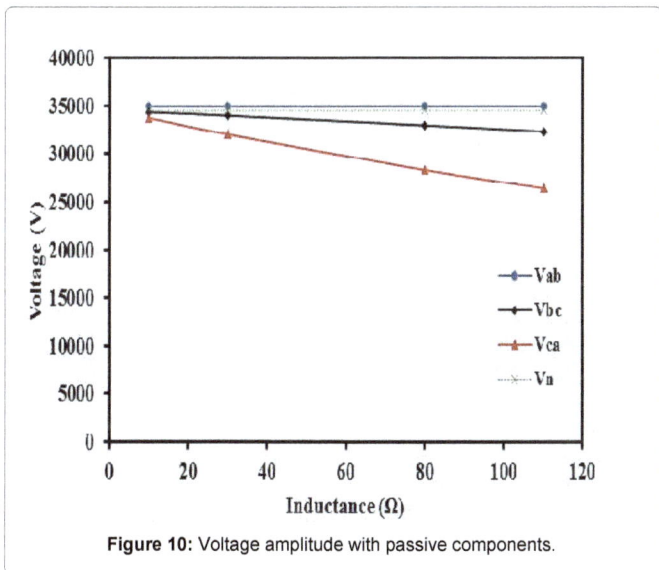

Figure 11: Factor unbalance with passive components.

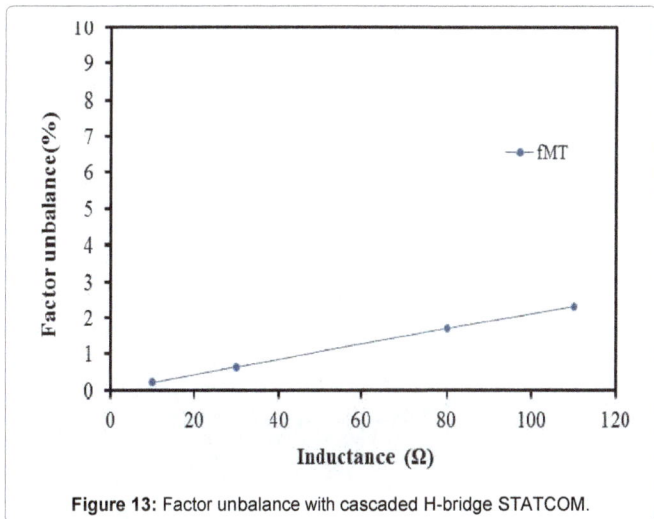

Figure 12: Voltage amplitude with H-bridge STATCOM cascaded: the system's three phase-to-phase voltages with cascaded H-bridge STATCOM.

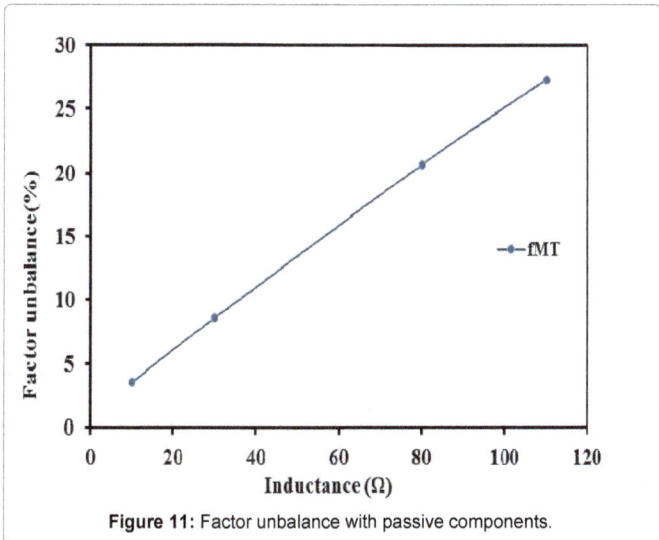

Figure 13: Factor unbalance with cascaded H-bridge STATCOM.

Conclusion

In this paper, we investigated the impact of the use of H-bridge cascaded STATCOM to reduce unbalance and voltage fluctuation on the three-phase shield wire. The unbalance and voltage fluctuation are created by the impedance difference.

The results show that the use of a STATCOM improves significantly the quality of the voltage of three-phase shield, with a very similar unbalance factor at acceptable limit compared to case when the passive components (capacitor bank) are used.

The variation of the inductance has a greater effect on the unbalance as that of the resistor. The method of unbalance reduction by H-bridge cascaded STATCOM is more efficient than passive components. From this study, we can conclude that the use of STATCOM can provide technical means to limit excessive voltage variation in SWS systems to ensure service's quality.

References

1. Iliceto F, Gatta FM, Cinieri E (1994) Rural Electrification of Developing Countries using the insulated Shield Wires of HV lines. New design criteria and operation experience. In Proc. CIGRE General Session.

2. Iliceto F, Cinieri E, Casely-Hayford L, Dokyi G (1989) New Concepts on MV Distribution from Insulated Shield Wires of HV lines. Operation Results of an

experimental system and Applications in Ghana. IEEE Transactions on Power Delivery 4: 2130-2144.

3. Iliceto F, Gatta FM, Lauria S, Dokyi G (1999) Three-Phase and Single-Phase Electrification in Developing Countries Using the Insulated Shield Wires of HV.

4. Cinieri E (1999) A new lossless circuit balancing the MV distribution systems from insulated shield wires of HV lines. Transmission and Distribution Conference, IEEE, New Orleans, LA 2: 722-728.

5. Tao Y, Wen-xia S, Cai-wei Y, Qing Y (2008) Experimental Investigation on the Impulse-Current Distribution of Grounding Electrodes with Various Structures. High Voltage Engineering and Application. International Conference on Chongqing: 269-272.

6. Alidou K (1991) Study of amplitude and phase imbalances of the tensions between networks Guard Insulated Cables electrical power lines Electrical Engineering. Brief, University of Quebec at Trois Rivières.

7. OTT R (2002) Qualité de la tension. Techniques de l'ingénieur, novembre.

8. Gatta FM, Iliceto F, Lauria S, Masato P (2005) Balancing methods of the 3-phase shield wire schemes. Browse Conference Publications, Power Tech, IEEE Russia, St. Petersburg.

9. Schéma directeur du réseau électrique, Archive de la Société Nationale d'Électricité du Burkina.

10. Melin P, Espinoza J, Guzman J, Rivera M (2013) Analysis and design of a Cascaded H-Bridge topology based on current-source inverters. Industrial Electronics Society, IECON-39th Annual Conference of the IEEE, Vienna.

11. Balde M, Doumbia ML, Cheriti A, Benachaiba C (2011) Comparative Study of NPC and Cascaded Converters Topologies. International Conference on Renewable Energies and Power Quality fevrier, Las Palmas de Gran Canaria, Spain.

12. Subbarao KRNV, Srinivas AVV, Krishna BV (2014) A level shifted PWM technique For controlled string source MLI. Power India International Conference(PIICON), 6th IEEE, Delhi.

13. Pillay P, Manyage M (2001) Definitions of voltage unbalance. IEEE Power Engineering Review 5: 50-51.

14. Shukla A, Nami A (2015) Multilevel Converter Topologies for STATCOMs, Springer Science.

15. Ibrahim ZB, Hossain L, Talib MHN, Mustafa R (2014) A five level cascaded H-bridge inverter based on space vector pulse width modulation technique. Energy Conversion (CENCON), IEEE Conference on, Johor Bahru.

5

An Electrically Tunable Liquid Crystal Lens for Fiber Coupling and Variable Optical Attenuation

Michael Chen, Chyong-Hua Chen, Yin-Chieh Lai, and Yi-Hsin Lin*

Department of Photonics, National Chiao Tung University, Hsinchu, 30010, R O C, Taiwan

Abstract

An electrically tunable Liquid Crystal (LC) lens for both of fiber coupling and variable optical attenuation is demonstrated. The LC lens modulates the beam waist coupling to the fiber by electrically changing the lens power. When the modulated beam waist is close to the core size of the fiber, the LC lens is operated as a lens coupler. When the beam waist increases by reducing the lens power, the LC lens is operated as a Variable Optical Attenuator (VOA) as a result of the corresponding coupling coefficient variation of the transformed beam into a multimode fiber. The study provides a way to design an optical device for fiber coupling and variable optical attenuation based on electrically tunable focusing optical component.

Keywords: Liquid Crystal (LC) lens; lens coupler; Variable Optical Attenuator (VOA)

Introduction

Variable Optic Attenuators (VOAs) are indispensable components for optical power control in the applications of telecommunication systems, high speed data networks and optical sensor systems. Several technologies have been applied to implement VOAs based on the optical mechanisms of reflection, diffraction, interference, absorption, or scattering. Important examples include optical Micro-Electromechanical System (MEMS) technology [1], acousto-optics [2], Planar Lightwave Circuits (PLC) [3,4], and Liquid Crystal (LC) technology [5-8]. In fiber systems, VOA is usually used with another component, a lens coupler, which function is to collimate the light coming out of a fiber or to couple the light into a fiber. A conventional lens coupler usually consists of lenses with a fixed focal length. Tunable focusing properties of liquid lenses with an electrically adjusting curved membrane or interface can be used for fiber coupling or waveguide coupling [9,10]. However, Fresnel reflection from the interface between two fluids, the thickness of the liquid lens, the gravity issue, and the difficulty of encapsulation still limit their applications. Up to now, the literatures only discussed the function of the lens coupler by using a tunable focusing liquid lens in a fiber system [9,10].

To make fiber systems more compact, an electrically tunable device with combinative functions of VOA and lens coupler is very advantageous. With the superior property of modulating lights through electrically tunable molecular orientations, Liquid Crystal (LC) is a great candidate for electrically tunable devices. Several scattering-type VOAs have been proposed to reduce the input power by using polymer-stabilized liquid crystals [6-8]. However, all these developed VOAs based on LC still require a lens coupler to couple the light into the fiber when applied to the fiber-connected optical systems. Our motivation is to design an electrically tunable LC device with functions of VOA and lens coupler for fiber-connected optical systems, which so far has not been realized. The LC phase modulator modulates the phase of the incident light. The lensing effect of the LC phase modulators is caused by the inhomogeneous spatial phase distribution originated from the inhomogeneous orientation distribution of the LC molecules, which can be controlled by the applied electric fields. The focal length of such a LC phase modulator or LC lens can be tunable as a positive lens or a negative lens depending on the electric fields [10,11]. LC lenses have various applications including image systems [12], pico-projector systems [13], optical zoom systems [14], concentrated photovoltaic systems [15], holographic systems [16], and ophthalmic lens [17].

In this paper, an electrically tunable LC lens for fiber coupling and variable optical attenuation is proposed and demonstrated. The LC lens electrically changes the lens power in order to adjust the beam size incident on the fiber. When the beam size is closed to the core size of the fiber, the LC lens is operated as a lens coupler. When the beam size increases by reducing the lens power, the LC lens is operated as a VOA. After introducing the operating principles, we use a LC lens to experimentally demonstrate the design concept and the results show that the maximum coupling efficiency can be 0.75 and the maximum attenuation can be as large as18 dB. The concept proposed in this paper is not only for LC lenses, but also for other optical components with electrically tunable focusing properties. The study provides a new way to design an optical device for fiber coupling and variable optical attenuation based on electrically tunable focusing optical components.

Mechanism and Operating Principles

Figure 1a and 1b depict a typical structure of the LC lens with positive and negative lens operations. We use the hole-patterned LC lens to demonstrate the concept [11,12]. In fact, the concept we proposed here can be applied to other structures of LC lenses. The components of the LC lens included three Indium Tin Oxide (ITO) glass substrates, two Polyvinyl Alcohol (PVA) layers for LC alignment, an insulating layer (Norland product Inc. NOA81), and a 50 μm nematic LC layer (Merck MLC-2070, $\Delta n=0.2609$ at 20°C at $\lambda=589.3$nm). In order to generate an inhomogeneous electric field, the ITO layer in the middle of the glass substrate was etched with a hole and connected to an applied Alternating Current (AC) voltage of V_1. PVA layers were coated on the glass substrates and mechanically rubbed in one direction to align the LC molecules. As a result, the LC molecules are aligned parallel to the glass substrate (x-direction in Figure 1) as $V_1=V_2=0V_{rms}$ and provide zero lens power (i.e. inverse of the focal length) for x-linearly polarized light.

*Corresponding author: Yi-Hsin Lin, Department of Photonics, National Chiao Tung University, Hsinchu, 30010, R O C, Taiwan
E-mail: yilin@mail.nctu.edu.tw

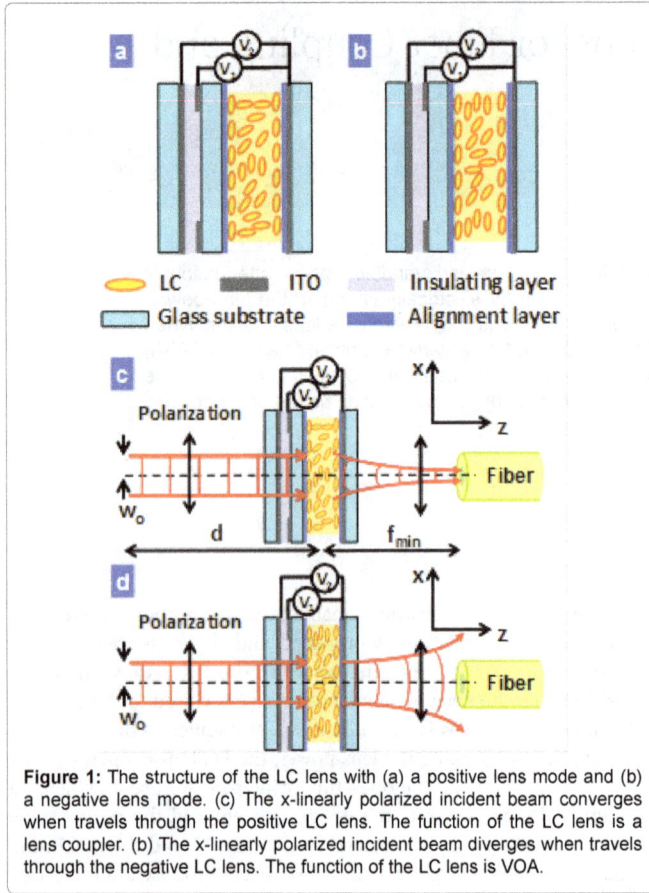

Figure 1: The structure of the LC lens with (a) a positive lens mode and (b) a negative lens mode. (c) The x-linearly polarized incident beam converges when travels through the positive LC lens. The function of the LC lens is a lens coupler. (b) The x-linearly polarized incident beam diverges when travels through the negative LC lens. The function of the LC lens is VOA.

As $V_1 > V_2$, the electric field near the peripheral region is stronger than in the center of the LC lens. As a result, the LC molecules at the center are more parallel to the x-direction and LC molecules near the edge are more parallel to the z-direction. Under this condition, the speed of x-linearly polarized light is slow in the center of the aperture and is fast in the peripheral region of the aperture. As a result, an incident plane wave of x-linearly polarized light is converted to a converged parabolic wave. The LC lens then provides a positive lens power with a parabolic spatial phase distribution, as illustrated in Figure 1a. Similarly, LC lens provides a negative lens power when $V_1 < V_2$, as shown in Figure 1b. As long as the aperture size (D) and the thickness (d_{LC}) of LC layer are fixed, the refractive indices difference (Δn) between the center and the peripheral region can decides the lens power of the LC lens (P_{LC}), which can be written in the form of [10]:

$$P_{LC}(V_1, V_2) = \frac{8 \cdot \Delta n(V_1, V_2) \cdot d_{LC}}{D^2} \quad (1)$$

When $V_1 > V_2$, the difference of refractive indices (Δn) is positive and the LC lens is operated as a positive lens. In addition, the lens power decreases as V_2 increases while V_1 is fixed. On the other hand, the LC lens is operated as a negative lens when $V_1 < V_2$ and the lens power gradually increases while V_1 increases and V_2 is fixed.

Figure 1c and 1d illustrate the operating principle of a LC lens for both fiber coupling and variable optical attenuation. A Gaussian beam with a beam waist of w_o impinges to a multimode fiber after propagating a distance of f_{min} (i.e. the minimum focal length). The distance between the beam waist (w_o) and the LC lens is d and the lens power of the LC lens is denoted as P_{LC}. Once we know the ABCD matrix of the system in Figure 1c and 1d, the beam waist (w) coupling to the fiber

can be expressed as $w = \sqrt{\lambda \times (A^2 + a \cdot B^2)/(a \times \pi \times (A \times D - B \times C))}$, where A,B,C and D are the elements in the ABCD matrix, λ is the optical wavelength, and $a = \lambda (\pi \times w_o^2)$. After we replace each element with the corresponding parameters mentioned above, the beam waist (w) can be written as a function of P_{LC}:

$$w = \sqrt{\frac{\pi^2 \cdot w_o^4 \cdot (1 - f_{min} \cdot P_{LC})^2 + \lambda^2 \cdot [f_{min} + d \cdot (1 - f_{min} \cdot P_{LC})]^2}{\pi \cdot w_o}} \quad (2)$$

For a lens coupler whose Numerical Aperture (NA) is much smaller than NA of the fiber, the influence of incident angle can be neglected while calculating the coupling efficiency. In our experiments, NA of the fiber (~0.29) is much larger than NA of the LC lens (<0.02). Therefore, we assume that the coupling efficiency (η) is only determined by the ratio of the fiber core radius (ρ) to the beam waist (w), which is equal to the following [18]:

$$\eta = \left[1 - \exp\left(-\frac{2 \cdot \rho^2}{M^2 \cdot w^2} \right) \right] \quad (3)$$

In Eq. (3), M^2 is the beam propagation factor defined as $\pi \times w \times \theta / \lambda$, where θ is the beam divergence of the incident wave. From Eq. (2) and Eq. (3), the coupling efficiency is determined by the incident beam waist (w), which can be continuously controlled by the electrically tunable lens power of the LC lens. As the lens power is relatively large (Figure 1c), the beam radius is similar to the fiber core radius, resulting in high coupling efficiency. The LC lens then works as a lens coupler. While the lens power decreases, the beam waist (w) is much larger than the radius of fiber core which reduces the optical power of the propagating light in the fiber. This means the optical attenuation can be adjusted by the LC lens. In Figure 1d, when a negative lens power is provided by the LC lens, the beam size is further enlarged and the optical attenuation range is also extended. Therefore, the LC lens can be a lens coupler and VOA at the same time. The mechanism illustrated above is applicable for either LC lenses with different structures or any tunable focusing devices whose lens power can be adjusted continuously from positive mode to negative mode.

Experiments and Discussions

In order to measure the lens power within an aperture of 1.33 mm under different applied voltages, the LC lens was placed between two crossed polarizers. The angles between the transmission axes of the polarizers and the rubbing direction were 45 degree. Therefore, the phase profiles with ring patterns were observed on the imaging plane as Figure 2a, 2b, and 2c. Figure 2a, and 2b show that the LC lens can either be a positive lens at $(V_1, V_2) = (85V_{rms}, 0V_{rms})$ or a negative lens at $(V_1, V_2) = (0V_{rms}, 40V_{rms})$. Figure 2c also displays a uniform phase profile at $(V_1, V_2) = (85V_{rms}, 85V_{rms})$ as the lens power of the LC lens was equal to zero. Two adjacent concentric rings in the phase profiles represent a phase difference of 2π radians. Based on Eq. (1), P_{LC} can be rewritten as $8 \times N \times \lambda / D^2$, where N is the number of concentric rings, λ is the operating wavelength, $\Delta n \times d_{LC}$ equals $N \times \lambda$, and D is the aperture size of the LC lens. Consequently, P_{LC} under different applied voltages can be calculated from the corresponding phase profiles, as shown in Figure 2(d). The LC lens is a negative lens at $V_2 = 40V_{rms} > V_1$ while it works as a positive lens at $V_1 = 85V_{rms} > V_2$. For $\lambda = 633$ nm, the evaluated lens power ranges from -8.9 m^{-1} to 30 m^{-1}. As to the lens power in the infrared region, we use Cauchy's equation, which indicates the dispersion relation of liquid crystal materials, to estimate the refractive indices of LC for the long optical wavelength [19]. Typically, the refractive indices of extraordinary (n_e) and ordinary (n_o) waves can be both expressed in the form of $n_{o,e} = A_{o,e} + B_{o,e} \cdot \lambda^{-2} + C_{o,e} \cdot \lambda^{-4}$ with coefficients (A_e, B_e, C_e) and (A_o, B_o, C_o), respectively. The corresponding coefficients of MLC-2070 measured by Abbe refractometer are $A_e = 1.7180$, $B_e = 0.0119$

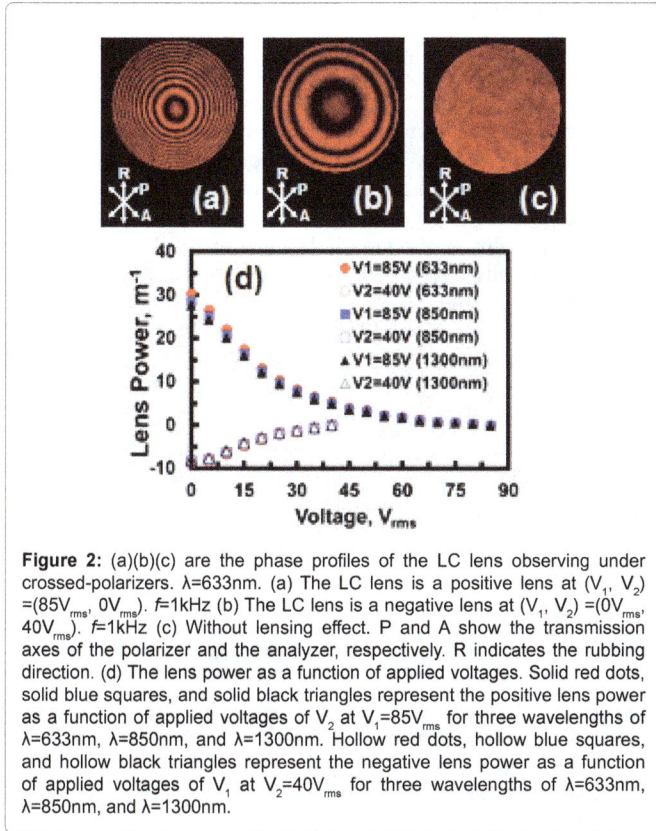

Figure 2: (a)(b)(c) are the phase profiles of the LC lens observing under crossed-polarizers. λ=633nm. (a) The LC lens is a positive lens at (V_1, V_2) =(85V_{rms}, 0V_{rms}). f=1kHz (b) The LC lens is a negative lens at (V_1, V_2) =(0V_{rms}, 40V_{rms}). f=1kHz (c) Without lensing effect. P and A show the transmission axes of the polarizer and the analyzer, respectively. R indicates the rubbing direction. (d) The lens power as a function of applied voltages. Solid red dots, solid blue squares, and solid black triangles represent the positive lens power as a function of applied voltages of V_2 at V_1=85V_{rms} for three wavelengths of λ=633nm, λ=850nm, and λ=1300nm. Hollow red dots, hollow blue squares, and hollow black triangles represent the negative lens power as a function of applied voltages of V_1 at V_2=40V_{rms} for three wavelengths of λ=633nm, λ=850nm, and λ=1300nm.

μm², C_e=0.0031μm⁴, A_o= 1.4899, B_o=0.0095 μm², and C_o=0.0002 μm⁴. Therefore, the lens power for λ=850 nm and λ=1300 nm can also be calculated from the results in visible region, which ranges are -8.3 m⁻¹~28.5 m⁻¹ and -8.1 m⁻¹~27.7 m⁻¹, respectively. As we can see in Figure 2d, the tunable lens power range decreases with the increasing optical wavelength because of the normal dispersion of MLC-2070.

In our experiments, we used three different laser sources for wavelengths of 633 nm, 850 nm, and 1300 nm: Melles Griot 05-srp-812, Power Technology APMT65 (850-100), and Agilent/HP 81552SM. Each laser passed through a single mode fiber (Thorlabs SM600 for visible light and Thorlabs SMF 28e+ for infrared light) and a lens collimator (Thorlabs F230FC-B for visible range, and Thorlabs F230FC-C for infrared range), which result in collimated incident beams with a diameter of 0.8 mm. In addition, a linear polarizer, whose transmission axis was parallel to the rubbing direction of the LC layer, was placed between each collimated laser beam and the LC lens. The incident light then coupled to a 30 cm long multimode fiber (Newport, F-MLD-500, core/cladding: 100 μm/140 μm diameter, NA=0.29) after passed through the LC lens. The distances between the LC lens and the fiber were decided by the minimum focal lengths at each wavelength, which were 3 cm as λ=633 nm, 3.3 cm as λ=850 nm and 3.4 cm as λ=1300 nm. The optical powers before and after transmitting through the fiber were detected by a power meter (OPHIR AN/2 for λ=633 nm and Newport 1916-R for λ=850 nm and λ=1300 nm). Therefore, we obtained the coupling efficiency (η) as a function of the lens power from the ratio of the measured powers, as shown in Figure 3a. Without the lens power of the LC lens, the coupling efficiencies are 0.027 for λ=633 nm, 0.025 for λ=850 nm and 0.031 for λ=1300 nm. When the lens power of the LC lens increases positively, the coupling efficiency also increases. The maximum coupling efficiencies are 0.75, 0.68, and 0.5 for wavelengths of 633 nm, 850 nm, and 1300 nm, respectively. This

indicates the LC lens is operated as a lens coupler because more lights are collected into the fiber. When the lens power of the LC lens increases negatively, the coupling efficiency decreases because of the divergence of the beam coupled to the fiber. The coupling efficiencies are different for the same lens power at different wavelengths, simply because the coupling efficiency depends on not only the lens power but also the wavelength. We further plotted the optical attenuation as a function of the lens power, as shown in Figure 3b. The optical attenuation is defined as 10×log₁₀ (η). In Figure 3b, the optical attenuation increases with a decrease of the lens power of the LC lens. This is because the beam size coupled into the fiber is larger as the lens power is smaller and then the optical attenuation is enhanced. The maximum attenuations are 17.3 dB at 633nm, 17.7 dB at 850 nm, and 16.7 dB at 1300 nm. Figure 3a and Figure 3b show that the LC lens can not only be an electrically tunable lens coupler but also an electrically tunable VOA.

Under the assumption of an ideal Gaussian beam (M^2=1), we can evaluate the beam waist (w) at different lens power from the known fiber core radius ρ=50 μm and Eq. (3). The results are plotted in Figure 4. Without focusing effect of the LC lens, the incident beams simply

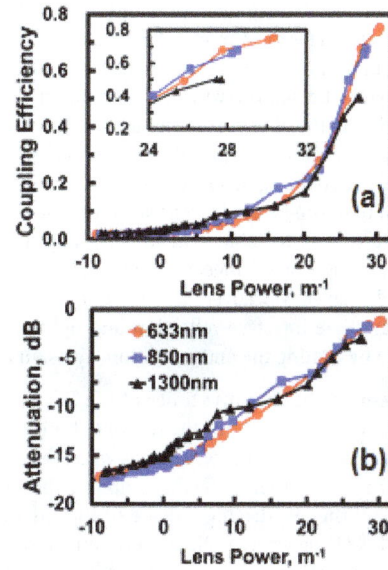

Figure 3: (a) Coupling efficiency as a function of the lens power at different wavelengths. Red dots, blue squares and black triangles stand for λ=633nm, λ=850nm and λ=1300nm, respectively. (b) Attenuation as a function of the lens power at different wavelengths.

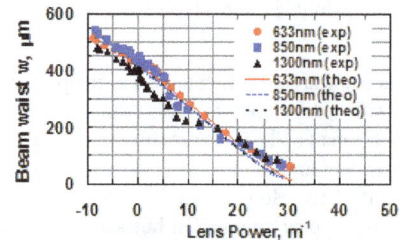

Figure 4: Beam waist (w) as a function of lens power. Solid red dots, solid blue squares, and solid black triangles represent experimental results at different wavelengths of 633 nm, 850 nm, and 1300 nm. Red solid line, blue dotted line, and black dotted line represent the theoretical results at different wavelengths of 633 nm, 850 nm, and 1300 nm.

pass through and maintain their beam waists (~400 μm). Because the beam waists coupling the fiber are much larger than the core radius (50 μm), the coupling efficiency is low (~0.027) and the optical attenuation is high (~15.7dB), as shown in Figure 3a and 3b. When the lens power increases, the beam waist turns out smaller which results in higher coupling efficiency and lower optical attenuation. The smallest beam waists for wavelengths of 633 nm, 850 nm and 1300 nm are 60 μm, 68 μm and 85 μm, respectively. At this moment, the beam waists are similar to the core radius of the fiber which causes maximum coupling efficiency as well as minimum optical attenuation. Therefore, an electrically tunable liquid crystal lens can be operated as an electrically tunable fiber coupling device with both functions of lens coupler and VOA by manipulating the beam waist coupling to the fiber. We also plotted the theoretical values of beam waists in Figure 4 based on an assumption of perfect thin lens and the relative experimental parameters in Eq (2): w_0=0.4 mm, f_{min}=33.0 mm for λ=633 nm, f_{min}=35.1 mm for λ=850 nm, f_{min}=36.1 mm for λ=1300 nm, d=12.5 cm for λ=633 nm, d=11.0 cm for λ=850 nm, and d=8.7 cm for λ=1300 nm. The theoretical minimum beam waists for 633 mm, 850 nm and 1300 nm are 16 μm, 24 μm and 37 μm, respectively. As a result, the theoretical minimum beam waists are smaller than the experimental values for all three wavelengths. This indicates that the phase profiles of the LC lens with large lens power are not identical to the parabolic distribution which is used in our theoretical model. Furthermore, scattering effect originated from inhomogeneous LC layer may also enlarge the beam waist. This explains the reason why the experimental beam waists cannot be smaller than the fiber core radius at the maximum lens power and restricts the maximum coupling efficiency to be lower than 0.8. In order to increase the coupling efficiency as well as optical attenuation, we not only need to optimize the LC lens structure for parabolic phase distribution, but also have to enlarge the tunable range in negative lens operation by using high bi-refringent LC materials or increasing the thickness of the LC layer. Moreover, the insertion loss of the lens was ~0.7dB due to multiple interface reflection and light scattering. This might be reduced by coating the anti-reflection layers on the LC lens.

We also measured the response times of the LC lens. As the lens power switched from 0 m^{-1} to 30 m^{-1}, the response time was 762 ms. On the other hand, the switching time was 382ms when the lens power was turned back to 0 m^{-1}. In order to further improve the response time, we can either adopt the overdriving scheme of applied voltages or use polymer network LC materials [19]. In addition, polymer-stabilized blue phase liquid crystals have been used to realize a LC lens with submilliseconds response time, which shows great potential for this study [20]. The LC lens in this paper still requires high applied voltages. This can be solved by using many other structures of LC lens operated with a driving voltage below 10 V$_{rms}$ [10,21]. Moreover, the optical efficiency can be enhanced by using polarization independent LC phase modulator, including double layered type, residual phase type or mixed type [17,20,22-30].

Compared to prior arts, the attenuation in the prior arts is around 30dB and polarization dependent loss less than 4dB with the response time of sub-ms~80 ms [1-7]. The VOA function we demonstrated is polarization dependent and the maximum attenuation is 15.7 dB with a switching time of ~1 sec. However, LC lens can provides the function of not only variable optical attenuation but also fiber coupling which differs from prior VOAs. Compared to typical lens couplers, such as GRIN lens coupler with a fixed focal length (low coupling loss ~0.1 dB), the LC lens not only serves as a lens coupler but also is able to control the attenuation level of optical power as a VOA due to an electrically tunable focal length of the LC lens. Therefore, the device we proposed

outperforms other types of VOA because of its dual functionality in a single device.

Conclusion

A mechanism for fiber coupling and variable optical attenuation via an electrically tunable LC lens is demonstrated. The beam waist coupling to the fiber can be controlled by adjusting the lens power of the LC lens. Therefore, the coupling coefficient and optical attenuation can be changed due to the tunable lens power. We have shown the experimental results for three wavelengths from visible to near infrared range. For broad band fiber systems, compact LC lens is a good candidate as lens coupler and VOA. Since polarization dependency of the LC lens, polarization-maintaining optical fibers are required in the practical applications. This concept is also applicable for other optical devices with tunable focusing ability. This study has several potential applications, such as gain-flattened amplifying systems and optical power equalization systems.

Acknowledgements

This research was supported by the National Science Council (NSC) in Taiwan under the contract no. 101-2112-M-009-011-MY3.

References

1. Marxer C, Griss P, de Rooij NF (1999) A variable optical attenuator based on silicon micromechanics. IEEE Photon Technol Lett 11: 233-235.

2. Mughal MJ, Riza NA (2002) Compact acoustooptic high-speed variable attenuator for high-power applications. IEEE Photon Technol Lett 14: 510-512.

3. Hurvitz T, Ruschin S, Brooks D, Hurvitz G, Arad E (2005) Variable optical attenuator based on ion-exchange technology in glass. J Lightwave Technol 23: 1918-1922.

4. Nishi H, Tsuchizawa T, Watanabe T, Shinojima H, Yamada K, et al. (2010) Compact and polarization-independent variable optical attenuator based on a silicon wire waveguide with a carrier injection structure. Jpn J Appl Phys 49: 04DG20.

5. Riza NA, Khan SA (2004) Liquid-crystal-deflector based variable fiber-optic attenuator. Appl Opt 43: 3449-3455.

6. Du F, Lu YQ, Ren HW, Gauza S, Wu ST (2004) Polymer-stabilized cholesteric liquid crystal for polarization-independent variable optical attenuator. Jpn J Appl Phys 43: 7083-7086.

7. Chen KM, Ren HW, Wu ST (2009) PDLC-based VOA with a small polarization dependent loss. Opt Commun 282: 4374-4377.

8. Wu YH, Liang X, Lu YQ, Du F, Lin YH, et al. (2005) Variable optical attenuator with a polymer-stabilized dual-frequency liquid crystal. Appl Opt 44: 4394-4397.

9. Shaik RP, Monch W, Krause H, Zappe H (2008) Reconfigurable liquid micro-lens system for variable fiber coupling. International Conference on Optical MEMs and Nanophotonics, Freiburg, Germany.

10. Ren HW, Wu ST (2012) Introduction to Adaptive Lenses. Wiley, Hoboken, NJ, USA.

11. Lin HC, Chen MS, Lin YH (2011) A review of electrically tunable focusing liquid crystal lenses. Trans Electr Electron Mater 12: 234-240.

12. Lin HC, Lin YH (2010) A fast response and large electrically tunable-focusing imaging system based on switching of two modes of a liquid crystal lens. Appl Phys Lett 97: 063505-063505-3.

13. Lin HC, Lin YH (2010) An electrically tunable focusing pico-projector adopting a liquid crystal lens. Jpn J Appl Phys 49: 102502-102502-5.

14. Lin YH, Chen MS, Lin HC (2011) An electrically tunable optical zoom system using two composite liquid crystal lenses with a large zoom ratio. Opt Express 19: 4714-4721.

15. Tsou YS, Lin YH, Wei AC (2012) Concentrating photovoltaic system using a liquid crystal lens. IEEE Photon Technol Lett 24: 2239-2242.

16. Lin HC, Collings N, Chen MS, Lin YH (2012) A holographic projection system with an electrically tuning and continuously adjustable optical zoom. Opt Express 20: 27222-27229.

17. Lin Y, Chen HS (2013) Electrically tunable-focusing and polarizer-free liquid crystal lenses for ophthalmic applications. Opt Express 21: 9428-9436.

18. Niu J, Xu J (2007) Coupling efficiency of laser beam to multimode fiber. Opt Commun 274: 315-319.

19. Yang DK, Wu ST (2006) Fundamentals of Liquid Crystal Devices. John Wiley & Sons Ltd, Chichester, UK.

20. Lin YH, Chen HS, Lin HC, Tsou YS, Hsu HK, et al. (2010) Polarizer-free and fast response microlens arrays using polymer-stabilized blue phase liquid crystals. Appl Phys Lett 96: 113505-113505-3.

21. Lin HC, Lin YH (2012) An electrically tunable-focusing liquid crystal lens with a low voltage and simple electrodes. Opt Express 20: 2045-2052.

22. Ren HW, Lin YH, Fan YH, Wu ST (2005) Polarization-independent phase modulation using a polymer-dispersed liquid crystal. Appl Phys Lett 86: 141110.

23. Lin YH, Ren HW, Wu ST (2009) Polarization-independent liquid crystal devices. Liq Cryst Today 17: 2-8.

24. Lin YH, Ren HW, Fan YH, Wu YH, Wu ST (2005) Polarization-independent and fast-response phase modulation using a normal-mode polymer-stabilized cholesteric texture. J Appl Phys 98: 043112.

25. Ren HW, Lin YH, Wen CH, Wu ST (2005) Polarization-independent phase modulation of a homeotropic liquid crystal gel. Appl Phys Lett 87: 191106.

26. Lin YH, Ren HW, Wu YH, Zhao Y, Fang JY, et al. (2005) Polarization-independent liquid crystal phase modulator using a thin polymer-separated double-layered structure. Opt Express 13: 8746-8752.

27. Ren HW, Lin YH, Wu ST (2006) Polarization independent and fast-response phase modulators using double-layered liquid crystal gels. Appl Phys Lett 88: 061123.

28. Huang Y, Wen CH, Wu ST (2006) Polarization-independent and submillisecond response phase modulators using a 90° twisted dual-frequency liquid crystal. Appl Phys Lett 89: 021103.

29. Lin YH, Tsou YS (2011) A polarization independent liquid crystal phase modulation adopting surface pinning effect of polymer dispersed liquid crystals. J Appl Phys 110: 114516.

30. Lin YH, Chen MS, Lin WC, Tsou YS (2012) A polarization-independent liquid crystal phase modulation using polymer-network liquid crystals in a 90 degree twisted cell. J Appl Phys 112: 024505.

6

The Need for a Whole Life Framework in Electrical Power System Asset Management and the Problems with Individual Silo like Asset Management System Contributions

James Bruiners*

Robert Gordon University, London, London UK

Abstract

The asset management of electrical power systems both in industry and academia currently offers up a wide spectrum of engineering practices and guidance to detect, diagnose and to combat asset deterioration. In doing so research largely aims to better enable enhanced decision or prediction making in single engineering applications such as partial discharge or dielectric condition measurements. This paper reviews the current state of the art as related to electrical asset management and sets out how todays engineering and maintenance heavy approaches are insufficient to meet the nature of the complex adaptive (Generation, Transmission & Distribution) systems. Furthermore this paper explains the need for a truly holistic governance framework capable of managing such complexities for power companies by using complex adaptive system science whilst grounded in the engineering, business and socio-technical attributes that applied engineering makes possible. In addition this paper outlines how utilising retroductive case study with hypothesis framework represents the best approach in creating such a fully holistic asset management capability for power companies.

Keywords: Framework; Electrical; Power; System; Asset

Introduction

This paper is concerned with exploring the topic of asset management for power companies and in doing so will forge the basis of justification for research in applied physical asset management of power systems assets in more holistic capabilities. It will also serve as means of representing the research methodologies utilised across the scientific community whilst including other important factors such as relevant industry and market conditions. Thus presenting the relevant findings and highlighting gaps in research, industry and furthermore present such information in light of the current state of the art as compared to the aims, objectives and proposal subsequently outlined in this paper. The overarching literature considerations are purposely designed to take a blended approach to the wider contributions in Asset Management, utilising a mixture of system engineering practices and making use of cognitive and requirements alignment techniques as described by the likes of Kirova et al. [1] and Mazak et al. [2] in providing a solid mechanism for a 'traceability matrix'. In addition to other inputs such as from complex adaptive system ideologies and concepts [3-5].

The specific aims and objectives of this paper are to provide a holistic management framework that will ensure a power company can effectively manage its physical assets across the entire lifecycle. That consists of the ability to manage in a more agnostic manner, whilst not ignoring prevailing and evolving organisational contributions and advancements. Furthermore this paper will highlight the necessary legacy components and outline how to aid the power company to continually improve towards optimised critical factors such as Risk, Cost and Human factors in doing so.

In March of 2016 an analysis of related scholarly papers and articles related to asset management was performed, this was taken across a ten year period from 2005 to 2015. Two key search terms "Holistic Asset Management" and "Power Systems Asset Management" were performed. In the first instance a mere thirty two relevant results were presented, of which it must be stated not all were entirely relevant. The latter search and more directly concerned to power systems asset management resulted in an even lower yield of just twelve results. A third search term that yielded zero results which was "Holistic Power Systems Asset Management", for brevity this was omitted from the graph below (Figure 1).

The low yield of papers and relevant academic platform to utilise or reference highlights in part a requirement for further investigation, especially in light of such a heavy scholarly reliance on direct activities in managing power systems assets such as maintenance programs/schedules or maintenance based optimisation [6] and although valid in part such maintenance prominent techniques only represent a

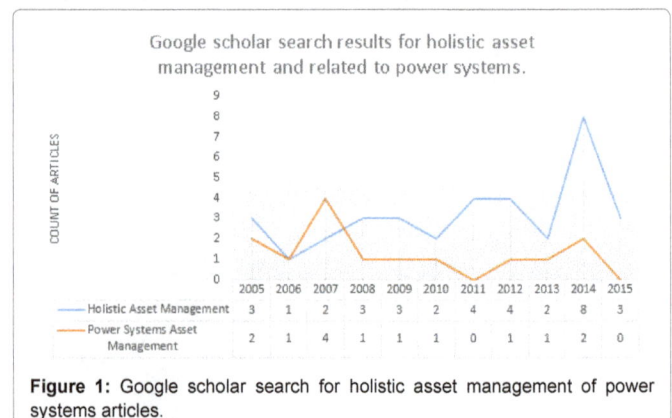

Figure 1: Google scholar search for holistic asset management of power systems articles.

***Corresponding author:** James Bruiners, Robert Gordon University, London, London UK, E-mail: jabruiners@outlook.com

single minded view on how to best manage assets through singular engineering disciplines.

Given that the aims and objectives of this research is to provide much needed guidance and method of assisting a power companies optimised approaches towards asset management it is important to build credible, reliable and relevant body of knowledge across all disciplines that apply to the management of power systems assets across the whole life cycle of the assets.

By reviewing the current state of the art in conjunction with the proposed over-developed prominence in singular asset management engineering disciplines enables the gaps in current thinking to be better established. Evaluating these existing gaps in holistically managing power system assets across the lifecycle will be better suited to evaluate the need for creating such holistic management governance through a synthesised framework. Since literature has not dealt with research design in [Asset Management] it would be beneficial to review the literature on research methodology and provide some guidelines [7]. Therefore when approaching this topic of research method relevance it is important to examine a wide platform of necessary facets (research design, literature review, industry and complexities etc.) all embedded in this literature review. This will ensure that the Asset Management as a total approach to managing physical 'things' for the power company can be better established.

By taking a broader approach the research can better establish a solid basis and fundamental framework/approach that intern can be founded both in academia and applied capability (Figure 2).

Although Figure 2 offers the guiding principle and subsequently the most mature approach it isn't sufficient enough for whole life cycle power systems management. This is important as applied to power systems given the long operating time frames for certain assets, such as Power Transformers, Cables, Towers, Control Systems etc., some of the major difference in this case pivots on the application and interpolation of three other major areas:

- Whole-life-cycle considerations for assets, not only in capitally intensive environments.

- Application to electrical power systems assets.

- The complex adaptive system properties that enable a more socio-technical and socio-economic capability.

These additions and changes are represented by the research schematic workflow below and as applied for a suitable approach in researching asset management for power companies across asset whole life cycles. Although the process of research will lend similarly to the framework, it still requires entirely different, more applied and include elements of complex adaptive systems (when desribing low load or distrubuted generation etc.) to the power systems engineering and mangemetn fratenerties, this be more robugstly explored throughout this paper (Figure 3) [7].

Asset Management: General

An overwhelming focus and attention of the management of physical Assets pivots on and around maintenance centred management [6]. A vast number of papers and research publications in the areas of Asset Management often highlights specific and individual areas such as Maintenance and Risk [7], Procurement [8], Deterioration Mechanics and Sciences surrounding diagnosis [9]. In particular publications point more heavily towards singular fields of engineering disciplines and quite often based on maintenance strategies and implementations of maintenance based asset integrity management [10].

Other asset focused elements such as planning [11]; asset condition and life assessment [12]; renewal/replacement [13]; outsourcing maintenance or contracts [14]; predictive plant maintenance [15]; performance strategy and performance management and measurements; information management amongst others. In large these asset management contributions are mostly conducted in specific asset related activities rather than a wider more enterprise and entire life-cycle systems approach [14-16].

The complex nature of infrastructure and power companies can be likened to that of complex adaptive systems [5], although major acknowledgements in the area of complex adaptive systems in infrastructure are predominately concerned with supply chain Brown et al. [11], and lacking the capability to influence in a system wide focus, especially in regards to whole-life-cycle and holistic asset management. These singular vein approaches of dealing with asset management could be therefore considered a complex problem such as indicated by McArthur et al. [17], examples include interlinked mechanical failures, human judgment, engineering design etc.

As the management of systems through systems design is generally

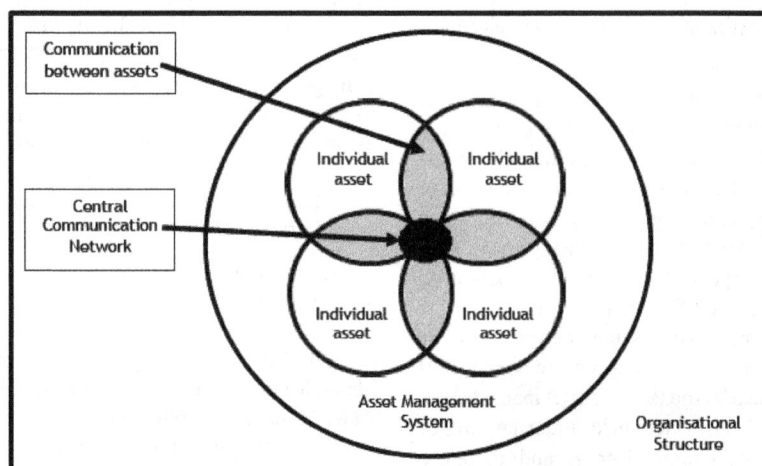

Figure 2: Research workflow for Capital Intensive asset management organisations [5].

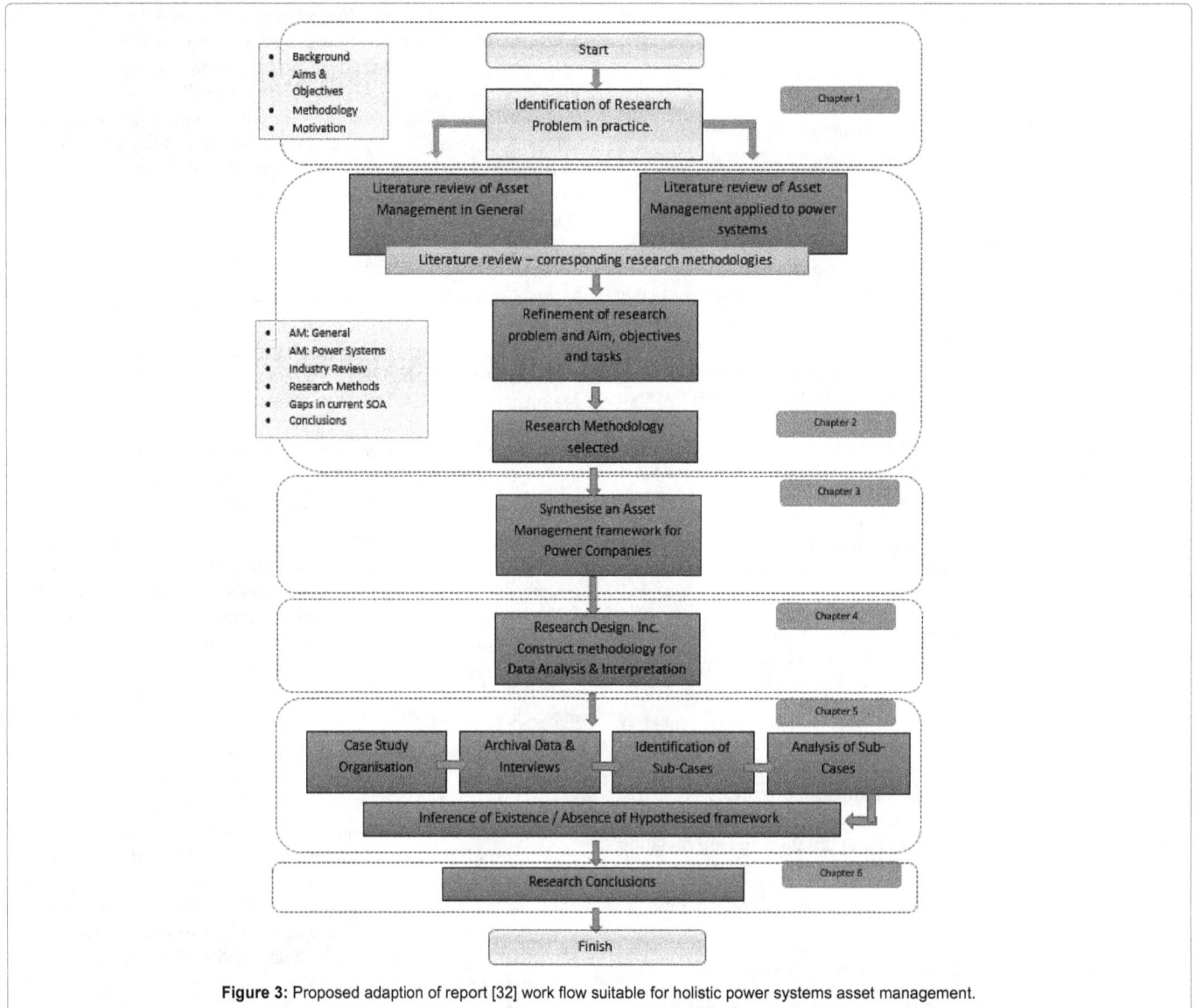

Figure 3: Proposed adaption of report [32] work flow suitable for holistic power systems asset management.

lacking in broadness and definition, likely born out of the seed of a lack of scholarly depth and the complexities [16,18-21] enhances the reason for the needed synthesis of a framework to deal with asset management in more holistic manners.

The Asset Management Council states that asset management is "The life cycle management of physical assets to achieve the stated outputs of the enterprise." Asset Management is "The system that plans and control the Asset-related activities" again upholding the concept that power system asset management represents a more complex system, given this argument it appears that asset management is to be considered a collection of interdependent and connected activities, not just singular activities concatenated [7]. Both go further than most to describe in capital intensive organisations the requirements for additional definition of what asset management in a more holistic life-cycle capacity truly means, but concedes the fact that asset management lacks understanding [7] as there is no widely held understanding of its role in organisational strategy, competitiveness and therefore success. The lack of definition of holistic asset management has led in part to various organisations and groups such as the Institute

of Asset Management to form frameworks and incorporate more multidisciplinary sets of interpretations, in essence trying to holistically approach this already broad and undefined term known as Asset Management, Mehairjan et al. [7] as problematic in that asset management is "increasingly a focus of attention in various industries to address interrelationships between the lifecycle of asset and the risks associated with the performance of asset related activities." However, the study falls short as it then proceeds to narrow the scope to only that of which greatly influences strategic elements of capital intensive organisations [22-27].

Asset Management has been viewed historically in a technical activity driven often by engineering functions that represented reliability centred maintenance. Literature appears not broad enough in focus, therefore highlighting a lack of Asset Management in whole life cycle approaches. The maintenance has an important stake hold in Asset Management but is only one part of a much larger 'variable' in effective management of assets across the asset whole life [19].

The holistic requirement and need for an approach in strategically measuring and applying effective asset management across the entire of lifecycle of an electrical asset is partially defined as above, yet still not fully developed; Developed meaning the fully integrated and enterprise wide management capability that enables all elements of the company to work together in a well-coordinated and optimised asset management manner.

The term 'holistic' is also an important one, industry specifications such as PAS 55:2008 defines the total life cycle of assets, "The time interval that commences with the identification of the need for an asset and terminates with the decommissioning of the asset or any liabilities there after". Brown et al. [11] demonstrate a key consideration towards holistic asset management, however, as above narrowly focus the efforts towards maintenance, only an element of the asset underlining lifecycle. This is confirmed through the works of Tse et al. [28] stating that "Asset Management is a holistic and interdisciplinary approach that covers the context of engineering asset the whole life cycle of the asset, from the acquisition to the disposal of the asset." Therefore any future research related to the holistic understanding of asset management must fully deal with the entire asset life cycle to be considered holistic.

Asset Management: Power Systems

The corresponding research that can be applied to holistic asset management and furthermore specifically in the case of power systems is somewhat opaque and weak from an academic viewpoint. This is likely due to the nature of holistic asset management only being discussed in scholarly works in the past decade as and as explained earlier.

The report in ref. [18] goes much further to create a compendium of works and practise that can assist in engineering fraternities, contributing to better understanding of individual realms as related to asset management but offers again no holistic framework or presence for power assets to be holistically represented in sufficient detail.

Again the overwhelming focus leans to a number of individually targeted and often engineering/technical related papers to power systems asset management. Trend analyser's Bonifay this prominence in technical direction [18]. Like in general asset management terms this engineering or technical focus plagues power asset management with the same inability to work holistically across the entire life cycle of the electrical systems assets. Multiple works can be cited in silo type engineering or maintenance/condition monitoring [4], wind power reliability engineering [21], Health Optimisation Modelling [22] and even though they attempt to add value to singular issues within power systems they do not approach the problem in light of a truly holistic capability, nor do they reference higher level reference architectures of which could be derived from and in as much seriously undermining the ability to provide aligned or synchronised asset management capabilities.

This once again adds to the complexity and difficulty as not addressing that asset management is not a single isolated existing entity within an organisation [7]. Furthermore in the power engineering landscape this appears to be an even more underdeveloped. Research topics based majorly on condition monitoring techniques such as Partial discharge [12] or leakage current monitoring. Again only tackling an engineering variable in and amongst the total holistic picture of asset management for power system assets. Reports in ref. [14,17] start to expand on the need for better planning of power systems in light of greater asset management implications through their approach that "Asset management is the art of balancing cost, performance, and risk." However their study dropped back into the engineering lines of

reliability centred approaches, likewise inhibiting the ability to develop applied holistic asset management to power systems.

Considering the technical prominence on maintaining assets in light of both general asset management and as related to power systems raises additional gaps, decision making processes "should be incorporated into an overall asset management concept such as described in PAS 55 or ISO 55000" yet offers no relevant framework applicable to power systems life cycle management [21,22]. The lack of both academic and applied works to achieve these goals (goals, meaning the overall asset management framework for power companies) underlines the troubles in power company's abilities to effectively apply whole life cycle asset management for its power system assets.

It can also be seen that the emerging research around power systems in light of the term 'Asset Management' lends towards power transformer assets in isolation, reports such as refs. [3,23], in these cases an advancement in engineering technological approaches both fall short of finding the relevant business approaches to supporting the holistic life cycle of the portfolio of assets.

Although, Mehairjan et al. [7] goes some way to initiate the conversation of Asset Management holistically the study falls grossly short in the ability to apply models out of strategic and or capital investment realms, furthermore the study applies it solely to capital intensive organisations and limited at large to strategic only capacities. For example he invites the reader to accept challenges in the life cycle of asset management ownership but reduces the scope by only dealing with capitally intensive situations; this does not apply well when considering power system assets as explained by this work earlier, inevitably leaving a significant gap in knowledge.

Furthermore these difficulties in finding the most fitting adaptation of asset management compound the problem. This especially when considering power engineering as an activity under a broader management, strategic framework or network of interconnected activities to build up best asset management practices. However this fits with the prognosis set forth earlier especially when considering that the topic of holistic asset management has only been developed in literature over the past twelve to thirteen years. Prior research and detailed reports rely heavily on specific related activities, following the relevant and logical adaptations of operations management, engineering and or system engineering.

However, as quantitative and positivist methodologies previous methods maybe there is little in the way of affirming the total complexity of asset management from 'Concept to Disposal' for the power industry. A few recognise the role of qualitative approaches in developing solid research, this needs to be considered when approaching the applicability of research topic in light of the state of the art in regards to asset management as a system and as asset management related to power companies [8].

Two prominent facets apply when approaching the theory of Asset Management which can be considered the discrete contributing activities and the interrelated and correlated system activities that constitutes Asset Management [5]. Before this paper goes someway to prescribing an optimised pathway for these interconnected activities it is important to also highlight that a combination of optimised strategies' must be deployed, therefore any research concerned with bridging this gap must go some way to understand, in the fullest, major and discrete business perspectives in doing so.

Notable Considerations

In this section the Asset Management research considerations will be exposed in greater detail, this is especially important when reviewing the general lack of clarity in research towards holistic asset management. This chapter will build on the previous sections in evaluating the nature of asset management, the resulting research and the approaches in Asset Management when considering guiding approaches. Detailing throughout principles that will govern research design in such criteria for example as quantitative and qualitative approaches. Furthermore this chapter will deal with the rationalisation behind case study research as an appropriate research method towards building a stronger body of knowledge through a more refined and better contribution of research [7].

What is Industry Saying About Asset Management

At times key motivators in developing a research topic are geared toward applying the findings in real life situations [9], often this is in light of the ability to increase decision making capabilities. This effect stipulates that all management research should be applied in essence [26]. Thus the ability to make better decisions in asset management capabilities such as planning, analysis, contracts and asset creation etc. Difficulties in applying academic works with its applied industrial counterpart by trying to close the gap between these two fields [27]. As report [25] acknowledges that, "Applied research is by nature much more problem oriented and could potentially alleviate the problem to value the applicability of envisioned research output". This gap between the research topics as highlighted in previous sections and as related to singular silos of asset management activities has widely led to the creation of arbitrary consultative frameworks as prescribed by the likes of the Institute of Asset Management, the Asset Management Council and the Asset Institute. The lack of holistic asset management frameworks in scholarly terms represents a truly challenging situation for the industries that could be suited to implementing such frameworks. ISO 55000 and PAS 55:2008 represent such standards or frameworks to assist, but themselves are not strongly implemented guidelines. Although these frameworks go some way to assist companies in better managing the holistic picture they fall short of providing the necessary guidance or rigorous testing of a dedicated research thesis.

Research Methods

This section will review the effectiveness of case study research in light of previous works and make critical analysis towards how to best

approach the proposed value as in light of this thesis. Due to the lack of maturity in the research of holistically applied asset management or indeed as applied to power systems creates a requirement to evaluate the mixtures of research methodologies. However, it is clear that the reports in refs. [7,27,28] do employ case study with hypothesised model.

Fortunately there is a wealth of new and historic material, research and information regarding principles of case study research methodologies as applied to engineering, power and related business systems. The importance of stipulating that none of the methods evaluated are thought to build a singular body of knowledge, considering the infancy in scholarly remit, the 'blended approach' to research methods will deliberately expand the capability to look further in building the most effective case study research possible for the hypothesis [19,29]. This will manifest itself as an applied mixture of the best suited research methods in essence to facilitate an effective, broader ability to utilise case study research. As ref. [5] prescribes a retroductive strategy with case study method to yield an appropriate manner in which to approach the research method. As the prior scholarly evidence demonstrates initial gaps in research, both as applied to asset management and asset management as applied to power systems assets. This is especially prevalent when considering the need to utilise, at times best practices and or established techniques from other research as part of any holistic asset management framework [28-30]. When considering the concept of retroduction in this manner it becomes apparently clear in light of the argument that "the dualism between pure inductive and deductive research processes can be overcome by introducing retroduction [30-32]. Retroduction makes this research possible by the process that characterized the linking of evidence (induction) and social theory (deduction) in a continually evolving, dynamic process." Thus report in ref. [28] approach is supported by the fact that Power Companies comprise of socio-technical diversity highlights and representative of complex adaptive systems (Figure 4) [11].

However as reported in ref. [29] this does not go far enough in holistically explaining the framework that builds up effective asset management and all the phenomenon, especially not nearly enough to apply to power companies. Given the explanation offered up by Smith and Stirling [33] that a case study represents an empirical inquiry that investigates a contemporary phenomenon in depth and within its real-life context, especially when the boundaries between phenomenon and context are not clearly evident, would therefore serve an appropriate research method for asset management. Taking a hypothesised model

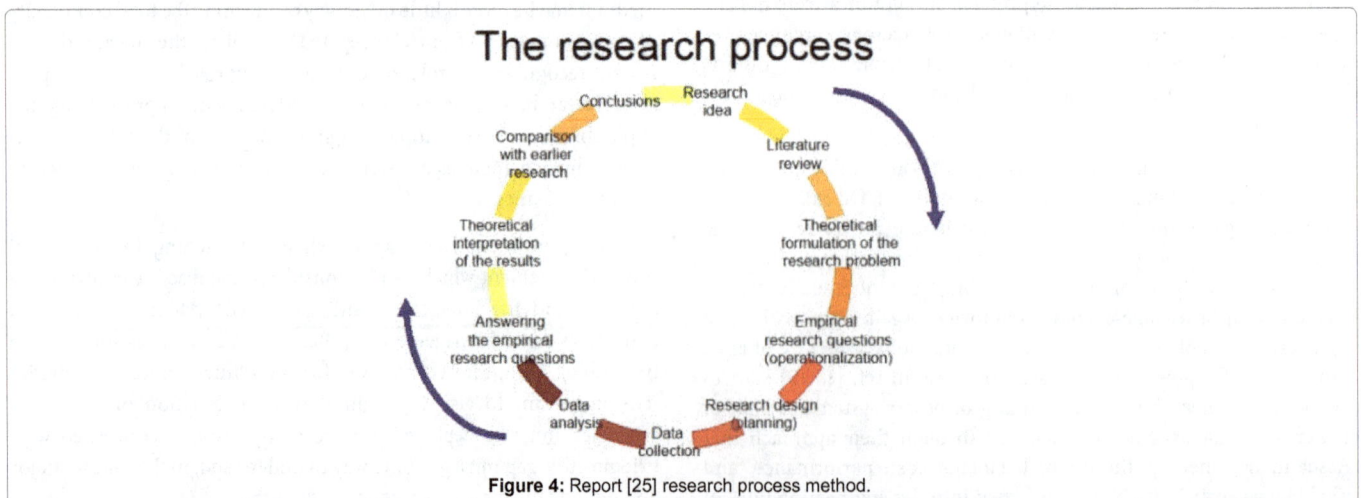

Figure 4: Report [25] research process method.

and merging with case study approach accordingly to Smith and Stirling [33] empowers the research in

- Coping with the technically distinctive situation in which there will be many more variables of interested than data points, and as one result.

- Relies on multiple sources of evidence, with data needing to converge in a triangulating fashion, and as another result.

- Benefits from the prior development of theoretical propositions to guide data collection and analysis.

An approach grafted from hypothesised model and tested through case study method works well [14,31]. That is that the combination of hypothesised model, as expressed as a holistic framework created by the author and tested through a case study offers a well attuned mechanism to research the holistic asset management picture in light of the real world context [33-36]. Although the research method shown above is very much applicable for the purpose of power companies operating in capital investment only activities, it is not inclusive enough to ensure whole-life-cycle optimisation of power systems, assets and or power companies throughout the entire life cycle. Entire life cycle in this case as specified by the Figure 5 and as can be schematically seen in the diagram.

Therefore the studies breadth and specificity towards power assets must therefore be extended to sufficiently align with the aim and objectives of this thesis. Figure 4 shows a method on researching the role of asset management for capital Intensive organisations highlights the limited scope of asset management and organisational strategy. Therefore although the method is useful and applicable the application must be suited to include those companies operating, owning or engaging in works directly related to power systems assets in greater detail. Hence the retrofit approach towards reviewing both Asset

Management in general academic application and as applied to power systems companies.

Gaps in the Current State of the Art

Given the clear indications and initial evidence for lack of knowledge regarding holistic asset management highlights a dual natured challenge, one being asset management as a holistic capability for companies and secondarily as applied directly to power companies. This literature review and research analysis highlights multiple gaps in both the asset management in general and as applied in the power systems fraternities of which can be summarised into two major elements.

Firstly as highlighted by the relative lack of holistic capabilities also secondarily compounded by an inability to apply credible holistic asset management across the asset lifecycle to power systems assets [37,38]. These gaps are to be expected given the relative lack of research conducted over the past decade and given such predominate previous research leaning towards engineering disciplines [26].

In addition and when considering the application of power systems asset management clearly identifies part of the problem when approaching power systems holistic asset management [34]. That is the lack of a corresponding holistic framework to utilise or apply. Some have attempted to morph a specific industrial concept to loosely fit elements of the power industry as demonstrated in report [35] yet again nothing specifically applied for power system assets.

In part this could be due to the complexity that surrounds decision making processes when concerning power system assets and as in the real world system wide setting. Given for example that power assets can represent multi-faceted business/government requirements and act as a provision of power to provide commerce and as a national/international infrastructure asset. As report in ref. [3] identifies this

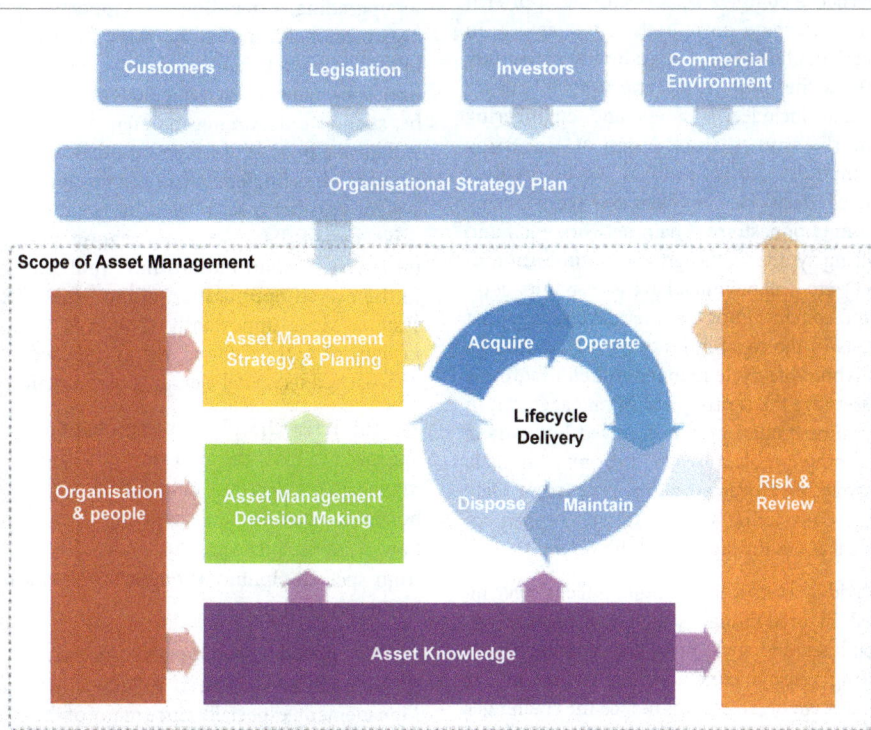

Figure 5: The Institute of Asset Management conceptual model.

challenge by calling out the complexity, "infrastructure management decision making is inherently an integrated process that requires the assimilation of a multitude of data, processes, and software systems." In consideration it can be said due to the applied nature of asset management of power system assets any suitable framework must inherit from an overarching capability that is holistically asset management driven.

Therefore the contribution of this research will be to fill two major challenges facing power systems assets across the lifecycle, that is:

- To ensure that a holistic full life-cycle asset management framework is developed and rigorously tested that,

- Applies to power companies.

Conclusion

The gaps in literature related to holistic asset management must take a complex system approach in prescribing a framework that reaches across the life cycle of power assets. That is the literature reviewed clearly identifies how in engineering terms individual areas of asset management can be approached and the manners in which these individual like contributions are focused in 'silo' like approaches. However, without the relevant overarching systems framework dealing with the combined approach to holistic asset management the optimisation of assets across the entire power asset portfolio cannot be fully realised.

In summary of the above paragraphs it is considered that effective Asset Management and the necessary controls are not represented solely by singular or independent activities, departments and or processes within an organisation. As highlighted in previous research these technical elements of asset management fall short on multiple fronts as the wider complex adaptive system natures are not taking widely into consideration as governance. Given the complex nature of the system concerned a complex socio-technical approach is the preferred approach in securing a more holistic framework, in achieving this end it has been identified that a retroductive case study and synthesised framework as the most fitting approach to ensure an integrated framework that included business and engineering elements to reach the aim. The synthesis and design of processing networks is a complex and multidisciplinary problem, which involves many strategic and tactical decisions at business (considering financial criteria, market competition, supply chain network, etc.) and engineering levels (considering synthesis, design and optimization of production technology, R&D, etc.), all of which have a deep impact on the profitability [30]. It is the conclusion of this literature review that a clear and evident gap exists in the capabilities of power companies to manage assets in a truly whole-life-cycle manner and although it is true to say that Physical Assets are the centre piece of management in this thesis ignoring the social or human presence geared in business environments is an underlining problem when creating a holistic framework for whole-life-cycle for power assets in the current state of the art. Especially when considering the nature of the human interaction with these physical assets throughout the life-cycle.

Given the extensive individual electrical and industrial engineering contributions have resulted in predominately linear improvements in the corresponding disciplines and in light of many a decade of 'industrial age like engineering' research concludes that science has yet to build a capacity of power system organisations to face the challenges of the present and future with sufficient capability, furthermore highlighting the requirement of a new research project.

Therefore the major gaps as seen by this review in scholarly terms clearly identify two major areas of which to increase in asset management capability, knowledge and research. That is in constructing a relevant asset management framework consisting largely of two major elements:

- Construct a holistic framework according to general terms applied in Asset Management whilst ensuring,

- Appropriateness of the framework to optimise asset management across power systems assets specifically.

The ability to do this constitutes a wide multi-disciplinary, multi-faceted, inter-connected complex system and to include understanding of research and effects across the model proposed during research, this is due to the expansive nature of Asset Management in general when creating a holistic framework. Often asset life cycle is separated into what might be considered manageable technical segments such as Maintenance or Finance, outlined above in the problem statements. Definitions from various commercial sector bodies highlight such elements as design, procurement, build and disposal as independent activities that make up a larger framework. The effective management of these cycles constitutes the broader pathway to effective strategic asset management [24].

It is true that organisations concerned with Asset Management require a decisive capability in approaching decisions, strategies, projects, organisation, information technological implementations, replacements, or end of life capabilities if optimised asset management in such settings is to be realised [17]. In this approach it is clear that asset management as applied in a system wide context is highly complicated and further represents cross collaboration of system wide expressions and interactions.

When considering how organisations may approach these asset management facets we need to expand on the 'how' or 'ways' of approaching such enhancements. For example on the approach to upgrading, expanding on supportive processes, business and or organisational capacity in ways to help the delivery of enhanced underlying asset management systems, naturally these decisions will have consequences on other areas of the system. Another example could be, such as outsourcing maintenance of Utility work to third parties will have a myriad of effects on other areas of the power company. Or perhaps when implementing a computerised maintenance management system or other asset related information systems to compliment the drive towards ISO 55000 information management compliance. Again these system contributions have multiple touch points and influence in the overall optimisation of any whole life cycle asset management system. More importantly though is to correlate the enhancement in light of both of the above as opposed to 'silo' and independently approached technical elements of maintaining assets only.

This research will aid in exposing the criticality of taking the broad approach in from the total view of disciplines and applications of asset management decision enhancement as a complex adaptive system, however, returning and applying throughout to the management of power assets. Subsequently, drawing on the experiences and insights from such disciplines as engineering, risk management, finance, and business development [36].

An applied research pattern is required to help secure the widest possible understanding of influence when considering holistic Asset Management, especially since many of the challenges appear contextual and to be seen in light of their activity, or field of activities is required in understanding the total combined picture – a case where the total is worth more than the sum of the parts and suggests a strong comparison to the likes of complex adaptive systems.

Therefore this requires that both researcher and the party responsible in the identification of the problem must be placed in the market itself, in real terms meaning a company that works with power systems in critical remits or as part of its core requirement to function. Both to fully rationalise the potential benefits and rewards from such a holistic framework.

This literature review concludes that substantial academic and industry value will be generated from applying a synthesised framework to assist the system of asset management, whilst helping rationalise assets and the management of assets throughout, supporting in essence the evolution towards optimised asset management for power companies. This approach will also inevitably educate-on-route rather than just collect, analyse and cast aspersions of theoretical nature solely. In taking this approach a truly representative model will absorb the complexity required whilst ensure the increases in and enhancements through the organisational management of the entire asset management system by a system and by a new synthesised model [32].

References

1. Kirova V, Kirby N, Kothari D, Childress G (2008) Effective requirements traceability: Models, tools, and practices. Bell Labs Technical Journal 12: 143-157.

2. Mazak A, Kargl H (2012) Cognitive Engineering meets Requirements Engineering, Bridging the Traceability Gap. ICSEA 2012, The Seventh International Conference on Software Engineering Advances.

3. Shahidehpour M, Ferrero R (2005) Time management for assets: chronological strategies for power system asset management. Power and Energy Magazine 3: 32-38.

4. Jahromi A, Piercy R, Cress S, Service J, Fan W (2009) An approach to power transformer asset management using health index. Electrical Insulation Magazine 25: 20-34.

5. Amin SM, Wollenberg BF (2005) Toward a smart grid: power delivery for the 21st century. Power and Energy Magazine, IEEE 3: 34-41.

6. Brown T, Beyeler W, Barton D (2004) Assessing infrastructure interdependencies: the challenge of risk analysis for complex adaptive systems. International Journal of Critical Infrastructures 1: 108-117.

7. Mehairjan RPY, Smit JJ, Djairam D (2014) Trends in Risk-Based Substation Asset Management & Lifetime Monitoring. International workshop on power transformers.

8. Calder BJ, Phillips LW, Tybout AM (1981) Designing research for application. Journal of consumer research, pp: 197-207.

9. Goel L, Aparna VP, Wang P (2007) A framework to implement supply and demand side contingency management in reliability assessment of restructured power systems. Power Systems, IEEE Transactions 22: 205-212.

10. Hastings NAJ (2009) Physical asset management. Springer Science & Business Media.

11. Brown R, Spare J (2004) Asset management, risk, and distribution system planning. Power Systems Conference and Exposition.

12. Halfawy MR (2008) Integration of municipal infrastructure asset management processes: challenges and solutions. Journal of Computing in Civil Engineering 22: 216-229.

13. Sather B (1999) Retroduction: an alternative research strategy? Business Strategy and the Environment 7: 245-249.

14. Campbell JD, Jardine AKS, McGlynn J (2011) Asset management excellence: optimizing equipment life-cycle decisions. CRC Press, Boca Raton, Florida, United States.

15. Schuman CA, Brent AC (2005) Asset life cycle management: towards improving physical asset performance in the process industry. International Journal of Operations & Production Management 25: 566-579.

16. Baker MJ (2000) Selecting a research methodology. The marketing review 1: 373-397.

17. McArthur SDJ, Davidson EM, Catterson VM, Dimeas AL, Hatziargyriou ND, et al. (2007) Multi-agent systems for power engineering applications- Part II: technologies, standards, and tools for building multi-agent systems. Power Systems, IEEE Transactions 22: 1753-1759.

18. Pagani GA, Aiello M (2013) The power grid as a complex network: a survey. In: Physica A: Statistical Mechanics and its Applications. Elsevier, pp: 2688-2700.

19. Blanchard BS (2009) Maintenance and support: a critical element in the system life cycle." Systems Engineering and management for Sustainable Development-Volume 2.

20. Madu CN (2000) Competing through maintenance strategies. International Journal of Quality & Reliability Management 17: 937-949.

21. Puletti F, Olivieri M, Cavallini A, Montanari GC (2006) Risk management of HV polymeric cables based on partial discharge assessment. In: Proc. IEEE PES Transmission Distribution Conference, pp: 626-633.

22. Vladimir Frolov, David Mengel, Wasana Bandara, Yong Sun, Lin Ma (2010) Building an ontology and process architecture for engineering asset management. Engineering Asset Lifecycle Management. Springer London, pp: 86-97.

23. Choubey SK (2011) System and method to calculate procurement of assets. US Patent No 8032401.

24. Cigre (2014) Contracts for Outsourcing Utility Maintenance Work.

25. Abu-Elanien AEB, Salama MMA (2010) Asset management techniques for transformers. Electric power systems research 80: 456-464.

26. Runeson P, Höst M, Sjoberg D (2009) Guidelines for conducting and reporting case study research in software engineering. Empirical software engineering 14: 131-164.

27. Dooley KJ (1997) A complex adaptive systems model of organization change. Nonlinear Dynamics, Psychology, and Life Sciences 1: 69-97.

28. Tse PWT, Mathew J, Wong K, Lam R, Ko CN (2013) Engineering Asset Managment - Systems, Professional Practices and Certification. Springer.

29. Baughman ML, Siddiqi SN (1991) Real-time pricing of reactive power: theory and case study results. Power Systems, IEEE Transactions 6: 23-29.

30. Quaglia A, Sarup B, Sin G, Gani R (2012) Integrated business and engineering framework for synthesis and design of enterprise-wide processing networks. Computers & Chemical Engineering 38: 213-223.

31. Skinne M, Kirwan A, William J (2011) Challenges of developing whole life cycle cost models for Network Rail's top 30 assets. IET and IAM Asset Management Conference.

32. Starkey K, Paula M (2001) Bridging the relevance gap: Aligning stakeholders in the future of management research. British Journal of management 12: S3-S26.

33. Smith A, Stirling A (2008) Social-ecological resilience and sociotechnical: critical issues for sustainability governance. STEPS Centre.

34. El-Akruti KO (2012) The Strategic role of engineering asset management in capital intensive organisations. University of Wollongong, Northfields Ave, Wollongong, Australia.

35. Faiz RB, Edirisinghe EA (2009) Decision making for predictive maintenance in asset information management. Interdisciplinary Journal of Information, Knowledge, and Management, pp: 23-26.

36. Schneider J, Gaul A, Neumann C, Hogräfer J, Wellßow W (2006) Asset management techniques. International Journal of Electrical Power & Energy Systems 28: 643-654.

37. Hooper R, Armitage R, Gallagher A, Osorio T (2009) Whole-life infrastructure asset management: good practice guide for civil infrastructure. CIRIA.

38. El-Akruti Khaled O, Dwight R (2010) Research methodologies for engineering asset management. ACSPRI Social Science Methodology Conference.

7

Electromagnetic Waves Propagation in Graphene Multilayered Structures

Mohammed M Shabat and Muin F Ubeid*

Department of Physics, Faculty of Science, Islamic University of Gaza, Palestine

Abstract

In this work, the reflected, transmitted, and loss powers due to the interaction of electromagnetic waves with graphene-dielectric structures are analyzed theoretically and numerically. The properties of the graphene material are given in detail and the required equations for its main parameters are defined. The formulations for the transverse electric waves case are provided. A recursive method is used to calculate the overall reflection and transmission coefficients of the structure. The reflected, transmitted, and loss powers are determined using these coefficients. In the numerical results, the mentioned powers are computed as a function of wavelength, angle of incidence, and dielectric thickness when graphene thickness changes.

Keywords: Graphene; Multilayered structure; Reflection; Transmission; Wavelength

Introduction

Graphene film is a monolayer of carbon atoms packed into a dense honeycomb crystal structure that can be viewed as an individual atomic plane extracted from graphite [1,2]. Graphene has high mobility and optical transparency, in addition to flexibility, robustness and environmental stability. It is remarkably strong for its very low weight (100 times stronger than steel), and it conducts heat and electricity. Because it is virtually two-dimensional, it interacts oddly with light and with other materials [3]. While scientists had theorized about graphene for decades, it was first produced in the lab in 2004 [4]. Despite its short history, graphene has already revealed a cornucopia of new physics and potential applications [5-8]. Andre Geim and Konstantin Nevoselov at the university of Manchester won the Nobel prize in physics in 2010 "for groundbreaking experiments regarding the two-dimensional material graphene" [9].

Researches on graphene-based structures have been developed quickly in both theoretical and experimental applications. Lee et al. [10] have investigated dynamic behavior of multilayer grapheme via supersonic projectile penetration. Liu et al. [11] have shown that an enhancement of graphene absorption is observed when the graphene monolayer is placed on the top or within dielectric mirrors. Zhu et al. [12] have studied optical transmittance of multilayer graphene. Singh et al. [13] have reviewed optomechanical coupling between a multilayer graphene mechanical resonator and a superconducting microwave cavity. Khan et al. [14] have demonstrated that, the optimized mixture of graphene and multilayer graphene, produced by the high-yield inexpensive liquid-phase-exfoliation technique, can lead to an extremely strong enhancement of the cross-plane thermal conductivity of the composite. Iorsh et al. [15] have proposed a new class of hyperbolic metamaterials for THz frequencies based on multilayer graphene structures. Min et al. [16] have shown the electronic structure of multilayer graphene. Rast et al. [17] have numerically analyzed a composite layered structure for, tunable, low-loss plasmon resonators, which consists of a noble metal thin film coated in graphene and supported on a hexagonal boron nitride substrate.

This paper is interested in transmission and reflection of electromagnetic waves by a graphene/dielectric periodic structure consisting of N periods. We consider the structure is embedded in vacuum and a monochromatic s-polarized plane electromagnetic wave is obliquely incident on it. The electric and magnetic fields are determined in each region using Maxwell's equations. Then Snell's law is applied and the boundary conditions are imposed at each interface to obtain the reflection and transmission coefficients. The reflected and transmitted powers of the structure are presented in terms of these coefficients. In the numerical analysis a recursive method [18,19] is used to calculate the mentioned powers as a function of wavelength, angle of incidence and the slab thickness when the graphene thickness changes. To check the results of the analysis used in these calculations, the conservation law of energy given in [20,21] is checked and it is clear that it is satisfied for all examples.

Theory

The considered waveguide structure consists of a pair of graphene (ε_2, μ_0) and dielectric (ε_3, μ_0) materials bounded by two half free spaces (ε_0, μ_0). The letter ε stands for permittivity of the related materials and the subscripts 2, 3 refer to region 2 and 3 and μ_0 is the permeability of free space. A perpendicular polarized plane wave in region 1 is incident on the plane z = 0 at some angle θ relative to the normal to the boundary (Figure 1).

The electric field in each region is [22, 23]:

$$\vec{E}_\ell = \left(A_\ell e^{ik_{\ell z}z} + B_\ell e^{-ik_{\ell z}z} \right) e^{i(k_{\ell x}x - \omega t)} \, \hat{y} \tag{1}$$

To find the corresponding magnetic field $\vec{H}_\ell$, we start with Maxwell's equation $\vec{\nabla} \times \vec{E}_\ell = -\partial\vec{B}/\partial t$, substituting $\vec{B} = \mu_\ell \vec{H}_\ell$ and solving for $\vec{H}_\ell$ yield:

$$\vec{H}_\ell = \frac{1}{\mu_\ell \omega}\left[(A_\ell k_{\ell x}e^{k_{\ell z}z} + B_\ell k_{\ell x}e^{-k_{\ell z}z})\hat{z} + (-A_\ell k_{\ell z}e^{k_{\ell z}z} + B_\ell k_{\ell z}e^{-k_{\ell z}z})\hat{x} \right]e^{i(k_{\ell x}x - \omega t)} \tag{2}$$

Where A_ℓ and B_ℓ are the amplitude of forward and backward traveling waves ($\ell = 1, 2, 3, 4$), $k_\ell = n_\ell \omega/c$ is the wave number inside the material and n_ℓ is the refractive index of it. Also the subscripts x and z represent the x- and z-components of the related wave number, respectively.

***Corresponding author:** Muin F. Ubeid, Department of Physics, Faculty of Science, Islamic University of Gaza P.O. 108, Gaza, Gaza Strip, Palestinian Authority
E-mail: mubeid@mail.iugaza.edu

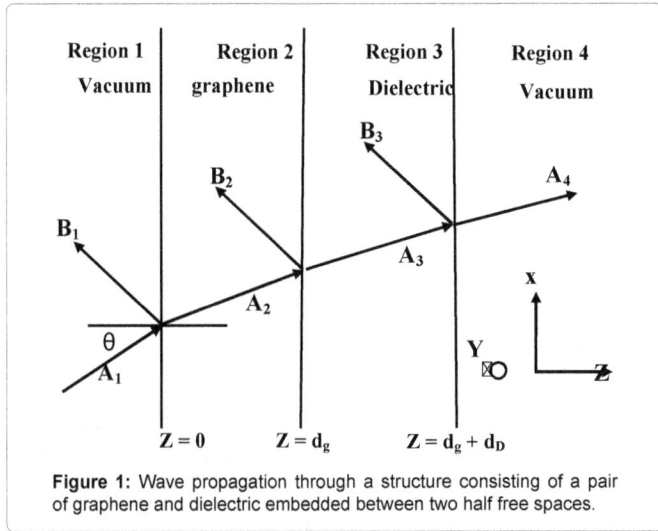

Figure 1: Wave propagation through a structure consisting of a pair of graphene and dielectric embedded between two half free spaces.

Matching the boundary conditions for $\vec{E}$ and $\vec{H}$ fields at each layer interface, that is at z=0, $E_{1y} = E_{2y}$ and $H_{1x} = H_{2x}$ and so on. This yields six equations with six unknown parameters [22,23]:

$$A_1 + B_1 = A_2 + B_2 \tag{3}$$

$$\frac{k_{1z}}{\mu_1}(A_1 - B_1) = \frac{k_{2z}}{\mu_2}(A_2 - B_2) \tag{4}$$

$$A_2 e^{ik_{2z}d_2} + B_2 e^{-ik_{2z}d_2} = A_3 e^{ik_{3z}d_2} + B_3 e^{-ik_{3z}d_2} \tag{5}$$

$$\frac{k_{2z}}{\mu_2}\left(A_2 e^{ik_{2z}d_2} - B_2 e^{-ik_{2z}d_2}\right) = \frac{k_{3z}}{\mu_3}\left(A_3 e^{ik_{3z}d_2} - B_3 e^{-ik_{3z}d_2}\right) \tag{6}$$

$$A_3 e^{ik_{3z}(d_2+d_3)} + B_3 e^{-ik_{3z}(d_2+d_3)} = A_4 e^{ik_{4z}(d_2+d_3)} \tag{7}$$

$$\frac{k_{3z}}{\mu_3}\left(A_3 e^{ik_{3z}(d_2+d_3)} - B_3 e^{-ik_{3z}(d_2+d_3)}\right) = \frac{k_{4z}}{\mu_4} A_4 e^{ik_{4z}(d_2+d_3)} \tag{8}$$

Where $k_{1x} = k_{2x} = k_{3x} = k_{4x} \equiv$ Snell's law and:

$$k_{\ell z} = \frac{\omega}{c}\sqrt{n_\ell^2 - n_1^2 \sin^2 \theta} \tag{9}$$

Fresnel coefficients (interface reflection and transmission coefficients r, t respectively) for perpendicular polarized light are given by [24]:

$$r_{ij} = \frac{\mu_j k_{iz} - \mu_i k_{jz}}{\mu_j k_{iz} + \mu_i k_{jz}} \tag{10}$$

$$t_{ij} = \frac{2\mu_j k_{iz}}{\mu_j k_{iz} + \mu_i k_{jz}} \tag{11}$$

Where i, j correspond to any two adjacent media.

The reflection and transmission coefficients R and T respectively of the structure are given by [25,26]:

$$R = \frac{B_1}{A_1} = \frac{r_{12} + r_{12}r_{23}r_{34}e^{i2k_{3z}d_3} + r_{23}e^{i2k_{2z}d_2} + r_{34}e^{i2(k_{2z}d_2+k_{3z}d_3)}}{1 + r_{23}r_{34}e^{i2k_{3z}d_3} + r_{12}r_{23}e^{i2k_{2z}d_2} + r_{12}r_{34}e^{i2(k_{2z}d_2+k_{3z}d_3)}} \tag{12}$$

$$T = \frac{A_4}{A_1} = \frac{t_{12}t_{23}t_{34}e^{i(k_{2z}d_2+k_{3z}d_3)}}{1 + r_{23}r_{34}e^{i2k_{3z}d_3} + r_{12}r_{23}e^{i2k_{2z}d_2} + r_{12}r_{34}e^{i2(k_{2z}d_2+k_{3z}d_3)}} \tag{13}$$

The reflectance R' and transmittance T' of the structure are given by:

$$R' = RR^*, \quad T' = \frac{k_{4z}}{k_{1z}}TT^* \tag{14}$$

Where R^* and T^* are the complex conjugate of R and T respectively. The law of conservation of energy is given by [20,21]:

$$R' + T' = 1 - P_{loss} \tag{15}$$

where P_{loss} is the loss power due to losses in the graphene slab.

For n'-layers structure shown in Figure 2 R and T are calculated as follows [26,27]:

$$R_{n'} = r_{n'-1,n'} \tag{16}$$

$$R_{n'-1} = \frac{r_{n'-2,n'-1} + R_{n'}e^{i2k_{(n'-1)z}d_{n'-1}}}{1 + r_{n'-2,n'-1}R_{n'}e^{i2k_{(n'-1)z}d_{n'-1}}} \tag{17}$$

$$R_{n'-2} = \frac{r_{n'-3,n'-2} + R_{n'-1}e^{i2k_{(n'-2)z}d_{n'-2}}}{1 + r_{n'-3,n'-2}R_{n'-1}e^{i2k_{(n'-2)z}d_{n'-2}}} \tag{18}$$

$$\vdots$$
$$\vdots$$

Continue on the same procedure until R_2 is reached which is the reflectance of the structure as a whole.

$$R_2 = \frac{r_{12} + R_3 e^{i2k_{2z}d_2}}{1 + r_{12}R_3 e^{i2k_{2z}d_2}} \tag{19}$$

The same procedure is performed for T_2:

$$T_{n'} = t_{n'-1,n'} \tag{20}$$

$$T_{n'-1} = \frac{t_{n'-2,n'-1}T_{n'}e^{i2k_{(n'-1)z}d_{n'-1}}}{1 + r_{n'-2,n'-1}R_{n'}e^{i2k_{(n'-1)z}d_{n'-1}}} \tag{21}$$

$$T_{n'-2} = \frac{t_{n'-3,n'-2}T_{n'-1}e^{i2k_{(n'-2)z}d_{n'-2}}}{1 + r_{n'-3,n'-2}R_{n'-1}e^{i2k_{(n'-2)z}d_{n'-2}}} \tag{22}$$

$$\vdots$$
$$\vdots$$

$$T_2 = \frac{t_{12}T_3 e^{i2k_{2z}d_2}}{1 + r_{12}R_3 e^{i2k_{2z}d_2}} \tag{23}$$

Where d_2, $d_{n'-1}$ and $d_{n'-2}$ are thicknesses of layers 2, n'-1and n'-2, respectively.

For graphene in regions 2, 4, 6 …, the complex refractive index is given by [28-30]:

$$n_4(\lambda) = 3 + \frac{iC\lambda}{3} \tag{24}$$

where C is 5.446 μm^{-1}, λ is the incident wavelength and $i = \sqrt{-1}$.

Numerical results and applications

In this section, the reflected, transmitted and loss powers of the strucure described in Figure 2 are calculated numerically as a function of wavelength, angle of incidence and dielectric thickness for changing graphene layer thickness. We have used the graphene described in eq. (24) and Fluorite (CaF$_2$) of refractive index 1.434 as a dielectric in each period. Three values the graphene thickness are considered [d_g = .34 nm, .68 nm (2 x .34 nm), .68 nm (3 x .34 nm)]. These thicknesses are reported in [31]. The central wavelength is assumed to be λ_0 = 600 nm, the thickness of Fluorite is $\lambda_0/2$ and the number of periods N =

7. Regions 3, 5, 7 … given in Figure 2 are assumed to be loss-less and the permeabilities of them are equal to the permeability of free space, μ_0. These materials do not affected by the magnetic field of incident radiations.

Figure 3 shows the reflected (reflectance), transmitted (transmittance) and loss powers as a function of the wavelength at the incidence angle of 30° when the graphene thickness changes. The wavelength is changed between 100 nm and 1700 nm, this range includes ultraviolet, visible and near infrared. We can see that,

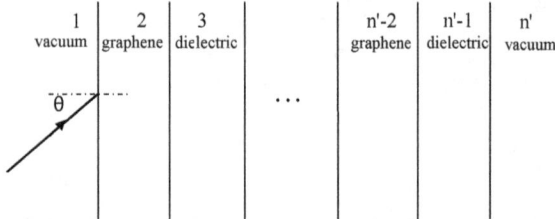

Figure 2: Oblique incidence of electromagnetic waves on graphene/dielectric periodic structure embedded in vacuum.

the powers generally have oscillatory characteristics in the given wavelength range. The degree of ripples decreases with wavelength for all values of the graphene thicknesses. Moreover, the reflected and transmitted powers decrease while the loss power increases with the graphene thickness.

Figure 4 illustrates the variation of the reflected, transmitted, and loss powers with the angle of incidence for 600 nm wavelength under three values of the graphene thickness. The angle of incidence is changed between 0° and 90° to realize all possible angles of incidence. Clearly the reflected power behaves as an oscillatory increasing function while the transmitted and loss powers show an oscillatory decreasing behavior with the angle of incidence. At 90° the reflected (the transmitted and loss) powers is maximum (minimum) at that angle for any value of the graphene thickness. The role of the graphene is clear at angles below 90°. The reflected and transmitted powers decrease while the loss power increases with the graphene thickness for any angle below 90°.

Figure 5 presents the reflected, transmitted and loss powers against the dielectric thickness at the incident angle of 30°. The dielectric thickness is changed from 0 nm to 400 nm. As it is confirmed from

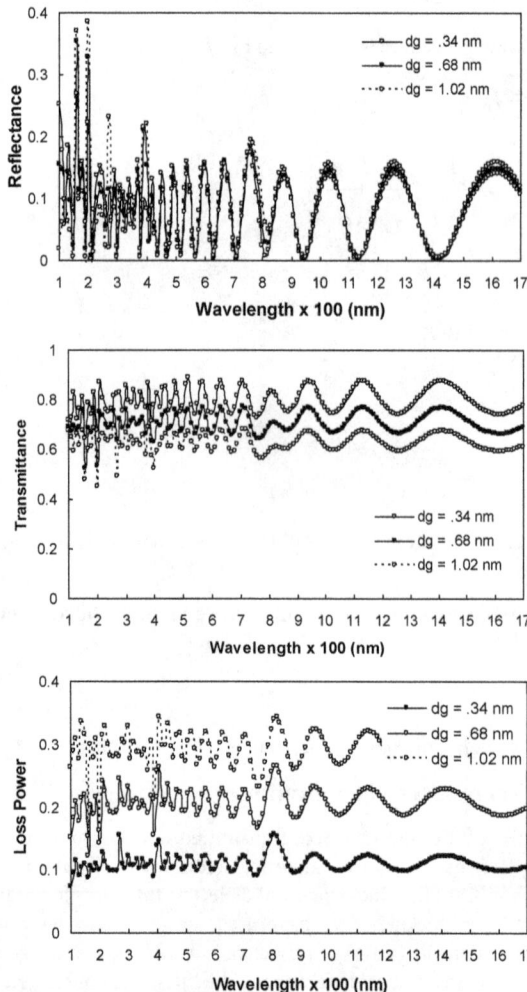

Figure 3: The reflected, transmitted and loss powers as a function of wavelength.

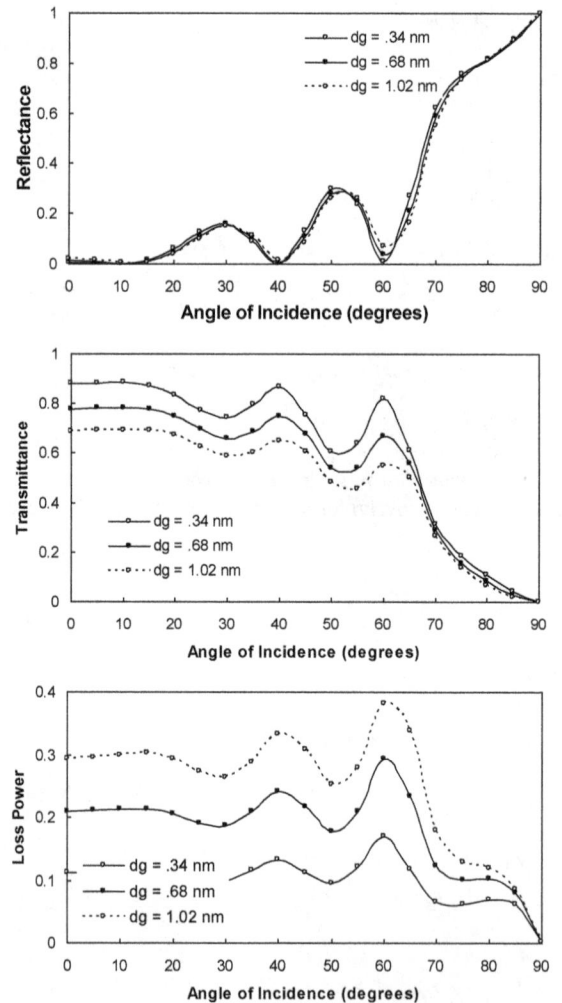

Figure 4: The reflected, transmitted and loss powers versus the angle of incidence.

the figure, the powers change periodically with thickness for any value of graphene thickness. The reflected and transmitted powers decrease while the loss power increases with graphene thickness.

Conclusions

In this paper, the reflection and transmission characteristics of the electromagnetic radiation propagation through a graphene/dielectric (Fluorite-CaF$_2$) periodic structure are studied in detail with the effect of the graphene material. The required equations for the electric and magnetic fields in each region are derived by Maxwell's equations. Then Snell's law is applied and the boundary conditions are imposed to calculate the reflection and transmission coefficients of the structure. Recursive method is used to solve the problem of electromagnetic wave propagation through the structure to obtain the reflected, transmitted, and loss powers. Finally, the mentioned powers as a function of wavelength, angle of incidence and the dielectric thickness etc. are studied numerically to observe the effect of the graphene on them. As it can be seen from the theoretical and the numerical results, if the grapheme thickness changes, the characteristic of the powers will be affected by this change. Numerical examples are already presented to illustrate the paper idea and to prove the validity of the obtained results. Moreover the law of conservation of energy is satisfied throughout the performed computations for all examples.

The results obtained could lead to design new devices, apparatus, components at the millimeter wave, optical, and microwave regimes. Furthermore, these results open a way to think how the availability of the graphene will change the functionality of future devices through the graphene/dielectric structure.

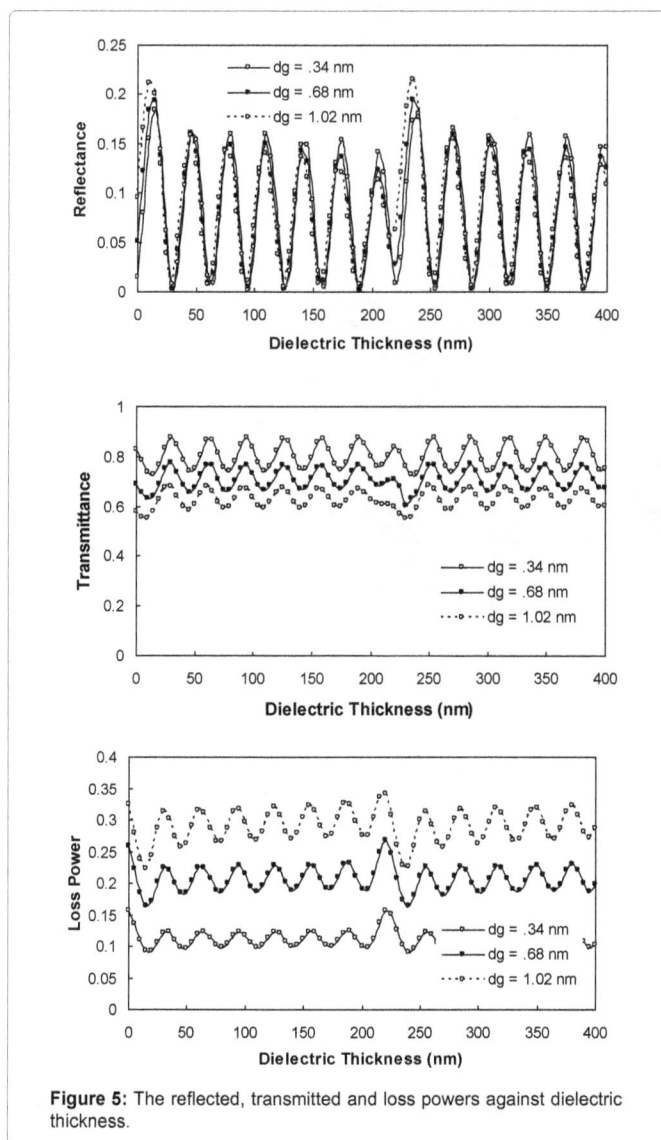

Figure 5: The reflected, transmitted and loss powers against dielectric thickness.

References

1. Geim AK, Novoselov KS (2007) "The rise of graphene", Nat. Mayyer 6: 191.

2. Geim AK (2009) "Graphene status and prospects", Science 324: 1530-1534.

3. Weber JW, Calado VE, Van de Sanden MC (2010) Optical constants of graphene measured by spectroscopic ellipsometry, Applied Physics Lett 97: 091904.

4. (2009) "This month in physics history: October 22, 2004; Discovery of graphene" APS news, series II 18: 2.

5. Riedle C, Coletti C, Iwasoki T, Zakharou AA, Starke U (2009), "Quasi-Free standing epitaxial graphene on Sic obtained by hydrogen interaction". Phys Rev Lett 103: 246804.

6. Prezzi D, Varsano D, Ruini A, Marini A, Molinari E (2008) "Optical properties of graphene nanoribbons: the role of many-body effects". Phys Rev B 77: 041404.

7. Zhu X, Su H (2010) "Excitons of edge and surface functionalized grapheme nanoribbons". J Phys Chem C 114: 17257-17262.

8. Blmatov D, Mou CY (2010) "Josephson effect in graphene SNS junction with a single localized detect" Physica B 2896.

9. The Nobel foundation, The Nobel prize in physics (2010).

10. Lee JH, Loya PE, Lou J, Thomas EL (2014) "Dynamic mechanical behavior of multilayer grapheme via supersonic projectile penetration". Science 346: 1092-1096.

11. Liu JT, Liu NH, Li J, Li XJ, Huang JH (2012) "Enhancement absorption of graphene with one dimensional photonic crysyal". App Phys Lett 101: 052104-052103.

12. Zhu SE, Yuan S, Janssen GCAM (2014) Optical transmittance of multilayer graphene, Europhysics Letters, 108.

13. Singh V, Bosman SJ, Schneider BH, Blanter YM, Castellanos-Gomes A, et al. (2014) "Optomechanical coupling between a multilayer graphene mechanical resonator and a superconducting microwave cavity", Nature Nanotechnology 9: 820-824.

14. Shahil KMF, Alexander AB (2012) Graphene-multilayer graphene nanocompsite as highly efficient thermal interface materials. Nano Lett 12.

15. Iorsh I, Mukhin I, Shadrivov I, Belov P, Kivshar Y (2013) "Novel hyperbolic metamaterials based on multilayered graphene structures". Phys Rev B 87: 075416.

16. Min H, MacDonald AH (2008) "Electronic structure of multilayer graphene", Progress of theoretical physics supplement No. 176. 277 (USA).

17. Rast L, Sulivan TI, Tewary VK (2013) "Stratified graphene/noble metal systems for low-loss plasmonics applications," Phys Rev B 87: 045428.

18. Vigoureux JM (1991) Polynomial formation of reflection and transmission by stratified planar structure. J Opt Soc Am 8: 1697-1701.

19. Kong JA (2002) Electromgnetic wave intraction with stratified negative isotropic media. Prog. Electromagn Res PIER 35: 1-52.

20. Sabah C, Uckun S (2007) Electromagnetic wave propagation through frequency-dispersive and loss double-negative slab. Opto-Elec Rev 15: 133-143.

21. Stancil DD, Prabhakar (2009) Spin Waves, New York: Springer.

22. Ubeid MF, Shabat MM, Sid-Ahmed MO (2012) numerical study of a structure containing left-handed material waveguide. Indian Journal of Physics (Springer) 86: 125-128.

23. Ubeid MF, Shabat MM (2014) Antireflection coating at metamaterial waveguide structures for solar energy applications. Energy Procedia (Elsevier) 50: 314-321.

24. Ubeid MF, Shabat MM, Sid-Ahmed MO (2012) Maximum and minimum transmittance of a structure containing N identical pairs of left- and right-handed materials. Journal of Nano- and Electronic Physics 4: 04007(1-5).

25. Ubeid MF, Shabat MM (2013) Transmitted powers of waves through superconductor-dielectric photonic crystal. Lecture Notes on Photonics and Optoelectronics 1: 35-39.

26. Fujiwara H (2007) Spectrometric Ellipsometry, USA: John Wiley and Sons, 46.

27. Vigoureux JM (1991) Polynomial formation of reflection and transmission by stratified planar structure, J Opt Soc Am 8: 1697-1701.

28. D Feng, Liu G, Zhang M, Jia D (2013) "D-shaped fiber optic SPR biosensors based on a metal-graphene structure". Chinese Optics Letters 11: 110607.

29. Wu L, Koh HS, Li EP (2010) Highly sensitive graphene biosensors based on surface plasmon resonance. OPTICS EXPRESS 18: 14395-14400.

30. Ubeid MF, Shabat MM (2010) Numerical investigation of a D-shape optical fiber sensor containing grapheme. Appl Phys A 118: 1113-1118.

31. Bruna M, Borini S (2009) Optical constant of graphene layers in the visible range. Applied Phys Lett 94: 031901.

8

The Main Approaches of Studying the Mechanisms of Action of Artificial Electromagnetic Fields on Cell

Yuriy Shckorbatov*

Institute of Biology, Kharkiv National University, Ukraine

Abstract

The effects of Electromagnetic Fields (EMFs) on health are discussed. These effects produce public interest to biological action of EMFs. The cell effects of EMFs have to be investigated to understand the mechanism of the biological action of EMFs. Among such mechanisms three main points are discussed: EMF action on cell membranes; EMF action on free radical concentration in cell; the action of EMF on intracellular regulatory systems. The analysis of experimental results proves the multilateral nature of the impact of EMFs on cell.

Keywords: Cell; Cell nucleus; Cell membrane; Microwaves; Magnetic field

Introduction

The problem of the influence of Electromagnetic Fields (EMFs) on biological objects has a long history. Now this problem attracts keen public interest in connection to increase of "electromagnetic pollution" of environment. In this century by Banik S, Bandyopadhyay S, Ganguly S; Vorst AV, Rosen A, Kotsuka Y; Barnes FS, Greenebaum B; Furse C, Christensen DA, Durney CH are reviewed the main facts of EMFs effects on cells [1-4]. The medical applications of EMFs are reviewed by Rosch PJ, Markov MS; Andrä W, Nowak H; Markov MS [5-7]. Some of theoretical models proposed to explain the basic phenomena associated with cell exposure to EMF are presented in review articles [8-21].

Now some new data are emerging and the approaches to the problem of the action of electromagnetic factors on cells are changing in accordance to the new developments in biology as a whole. I would like to present the brief description of the main directions of research in this field.

The most important, in our opinion, are three primary biological mechanisms of action of EMFs on cell. First, the action on cell membranes, second, the action on free radical concentration in cell, and the third, the action of EMF on intracellular regulatory systems. The known effects of EMFs on numerous isolated enzyme reactions may acquire the biological meaning in their connection to the mentioned above mechanisms.

Traditionally, the biological effects of EMFs are divided in two groups: "thermal" and "non-thermal". For microwave radiation the EMF irradiation with the level of surface power density less than 10 mW/cm² is assumed as not inducing the significant thermal effects in the biological objects and the effects of microwaves with surface power density more than 10 mW/cm² as "thermal" [22]. This classification is provisional, but it enables to separate the effects of EMFs related mainly with heating of biological tissue. In this work we will discuss non-thermal effects of EMFs. First, in order to show the importance of the issues discussed for humans we consider some of the effects of electromagnetic fields on human organism.

Effects of Exposure to Anthropogenic EMFs on Human Health

The environmental and health effects of electromagnetic radiation produced by mobile phones, mobile phone base stations, TV transmitting towers, and some medical equipment induce public concern. Some investigations analyzing the health effects of the electromagnetic radiation produced by these sources of electromagnetic radiation are observed below.

Electromagnetic fields of Radiofrequency (RF) range cover wide spectrum of frequencies. The ionosphere very effectively shields the earth's biosphere from radiations of this type originating in space. Electromagnetic fields and radiation of high intensity may be generated by natural electrical phenomena such as those accompanying thunderstorms. However, in the frequency range of 100 kHz to 300 GHz, the intensity of natural fields and radiation is low. Exposure of the urban population in the USA to man-made microwave sources was found by Janes (1979) to vary from a very low value to as high as 100 μW/cm². The median exposure to the total microwave flux from external sources for this population was calculated to be 0.005 μW/cm². Although concern about microwave and RF effects and possible hazards arose first in highly developed countries, the problem is universal [23].

The mobile telephony is widespread in the modern world. The data about the usage of mobile telephony are presented in review [24]: "In 2010, cell phone subscribers in the U.S. numbered 287 million, Russia 220 million, Germany 111 million, Italy 87 million, Great Britain 81 million, France 62 million, and Spain 57 million". The "microwave sickness syndrome" among people working with microwaves was first described in Soviet and Polish medicine, this syndrome includes fatigue, dermographism, headaches, insomnia, changes in blood pressure, tumors, impotence, skin symptoms and memory impairment, among others. The skepticism about this syndrome was expressed in the West, but now is admitted the possible neurological basis for this syndrome [24,25]. It was proved that there was a significant relation of some symptoms to measured EMF power density; this was highest for headaches. Authors express their opinion that despite very low exposure

*Corresponding author: Yuriy Shckorbatov, Institute of Biology, Kharkiv National University, pl. Svobody, 4, Kharkiv, 61077, Ukraine
E-mail: yuriy.shckorbatov@gmail.com

to HF-EMF, effects on wellbeing and performance cannot be ruled out, however, mechanisms of action at low EMF levels are unknown [26].

In many investigations the frequency of micronuclei and chromosomal aberrations tests which indicate the level of genetic damage (genotoxic effects) are used. Micronuclei are emerging in cells after action of different factors (mutagenic factors) which influence the process of cell division or the process of origin of mutations. Chromosome aberrations are the impairments in chromosome structure visible in light microscope. The results of one of such investigations show that there was no significant difference of micronucleus frequency and chromosomal aberrations in human blood cells between the groups of people living around mobile phone base stations and control group [27].

Some investigations indicate an increased risk for cancer - acoustic neuroma and glioma after >10 years of mobile phone usage. Authors conclude that current standard for exposure to microwaves during mobile phone use is not safe for long-term exposure and needs to be revised [16,28]. But the increased cancer risk among mobile telephone users is not detected by some investigators. It was observed no increased risk for basal cell carcinoma, squamous cell carcinoma, or melanoma of the head and neck, little evidence of an increased skin cancer risk was observed among mobile phone users after at least 13 years of mobile telephone usage [29].

The association between childhood acute lymphoblastic leukemia and power lines was revealed. This study emphasizes on risk of acute lymphoblastic leukemia following living close to overhead high voltage power lines. Authors propose consider legal limitation for building constructions in at least 600 meters from high voltage power lines [30].

To investigate the risk of early childhood cancers associated with the mother's exposure to radiofrequency radiation from mobile phone base stations (masts) during pregnancy. It was demonstrated that is no association between the risk of early childhood cancers and mother's exposure to mobile phone base stations during pregnancy [31].

The epidemiological evidence suggests an association between occupational exposure to ELF-EMF and Alzheimer disease. However, some limitations affecting the results from this meta-analysis should be considered [32].

The possible health hazardous effects of magnetic resonance imaging (MRI) were also explored. While the whole data does not confirm a risk hypothesis, it suggests a need for further studies and prudent use in order to avoid unnecessary examinations, according to the precautionary principle [33]. In 2003, the FDA declared non-significant risk status for MRI clinical systems generating static fields up to 8T [34].

In spite of existing of some investigations indicating the negative health impact of EMF the general estimation of the EMF-induced health hazard was optimistic. The International Commission on Non-Ionizing Radiation Protection makes such conclusion based on the analysis of many epidemiologic investigations. Results of epidemiological studies to date give no consistent or convincing evidence of a causal relation between RF exposure and any adverse health effect. On the other hand, these studies have too many deficiencies to rule out an association. In the last few years the epidemiologic evidence on mobile phone use and risk of brain and other tumors of the head has grown considerably. In opinion of members of International Commission on Non-Ionizing Radiation Protection, overall the studies published to date do not demonstrate a raised risk within approximately ten years of use for

any tumor of the brain or any other head tumor [35]. Nevertheless, the WHO International Agency for Research on Cancer (IARC) evaluated the extremely low frequency magnetic fields as possibly carcinogenic to humans (Group 2B). Static magnetic fields and static and extremely low frequency electric fields could not be classified as to carcinogenicity to humans [36]. IARC has classified radiofrequency electromagnetic fields as possibly carcinogenic to humans (Group 2B), based on an increased risk for glioma, a malignant type of brain cancer, which is associated with wireless phone use [37].

The problem of adequate assessment of EMF influence on human organism is very important. We believe that the state of individual cell in our organism is directly linked to the state of the whole organism. Stress-related changes in the cell nucleus are generally associated with transitions of main cell nucleus component chromatin from the diffused state euchromatin to more condensed state — heterochromatin. We propose to use euchromatin → heterochromatin transitions in cells of human buccal epithelium (cells from the surface of the cheek's mucosa) to assess the EMF-induced effects. As a criterion for determination of changes in the state of human buccal epithelium cell under the EMF action we proposed the Heterochromatin Granule Quantity (HGQ) in the cell nucleus [38].

EMF Action on Cell Membranes

The interest of investigators to EMFs effects on the cell membranes was stimulated by the theoretical work [39]. The author drives a conclusion very interesting to biologists. It should be remembered that cells have a membrane of about 10^{-6} cm thickness which maintains a very strong dipolar layer. In the (from the point of view of physics) complicated shape of a cell surface, local vibrations of a part of the cell membrane are feasible such that the positive and the negative part of a particular section of the membrane vibrate against each other leading to an oscillating electric dipole. Its frequency is of the order 10^{11}-10^{12} sec^{-1} if a sound velocity in the layer of order 10^{5}-10^{6} cm/sec perpendicular to the surface is assumed [39]. Indeed in many experimental works the effects of microwaves upon biological membranes were observed. The increase of cell membrane permeability to sodium and potassium ions was demonstrated [40-42]. Interestingly, that the microwave-induced increase of permeability to sodium ions was found to be reversible and returns to normal level within 60 min [42]. As it was supposed that the increase of permeability to ions is due to "micro-thermal" effects in cell membranes induced by microwaves [40]. Now the corresponding calculations of EMF-induced thermal gradients in membranes are done. The calculation on the impact of electric field induced thermal gradients across the plasma membrane may be applied to either Pulsed Electric Fields (PEFs) or Alternating Current (AC) fields [43]. The so-called "resonance" effects of microwaves on membrane permeability, i.e. the effect may be induced by EMF of the strictly determined wavelength and not induced at near frequencies, also were discussed [44]. It was demonstrated that in cells of algae *Nitellopsis obtusa* exposed to microwaves at frequencies of microwaves of 49, 70 and 76 GHz the chloride transmembrane current increased by 200 - 400%, at frequencies of 41.5 but at 71 GHz it decreased [45].

The microwave-induced effects (frequency of 37,5; 18,75 and 36.64 GHz) of increase of permeability of cell membranes to cytological stains (vital dyes) in living human buccal epithelium cells were reported [46,47]. Interestingly, that microwave-induced change in permeability to vital dye may recover in 2 hour period [47].

The effects of exposure to a 50 Hz magnetic field (maximum of 41.7 to 43.6 mT) on the membrane protein structures of living HeLa cells were studied using attenuated total reflection infrared spectroscopy. One min of such exposure shifted peak absorbance of the amide I band to a smaller wave number, reduced peak absorbance of the amide II band, and increased absorbance at around 1600 cm^{-1}. These results suggest that exposure to the ELF magnetic field has reversible effects on the peptide linkages, and changes the secondary structures of cell membrane proteins [48].

The experiments were also held in model membranes [49]. The suspension of egg lecithin multilamellar vesicles (liposomes) was exposed to 900 MHz microwave radiation for 5 h. Specific Absorption Rate (SAR) of the radiation for the investigated liposome sample was 12 ± 1 W/kg. Liposomal changes were monitored using a light scattering technique. Optical anisotropy of the liposome sample decreased dramatically upon exposure to microwave radiation, indicating structural changes in acyl chain packing. Infrared (IR) and Nuclear Magnetic Resonance ($^{(1)}$H NMR) studies showed an increased damage upon exposure to microwaves. The changes observed in the $^{(1)}$H NMR spectrum of the microwave exposed sample indicated hydrolysis of carboxylic and phosphoric esters. IR study showed conformational changes in the acyl chains of the lipids upon microwave exposure. However, both IR and $^{(31)}$P NMR did not show any appreciable changes in the head group part of the lipids [49].

A critical evaluation of three theories that describes the effects of weak electromagnetic fields on channel proteins in the cell membrane was done [50].

EMF Action on Free Radical Concentration in Cell

A free radical is a molecule or atom that has unpaired valence electrons. Free radicals are formed in cells and are involved in many intracellular processes. These compounds are capable to react with the cell structures causing their deterioration, so their excess is unfavourable. In particular, the free radical excess induces mutations and apoptosis – the programmed cell death. The "radical pair mechanism" is proposed to explain the phenomenon of "compass" in birds [51,52].

The role of Static Magnetic Field (SMF) in production of free radicals in cell is discussed in the review [11]. Authors propose that the primary cause of changes in cells after incubation in external SMF is disruption of free radical metabolism and elevation of their concentration. Such disruption causes oxidative stress and, as a result, damages ion channels, leading to changes in cell morphology and expression of different genes and proteins and also changes in apoptosis and proliferation [11]. The review of microwave effects on free radical formation in connection to male reproductive system one can find in [13]. Many investigations are analysing the microwave-induced effects on free radical concentration in cell, the problem of so-called oxidative stress. This problem attracts public concern in connection with interest to possible hazards of using mobile telephones. The aspects of the problem of EMF-induced free radicals in connection with neurodegenerative diseases are reviewed in review [53]. Some of experimental investigations studying the microwave-induced changes in the free radical situation in the cell are presented below.

Rats were exposed to 900 MHz EMF for 7 days (1 h/day). In other experimental groups, rats were exposed to EMF and pretreated with of Ginkgo biloba extract (Gb). Subsequently, oxidative stress markers and pathological changes in brain tissue were examined for each group. Oxidative damage was registered in the experimental group but these alterations were prevented by Gb treatment. Furthermore, Gb prevented the EMF-induced cellular injury in brain tissue which was demonstrated by histopathology methods [54].

The exposure of rats to electromagnetic radiation produced by GSM mobile phone induced in plasma a significant decrease in enzymes that defend cell from the elevation of free radical level: Catalase (CAT) and Superoxide Dismutase (SOD). The effect of acute doses of EMF on the rat's antioxidant status is significantly higher than that of fractionated doses of the same type of radiation [55].

In experiment on male guinea pigs exposed from 890- to 915-MHz EMF (217-Hz pulse rate, 2-W maximum peak power, SAR 0.95 w/kg) of a cellular phone for 12 h/day (11-h 45-min stand-by and 15-min spiking mode) for 30 days the changes in free radical marker malonic aldehyde (MDA) and antioxidant substances were shown. It was found that the MDA level increased (P<0.05), glutathione level and CAT enzyme activity decreased (P<0.05), and vitamins A, E and D(3) levels did not change (P>0.05) in the brain tissues of EMF-exposed guinea pigs. It was concluded that electromagnetic field emitted from cellular phone might produce oxidative stress in brain tissue of guinea pigs [56].

In experiments on Wistar female rats exposed to PC irradiation of monitor of Cathode Ray Tube (CRT) type the changes in eye corneal and lens tissues were investigated. In corneal tissue, MDA levels and CAT activity were found to increase in the computer group compared with the control group. Regarding lens tissue, MDA levels SOD activity and glutathione peroxidase activity (GSH-Px – the enzyme protecting cell from oxidative damage) were found to increase, as compared to the control group. The authors suppose that results of this study suggest that computer monitor radiation leads to oxidative stress in the corneal and lens tissues, and that vitamin C may prevent oxidative effects in the lens [57].

If the human Peripheral Blood Mononuclear Cells (PBMC) were exposed to the EMF of 900 MHz radiofrequency at a Specific Absorption Rate (SAR) of ~0.4W/kg longer than two hours, the apoptosis (the programmed cell death) is induced. The authors suppose that the activation of free radicals - Reactive Oxygen Species (ROS) is triggered by the conformation disturbance of lipids, protein, and DNA induced by the cell exposure [58].

In experiments with Wister rats located in the vicinity of the base station, estimated distance was less than 10 m the changes in free radical concentration were investigated. The frequency of radiation was 1800 MHz, SAR 0.95-2 W/kg for 40 and/or 60 days continuously. The study demonstrates slight decrease in the activity of glutathione reductase, lipid peroxidation as measured by MDA and total cholesterol in all tissues investigated when rats were exposed to radiation emitted from base station in 40 days. These parameters decreased when the period of exposure was extended to 60 days [59].

In experiments with the primary cultured cortical neurons exposed to pulsed electromagnetic fields at a frequency of 1800 MHz modulated by 217 Hz at an average SAR of 2 W/kg the oxidative damage was demonstrated [60]. At 24 h after exposure the radiation induced a significant increase in the levels of 8-hydroxyguanine (8-OHdG), a common biomarker of DNA oxidative damage, in the mitochondria of neurons. Concomitant with this finding, the copy number of mitochondrial DNA (mtDNA) and the levels of mitochondrial RNA (mtRNA) transcripts showed an obvious reduction after RF exposure. Each of these mtDNA disturbances could be reversed by pretreatment with melatonin, which is known to be an efficient antioxidant in the brain. Together, these results suggested that 1800 MHz RF radiation could cause oxidative damage in mtDNA in primary cultured neurons. In authors' opinion the oxidative damage of mtDNA may account for the neurotoxicity of RF radiation in the brain [60].

The investigation of EMF of mobile telephone on oxidation situation in rat sperm was done [61]. The frequency of the cell phone was 900 MHz, pulse GSM mode. The specific absorption rate was estimated to be 0.9 W/kg. The MDA concentration, SOD, CAT, and GPx activity were determined in sperm samples. The result shows a significant MDA increase in the mobile phone-exposed group as compared with the control ones. The reduction in GPx and SOD activity and an increase in CAT activity were observed after animal exposure to microwaves. Authors conclude that overproduction of reactive oxygen species (ROS) under microwave field exposure. Authors propose hypothetical scheme of action of EMF on enzyme histone kinase activity, micronuclei, cell cycle, and antioxidant enzymes via increase of intracellular concentration of free radicals as a primary mechanism [61].

In male Fischer-344 rats exposed to 900 MHz microwave radiation (SAR = 5.9 x 10^{-4} W/kg) and 1800 MHz microwave radiation (SAR = 5.8 x 10^{-4} W/kg) for 30 days (2 h/day) the significant impairment in cognitive function and induction of oxidative stress in brain tissues was registered [62,63].

As one can see, the microwave irradiation produces the oxidation stress in different tissues of laboratory animals and antioxidants added to animal nutrition may to definite extent prevent such consequences of microwave exposure. As antioxidants were used the extract of *Ginkgo biloba* [54], vitamin E and vitamin A [55], vitamins A, E and D [56], vitamin C [57], and even garlic [64].

The low-energy EMFs of Extremely Low Frequency (ELF) also induce the oxidative stress [65-68], but such effect is not always registered [69,70]. The prolongation of ELF exposure can result in adaptation to this factor [70].

It was supposed that the cell membranes may be the target of ROS [71] so the membrane permeabilization may be induced by the EMF-resulted ROS overproduction. At the same time the highly significant relationships was established between SAR, the oxidative DNA damage bio-marker, 8-OH-dG, and DNA fragmentation after microwave (1.8 GHz SAR from 0.4 W/kg to 27.5 W/kg) exposure of human spermatozoa [72]. About the analogous oxidative damage induced by EMF in rats mitochondrial DNA it was mentioned above [60]. So, the oxidative stress produced by EMFs may be the cause different negative consequences of EMF – micronuclei formation [61,73-75] and DNA strand breaks [76-78].

The Action of EMF on Intracellular Regulatory Systems

The fact that EMFs may regulate gene expression was demonstrated in many experimental works. Mainly the interest of researchers is concentrated on genes coding the proteins connected with the process of regulation of gene activity, for instance, the proteins c-jun, c-myc; or proteins connected with stress reaction of cells, first of all so-called heat shock proteins, for instance, hsp-70; or very important regulatory proteins p53 and p21. There are evidences of microwave exposure (1,71 GHz) influence on hsp70, c-jun, c-myc, and p21 levels in p53-deficient cells, but not in wild-type cells [79]. The EMF-induced changes of the transcript level of cell cycle regulatory and apoptosis-related genes were shown [80,81]. The 3.5 fold upregulation of mRNA stress-related transcription factor bZIP after exposure to 900 MHz microwaves was shown [82]. After the exposure of human lens epithelial cells to 1800-MHz GSM-like radiation for 2 h (SAR - 1.0, -3.5 W/kg) four proteins were upregulated. The heat-shock protein 70 (HSP70) and heterogeneous nuclear ribonucleoprotein K (hnRNP K) were among upregulated proteins [83]. The proteomic analysis revealed fourteen proteins specifically up-regulated by EMF (GSM 1800, 2 W/kg) in Jurkat cells. The examples of proteins specifically up-regulated by EMF are: heat shock protein 70, ubiquitin carboxyl-terminal hydrolase 14 and 26S protease regulatory subunit 6B [83]. Evidences of EMF-induced changes in transcriptiome (all transcripts synthesized by one cell or group of cells, including mRNA and non-coding RNA) and proteome (all proteins synthesized by one cell or group of cells) one may find in [84-86].

The EMF-induced changes in transcriptome and proteome are not registered in several studies. For instance, to seek alterations in gene transcription in bone marrow cells following *in vivo* exposure of juvenile mice to power frequency magnetic fields, young (21–24-day old) C57BL/6 mice were exposed to a 100 μT 50 Hz magnetic field for 2 h. A pilot experiment with 6 exposed (E) and 6 non-exposed (NE) mice identified four candidate responsive transcripts (two unknown transcripts (*AK152075* and *F10-NED*), phosphatidylinositol binding clathrin assembly protein (*Picalm*) and exportin 7 (*Xpo7*)). A larger experiment compared 19 E and 15 NE mice using two independent QRT-PCR assays and repeated microarray assays. No significant field-dependent changes were seen, although *Picalm* showed a trend to significance in one QRT-PCR assay [87].

The experiments in which the EMF-treated cells not revealed the specific changes in transcription or proteome changes are described in [88-96]. The differences in results in genomic and proteomic experiments with EMF-treated cells may be connected with methodological problems [97]. In our opinion the existence of results not revealing the EMF-induced changes in transcription not denies the results of experiments in which such changes were shown. What theoretic explanations of EMF-induced gene activity regulation are proposed?

In a series of works of R. Goodman and M. Blank and co-authors some mechanisms of the regulatory action of EMFs on gene activity were proposed. The activation of binding of both HSF and AP-1 to DNA is induced by magnetic fields [17]. The genes response to EMFs is regulated by with Electromagnetic Response Elements. A 900 base pair segment of the c-myc promoter, containing eight nCTCTn sequences, is required for the induction of c-myc expression by Electromagnetic (EM) fields. Similarly, a 70 bp region of the HSP70 promoter, containing three nCTCTn sequences, is required for the induction of HSP70 expression by EM fields. These sequences appear to act as Electromagnetic Field Response Elements (EMREs), since the ability of an EM field to induce stress proteins gradually disappears as the EMREs are mutated [18]. The EMF may interact directly with electrons in DNA and the electron transfer would result in gene expression [19]. Because of the low energy required, interaction with electrons in H-bonds may be the initial perturbation that leads double stranded DNA to come apart and begin the complex process of transcription to messenger RNA [20]. This model is detailed in [21]. Transfer of charge in electromagnetic field could contribute to separation of base pairs in DNA. An increase in local charge can cause separation of small groups of base pairs, and the low electronegativities of CTCT bases associated with the response to EMF increase the likelihood of electron displacement. EMF initiated DNA separation can set in motion the inter-connected biochemical signaling pathways that are activated in the stress response [21].

The EMF-induced effects on calcium concentration in cell are very interesting in connection with the role of calcium as an intracellular signal messenger [98] and the regulator of gene activity [98-101]. The calcium concentration increases in cell under the action of EMFs [102,103]. During the chronic exposure the calcium ions are released from the membrane, which are membrane bound and are released into the cytosol [104]. The influence of Extremely Low Frequency (ELF) magnetic fields on the transport of Ca^{2+} was studied in a biological system consisting of highly purified plasma membrane vesicles. Vesicles were exposed for 30 min at 32°C and the calcium efflux was studied using radioactive ^{45}Ca as a tracer. The plasma membrane vesicles were loaded with Ca^{2+}. After 30 min the vesicular Ca^{2+} content was decreased to approximately 50% of the initial value. Static magnetic fields ranging from 27 to 37 mT and time varying magnetic fields with frequencies between 7 and 72 Hz and amplitudes between 13 and 114 mT (peak) were used. At suitable combinations of static and time varying magnetic fields directly interact with the Ca^{2+} channel protein in the cell membrane, and it could be quantitatively confirm the model proposed in [105].

Calcium plays a leading role in EMF-induced cell differentiation, as it is shown in the works cited below. Exposure to 50 Hz EFs significantly enhanced proliferation in human neuroblastoma IMR32 (+40%) and rat pituitary GH3 cells (+38%). These data provide the direct evidence that EFs enhance the expression of voltage-gated Ca^{2+} channels on plasma membrane of the exposed cells. The consequent increase in Ca^{2+} influx is likely responsible for the EF-induced modulation of neuronal cell proliferation and apoptosis [106]. The differentiating neural stem/progenitor cells were exposed to EMNs (1 mT, 50 Hz), Cell exposure promotes neuronal differentiation of cells by upregulating Ca(v)1-channel expression and function [107].

The recent review [108] provides support for a pathway of the biological action of ultralow frequency and microwave EMFs, nanosecond pulses and static electrical or magnetic fields via EMF activation of voltage-gated calcium channels which leads to rapid elevation of intracellular Ca^{2+}, nitric oxide and in some cases at least, peroxynitrite. Potentially therapeutic effects may be mediated through the Ca^{2+}/nitric oxide/cGMP/protein kinase G pathway. Pathophysiological effects may be mediated through the Ca^{2+}/nitric oxide/peroxynitrite pathway.

In works of research group of Kharkiv National University the EMFs-induced condensation of chromatin in cell nuclei was demonstrated. Chromatin – the main component of cell nucleus in the period when cell is not dividing is presented in two forms – euchromatin (the main part of chromatin) and the heterochromatin (the condensed, non-active chromatin). The simple method for determination of the portion of heterochromatin by counting the number of heterochromatin granules in a probe of cells and determination of average number per cell was proposed for buccal epithelium cells – cells of cheek mucosa [109]. In a series of works it was demonstrated that the low energy EMF (10 -100 μW/cm²) and static magnetic field (25 mT) induce chromatin condensation (heterochromatinization) [38,47,110-114]. It is supposed that this phenomenon reflects the process of regulation of genetic activity at cytological level, because heterochromatinization is connected with decrease of functional activity of chromatin [115,116]. The EMF-induced chromatin condensation is the necessary stage of further thin process of regulation activity of individual genes [114]. It was shown the decrease of the functional activity of polytene (giant) chromosomes in EMF-exposed Drosophila. In larvae developed from the EMF-exposed eggs at the prepupal stage 3 of 8 chromosomal puffs tested (71CE, 82EF, and 83E) had significantly smaller dimensions than these in control [117]. In our opinion, the mechanism of EMF-induced chromatin condensation may include the primary effects connected with changes in DNA-protein interactions induced by EMF, and also the known mechanisms of chromatin remodeling during chromatin condensation. The increase of the intranuclear concentration of calcium ions may be involved in the process of EMF-induced chromatin condensation.

Conclusion

The problem of primary mechanism of EMF-induced effects on the cellular level is far from its solving. It is difficult to identify the leading cause of such effects. Sometimes the effects are registered in some laboratories, but not registered in others. This phenomenon may be partly explained by differences in the EMF sources and experimental objects applied. The cells of different genotype [118,119] and of the different of stage of differentiation [120] reveal different reactions to the same EMFs. So, these differences may be taken into consideration when experiments are panned and analyzed. The good established are facts of oxidative stress and gene regulation changes after EMFs action. But sometimes it is difficult to distinguish the primary cause of EMF-induced effect and the secondary consequence of EMF action. This consideration may be applied to the EMF-induced oxidation stress and to gene regulation pattern, because these events accompany the action of different unfavorable factors on cell. It is possible that such complex reaction as cell reaction on EMFs has many competing mechanisms and under different conditions to the fore is advanced one of them. In our opinion the primary reaction to electromagnetic field may be connected with the level of gene activity regulation. The reaction of chromatin condensation is one of the "safe" cell reactions to EMF which is always registered and which we can observe immediately after applying of EMF. The investigation of molecular mechanisms underlying this reaction can give new important information about the primary mechanism of EMF action on cell.

References

1. Banik S, Bandyopadhyay S, Ganguly S (2003) Bioeffects of microwave--a brief review. Bioresour Technol 87: 155-159.

2. Vorst AV, Rosen A, Kotsuka Y (2006) RF/microwave interaction with biological tissues (Wiley Series in microwave and optical engineering). Wiley-IEEE Press, New York, USA.

3. Barnes FS, Greenebaum B (2007) Bioengineering and biophysical aspects of electromagnetic fields. Handbook of biological effects of electromagnetic fields (3rd edn.), CRC Press, Boca Raton, New York, Washington, USA.

4. Furse C, Christensen DA, Durney CH (2009) Basic Introduction to bioelectromagnetics (2nd edn.), CRC Press, Boca Raton, New York, Washington, USA.

5. Rosch PJ, Markov MS (2004) Bioelectromagnetic medicine. Marcel Dekker Inc, New York, USA.

6. Andrä W, Nowak H (2007) Magnetism in medicine: a handbook (2nd edn.). Wiley-VCH Verlag GmbH & Co. KGaA.

7. Markov MS (2007) Magnetic field therapy: a review. Electromagn Biol Med 26: 1-23.

8. Takahashi A, Ohnishi T (2009) Molecular mechanisms involved in adaptive responses to radiation, UV light, and heat. J Radiat Res 50: 385-393.

9. Kovacic P, Somanathan R (2010) Electromagnetic fields: mechanism, cell signaling, other bioprocesses, toxicity, radicals, antioxidants and beneficial effects. J Recept Signal Transduct Res 30: 214-226.

10. Markov MS (2014) Electromagnetic fields and life. J Electr Electron Syst 3: 119.

11. Ghodbane S, Lahbib A, Sakly M, Abdelmelek H (2013) Bioeffects of static magnetic fields: oxidative stress, genotoxic effects, and cancer studies. Biomed Res Int 2013: 602987.

12. Szmigielski S (2013) Reaction of the immune system to low-level RF/MW exposures. Sci Total Environ 454-455: 393-400.

13. Desai NR, Kesari KK, Agarwal A (2009) Pathophysiology of cell phone radiation: oxidative stress and carcinogenesis with focus on male reproductive system. Reprod Biol Endocrinol 7: 114.

14. Kesari KK, Siddiqui MH, Meena R, Verma HN, Kumar S (2013) Cell phone radiation exposure on brain and associated biological systems. Indian J Exp Biol 51: 187-200.

15. Bhat MA, Kumar V, Gupta GK (2013) Effects of mobile phone and mobile phone tower radiations on human health. International Journal of Recent Scientific Research 4:1422- 1426.

16. Hardell L, Carlberga M, Hansson Mild K (2013) Use of mobile phones and cordless phones is associated with increased risk for glioma and acoustic neuroma. Pathophysiology 20: 85-110.

17. Lin H, Han L, Blank M, Head M, Goodman R (1998) Magnetic field activation of protein-DNA binding. J Cell Biochemis 70: 297-303.

18. Lin H, Blank M, Rossol-Haseroth K, Goodman R (2001) Regulating genes with electromagnetic response elements. J Cell Biochemis 81: 143-148.

19. Goodman R, Blank M (2002) Insights into electromagnetic interaction mechanisms. J Cell Physiol 192: 16-22.

20. Blank M, Goodman R (2004) Initial interactions in electromagnetic field-induced biosynthesis. J Cell Physiol 199: 359-363.

21. Blank M, Goodman R (2008) A mechanism for stimulation of biosynthesis by electromagnetic fields: charge transfer in DNA and base pair separation. J Cell Physiol 214: 20-26.

22. Michaelson SM (1980) Microwave biological effects: An overview. Proceedings of the IEEE 68: 40-49.

23. World Health Organization (1981) International programme on chemical safety environmental health criteria. Radiofrequency and microwaves, Geneva.

24. Levit BB, Lai H (2010) Biological effects from exposure to electromagnetic radiation emitted by cell tower base stations and other antenna arrays. Environ Rev 18: 369-395.

25. Hocking B (2001) Microwave sickness: a reappraisal. Occup Med (Lond) 51: 66-69.

26. Hutter H-P, Moshammer H, Wallner P, Kundi M (2006) Subjective symptoms, sleeping problems, and cognitive performance in subjects living near mobile phone base stations. Occup Environ Med 63: 307-313.

27. Yildirim M, Yildirim A, Zamani AG, Okudan N (2010) Effect of mobile phone station on micronucleus frequency and chromosomal aberrations in human blood cells. Genet Couns 21: 243-251.

28. Hardell L, Carlberg M, Hansson Mild K (2009) Epidemiological evidence for an association between use of wireless phones and tumor diseases. Pathophysiology 16: 113-122.

29. Poulsen AH, Friis S, Johansen C, Jensen A, Frei P, et al. (2013) Mobile phone use and the risk of skin cancer: a nationwide cohort study in Denmark. Am J Epidemiol 178: 190-197.

30. Sohrabi M-R, Tarjoman T, Abadi A, Yavari P (2010) Living near overhead high voltage transmission power lines as a risk factor for childhood acute lymphoblastic leukemia: a case-control study. Asian Pac J Cancer Prev 11: 423-427.

31. Elliott P, Toledano MB, Bennett J, Beale L, de Hoogh K, et al. (2010) Mobile phone base stations and early childhood cancers: case-control study. BMJ 340: c3077.

32. García AM, Sisternas A, Hoyos SP (2008) Occupational exposure to extremely low frequency electric and magnetic fields and Alzheimer disease: a meta-analysis. Int J Epidemiol 37: 329-340.

33. Hartwig V, Giovannetti G, Vanello N, Lombardi M, Landini L, et al. (2009) Biological effects and safety in magnetic resonance imaging: A review. Int J Environ Res Public Health 6: 1778-1798.

34. Zaremba LA (2003) Guidance for industry and FDA staff: criteria for significant risk investigations of magnetic resonance diagnostic devices. U.S. Department of Health and Human Services Food and Drug Administration. Silver Spring, MD, USA.

35. Vecchia P, Matthes R, Ziegelberger G, Lin J, Saunders R, Swerdlow A (2009) Exposure to high frequency electromagnetic fields, biological effects and health consequences (100 kHz-300 GHz). Review of the scientific evidence on dosimetry, biological effects, epidemiological observations, and health consequences concerning exposure to high frequency electromagnetic fields (100 kHz to 300 GHz). International Commission on Non-Ionizing Radiation Protection.

36. WHO (2002) IARC Monographs on the evaluation of carcinogenic risks to humans IARC Monograph, Volume 80. Non-Ionizing radiation, Part 1: Static and Extremely Low-Frequency (ELF) electric and magnetic fields.

37. WHO Press Release (2011) IARC classifies radiofrequency electromagnetic fields as possibly carcinogenic to humans.

38. Shckorbatov Y (2012) The state of chromatin as an integrative indicator of cell stress. New Developments in Chromatin Research 6: 123-144. Nova Publishers, New York, USA.

39. Fröhlich H (1968) Long-range coherence and energy storage in biological systems. Int J Quantum Chem 2: 641-649.

40. Ismailov ESh (1971) Mechanism of the effect of microwaves on the permeability of erythrocytes for potassium and sodium ions. Nauchnye Dokl Vyss Shkoly Biol Nauki 3: 58-60 (in Russian).

41. Alekseev SI, Ziskin MC (1995) Millimeter microwave effect on ion transport across lipid bilayer membranes. Bioelectromagnetics 16: 124-131.

42. Liburdy RP, Vanek PFJr (1985) Microwaves and the cell membrane: II. Temperature, plasma, and oxygen mediate microwave-induced membrane permeability in the erythrocyte. Radiat Res 102: 190-205.

43. Garner AL, Deminsky M, Neculaes VB, Chashihin V, Knizhnik A, et al. (2013) Cell membrane thermal gradients induced by electromagnetic fields. J Appl Phys 113: 214701-214701-11.

44. Golant MB (1989) Resonance effect of coherent millimeter-band electromagnetic waves on living organisms. Biofizika 34: 1004-1014 (in Russian).

45. Kataev AA, Aleksandrov AA, Tikhonova LL, Berestovskii GN (1993) Frequency-dependent effect of millimeter electromagnetic waves on ionic currents algae Nitellopsis. Non-thermal effects. Biofizika 38: 446-462 (in Russian).

46. Shckorbatov YG, Shakhbazov VG, Navrotskaya VV, Grabina VA, Sirenko SP, et al. (2002) Electrokinetic properties of nuclei and membrane permeability in human buccal epithelium cells influenced by the low-level microwave radiation. Electrophoresis 23: 2074-2079.

47. Shckorbatov YG, Pasiuga VN, Kolchigin NN, Grabina VA, Ivanchenko DD, et al. (2011) Cell nucleus and membrane recovery after exposure to microwaves. Proceedings of the Latvian Academy of Sciences 65: 13-20.

48. Ikehara T, Yamaguchi H, Hosokawa K, Miyamoto H, Aizawa K (2003) Effects of ELF magnetic field on membrane protein structure of living HeLa cells studied by Fourier transform infrared spectroscopy. Bioelectromagnetics 24: 457-464.

49. Gaber MH, Abd El Halim N, Khalil WA (2005) Effect of microwave radiation on the biophysical properties of liposomes. Bioelectromagnetics 26: 194-200.

50. Halgamuge MN, Perssont BRR, Salford LG, Mendis P, Eberhardt J (2009) Comparison between two models for interactions between electric and magnetic fields and proteins in cell membranes. Environ Eng Sci 26: 1473-1480.

51. Hill E, Ritz T (2010) Can disordered radical pair systems provide a basis for a magnetic compass in animals? J R Soc Interface 7: S265-S271.

52. Rodgers CT, Hore PJ (2009) Chemical magnetoreception in birds: The radical pair mechanism. Proc Natl Acad Sci USA 106: 353-360.

53. Consales C, Merla C, Marino C, Benassi B (2012) Electromagnetic fields, oxidative stress, and neurodegeneration. International Journal of Cell Biology 2012: 16 pages.

54. Ilhan A, Gurel A, Armutcu F, Kamisli S, Iraz M, et al. (2004) Ginkgo biloba prevents mobile phone-induced oxidative stress in rat brain. Clin Chim Acta 340: 153-162.

55. Elhag MA, Nabil GM, Attia AM (2007) Effects of electromagnetic field produced by mobile phones on the oxidant and antioxidant status of rats. Pak J Biol Sci 10: 4271-4274.

56. Meral I, Mert H, Mert N, Deger Y, Yoruk I, et al. (2007) Effects of 900-MHz electromagnetic field emitted from cellular phone on brain oxidative stress and some vitamin levels of guinea pigs. Brain Res 1169: 120-124.

57. Balci M, Namuslu M, Devrim E, Durak İ (2009) Effects of computer monitor-emitted radiation on oxidant/antioxidant balance in cornea and lens from rats. Mol Vis 15: 2521-2525.

58. Yao-Sheng Lu, Bao-Tian Huang, Yao-Xiong Huang (2012) Reactive oxygen species formation and apoptosis in human peripheral blood mononuclear cell induced by 900 MHz mobile phone radiation. Oxidative Medicine and Cellular Longevity 2012: 8 pages.

59. Achudume AC, Onibere B, Aina F (2009) Bioeffects of electromagnetic base station on glutathione reductase, lipid peroxidation and total cholesterol in different tissues. Biology and Medicine 1: 33-38.

60. Xu S, Zhou Z, Zhang L, Yu Z, Zhang W, et al. (2010) Exposure to 1800 MHz radiofrequency radiation induces oxidative damage to mitochondrial DNA in primary cultured neurons. Brain Res 1311: 189-196.

61. Kesari KK, Kumar S, Behari J (2011) Effects of radiofrequency electromagnetic wave exposure from cellular phones on the reproductive pattern in male Wistar rats. Appl Biochem Biotechnol 164: 546-559.

62. Megha K, Deshmukh PS, Banerjee BD, Tripathi AK, Abegaonkar MP (2012) Microwave radiation induced oxidative stress, cognitive impairment and inflammation in brain of Fischer rats. Indian J Exp Biol 50: 889-896.

63. Deshmukh PS, Banerjee BD, Abegaonkar MP, Megha K, Ahmed RS, et al. (2013) Effect of low level microwave radiation exposure on cognitive function and oxidative stress in rats. Indian J Biochem Biophys 50: 114-119.

64. Avci B, Akar A, Bilgici B, Tunçel ÖK (2012) Oxidative stress induced by 1.8 GHz radio frequency electromagnetic radiation and effects of garlic extract in rats. Int J Radiat Biol 88: 799-805.

65. Goraca A, Ciejka E, Piechota A (2010) Effects of extremely low frequency magnetic field on the parameters of oxidative stress in heart. J Physiol Pharmacol 61: 333-338.

66. Ciejka E, Kleniewska P, Skibska B, Goraca A (2011) Effects of extremely low frequency magnetic field on oxidative balance in brain of rats. J Physiol Pharmacol 62: 657-661.

67. Chu LY, Lee JH, Nam YS, Lee YJ, Park WH, et al. (2011) Extremely low frequency magnetic field induces oxidative stress in mouse cerebellum. Gen Physiol Biophys 30: 415-521.

68. Selaković V, Rauš Balind S, Radenović L, Prolić Z, Janać B (2013) Age-dependent effects of ELF-MF on oxidative stress in the brain of Mongolian gerbils. Cell Biochem Biophys 66: 513-521.

69. Hong MN, Han NK, Lee HC, Ko YK, Chi SG, et al. (2012) Extremely low frequency magnetic fields do not elicit oxidative stress in MCF10A cells. J Radiat Res 53: 79-86.

70. Glinka M, Sieroń A, Birkner E, Cieślar G (2013) Influence of extremely low-frequency magnetic field on the activity of antioxidant enzymes during skin wound healing in rats. Electromagn Biol Med 32: 463-470.

71. Avery SV (2011) Molecular targets of oxidative stress. Biochem J 434: 201-210.

72. De Iuliis GN, Newey RJ, King BV, Aitken RJ (2009) Mobile phone radiation induces reactive oxygen species production and DNA damage in human spermatozoa in vitro. PLoS One 4: e6446.

73. Zotti-Martelli L, Peccatori M, Maggini V, Ballardin M, Barale R (2005) Individual responsiveness to induction of micronuclei in human lymphocytes after exposure in vitro to 1800-MHz microwave radiation. Mutat Res 582: 42-52.

74. Karaca E, Durmaz B, Aktug H, Yildiz T, Guducu C, et al. (2012) The genotoxic effect of radiofrequency waves on mouse brain. J Neurooncol 106: 53-58.

75. Atlı Şekeroğlu Z, Akar A, Şekeroğlu V (2013) Evaluation of the cytogenotoxic damage in immature and mature rats exposed to 900 MHz radiofrequency electromagnetic fields. Int J Radiat Biol 89: 985-992.

76. Paulraj R, Behari J (2006) Single strand DNA breaks in rat brain cells exposed to microwave radiation. Mutat Res 596: 76-80.

77. Deshmukh PS, Megha K, Banerjee BD, Ahmed RS, Chandna S, et al. (2013) Detection of low level microwave radiation induced deoxyribonucleic acid damage vis-à-vis genotoxicity in brain of Fischer rats. Toxicol Int 20: 19-24.

78. Mihai CT, Rotinberg P, Brinza F, Vochita G (2014) Extremely low-frequency electromagnetic fields cause DNA strand breaks in normal cells. J Environ Health Sci Eng 12: 15.

79. Czyz J, Guan K, Zeng Q, Nikolova T, Meister A, et al. (2004) High frequency electromagnetic fields (GSM signals) affect gene expression levels in tumor suppressor p53-deficient embryonic stem cells. Bioelectromagnetics 25: 296-307.

80. Nikolova T, Czyz J, Rolletschek A, Blyszczuk P, Fuchs J, et al. (2005) Electromagnetic fields affect transcript levels of apoptosis-related genes in embryonic stem cell-derived neural. FASEB J 19: 1686-1688.

81. ZhaoT-Y, Shi-Ping Zou S-P, Knapp PE (2007) Exposure to cell phone radiation up-regulates apoptosis genes in primary cultures of neurons and astrocytes. Neurosci Lett 412: 34-38.

82. Vian A, Roux D, Girard S, Bonnet P, Paladian F, et al. (2006) Microwave irradiation affects gene expression in plants. Plant Signal Behav 1: 67-70.

83. Gerner C, Haudek V, Schandl U, Bayer E, Gundacker N, et al. (2010) Increased protein synthesis by cells exposed to a 1,800-MHz radio-frequency mobile phone electromagnetic field, detected by proteome profiling. Int Arch Occup Environ Health 83: 691-702.

84. Li HW, Yao K, Jin HY, Sun LX, Lu DQ, et al. (2007) Proteomic analysis of human lens epithelial cells exposed to microwaves. Jpn J Ophthalmol 51: 412-416.

85. Tsai MT, Li WJ, Tuan RS, Chang WH (2009) Modulation of osteogenesis in human mesenchymal stem cells by specific pulsed electromagnetic field stimulation. J Orthop Res 27: 1169-1174.

86. Hinsenkamp M, Collard JF (2011) Bone Morphogenic Protein--mRNA upregulation after exposure to low frequency electric field. Int Orthop 35: 1577-1581.

87. Kabacik S, Kirschenlohr H, Raffy C, Whitehill K, Coster M, et al. (2013) Investigation of transcriptional responses of juvenile mouse bone marrow to power frequency magnetic fields. Mutat Res 745-746: 40-45.

88. Morehouse CA, Owen RD (2000) Exposure of Daudi cells to low-frequency magnetic fields does not elevate MYC steady-state mRNA levels. Radiat Res 153: 663-669.

89. Yomori H, Yasunaga K, Takahashi C, Tanaka A, Takashima S, et al. (2002) Elliptically polarized magnetic fields do not alter immediate early response genes expression levels in human glioblastoma cells. Bioelectromagnetics 23: 89-96.

90. Shi B, Farboud B, Nuccitelli R, Isseroff RR (2003) Power-line frequency electromagnetic fields do not induce changes in phosphorylation, localization, or expression of the 27-Kilodalton heat shock protein in human keratinocytes. Environ Health Perspect 111: 281-288.

91. Coulton LA, Harris PA, Barker AT, Pockley, AG (2004) Effect of 50 Hz electromagnetic fFields on the induction of heat-shock protein gene expression in human leukocytes. Radiat Res 161: 430-434.

92. Thorlin T, Rouquette JM, Hamnerius Y, Hansson E, Persson M, et al. (2006) Exposure of cultured astroglial and microglial brain cells to 900 MHz microwave radiation. Radiat Res 166: 409-421.

93. Chauhan V, Mariampillai A, Bellier PV, Qutob SS, Gajda GB, et al. (2006) Gene expression analysis of a human lymphoblastoma cell line exposed in vitro to an intermittent 1.9 GHz pulse-modulated radiofrequency field. Radiat Res 165: 424-429.

94. Nylund R, Kuster N, Leszczynski D (2010) Analysis of proteome response to the mobile phone radiation in two types of human primary endothelial cells. Proteome Science 8: 52.

95. Kim KB, Byun HO, Han NK, Ko YG, Choi HD, et al. (2010) Two-dimensional electrophoretic analysis of radio frequency radiation-exposed MCF7 breast cancer cells. J Radiat Res 51: 205-213.

96. Sakurai T, KiyokoewaT, Eijiro Narita E, Suzyki Y, Taki M, et al. (2011) Analysis of gene expression in a human-derived glial cell line exposed to 2.45 GHz continuous radiofrequency electromagnetic fields. J Radiat Res 52: 185-192.

97. Vanderstraeten J, Verschaeve L (2008) Gene and protein expression following exposure to radiofrequency fields from mobile phones. Environ Health Perspect 116: 1131-1135.

98. Cerella C, Diederich M, Ghibelli L (2010) The Dual Role of Calcium as Messenger and Stressor in Cell Damage, Death, and Survival. Int J Cell Biol 2010: 546163.

99. Hardingham GE, Bading H (1998) Nuclear calcium: a key regulator of gene expression. Biometals 11: 345-358.

100. Alonso MT, García-Sancho J (2011) Nuclear Ca^{2+} signaling. Cell Calcium 49: 280-289.

101. Bengtson CP, Bading H (2012) Nuclear Calcium Signaling. Synaptic Plasticity, Advances in Experimental Medicine and Biology, Springer-Verlag, Wien, NY, USA.

102. Cho MR, Thatte HS, Silvia MT, Golan DE (1999) Transmembrane calcium influx induced by ac electric fields. FASEB J 13: 677-683.

103. Pazur A, Rassadina V (2009) Transient effect of weak electromagnetic fields on calcium ion concentration in Arabidopsis thaliana. BMC Plant Biology 9:47.

104. Paulraj R, Behari J (2012) Biochemical Changes in Rat Brain Exposed to Low Intensity 9.9 GHz Microwave Radiation. Cell Biochem Biophys 63: 97-102.

105. Bauréus Koch CLM, Sommarin M, Persson BRR, Salford LG, Eberhardt JL (2003) Interaction between weak low frequency magnetic fields and cell membranes. Bioelectromagnetics 24: 395-402.

106. Grassi C, D'Ascenzo M, Torsello A, Martinotti G, Wolf F, et al. (2004) Effects of 50 Hz electromagnetic fields on voltage-gated Ca^{2+} channels and their role in modulation of neuroendocrine cell proliferation and death. Cell Calcium 35: 307-315.

107. Piacentini R, Ripoli C, Mezzogori D, Azzena GB, Grassi C (2008) Extremely low-frequency electromagnetic fields promote in vitro neurogenesis via upregulation of Ca(v)1-channel activity. J Cell Physiol 215: 129-139.

108. Pall ML (2013) Electromagnetic fields act via activation of voltage-gated calcium channels to produce beneficial or adverse effects. J Cell Mol Med 17: 958-965.

109. Shckorbatov YG (1999) He-Ne laser light induced changes in the state of the chromatin in human cells. Naturwissenschaften 86: 452-453.

110. Shckorbatov YG, Pasiuga VN, Kolchigin NN, Batrakov DO, Kazansky OV, et al. (2009) Changes in the human nuclear chromatin induced by ultra wideband pulse irradiation. Central European Journal of Biology 4: 97-106.

111. Shckorbatov YG, Pasiuga VN, Kolchigin NN, Grabina VA, Batrakov DO, et al. (2009) The influence of differently polarized microwave radiation on chromatin in human cells. Int J Radiat Biol 85: 322-329.

112. Shckorbatov YG, Pasiuga VN, Goncharuk EI, Petrenko TPh, Grabina VA, et al. (2010) Effects of differently polarized microwave radiation on the microscopic structure of the nuclei in human fibroblasts. J Zhejiang Univ Sci B 11: 801-805.

113. Shckorbatov YG, Rudneva II, Pasiuga VN. Grabina VA, Kolchigin NN, et al. (2010) Electromagnetic fields effects on Artemia hatching and chromatin state. Central European Journal of Biology 5: 785-790.

114. Shckorbatov YG, Katrich VA, Pasiuga VN, Rudenko AO (2013) Cell Response to Electromagnetic Field: Nuclear and Membrane Mechanisms. Nova Biomedical, New York, USA.

115. Martin RM, Cardoso MC (2010) Chromatin condensation modulates access and binding of nuclear proteins. FASEB J 24: 1066-1072.

116. Biran A, Meshorer E (2012) Concise review: chromatin and genome organization in reprogramming. Stem Cells 30: 1793-1799.

117. Shakina LA, Pasiuga VN, Dumin OM, Shckorbatov YG (2011) Effects of microwaves on the puffing pattern of D. melanogaster. Central European Journal of Biology 6: 524-530.

118. Tenuzzo B, Chionna A, Panzarini E, Lanubile R Tarantino P, et al. (2006) Biological effects of 6 mT static magnetic fields: a comparative study in different cell types. Bioelectromagnetics 7: 560-577.

119. Ding G-R, Yaguchi H, Yoshida M, Miyakoshi J (2000) Increase in X-Ray-induced mutations by exposure to magnetic field (60 Hz, 5 mT) in NF-kB-inhibited cells. Biochem Bioph Res Co 276: 238-243.

120. Tsai MT, Li WJ, Tuan RS, Chang WH (2009) Modulation of osteogenesis in human mesenchymal stem cells by specific pulsed electromagnetic field stimulation. J Orthop Res 27: 1169-1174.

9

Wind Energy Conservation with Grid Levelling for Transient Loads

Heera K*

Electrical and Electronics Engineering, Francis Xavier Engineering College, Tirunelveli, India

Abstract

The paper discusses the maximum power point tracking in a grid connected PMSG based WECS. To the variable-speed wind turbine, if the rotor speed can always be adjusted to make the turbine operate under optimum tip speed ratio then it means that the turbine realizes the MPPT operation. For this purpose the P and O tracking algorithm is adopted. In addition to fully recognize the wind energy it is necessary to integrate it to the grid and hence grid parameters are regulated as well. The proposed system is developed in Matlab environment.

Keywords: Grid integration; MPPT control; PMSG; Wind energy systems

Introduction

Among different sorts of renewable vitality, wind vitality is dealt with as the most difficult one because of its free accessibility, arrangements encouraging, and the development of turbine systems. Renewable vitality particularly wind vitality conversion systems have attracted an expanding enthusiasm for the past years since they could be considered as affirmed choices for sustaining the consistent developing vitality needs [1,2]. The development of renewable energy in India is colossal and wind energy demonstrates to be the best answer for the problem of exhausting fossil powers, importing of coal, nursery gas emission, and ecological contamination and so on. In this way, wind vitality transformation innovation has turn into the examination centre of scientists everywhere throughout the world. Renewable Energy in India is colossal and Wind Energy demonstrates to be the best answer for the problem of exhausting fossil powers, importing of coal, nursery gas emission, and ecological contamination and so on. In this way, wind vitality transformation innovation has turn into the examination centre of scientists everywhere throughout the world [3].

In advanced wind vitality change framework two turbine structures are favoured: DFIG and PMSG Albeit, both of these structures highlight enhanced proficiency, diminished streamlined loads, and simplicity of dynamic and receptive force regulation, recent is significantly more dependable than the previous, considering the likelihood of wiping out of gearbox. In this way, an immediate commutes PMSG based WECS is considered in this paper. To collect more vitality from the variety winds, MPPT control ought to be incorporated in the power control framework. The diverse strategies for MPPT framework are characterized in [4-6]. However till date, there is no decisive proof is accessible as to which MPPT framework is prone to give a more effective and less costly in written works.

Wind Energy Conversion System Technology

A WECS is a structure that transforms the kinetic energy of the incoming air stream into electrical energy. Modern Wind Energy Conversion System (WECS) is shown in Figure 1 and the energy conversion chain is organised into four subsystems:

Aerodynamic subsystem

Consisting mainly of the turbine rotor, which is composed of blades and turbine hub, which is the support for blades.

Drive train

This is generally composed of low-speed shaft–coupled with the turbine, hub, speed multiplier and high-speed shaft–driving the electrical generator.

Electromagnetic subsystem

It mainly consists of the electric generator.

Electric Subsystem

Electrical subsystem includes the elements for grid connection and local grid. The circuit diagram of the proposed system as in Figure 1 includes variable speed wind turbine, permanent magnet synchronous generator, power electronic components which includes rectifier, inverter, boost converter and the control system which is a PI controller [7-9]. The description of the circuit diagram parameters are described in the following sections (Figure 2).

The tip speed ratio of a wind turbine is a variable expressing the ratio between the peripheral blade speed and the wind speed. It is denoted by λ and computed as below:

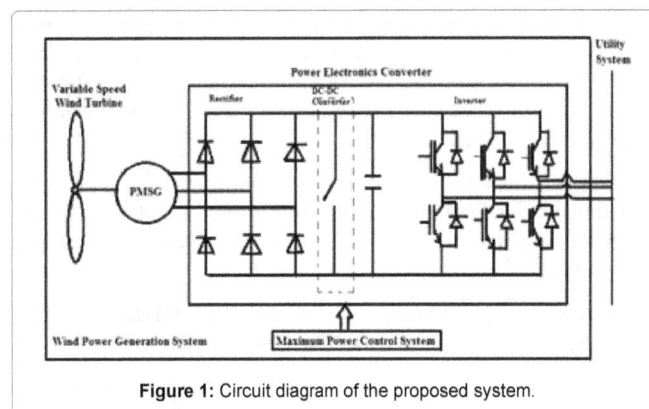

Figure 1: Circuit diagram of the proposed system.

***Corresponding author:** Heera K, Electrical and Electronics Engineering, Francis Xavier Engineering College, Tirunelveli, India, E-mail: heeraakn@gmail.com

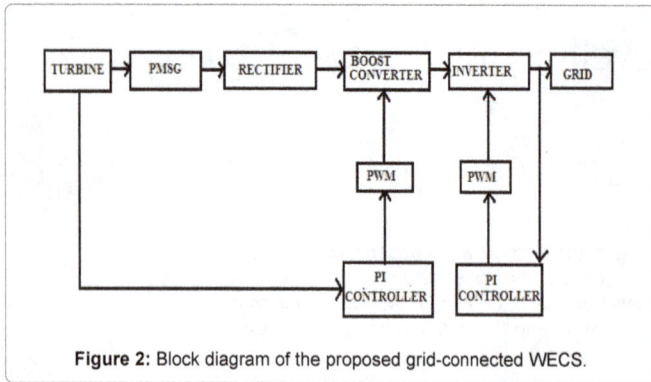

Figure 2: Block diagram of the proposed grid-connected WECS.

$$\lambda = \frac{R\omega}{v} \tag{1}$$

Where R is the blade length, ω is the rotor speed, v is the wind speed and the power extracted by a wind turbine whose blade length is R is expressed as:

$$P_T = \frac{1}{2}\rho\pi R^2 C_p(\lambda)v^3 \tag{2}$$

Therefore,

$$C_p = 4a(1-a)^2 \tag{3}$$

$$C_p = 4a(1-a)^2$$

The maximum value of Cp occurs for a=1/3 and hence Cpmax=0.59 known as the Betz limit and represents the maximum power extraction efficiency of a wind turbine [10].

Working

(a) When wind speed is below cut-in speed the machine does not produce power. If the rotor has a sufficient staring torque, it may start rotating below this wind speed. However, no power is extracted and the rotor rotates freely. In many modern designs the aerodynamic torque produced at the standstill condition is quite low and the rotor has to be started (by working the generator in the motor mode) at the cut-in wind speed.

(b) At normal wind speeds, maximum power is extracted from wind. The maximum power point is achieved at a specific (constant) value of the TSR. Therefore, to track the maximum power limit point, the rotational speed has to be changed continuously in proportion to the wind speed [11,12].

(c) At high winds, the rotor speed is limited to maximum value depending on the design limit of the mechanical components. In this region, the Cp is lower than the maximum and the power output is not proportional to the cube of the wind speed.

(d) At even higher wind speeds, the power output is kept constant at the maximum value allowed by the electrical components.

(e) At a certain cut-out or furling wind speed, the power generation is shut down and the rotation stopped in order to protect the system components.

The output power evolves according to Equation (2), proportionally with the wind speed cubed, until it reaches the wind turbine rated power. This output power from turbine is fed into the PMSG. PMSG is favoured more and more in developing new designs because of its higher efficiency, high power density, availability of high-energy permanent magnet material at reasonable price, and the possibility of providing smaller turbine diameter in direct drive applications [13].

Then power conversion for wind energy systems occurs in two stages. The first stage is rectification, where the alternating current (AC) is transformed into direct current (DC). The boost converter steps up the input DC voltage. The second stage is inversion where the direct current is transformed back into alternating current. PI controller is adopted in this system as this will optimize the conversion coefficient to maintain maximum power output. The inputs to the controller are the wind speed and voltage, current that are to be fed into the grid. The PI controller regulates the inputs and feeds the error signal to PWM. The PWM scheme is most commonly used because of the possibility of voltage regulation, but it will also cancel out multiples of the third harmonic to help improve output power quality. The inverter receives the switching signals from the PWM which in turn regulates the incoming DC link voltage and current and feeds it into the grid. The wind speed tracking is also shown.

The basic device in the wind energy conversion system is the wind turbine which transfers the kinetic energy into a mechanical energy. The wind turbine is connected to the electrical generator through a coupling device gear train or a direct drive system. The output of the generator is given to the electrical grid by employing a proper controller to avoid the disturbances and to protect the system or network. The detailed description of various blocks are already discussed in the above sections [14-16].

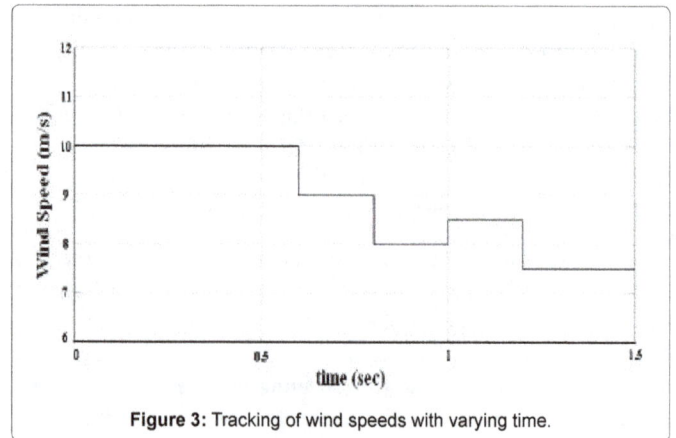

Figure 3: Tracking of wind speeds with varying time.

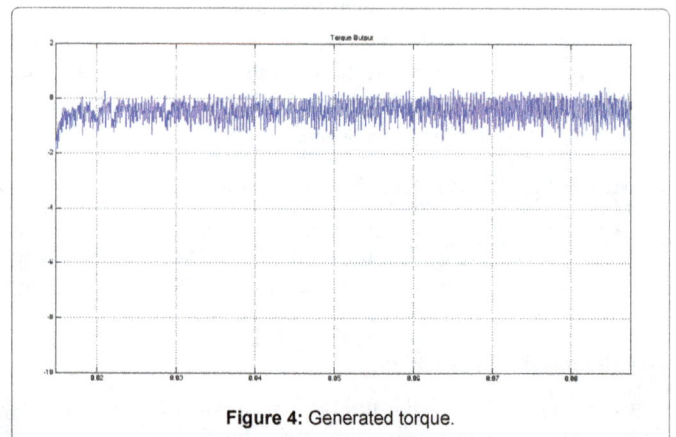

Figure 4: Generated torque.

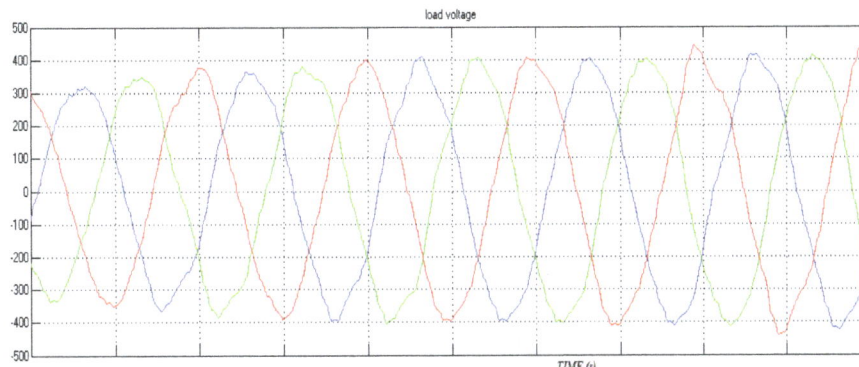

Figure 5: Input voltage to grid.

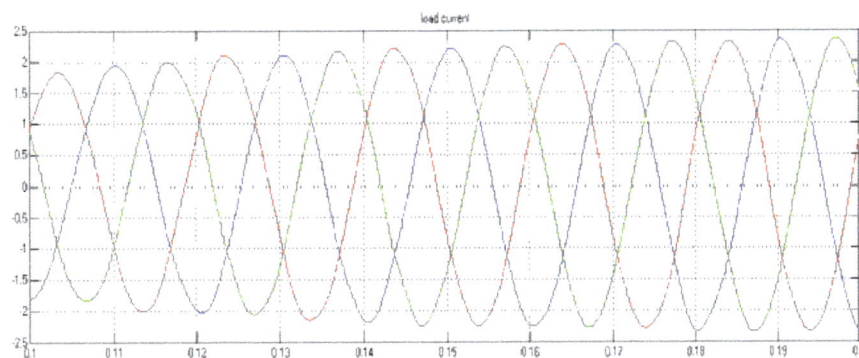

Figure 6: Input current to grid.

Simulation Results

This chapter presents the results of the proposed wind energy conservation system with grid levelling for transient loads. Simulation results are shown below in the following sections.

The intended contribution of this paper is to find out a relation between the MPPT speed and the transient loads (torque ripples). Hence a graph showing the waveform of the tracking of wind speed with time is shown in Figure 3 generated torques is shown in Figure 4.

The WTS controller outputs a torque command which contains the turbine dynamic information to the inverter, which is working in torque control mode. Because the PMSG is driven by this inverter, it will generate a torque that is equal to that of a real wind turbine. The validity of the wind turbine emulator has already been verified in the previous work. As mentioned earlier the output from the wind energy system is integrated with the grid in order to fully utilise its potential (Figure 5).

The voltage fed into the grid from the inverter. The current waveforms that are free from ripples which are obtained as outputs from the inverter are also given to the grid are shown in Figure 6.

The first plot is the output voltage, as can be seen, without the harmonic filters; the output is essentially a square wave due to the switching nature of the inverter. The second plot is of the output current, it is not in phase with the output voltage. Both the unclean output voltage and current lead to an unstable output active power (P) and a large, also unstable, reactive power (Q) output.

Conclusion

The main focus of this paper is on proposing a systematic study on the MPPT system to get a good compromise between the MPPT speed and the transient load. Furthermore, to confirm that the WECS can operate at the designed system bandwidth, P and O control method is proposed. The MPPT controller helps in tracking wind speeds varying with time. In addition, the system includes a PI controller to control the turbine speed and the grid voltage on the generator side and the grid side respectively. The controller further inputs pulses to the PWM inverter, the output of which are fed to grid.

References

1. Chen J, Gong C (2013) New overall power control strategy for variable-speed fixed-pitch wind turbines within the whole wind velocity range. IEEE Trans Ind Electron 60: 2652-2660.

2. Chen J, Chen J, Gong C (2013) Constant-bandwidth maximum power point tracking strategy for variable speed wind turbines and its design details. IEEE Trans Ind Electron 60: 5050-5058.

3. Mesemanolis A, Mademlis C, Kioskeridis I (2012) Maximum electrical energy production of a variable speed wind energy conversion system. IEEE Trans Ind Electron 1: 1029-1034.

4. Errami Y, Maaroufi M, Ouassaid M (2012) Control scheme and maximum power point tracking of variable speed wind farm based on the PMSG for utility network connection. International Conf on Complex Systems.

5. Mendis N, Muttaqi KM, Sayeef S, Perera S (2012) Standalone operation of wind turbine-based variable speed generators with maximum power extraction capability. IEEE Trans on Energy Convers 27: 822- 834.

6. SMR Kazmi, Goto H, Guo HJ, Ichinokura O (2011) A novel algorithm for fast

and efficient speed-sensorless maximum power point tracking in wind energy conversion system. IEEE Trans on Ind Electron 58: 29-36.

7. Khan SA, Hossain MI (2011) Intelligent control based maximum power extraction strategy for wind energy conversion systems. Electrical and Computer Engineering, Niagara Falls.

8. Agarwal V, Aggarwal RK, Patidar P, Patki C (2010) A novel scheme for rapid tracking of maximum power point in wind energy generation systems. IEEE Trans Energy Convers 25: 228-236.

9. Shirazi M, Viki AH, Babayi O (2009) A comparative study of maximum power extraction strategies in PMSG Wind Turbine System. IEEE Trans Electrical Power and Energy Conference, Montreal.

10. Xinyin Z, Minqiang H, Xiaohu C, Zaijun W (2008) The research on grid-connected wind-power generation system of variable speed permanent magnet synchronous wind generator. Electric utility deregulation and reconstructing and power technologies, Nanjuing.

11. Koutroulis E, Kalaitzakis K (2006) Design of a maximum power tracking system for wind-energy-conversion applications. IEEE Trans Ind Electron 53: 486-494.

12. Chinchilla M, Arnaltes S, Burgos JC (2005) Control of permanent magnet generators applied to variable-speed wind-energy systems connected to the grid. IEEE Trans Energy Convers 21: 130-135.

13. Knight AM, Peters GE (2005) Simple wind energy controller for an expanded operating range. IEEE Trans Energy Convers 20: 459-466.

14. Wang Q, Chang LC (2004) An intelligent maximum power extraction algorithm for inverter-based variable speed wind turbine systems. IEEE Trans on Power Electronics 19: 1242-1249.

15. Datta R, Ranganathan VT (2003) A method of tracking the peak power points for a variable speed wind energy conversion system. IEEE Trans on Energy Convers 18: 163-168.

16. Yaoqin J, Zhongqing Y, Binggang C (2002) A new maximum power point tracking control scheme for wind generation. Int Conf Power Syst Technol 1: 144-148.

10

Colpitts Oscillator: Design and Performance Optimization

Ankit Rana*

B. Tech. (ECE), Bharati Vidyapeeth's College of Engineering, A-4, Paschim Vihar, New Delhi, India

Abstract

From the very fundamental oscillator, a simple pendulum wherein there is a constant energy switch between potential and kinetic energy, oscillators have seen groundbreaking changes in setup, operation and their applicability. There are harmonic oscillators which produce a continuous sine wave output of certain frequencies as per the passive components involved. Additionally, are known Relaxation oscillator which yield triangular, square and sawtooth waves as output to name a few. The present paper deals with the details of how a fundamental Colpitts oscillatory circuit can be designed. Furthermore, we would take a look at optimizing its performance with change in several dependent characteristics in oscillation. We would conclude with an inference pertaining to the best customization with could be put to practical usage.

Keywords: Feedback; LC combination; Multisim; Tank circuit; VFO; Resonance; Magnetic energy leakage; Microfarad(uF); Oscillators; Sinusoidal

Introduction

Back in mid-1912, Edwin Armstrong while carrying out experiments with triodes wasn't aware of a spectacular phenomena due to a component which was going to serve as the basis of coupling and amplification amongst many other applications. Until then, the experiments had incorporated these devices as a detector for amplitude modulated waves. Having utilized them for this functionality, nobody was aware of the reasons as to why that was happening. By means of coupling one terminal of the device to another, it was observed that he could achieve large signal gain. We today term the phenomenon as positive feedback in circuits. He, with his invention of the radio had produced a unique oscillator. The uniqueness was justified since the limitation of other oscillators producing output in the kHz domain had now been extended to the MHz domain [1].

Ever since then, there has been a plethora of oscillator circuits that have been invented and employed industrially worldwide. Continuous Sine Wave oscillators or Square wave oscillators find their application in a myriad of fields in converting DC input to a variant A.C. output, in amplification of signals, synchronization purposes to name a few. One such circuit which utilizes storage and dissipation of magnetic energy, namely the Colpitts Oscillator has formed the basis of this study. A few of modifications in the conventional circuit have hereby been inferred.

The Colpitts Oscillator

Classification

The Colpitts Oscillator is known to work on feedback from the divider setup that is used in the circuit. The voltage divider is either made by 2 inductors or by using tapping on the single inductor. In either of the cases, if the desired application is that of a VFO, the usage is not as much preferred as with the case of Clapp Oscillator. In the latter, an extra capacitor is used for tuning to the optimum frequency and hence a better sustained waveform is achievable readily [2,3].

However, Colpitts form the basis of either of these circuitries, capacitor-tuned Clapp or the conventional circuit. And consequently, is believed in this experimental study to be encompassing the behavioral results for the counterpart circuits as well [2].

Frequency tuning and parameters that effect it

Since the Colpitts oscillator is a type of tank circuit (LC combination) and works on feedback of energy, the mathematical expression underlying its operation is the same as that of first-order LC circuit i.e.

$$f = (1/2\pi\sqrt{L*C}) \qquad (1)$$

And since, the Colpitts must have two capacitors for compensating purposes, the mathematical expression to obtain the frequency of operation is depicted as [4,5]:

$$f = 1/2\pi\sqrt{\frac{L(C1*C2)}{(C1+C2)}} \qquad (2)$$

It can be noted that the two capacitors in series with each other result in the expression in the denominator.

From the study of its parasitic elements and the transconductance (g_m) concept, it is known that a negative value of input resistance only would be able to sustain oscillations at the output [6,7].

Oscillations are obtained only for a large value of transconductance (g_m) and for smaller values of capacitor elements used.

Experimental Approach to Colpitts Oscillator Design

A industrially acclaimed simulator, namely Multisim was used to observe the influence of several parameters on a Colpitts oscillator design with multiple modifications and optimization aims in mind [5,7].

A first order Colpitts Oscillator has been drawn as a schematic over the simulator. This schematic circuit was introduced with several modifications viz. change in input resistance, changes in capacitors and the inductor coil's inductance. The parameters mentioned herewith are the fundamental governing dependencies in the performance of a Colpitts Oscillator. Increase in the value of any of these would leave

*Corresponding author: Ankit Rana, B. Tech. (ECE), Bharati Vidyapeeth's College of Engineering, A-4, Paschim Vihar, New Delhi-110063, India
E-mail: ankitrana1709@gmail.com

an impact on the output sinusoid produced. From a perfect sinusoid under optimum conditions (mentioned as conclusion) to a distorted wave output, the tank circuit's myriad of oscillatory behaviors have been incorporated [8,9].

The effects were simulated over Multisim and analyzed over time. Conclusions from the same were drawn (Figures 1-12).

The behavior as per the results obtained from this comparative study have been incorporated as conclusions to the customization activity.

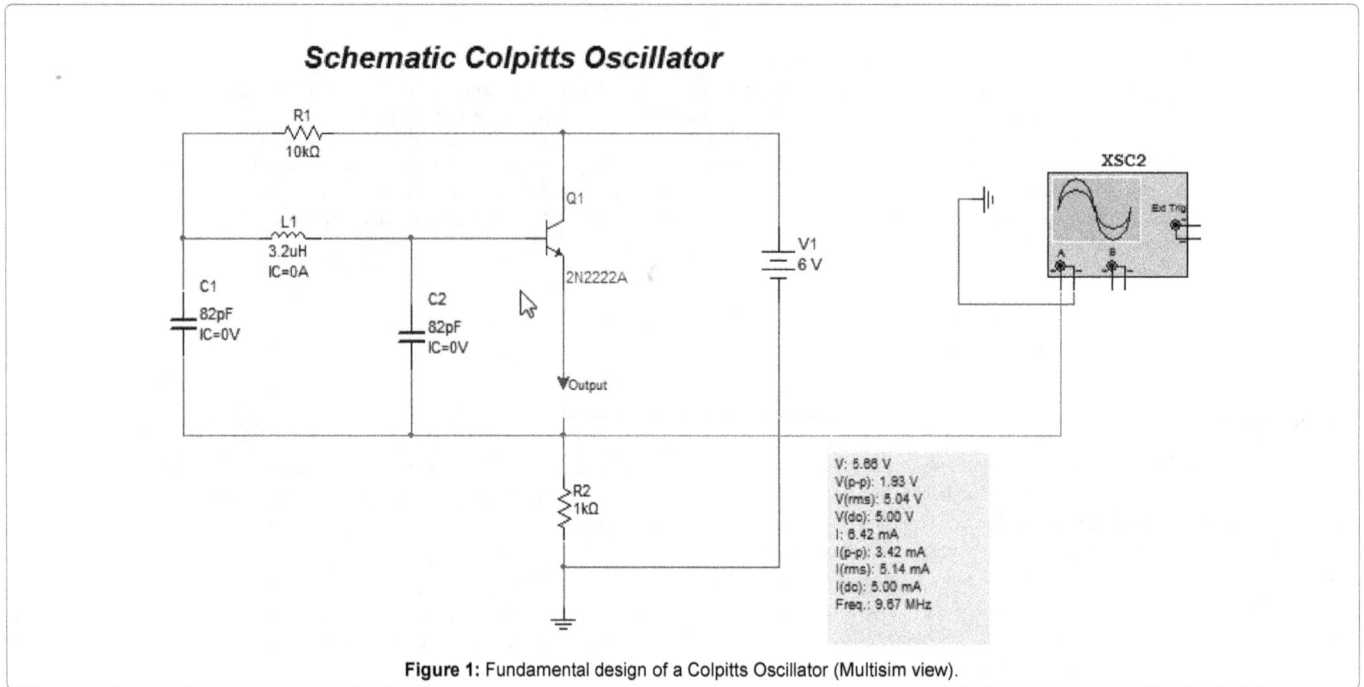

Figure 1: Fundamental design of a Colpitts Oscillator (Multisim view).

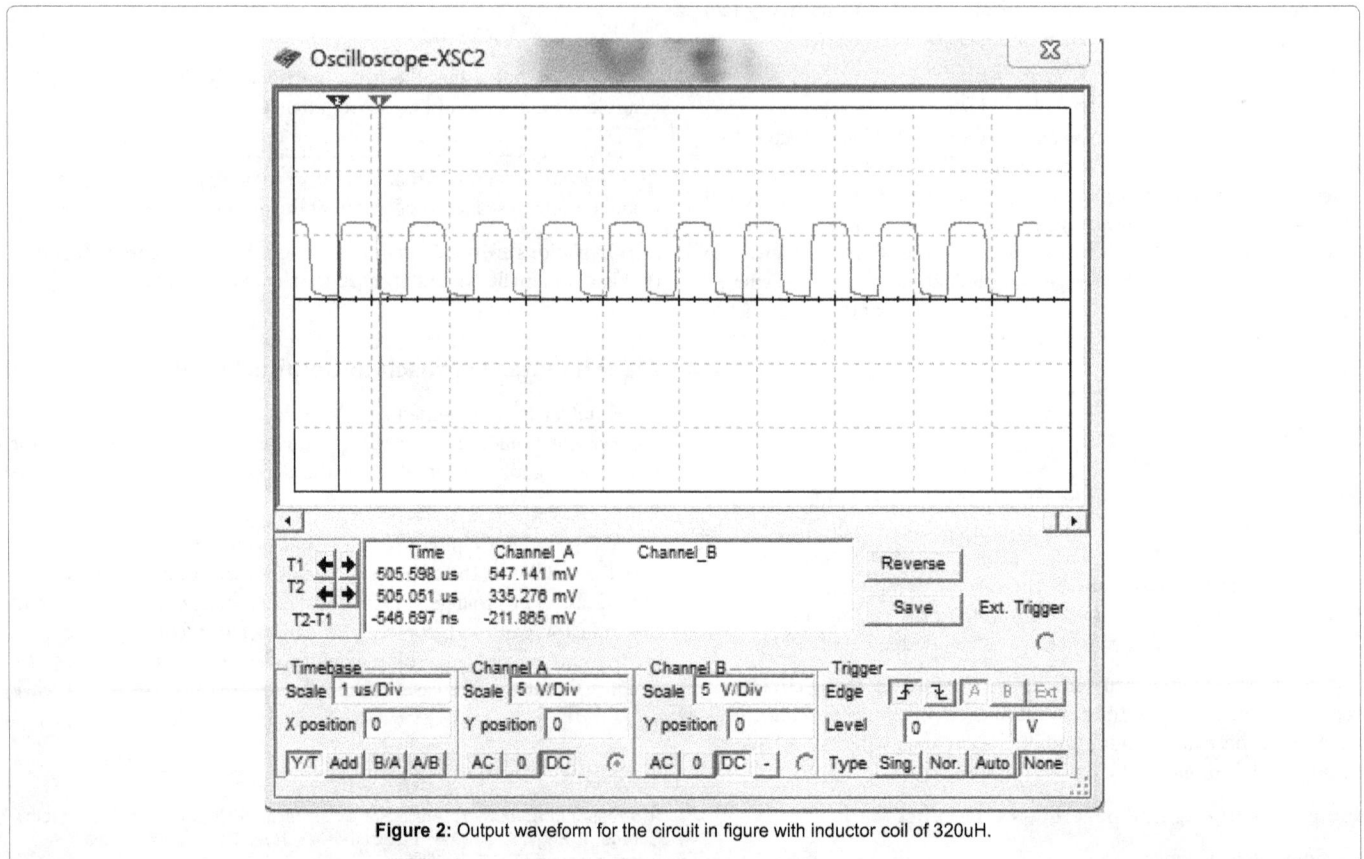

Figure 2: Output waveform for the circuit in figure with inductor coil of 320uH.

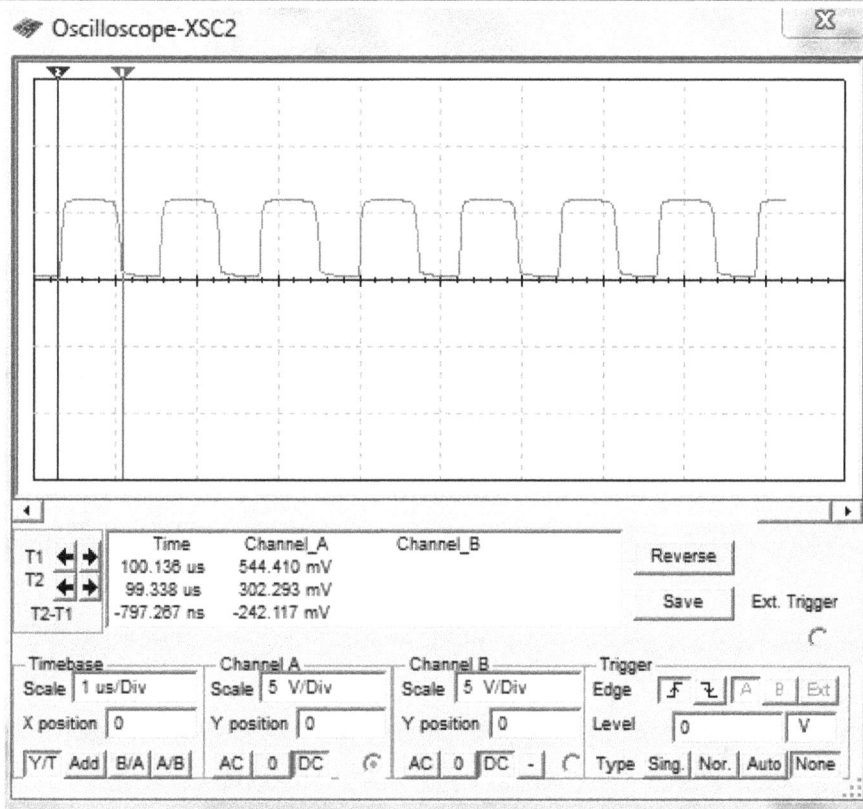

Figure 3: Output waveform for the circuit in figure with inductor coil of 640uH.

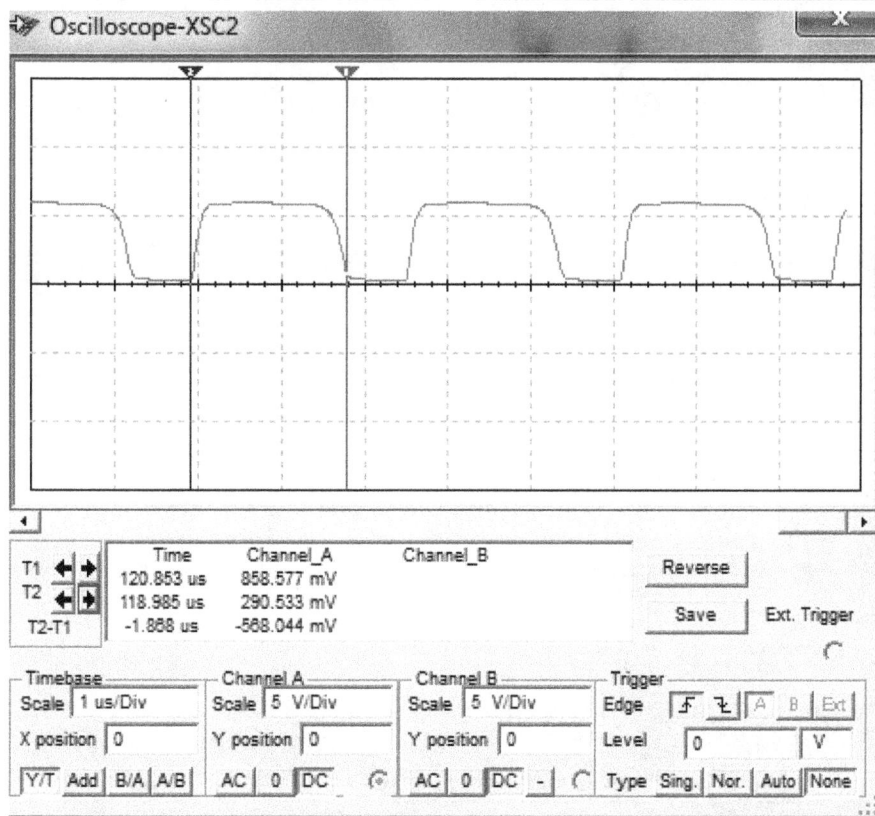

Figure 4: Output waveform for the circuit in figure with inductor coil of 3200uH.

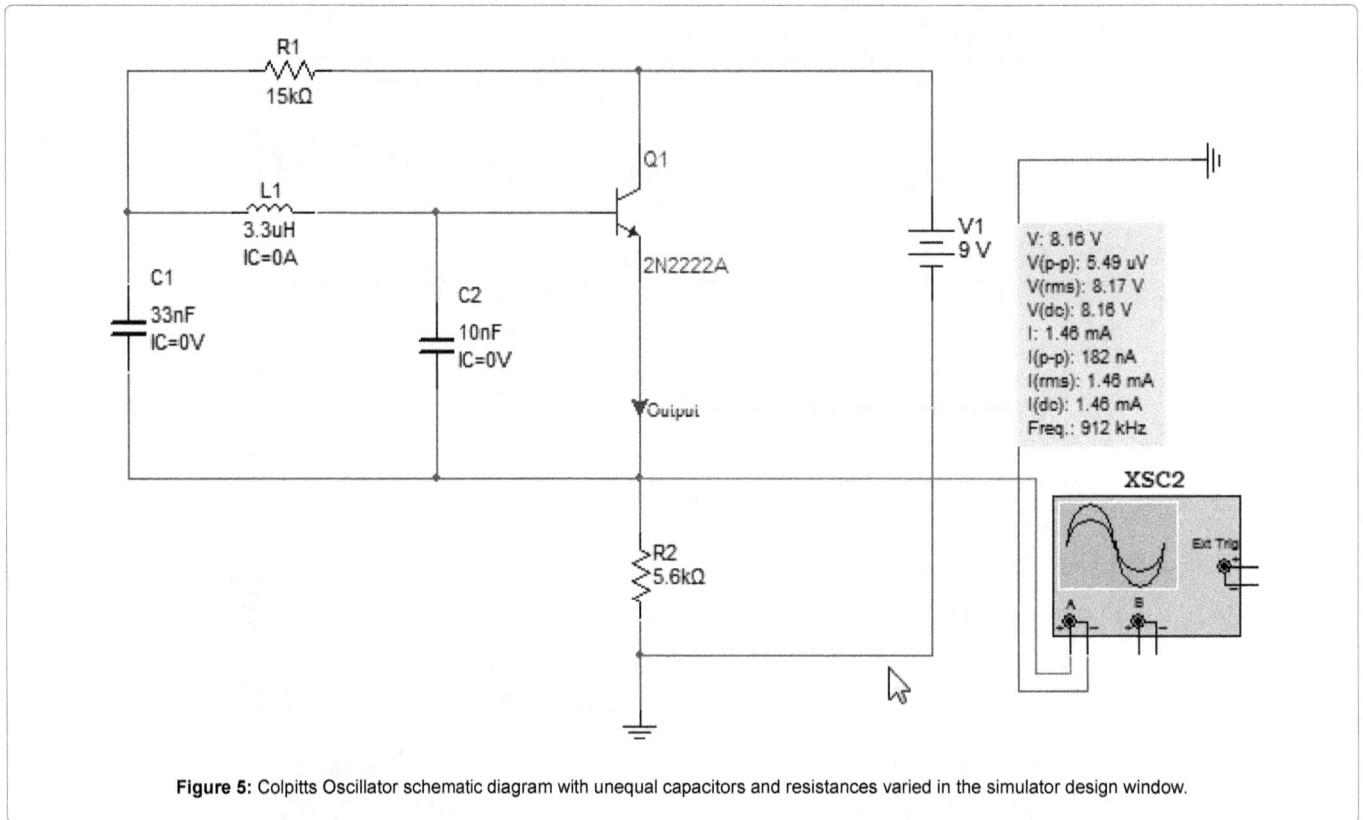

Figure 5: Colpitts Oscillator schematic diagram with unequal capacitors and resistances varied in the simulator design window.

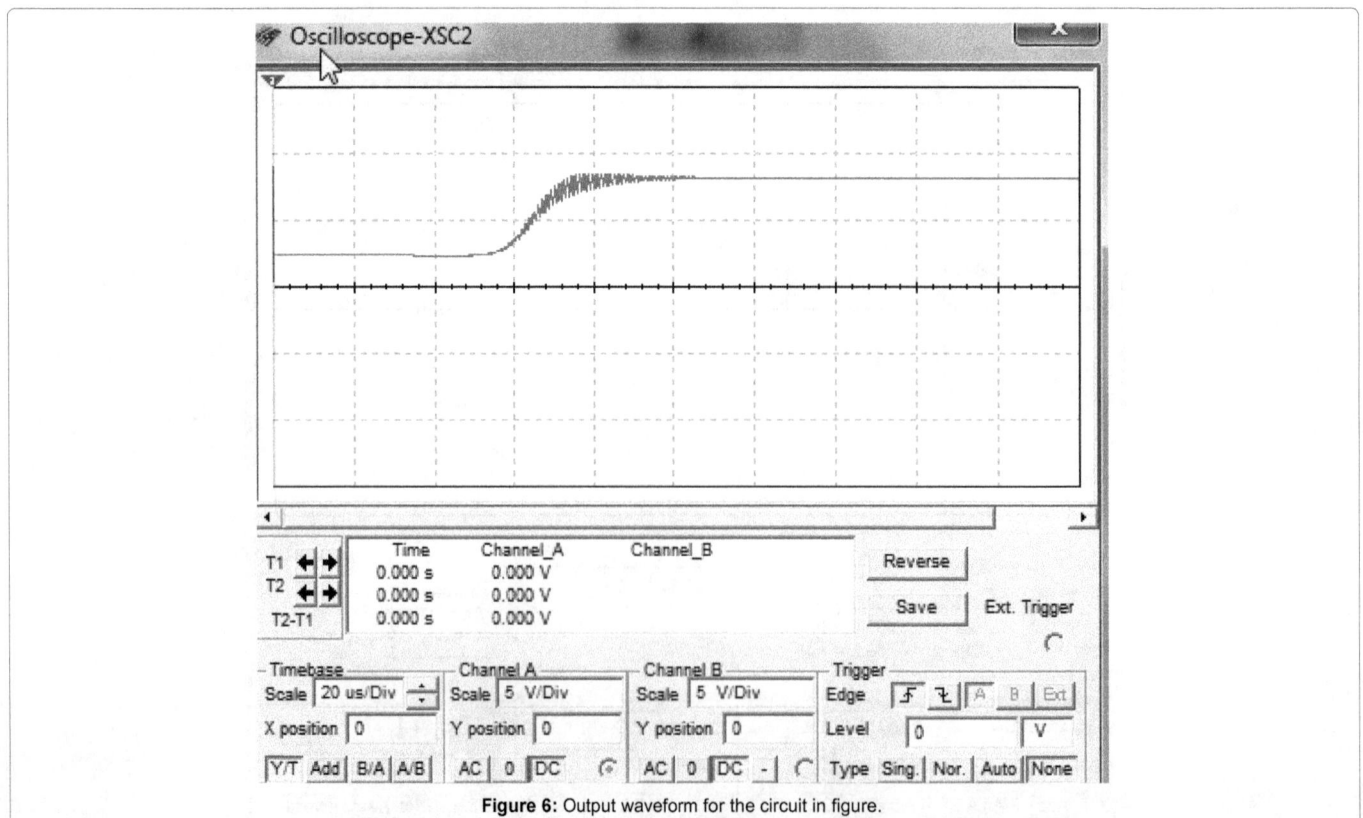

Figure 6: Output waveform for the circuit in figure.

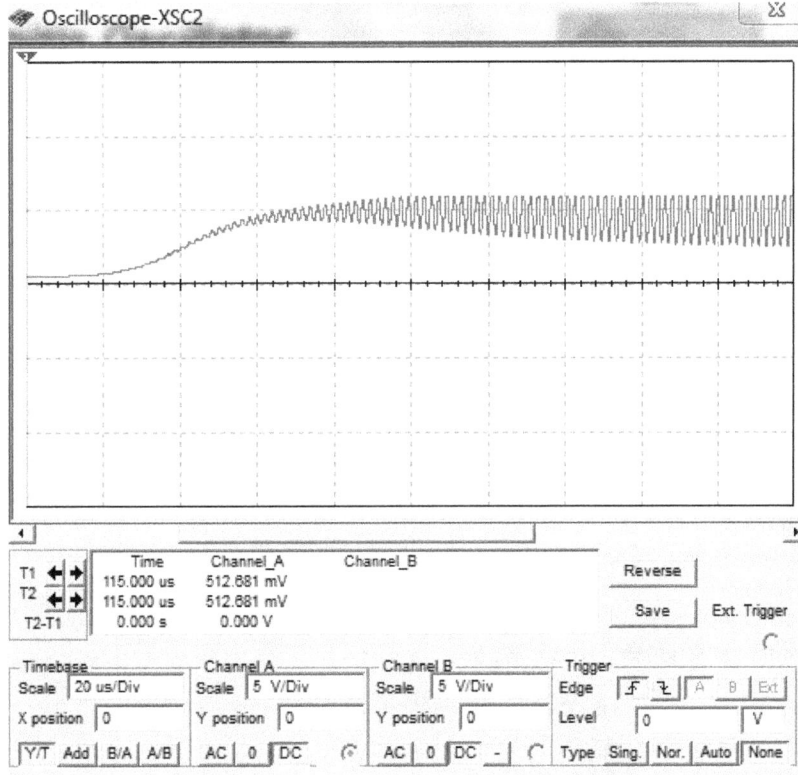

Figure 7: Output waveform for the circuit in figure when the capacitance had been decreased to 60nF each.`

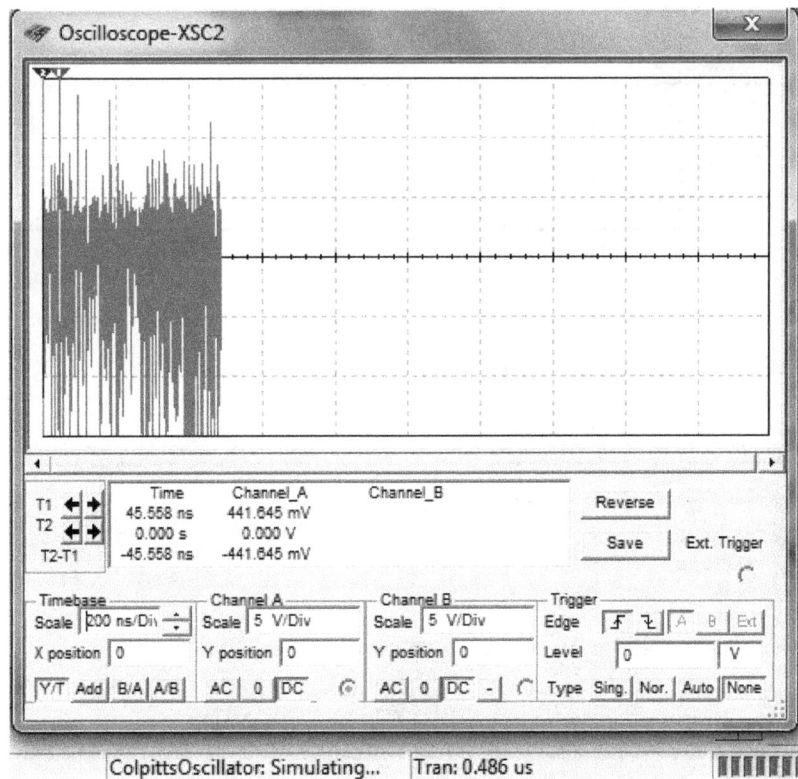

Figure 8: Output waveform for the circuit in figure when the inductance of the coil had been decreased to 2pH.

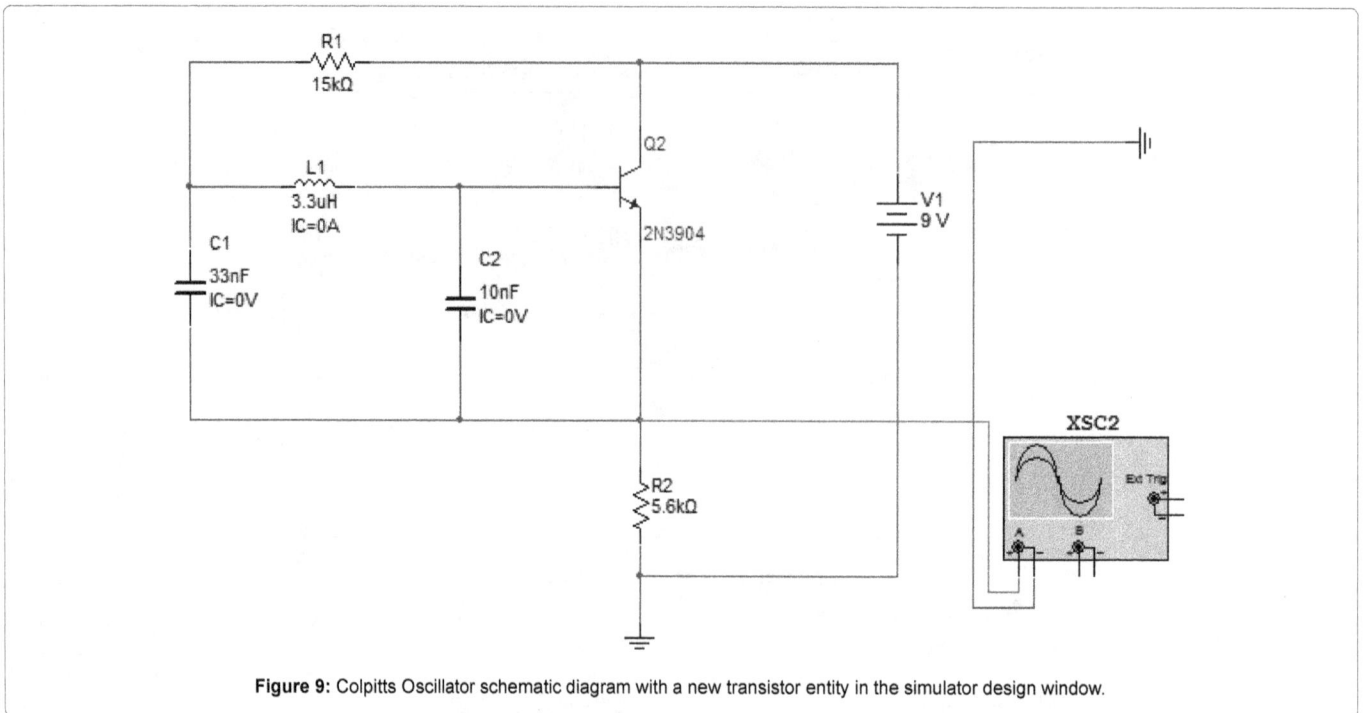

Figure 9: Colpitts Oscillator schematic diagram with a new transistor entity in the simulator design window.

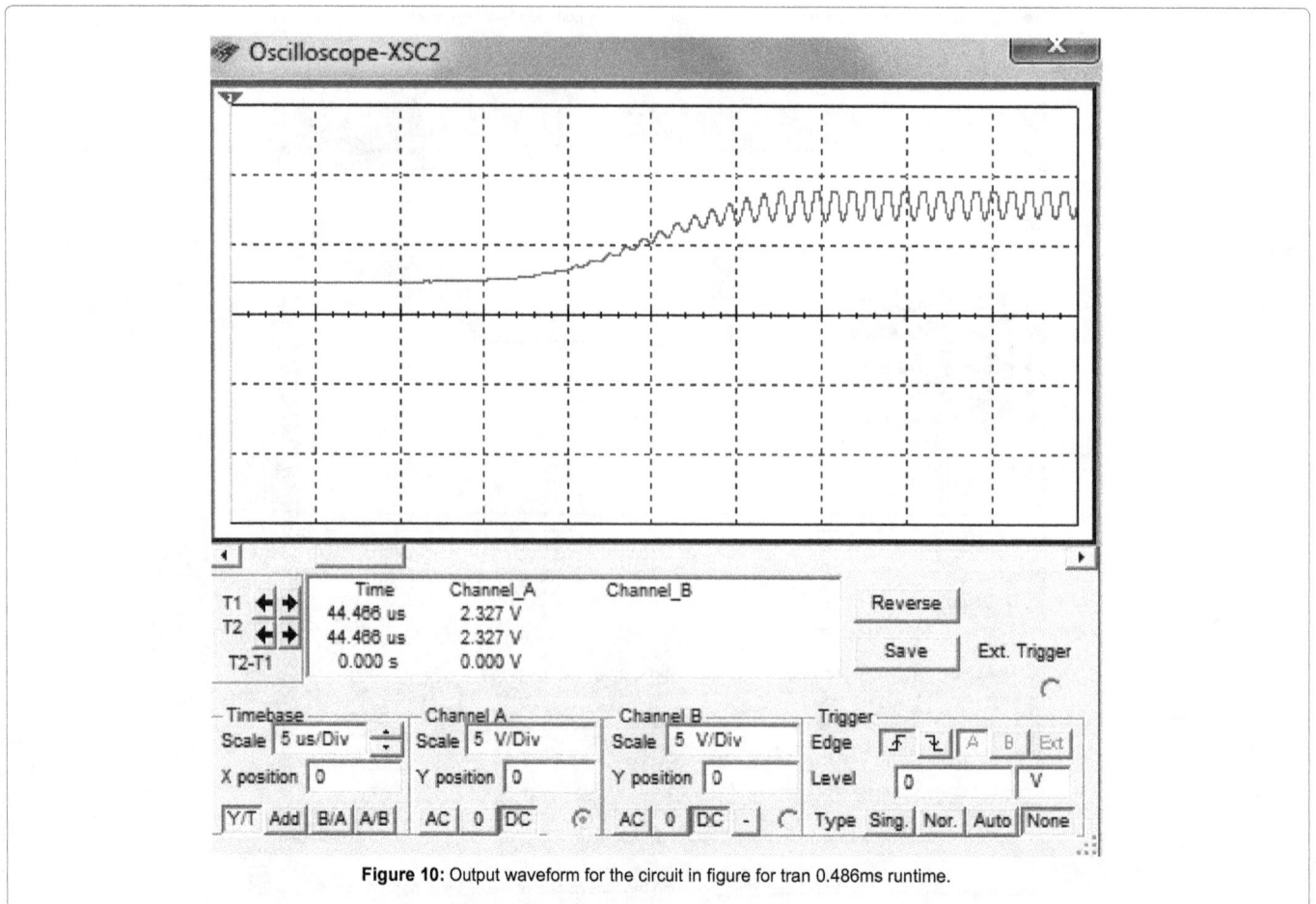

Figure 10: Output waveform for the circuit in figure for tran 0.486ms runtime.

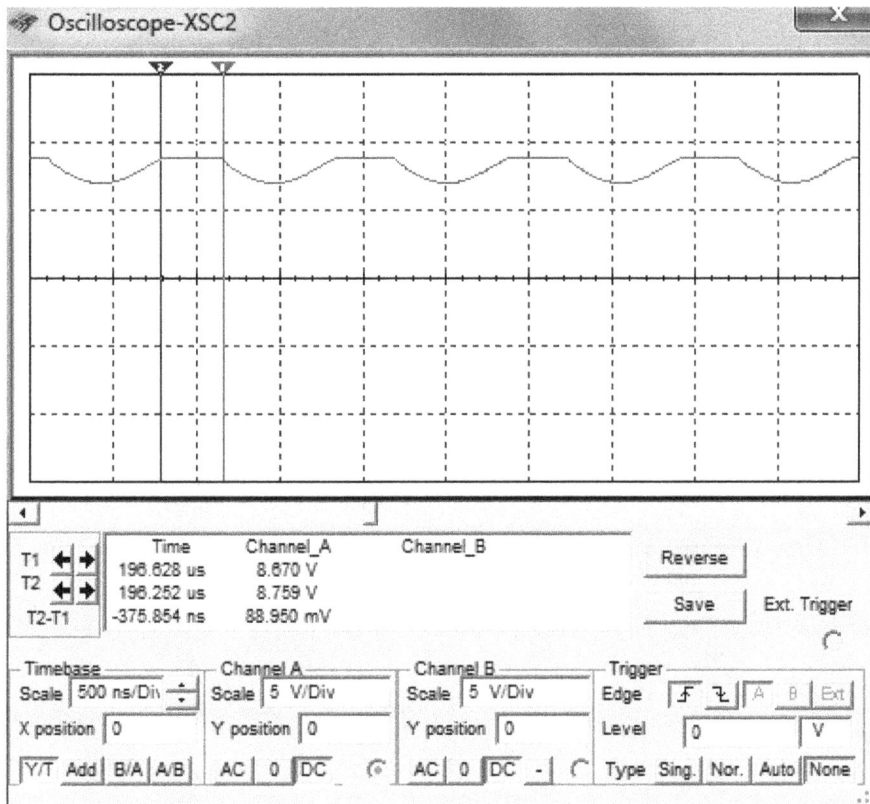

Figure 11: Output waveform for the circuit in figure for pulse flat-top duration in the same runtime.

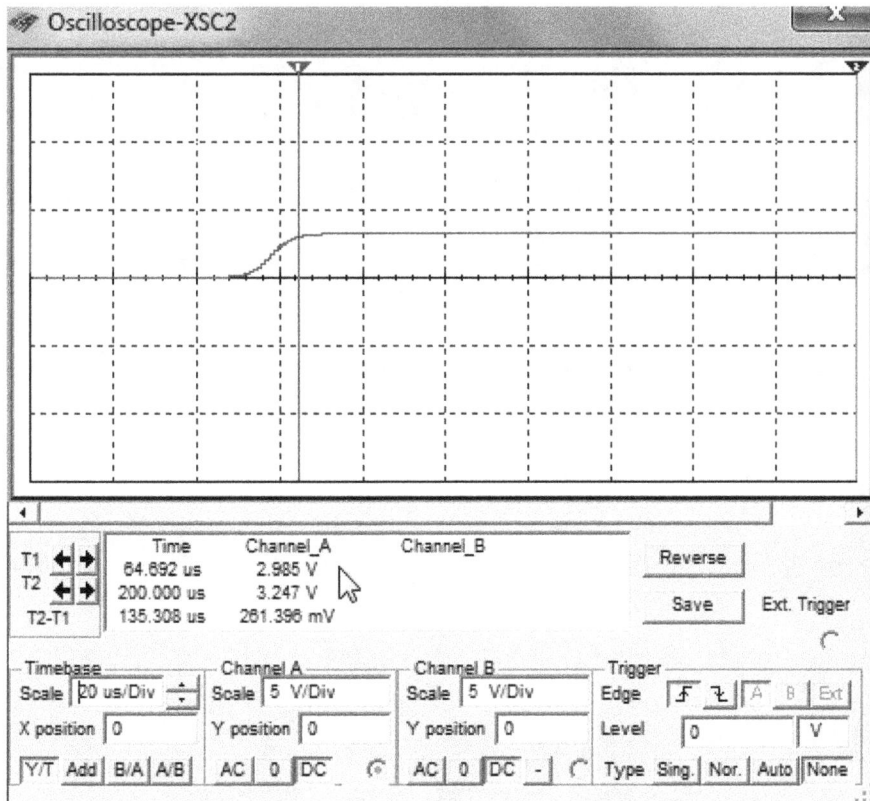

Figure 12: Output waveform for the circuit in figure for resistance value altered to 68 ohms.

Conclusion

As is the scenario with any other oscillator design, the amplification (positive or negative) of the active component must be relatively higher than the attenuation shown by the capacitive voltage divider implemented, to obtain a stable functionality or in other words, a smoother sine-wave output. Consequently, a Colpitts oscillator utilized as a variable frequency oscillator (VFO) shows at its best operation when a variant inductance is utilized for tuning the circuit, contrary to the case of tuning one of the two capacitors. In the case, tuning by means of a variable capacitor is required, it must be done by means of a third capacitive entity connected in parallel combination to the inductor coil (or in series). The amount of feedback depends on the values of capacitive entities with the smaller the values of capacitance the more will be the obtained feedback [3,9]. The same is optimally adjusted to attain un-damped oscillations. In the same scenario, the relative difference or the ratio of two capacitive entities is a big determinant since the two are ganged together. The amplification of the active component must be marginally larger than the attenuation seen due to the voltage divider (capacitive combination), to obtain the most stable operation possible.

References

1. Razavi B (2001) Design of Analog CMOS Integrated Circuits. McGraw-Hill.

2. Blanchard J (1941) The History of Electrical Resonance. Bell System Technical Journal, USA.

3. Gottlieb I (1997) Practical Oscillator Handbook, Elsevier, USA.

4. Huurdeman AA (2003) The worldwide history of telecommunications, Wiley-IEEE, USA.

5. Rohde UL, Poddar A, Bock G (2005) The Design of Modern Microwave Oscillators for Wireless Applications: Theory and Optimization, John Wiley & Sons.

6. Rohde UL, Rudolph M (2012) RF / Microwave Circuit Design for Wireless Applications (2ndedn) John Wiley & Sons.

7. http://www.howstuffworks.com/oscillator.htm

8. Sarkar S, Sarkar S, Sarkar BC (2013) Nonlinear Dynamics of a BJT Based Colpitts Oscillator with Tunable Bias Current. IJEAT.

9. http://www.ittc.ku.edu/~jstiles/622/handouts/Oscillators%20A%20Brief%20History.pdf

11

Implementation of a New Compact Inverter Structure Controlled by Numeric Sinusoidal Pulse Width Modulation for Photovoltaic Applications

Abounada A, Brahmi A, Chbirik G* and Ramzi M

Electrical Engineering Department, Faculty of Sciences and Technology, University of Soultan Moulay Sliman, Laboratory of Automatic, Energy Conversion and Microelectronic, P.B: 523 Mghila, Beni Mellal 2300, Morocco

Abstract

In order to cover energy requirement, the photovoltaic is one of the proposed solutions. However, according to the aim of its utilization, the direct current output voltage of photovoltaic source should be adjusted. The boost inverter is a recent power processing stage that can increase, filter and alternate direct current input voltage. So as to control it, there are various modulation types. Among them, the sinusoidal pulse width modulation is presented in a new digital form. The operating principle of the aforementioned inverter and command have been analyzed and verified by simulation and realization. Furthermore, frequency analysis of output voltage signal proves efficient results.

Keywords: Renewable energy; Boost; Inverter; SPWM; Control; Microcontroller

Introduction

Energy crises and environmental pollution caused by consumption of traditional energy sources lead scientists to think about renewable energies to satisfy the world needs of electricity [1]. So, nowadays renewable energies are the most important subjects that researches and studies discuss [2]. Among these energy types, the solar is predicted to cover the maximum need of energy in the future [3]. Thus, a lot of articles present different photovoltaic systems structures [2,4,5], which are based on many power processing stages. This paper will focus on the inverter stage and its command. The structure presented here is very recent and understudied; it allows a conversion in a single stage with AC output voltage higher than the input voltage [6]. It is the boost inverter. The inverter supply is provided by three series batteries of 24 V. Thus, the inverter is powered by 72 V. The inverter output is modulated by a digital sinusoidal pulse width modulation SPWM inspired from the analogical one.

To give more details on this structure, this paper will contain in the first part the construction of the inverter, its principle of operating, and a simulation which interpret its performance. In the second part, the experimental study is presented to prove simulation results.

System Description

The boost inverter discussed here allows producing a filtered and amplified alternative voltage. It contains two bidirectional boost converters connected deferentially to a load [6,7]. Controlled by sinusoidal pulse width modulation SPWM command, each converter produces a DC voltage and an alternative component. The alternative components are sinusoidal signals and in phase opposition, whereas the DC components have the same value. This structure is shown by Figure 1 [8].

The output voltages V_{01} and V_{02} for both converters are presented by Eq. (1) and Eq. (2) [9,10].

$$V_{01}(t) = V_{dc} + V_{max} \sin(\omega t) \tag{1}$$

$$V_{02}(t) = V_{dc} - V_{max} \sin(\omega t) \tag{2}$$

The output voltage of the inverter is in "Eq. (3)"

$$V_{02(t)} = V_{01}(t) - V_{02}(t) = 2 * V_{max} \sin(\omega t) \tag{3}$$

In order to explain the functioning of this inverter, we will consider one converter as shown in Figure 2 [9,10]:

The functioning of the converter is divided into two parts:

$[0, d*T_{c3}]$: The switch M_3 is closed while M_4 is open, I_{l3} increases linearly, D_4 is reverse polarized, C_3 provides the load with energy, which decreases V_{01}.

$[d*T_{c3}, T_{c3}]$: The switch M_4 is closed while M_3 is open, C_3 is charging by I_{l3}. So, the voltage V_{01} increases [9,10].

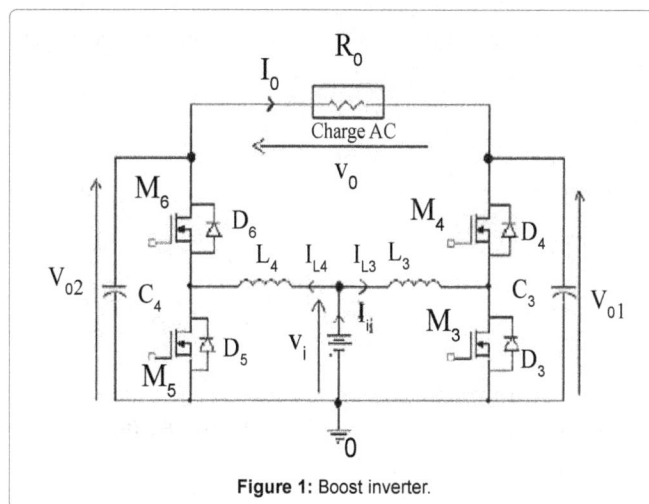

Figure 1: Boost inverter.

---

*Corresponding author: Chbirik G; Electrical Engineering Department, Faculty of Sciences and Technology, University of Soultan Moulay Sliman, Laboratory of Automatic, Energy Conversion and Microelectronic P.B: 523 Mghila Beni Mellal, Morocco, E-mail: ghizlane.fst@gmail.com

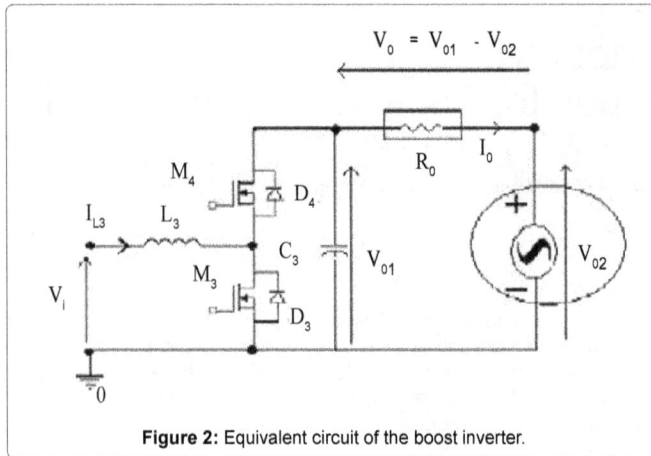

Figure 2: Equivalent circuit of the boost inverter.

Where d is the duty cycle, Tc is the switching period

When one converter is in boost operating the other should be in buck operating. The average voltage's expression for the first converter [11] is in Eq. (4) while that of the second is corresponding to Eq. (5):

$$V_{01} = \frac{Vi}{1-d} \qquad (4)$$

$$V_{02} = \frac{Vi}{d} \qquad (5)$$

Then the transfer function can be concluded in Eq. (6):

$$\frac{V_0}{Vi} = \frac{2d-1}{d(1-d)} \qquad (6)$$

(6) For $d_0 = 0.5$, $V_0 = 0$, then if we vary d near to d_0 we will have an AC output voltage. Via the transfer function we can conclude that the duty cycle is "Eq. (7)"

$$d = 0.5 - \frac{2 - \sqrt{4 + \left(\frac{V_0}{V_i}\right)^2}}{2 * \left(\frac{V_0}{V_i}\right)} \qquad (7)$$

A linearization of this equation around d_0 gives "Eq. (8)":

$$d = \frac{1}{2} + \frac{V_0}{8 * V_i} \qquad (8)$$

If we consider that V_0 is sinusoidal we will have "Eq. (9)"

$$d = \frac{1}{2} + \frac{V_{0max} \sin(2\pi ft)}{8 * V_i} \qquad (9)$$

Where f is the frequency

This equation presents the duty cycle's variation near to d_0. To elaborate the command SPWM, in order to have a sinusoidal V_0, we will use the aforementioned "Eq. (9)".

The Digital Sinusoidal Pulse Width Modulation

To produce a sinusoidal voltage using this inverter, we command it with a digital modulation SPWM inspired from the analogical one.

It applies the same concept which is logic comparison of a sinusoidal wave to another triangular. So as to command the interrupters, we make use of the intersections of these signals [11-13].

As Figure 3 presents, the triangular signal is produced by a binary counter. It counts till a maximum value, stored in a register, then counts down until a minimal value. The sinusoidal one is represented by a table of values obtained by sampling a sinusoidal signal; this table is stored in a memory. The period of sampling is the same as commutation period of the inverter, which is equal to the period of the triangular signal elaborated by the counter.

In each period of sinusoidal signal, a software comparison between these two signals is done so as to produce an impulsion in higher or lower state.

The sinusoidal signal period gives suit of periodic impulsions modeled in width according to a sinusoidal law.

The numeric command SPWM is implemented in the microcontroller 16 F876 (20 MHz). The resulted SPWM signal should command MOSFET transistors. Since the intensity and the amplitude of this signal are unable to commutate the MOSFET, we added a driver between the microcontroller and the transistors. It is IR2111 circuit.

We can also produce the timing sequence by Excel software or Matlab and store it on the chip. This will allow us to optimize the chip performance.

Finally, in order to protect the microcontroller we added an optocoupler before the driver.

The circuit which produces the SPWM signals is presented in Figure 4.

Simulation

This part present a simulation of the inverter structure and its command discussed in the beginning under a power of 300 W. The inverter characteristics simulated are: Inductor: 0.23 µH, The capacitor: 260 µF, MOSFET: W45NM50 (Table 1).

The following Figures 5-10 show the realized inverter circuit in Orcade software and its simulation results:

According to the simulation results, the output voltage waveforms are sinusoidal; they have a frequency of 50 Hz and amplitude of 110 V.

Realization

After verification and validation by simulation, now we present a practical implementation of the realized inverter structure and its command.

Magnetic Circuit	EI40
Wire section	5,27 mm²
Value of air gaps	2 mm
Number of coils	45

Table 1: Inverter characteristics.

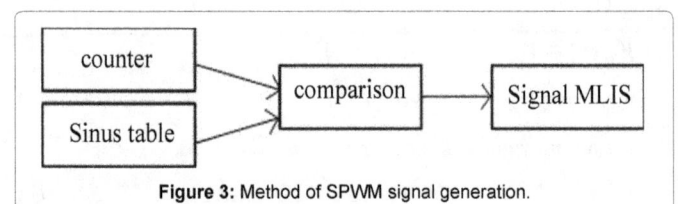

Figure 3: Method of SPWM signal generation.

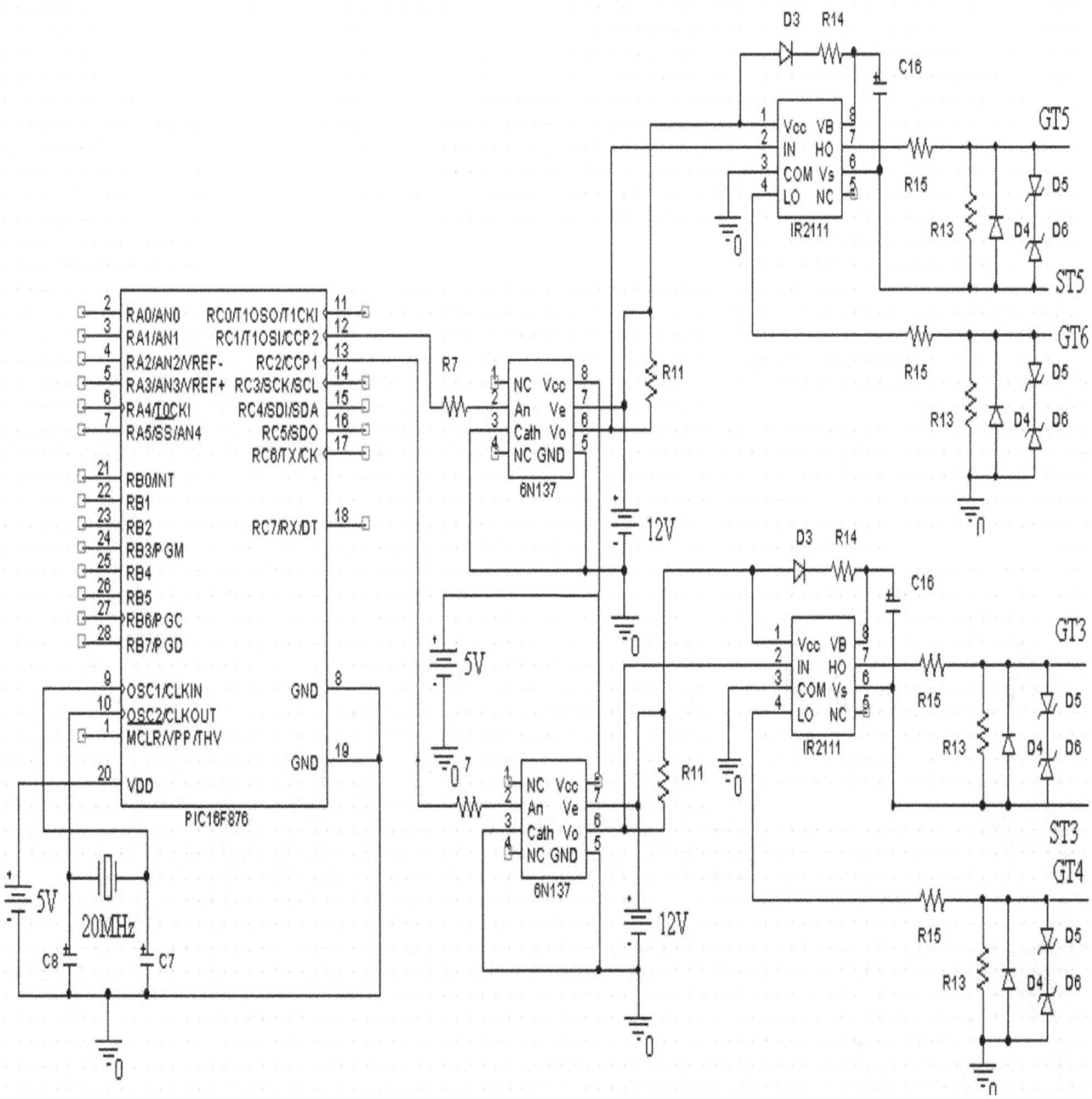

Figure 4: SPWM electroniccuit cir.

Inverter

In Figure 11 below shows the fabricated inverter.

The command card: The control card is built with the microcontroller PIC16F876 and three circuits. They are used for isolation, adaptation and interfacing with the microcomputer. This card has two inputs and five outputs, four of them are used to command the switches of the inverter.

In Figure 12 below shows the realized command card.

The microcontroller chosen to implement the SPWM technique has two PWM outputs and a memory. The outputs generate the SPWM signal whereas the memory accommodates the program. The maximum clock frequency that the microcontroller can accept is 20 MHz. The implementation of the mechanism SPWM in the microcontroller goes through three steps: Firstly, a configuration in pulse with modulation mode of two outputs "CCP1" and "CCP2" should be done.

Secondly, the period T_{C3} of PWM signals is defined according to "Eq. (10)":

$$T_{C3} = (PR_2 + 1) \times T_{OSC} \times T_P \qquad (10)$$

Where the period of the oscillator is T_{osc}, the predivisor of timer 2 is T_p and PR_2 is a register.

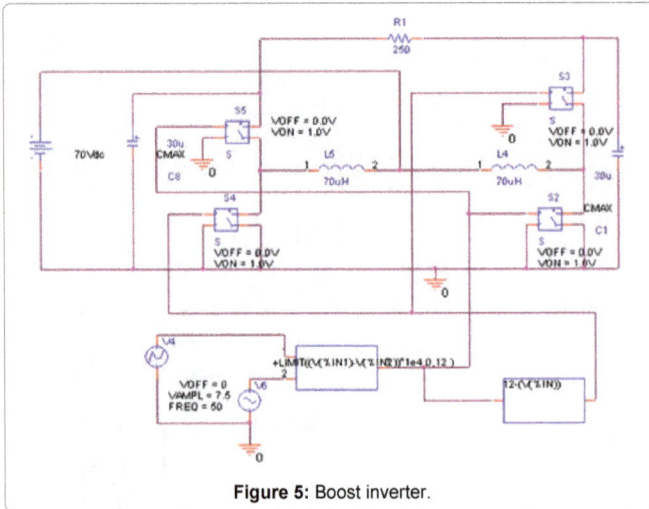

Figure 5: Boost inverter.

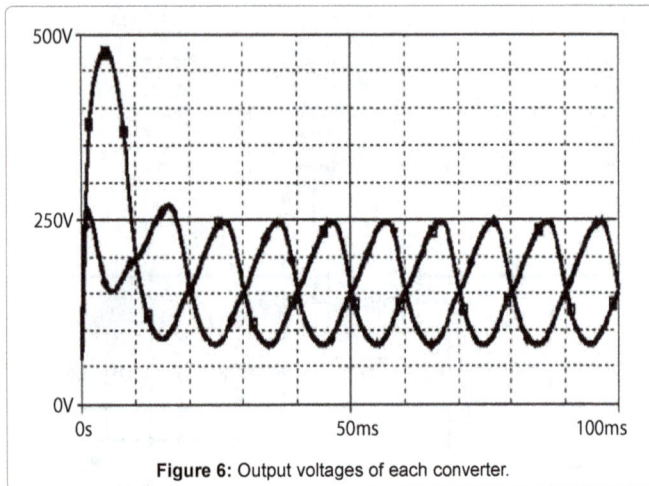

Figure 6: Output voltages of each converter.

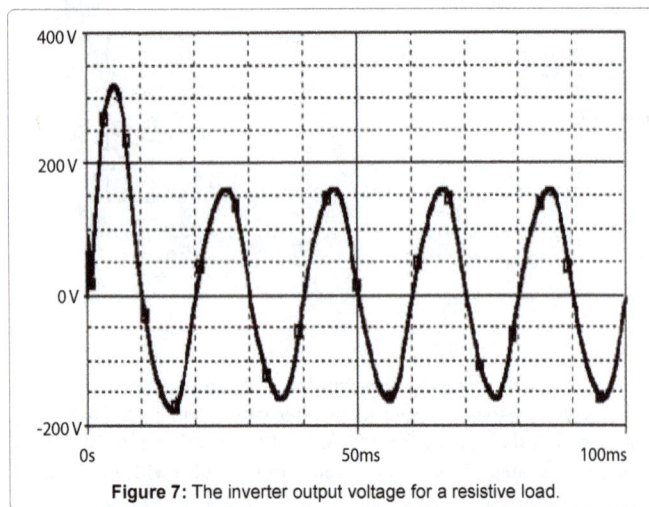

Figure 7: The inverter output voltage for a resistive load.

In order to fix the T_{C3} value to a required one, we should load the register PR_2 by a decimal value from "0" to "255".

Thirdly, the values K_i, which give the duty cycles D_i of number n, will be stored in a table and then loaded periodically according to a repetitive process. The loading procedure will be done by a timekeeper;

it can belong to one of registers CCPR1L or CCPR2L according to "Eq. (11)" or "Eq. (12)".

$$(CCPR1L) \times T_{OSC} \times T_P \tag{11}$$

$$(CCPR2L) \times T_{OSC} \times T_P \tag{12}$$

Depending on "Eq. (13)", the generation of table's values K_i can be done by using a software tool. Which help to obtain the duty cycle D_i depending on "Eq. (14)".

$$K_i = B + A \times Sin\left(\frac{2 \times \pi \times i}{n}\right) \tag{13}$$

The value of i goes from 1 until n

$$D_i = \frac{k_i}{PR_2} \tag{14}$$

The constants B, A and n will be determined as described subsequently.

The number n presents the number of duty cycles. It's given by the report T_{c3}/T_r.

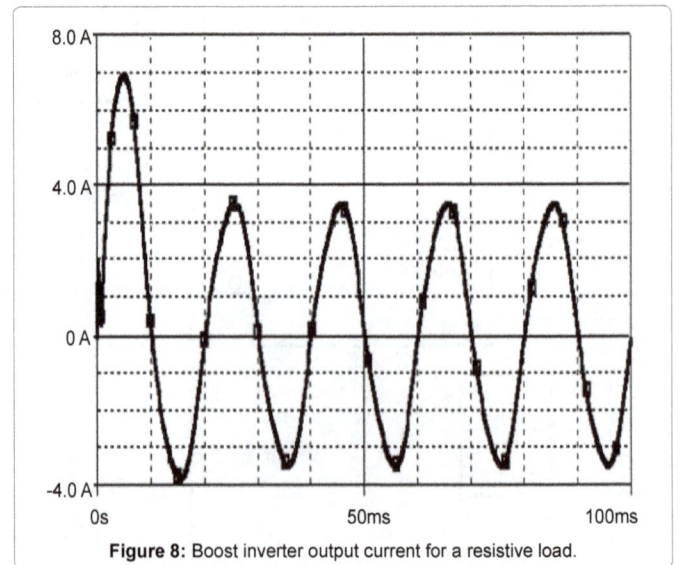

Figure 8: Boost inverter output current for a resistive load.

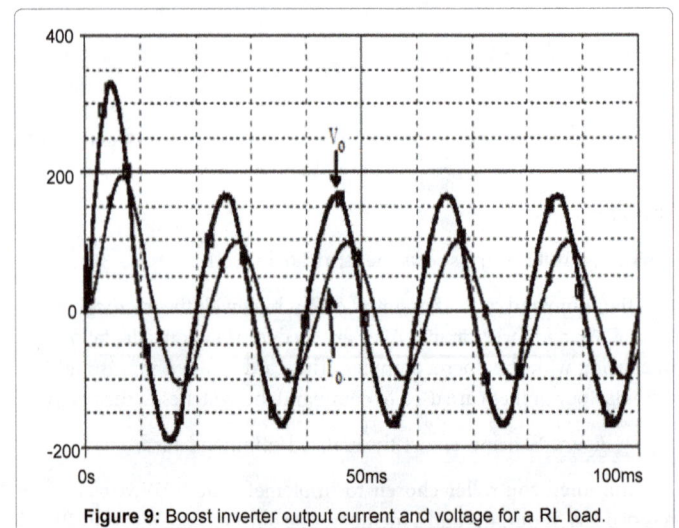

Figure 9: Boost inverter output current and voltage for a RL load.

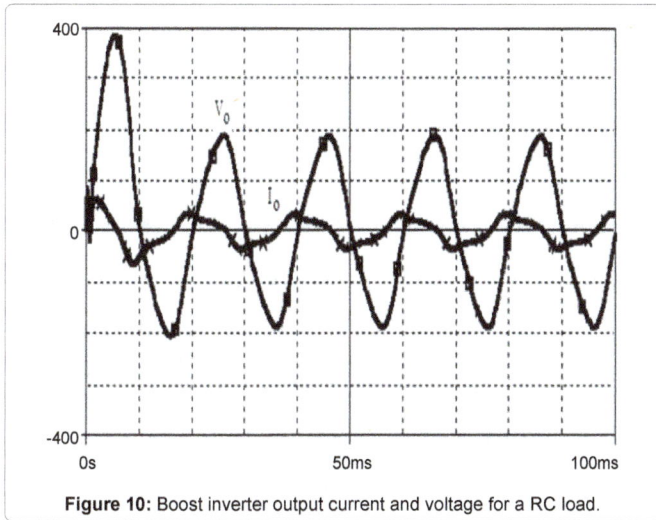

Figure 10: Boost inverter output current and voltage for a RC load.

Figure 11: Realized boost inverter.

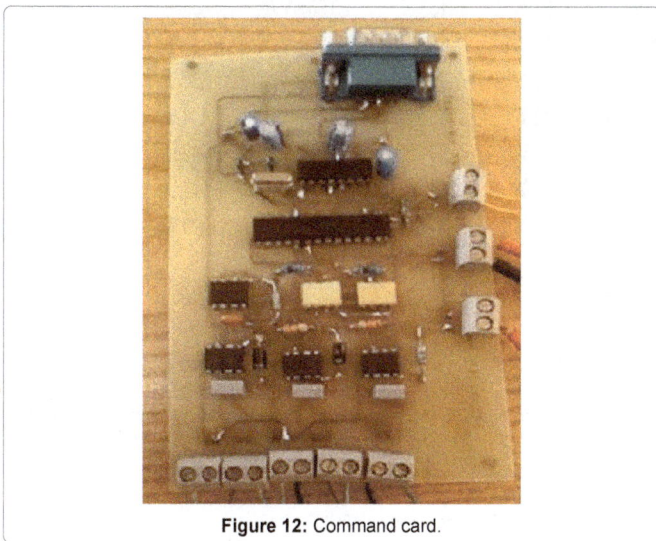

Figure 12: Command card.

To choose A and B, on the one hand, we should consider the size of both registers CCPR1L and CCPR2L of 8 bits, thus these constants are positive and decimal. On the other hand, they will be chosen in order to have the duty cycle value equal to 50% for t=0, equal to 95% for t=T$_r$/4 and equal to 5% for t=3 T$_r$/4.

The maximum duty cycle value 95% is obtained by "Eq. (15)":

$$\sin\left(\frac{2 \times \pi \times i}{n}\right) = 1 \qquad (15)$$

That will help to conclude the value of B+A

As the duty cycle value 50% is obtained depending on "Eq. (16)":

$$\sin\left(\frac{2 \times \pi \times i}{n}\right) = 0 \quad (16)$$

The value of B and then that of A will be easily deduced.

The implanted SPWM program undergoes from an interruption. This made the two values Ki and 255-Ki read and loaded simultaneously in both registers CCPR1L and CCPR2L. Then during every period T$_{C3,}$ two complementary motives SPWM are produced. This is caused by the over follow of timer 2. So we obtain the two desired signals SPWM during the period T$_r$.

The value of B and then that of A will be easily deduced.

The implanted SPWM program undergoes from an interruption. This made the two values Ki and 255-Ki read and loaded simultaneously in both registers CCPR1L and CCPR2L. Then during every period T$_{C3,}$ two complementary motives SPWM are produced. This is caused by the over follow of timer 2. So we obtain the two desired signals SPWM during the period T$_r$.

Experimental results

The output differential voltage of the realized boost inverter, applied on a resistive load, a capacitive load and inductive load, is presented in Figures 13-15.

These results show an output voltage highly near to be sinusoidal. It has a frequency of 59.94, a RMS voltage of 110.

Frequency analysis

Applying Fast Fourier Transform FFT on the inverter output voltage of every loads type offer various characteristics which are illustrated in Figure 16.

The fundamental has amplitude of 110 V; the 3rd harmonics remains the most dominant with low amplitude of 3.32 V. This value can be negligible compared to the amplitude of the fundamental.

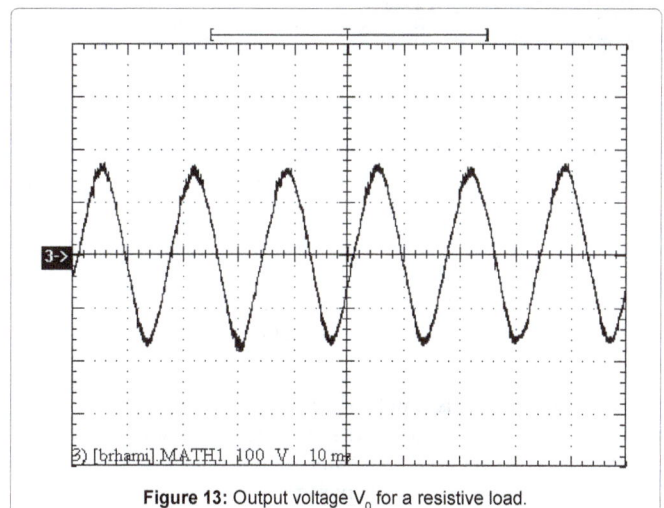

Figure 13: Output voltage V$_0$ for a resistive load.

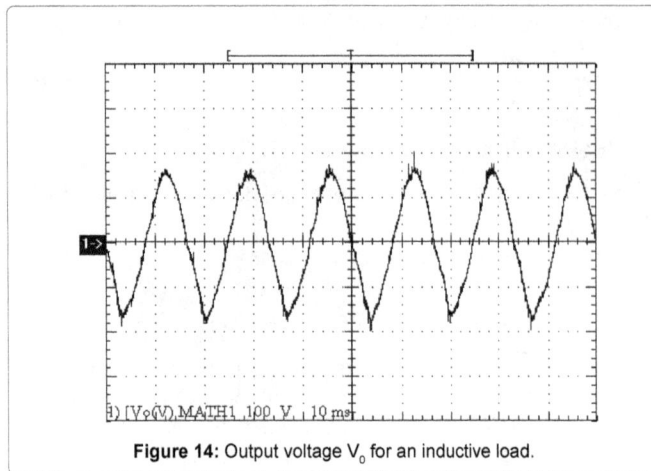

Figure 14: Output voltage V_0 for an inductive load.

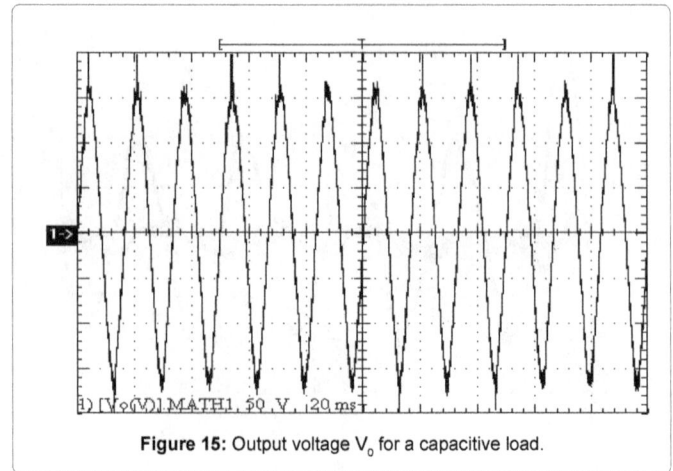

Figure 15: Output voltage V_0 for a capacitive load.

		Voltage = 110.09 V		Current = n/a		Voltage = 110.91 V			Voltage = 112.21 V		
		Voltage THD = 4.164 %		Current THD = n/a		Voltage THD = 4.588 %			Voltage THD = 4.622 %		
		Power Factor = n/a		Displacement Power Factor		Power Factor = n/a			Power Factor = n/a		
		Apparent Power = n/a		Reactive Power = n/a		Apparent Power = n/a			Apparent Power = n/a		

	Frequency	Voltage RMS	Voltage % of Fund.	Voltage Phase	Voltage RMS	Voltage % of Fund.	Voltage Phase	Voltage RMS	Voltage % of Fund.	Voltage Phase
Fundamental	59.940 Hz	109.60 V	100.000 %	0.0000	110.65 V	100.000 %	0.0000	110.80 V	100.000 %	0.0000
Harmonic 2	119.88 Hz	928.34m V	0.847 %	-62.551	2.5168 V	2.275 %	-69.076	3.0067 V	2.714 %	-93.327
Harmonic 3	179.82 Hz	3.3271 V	3.036 %	-1.2173	3.7867 V	3.422 %	15.665	3.8359 V	3.462 %	1.7724
Harmonic 4	239.76 Hz	898.29m V	0.820 %	144.65	1.0919 V	0.987 %	-151.30	543.63m V	0.491 %	151.75
Harmonic 5	299.70 Hz	1.2976 V	1.184 %	-172.00	1.4031 V	1.268 %	150.63	511.54m V	0.462 %	164.54
Harmonic 6	359.64 Hz	1.1826 V	1.079 %	-73.766	624.79m V	0.565 %	-17.426	508.27m V	0.459 %	151.08
Harmonic 7	419.58 Hz	981.44m V	0.895 %	41.093	390.88m V	0.353 %	-163.07	763.53m V	0.689 %	74.691
Harmonic 8	479.52 Hz	757.23m V	0.691 %	19.301	678.75m V	0.613 %	130.41	441.99m V	0.399 %	-155.89
Harmonic 9	539.46 Hz	1.3087 V	1.194 %	-167.52	264.53m V	0.239 %	-75.936	554.53m V	0.500 %	169.37
Harmonic 10	599.40 Hz	827.25m V	0.755 %	-23.389	378.94m V	0.342 %	99.914	492.39m V	0.444 %	18.786
Harmonic 11	659.34 Hz	467.90m V	0.427 %	29.883	290.74m V	0.263 %	60.180	54.044m V	0.049 %	-56.266
Harmonic 12	719.28 Hz	430.21m V	0.393 %	70.364	683.51m V	0.618 %	56.649	159.31m V	0.144 %	-44.136
Harmonic 13	779.22 Hz	343.88m V	0.314 %	-15.485	301.10m V	0.272 %	-91.870	20.104m V	0.018 %	140.58

Figure 16: FFT for different load types.

Practical results obtained show a favorable functioning of this inverter; its yield can be improved by an optimal choice of the interrupters used to establish it.

Conclusion

In this paper we have established and evaluated a new inverter structure which allows boosting, undulating and filtering a DC voltage in a single stage.

The modulation is implemented in a digital form using a microcontroller which guaranties a good functioning. And the realization done supports and shows the optimal functioning of the inverter with different linear and non linear loads.

References

1. Lahouar FE, Slama JBH, Hamouda M, Mustapha FB (2014) Comparative study between two and three-level topologies of grid connected photovoltaic converters. 5th International Renewable Energy Congress (IREC): 1-6.

2. Boumaaraf H, Talha A, Bouhali O (2015) A three-phase NPC grid-connected inverter for photovoltaic applications using neural network MPPT. Renewable and Sustainable Energy Reviews 49: 1171-1179.

3. Islam M, Mekhilef S, Hasan M (2015) Single phase transformer less inverter topologies for grid-tied photovoltaic system: A review. Renewable and Sustainable Energy Reviews 45: 69-86.

4. Islam M, Hasan M, Akter P, Rahman M (2014) A new transformerless inverter for grid connected photovoltaic system with low leakage current. International Conference on Electrical Information and Communication Technology (EICT), Khulna: 1-6.

5. Islam M, Mekhilef S (2014) An improved transformer less grid connected photovoltaic inverter with reduced leakage current. Energy Conversion and Management 88: 854-862.

6. Sanchis AU, GUBIA E, Marroyo L (2005) Boost DC–AC Inverter: A New Control Strategy. IEE Transaction of Power Electronics 20: 343-353.

7. Albea C, Gordillo F, De Wit CC (2008) Adaptive Control of the Boost Inverter with Load RL, 17th IFAC World Congress, Seoul: 3316-3321.

8. Mostafa SMG (2012) Designing a Boost Inverter to Interface between Photovoltaic System and Power Utilities. IOSR Journal of Electrical and Electronics Engineering 3: 01-06.

9. Albea C, Gordillo F, Aracil J (2006) Control of the Boost DC-AC Converter by Energy Shaping. The 32nd Annual Conference of IEEE Industrial Electronics society, Paris: 754-759.

10. Akhter R (2007) A new technique of PWM boost inverter for solar home application. BRAC University Journal 4: 39-45.

11. Meddah M, Bouchetata NB (2011) Summary static DC/AC for photovoltaic systems. Renewable Energy Reviews ICESD'11 Adrar: 101-112.

12. Tehrani KA (2010) Design, Synthesis and Application of a New Robust Control by PID for The Fractional Multilevel Inverters,Theses.

13. Berrezzek F (2006) Etude des Différentes Techniques de Commande des Onduleurs à MLI Associés à une Machine Asynchrone.

12

Wearable Real Time Health and Security Monitoring Scheme for Coal Mine Workers

Jayabharata S* and Marimuthu CN

Department of ECE, Nandha Engineering College, Erode, Tamil Nadu, India

Abstract

This paper deals with implementing a supervision system for coal mine and underground workers, which is essential to avoid the workers illness and death. The proposed recovery system consists of all primary aspects of the coal mine and underground areas. This system incorporates a sensor array, GSM, RF and controller modules. ARM7 (LPC2148) Microcontroller is fully automated measuring system. ARM7 processor is used for measuring the environment parameters with high reliability and accuracy and smooth control by using sensor networks. Consequently, advance detecting crucial conditions the microcontroller starts alerting the mine workers by the alarm system and sends the alert messages to fire and ambulance services by using GSM modem. In addition, the observed parameter's value will be displayed on a PC by using RF (CC2500) module, which is at the control station. At the hazardous situation, this system shows the shortest and available way out path for the workers to move away from the harmful environment.

Keywords: Coal mine; GSM; Sensors array; SMS; PC; RF

Introduction

Air pollution and gas explosion are increasing day by day and become foremost crisis in the coal mines and other industries. Safety of the human being is an essential aspect in any industry, especially in the field of mining and underground industry. The coal mine has been a very precarious activity which results in a number of detrimental effects on the environment such as suffocation, roof collapse, gas poisoning, gas explosions and hazardous greenhouse gases may be released into the air.

Air pollution leads to the personal and health impacts on lots of people causing illness and fatality. Explosive gas emission panics the human health, occupational safety of the coal mines workers [1]. Consequently these pollutions cause hazardous effects for the coal mines workers. There are a lot of works are carried out in the same manner in the case of coal mine monitoring. However existing and extant security monitoring systems cannot monitor all environmental parameters, therefore workers cannot monitor properly. The remedy for this issue is a wearable cost effective protection system with low power consumption and high performance [2]. This venture aims to increase the occupational safety and to protect the workers effectively in the hazardous air polluted environment.

The major accidents occurred in the coal mine and underground units are based on fire, natural gas and overheating of surroundings. Coal mine safety monitoring system is mainly based on wireless sensor network. Wireless network can provide a timely and accurately reveal energetic condition of staff in the underground regions to control center computer system. This protection system consists of all primary aspects of the coal mine and underground area. Therefore, to precocious systematic understanding of the health impacts of personal exposure to these pollutants, a miniaturized monitoring device is considered necessary for those individuals human can wear or carry to constantly examine their surrounding environment [3].

In this paper we describe Health Gear, a wearable real-time health monitoring system. Health Gear consists of a set of physiological sensors wirelessly connected via Bluetooth to a Bluetooth-enabled cell phone. We describe our experience using Health Gear with an oximeter to constantly monitor and analyze the user's blood oxygen level (SpO$_2$),

heart rate and plethysmographic signal in a light-weight fashion. In this paper, we describe one exemplary application of Health Gear for monitoring users in their sleep in order to detect sleep apnea events. We are currently working on a few other applications of Health Gear, including monitoring pilots during their flights to detect dangerous drops in blood oxygen levels due to altitude, and constant, daily monitoring to identify correlations between contextual information (i.e., date, current activity, location, etc.) and changes in heart-rate and/or blood oxygen levels [4].

System Design

The proposed protection system covers the most ameliorate and primary necessity aspects of the coal mine workers. This rescue system for coal mine workers is divided into two segments. Miner module and ground station unit. The wearable hardware device (miner module) is attached to the body of the mine workers (Figure 1).

Miner module

The miner's module consists of the sensor arrays, ARM7 microcontroller, RF transmitter and GSM modem The sensor module incorporates with various sensors to measures the real-time underground parameters like temperature, humidity, light, fire and position of worker furthermore multiple gas and it's concentrations in underground station [5]. During the hazardous situation of the mine, miner's module gives information to the control station through the RF transmitter. The control station will alert the workers after receiving the hazardous information from the ground station by using RF receiver.

ARM7 Microcontroller (LPC2148) is the heart of this module unit. It is used to take the necessary decision whenever the predefined comfort

***Corresponding author:** Jayabharata S, Department of ECE, Nandha Engineering College, Erode, Tamil Nadu, India, E-mail: jayabrathasomu@gmail.com

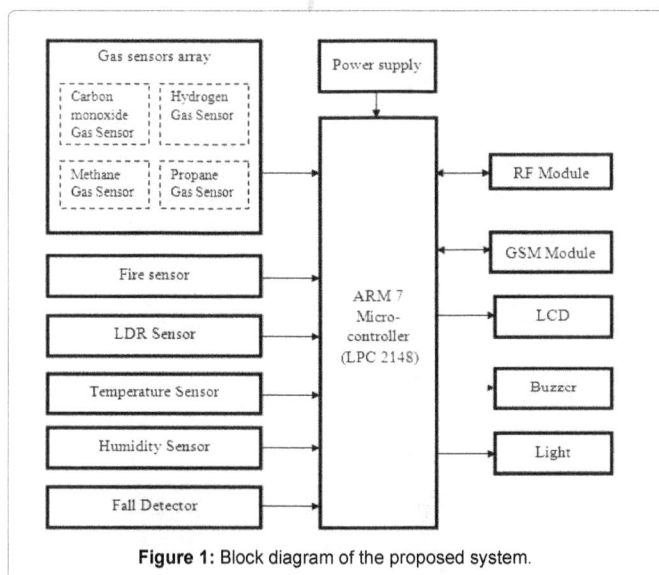

Figure 1: Block diagram of the proposed system.

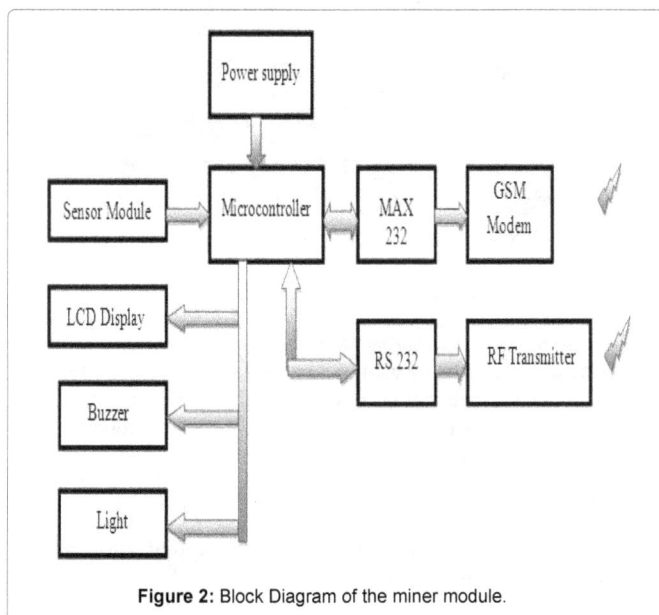

Figure 2: Block Diagram of the miner module.

level rises in the underground unit. A 2×16 character LCD module is interfaced to show all the parameters like temperature, humidity etc., at underground miner's module. GSM Modem is interfaced with the Module to send the message to the remote location similar to the fire station and the ambulance service (Figure 2).

Ground station module

Ground station control unit consists of RF receiver and Personal Computer to monitor the state of affairs in the coal mine (Figure 3). The hazardous conditions of the coal mine unit state will be displayed on the PC by using the serial to USB converters (RS. 232), which is used to take a necessary precaution action for the workers [6]. With the help of this alert system, workers will be alerted in the dangerous environment. Once the measured parameter values are more than the safety level pre-programmed at microcontroller, it decodes different type of beep alarms.

Hardware Description

Hardware description comprises of four modules namely sensor, controller, GSM and RF modules.

Sensor network

Sensor network incorporates several sensors to monitor the hazardous environment of the underground unit.

Gas sensor technologies: The most common gas sensor technologies are metal oxide sensors (MOS), non-dispersive infrared gas sensors (NDIR), catalytic sensors, photo Ionization detectors (PID), flame ionization detector (FID), thermal conductivity sensor, and electrochemical (EC) sensors. Due to inherent complexity and sophistication, PID, FID and NDIR sensors have high production cost and are difficult to miniaturize. Although catalytic sensors are less sensitive to ambient environment changes and have simple structure, they have very short lifetimes due to poisonings and can only measure combustible gases [7].

Among gas sensor technologies, MOS sensors stand out as candidates for wearable systems, especially because their structures are compatible with modern micro fabrication processes, therefore a suitable device for low cost miniaturization. MOS sensors have low power consumption, low-cost, good selectivity, and can sense a wide range of gases such as CH_4, NH_3, CO, NO, NO_2, H_2, SO_2, CO_2, H_2S and O_2 Therefore, MOS sensors were chosen in this system for multi-gas measurement [8-10].

Fire sensor

The Fire sensor works on the principle of IR rays or heat radiation detection for protection against fire. When the sensor is detecting any kind of fire it will give an interrupt signal to the microcontroller to alert the workers in underground regions.

Light Dependent Resistor

LDR used to sense the sense the darkness at the underground unit. The amount of light intensity level reduced in the working area then LDR triggered, circuit will turn ON the LED. For the convenient purpose, possibly the LED will be fixed with the workers helmet. Consequently to provide intensity the LED lights present on miner's module.

Temperature sensor (LM35)

LM35 is a precision IC temperature sensor. The electrical output voltage is proportional to the Centigrade temperature. It measures the temperature more accurately than a thermistor and no need of amplifying the output voltage.

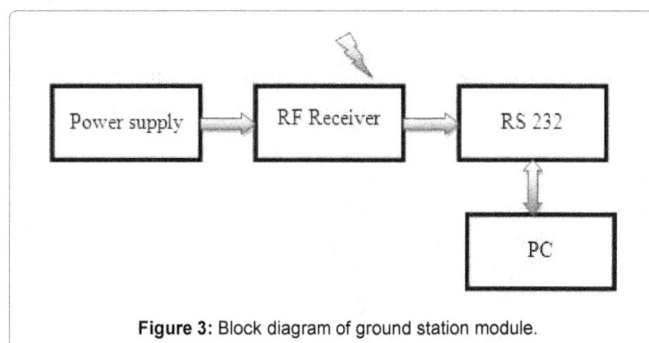

Figure 3: Block diagram of ground station module.

Humidity sensor (HSM-20G)

Humidity is the amount of water vapor in the air. The humidity sensor HSM-20G is of resistive type and converts relative humidity into standard voltage output. It is an analog humidity and temperature sensor that outputs analog voltage respects to relative humidity.

Fall detector (ADXL335)

Fall detector is sensitive to both linear acceleration and the local gravitational field. The ADXL335 is a low power, thin, small, complete 3-axis accelerometer with signal conditioned voltage outputs. The output signals are analog voltages that are proportional to acceleration. Its measurement range is of ±3 g. X-axis is connected with controller and continuously checks that- g value change.

Microcontroller (LPC 2148): The ARM 7 is a 32-bit RISC instruction set developed by ARM Holdings. It is also known as the Advanced RISC Machine. In this scheme, ARM 7 (LPC 2148) microcontroller is used. The sensor module is connected to LPC 2148 and it takes necessary decisions based on sensor module output [11].

Due to their tiny size and low power consumption, LPC2141/42/44/46/48 are ideal for applications where miniaturization is a key requirement, such as access control and point-of-sale. Serial communications interfaces ranging from a USB 2.0 Full-speed device, multiple UARTs, SPI, SSP to I2C-bus and on-chip SRAM of 8 kb up to 40 kb, make these devices very well suited for communication gateways and protocol converters, soft modems, voice recognition and low end imaging, providing both large buffer size and high processing power. Various 32-bit timers, single or dual 10-bit ADC(s), 10-bit DAC, PWM channels and 45 fast GPIO lines with up to nine edge or level sensitive external interrupt pins make these microcontrollers suitable for industrial control and medical systems.

GSM module: GSM module is a compact and reliable wireless module. GSM module is used to send the message to the ground or remote station. Plain Text message may be sent through the modem by interfacing only three signals of the serial interface of a modem with microcontroller (TxD, RxD and GND).

In this system, transmit and receive signal from a serial port of the microcontroller is connected respectively with the transmit signal (TxD) and receive signal (RxD) of the serial interface of GSM Modem. The LPC 2148 alerts GSM modem in the miner module to send SMS to fire and ambulance service stations in adverse situations and the acknowledgement SMS is sent to the miner's module from control station.

RF (CC2500) module: Effective communication system is an essential in a coal mine and underground area. Wired network communication is inconvenient and inefficient method compared to wireless communication system.

Consequently this coal mine recovery scheme communication is based on RF wireless transceiver module (Figure 4).

RF module provides easy and flexible wireless data transmission between devices. This is based on AVR Atmega8 with serial output which can be interfaced directly to PC. The CC2500 is a low-cost 2.4 GHz transceiver designed for very low-power wireless applications purpose. RF wireless transceiver module establishes communication between the underground and ground station units to communicate about the environmental changes in the underground unit.

Figure 4: GSM module.

Figure 5: Overall recovery system hardware setup.

Software Description

As this recovery system is based on embedded systems, the firmware enlargement is prepared by using Embedded C language. To simulate this design in hardware various software tools in set up needed for this system. Consequently the most significant among these tools are Keil IDE, Flash Magic Tera Term and HyperTerminal.

Keil IDE

Most of the embedded programmer uses the Keil software. Keil version 4 is used to compile the Embedded C code. A compiler is used to compile the code and convert the source code to hex files with the help of an IDE.

Flash magic and tera term

The Flash magic tool is used to flash the Programme in ARM7 microcontroller board. Tera term is used to communicate with the serial port. With the help of the Tera term, user can receive the information about underground unit in the control center PC with the help of RF receiver.

Hyper terminal

The Hyper Terminal tool is used to monitor the information about underground unit with the help of a Serial Ports in PC. RF receiver is used to send the information to a remote location.

Results

The Overall system's results are discussed in this section. The LPC2148 Evolution Board is heart of all functionalities in miner module i.e. Monitoring, Processing collected data and taking necessary action based on the limits given for individual sensors (Figure 5).

The following figure shows the abnormality levels of the parameter in the underground module. The predefined comfort value is programmed in the microcontroller, proposed protection system continuously monitor the underground unit (Figure 6).

Whenever the predefined value is more than the safety level pre-programmed at microcontroller, the controller decodes beep alarms through the speaker connected with controller. Corresponding parameters value displayed in the LCD display and get SMS alert through GSM Module. This above figure shows the emergency shortest and risky free way-out to the workers. With the help of this system worker will move fast and not panic to shift away from the environment (Figure 7).

On detection of Abnormal activity at miner module the core system alerts and sends SMS to either Fire station or Ambulance based on the Interrupt source (Figure 8).

The above figure shows the Hyper terminal screen i.e. Remote Station. In this station information will be send to a PC by using RF receiver to monitor the data in ground station (Figure 9). This statistics records will be displayed in pc, which provides the complete information of workers and statistics of all the parameters.

One such disorder is sleep apnea, which is an under-diagnosed, but common condition that affects both children and adults. It is characterized by periods of interrupted breathing (apnea) and periods of reduced breathing (hypoapnea). Both types of events are generally considered to be equal in terms of their impact on patients. The most common form of sleep apnea, called obstructive sleep apnea (OSA),

Figure 8: SMS alert system.

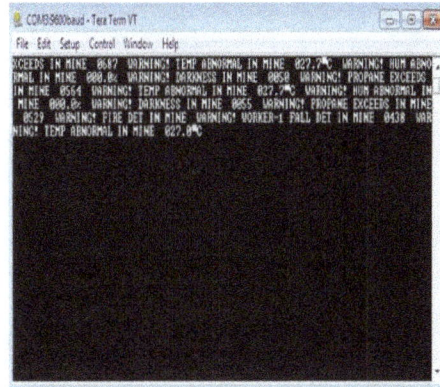

Figure 9: Remote section monitoring screen.

is caused by the partial or complete constriction of the patient supper airway. Regular sleep apnea leads to repeated hypoxemia, as phyxia and awakenings, and produces immediate symptoms such as increased heart rate and high blood pressure and long term symptoms such as extreme fatigue, poor concentration, a compromised immune system, slower reaction times and cardio/cerebrovascular problems.

Conclusion

This proposed system covered the most Important and Primary necessity aspect of any mine workers safety. More security is provided by GSM module, used to send the message to fire and ambulance service to evade the damages. Furthermore, this system perhaps extended with an ability to find the shortest exit path for the workers in case of fire accidents in the underground mines. Then the ground station alerts the workers using voice over IC fixed in the miner module about the shortest furthermore safest path in case of fire accidents in one or more underground units. All these sensors can be easily placed on Miner's Helmet that helps in continuous monitoring. The extant mine security system can be effectively replaced by this rescue safety system proposed in this paper.

References

1. Li H, Mu X, Yang Y, Mason AJ (2014) Low power multi-mode electrochemical gas sensor array system for wearable health and safety monitoring, IEEE Sensor Journal.

2. Zhang P (2012) Design of wireless mine gas monitoring and control system based on nRF2401, in Computer Science and Service System (CSSS), International Conference 1051-1054.

Figure 6: Miner's module alert system.

Figure 7: Emergency way-out indications.

3. Tsow F (2009) A wearable and wireless sensor system for real-time monitoring of toxic environmental volatile organic compounds, Sensors Journal IEEE 9: 1734-1740.

4. Tang KT (2014) A 0.5 V 1.27 mW nose-on-a-chip for rapid diagnosis of ventilator-associated pneumonial, in Solid-State Circuits Conference Digest of Technical Papers (ISSCC), IEEE International 420-421.

5. (2011) A CMOS single-chip gas recognition circuit for metal oxide gas sensor arrays, circuits and systems I: Regular Papers, IEEE Transactions 58: 1569-1580.

6. Krithika N, Seethalakshmi R (2014) Safety scheme for mining industry using zigbee module Indian Journal of Science and Technology 7: 1222-1227.

7. Qiang C, Ping SJ, Zhe Z, Fan Z (2009) Zig Bee based intelligent helmet for coal miners, Proc IEEE World Congress on Computer Science and Information Engineering 433-435.

8. Oliver N and Flores F. Health gear: A Real-time Wearable System for Monitoring and Analyzing Physiological Signals -Mangas 2.

9. Wei S, Li LL (2009) Multi-parameter Monitoring System for Coal Mine based on Wireless Sensor Network Technology, Proc. International IEEE Conference on Industrial Mechatronics and Automation 225-27.

10. Jin-ling S, Heng-wei G, Yu-jun S (2010) Research on Transceiver System of WSN Based on V-MIMO Underground Coal Mines, Proc. International Conference on Communications and Mobile Computing 374-378.

11. Kiran KV, Narasimha E, Shruthi Y (2013) Smart Helmet For Coal Miners Using Zigbee Technology International Journal for Research in Science and Advanced Technologies 2: 67-69.

13

Power Quality Improvement of Constant Frequency Aircraft Electric Power System Using Genetic Algorithm and Neural Network Control Based Control Scheme

Saifullah Khalid*

Department of Electrical Engineering, IET Lucknow, India

Abstract

Synchronous Reference Frame Strategy for extracting reference currents for shunt active power filters have been modified using Artificial Neural Network, Genetic Algorithm based controller and their performances have been compared. The acute analysis of Comparison of the compensation capability based on THD and speedwell be done, and recommendations will be given for the choice of technique to be used. The simulated results using MATLAB model are shown, and they will undoubtedly prove the importance of the proposed control technique of aircraft shunt APF.

Keywords: Aircraft electrical system; Shunt Active Filter (APF); Synchronous reference frame strategy; ANN; Genetic algorithm; THD

Introduction

More advanced aircraft power systems [1-3] have been needed due to increased use of electrical power on behalf of other alternate sources of energy. The subsystems like flight control, flight surface actuators, passenger entertainment, are driven by electric power, which flowingly increased the demand for creating aircraft power system more intelligent and advanced. These subsystems have extensive increased electrical loads i.e. power electronic devices, increased feeding of electric power, additional demand for power, and above to all of that great stability problems.

In peculiarity to standard supply system, the source frequency is of 50 Hz, whereas, aircraft AC power system works on the source frequency of 400 Hz [1-3]. Aircraft power utility works on source voltage of 115/200 V. The loads applicable to the plane a system differs from the loads used in 50 Hz system [1]. When we deliberate the generation portion; aircraft power utility will remain AC driven from the engine for the plane primary power. Novel fuel cell technology can be used to produce a DC output for ground power, and its silence process would match up to suitably with the Auxiliary Power Unit (APU). Though when considering the dissemination of primary power, whether AC or DC; each approach has its merits. In DC distribution, HVDC power distribution systems permit the most resourceful employ of generated power by antithetical loss from skin effect. This allows paralleling and loads sharing amongst the generators. In AC distribution, AC Flogging is very clear-cut at high levels too. Due to its high dependence on HVDC system, a wide range of Contactors, Relays can be exploited.

While talking about Aircraft Power Systems we also need to consider increased power electronics application in aircraft which creates harmonics, large neutral currents, waveform distortion of both supply voltage and current, poor power factor, and excessive current demand. Besides if some non-linear loads is impressed upon a supply, their effects are additive. Due to these troubles, there may be nuisance tripping of circuit breakers or increased loss and thermal heating effects that may provoke early component failure. This is a prodigious problem to every motor loads on the system. Hence, decent power quality of the generation system is of scrupulous attention to the Aircraft manufacturer. We discern that aircraft systems work on high frequency so even on the higher frequencies in the range of 360 to 900 Hz; these components would remain very significant.

Today, advanced soft computing techniques are used widely in the involuntary control system, and optimization of the system applied. Several of them are such as Fuzzy logic [4-8], optimization of active power filter using GA [9-12], power loss reduction using particle swarm optimization [13], Artificial neural network control [14-18] applied in together machinery and filter devices.

In this paper, ANN based and Genetic Algorithm optimized have been used together to mend the complete performance of active filter for the lessening of harmonics and other delinquents created into the aircraft electrical system because of the non-linear loads [1]. The simulation results clearly show their effectiveness. The simulation results acquired with the new model are much improved than those of traditional method.

The paper has been modified in a sequential manner. The APF outline and the load under contemplation are discussed in Section II. The control algorithm for APF converses in Section III. MATLAB/Simulink based simulation results are presented in Section IV, and finally Section V concludes the paper [19-23].

System Depiction

The aircraft power system is a three-phase power system with the frequency of 400 Hz. As exposed in Figure 1, Shunt Active Power Filter improves the power quality and compensates the harmonic currents in the system [22,24-28]. The shunt APF is comprehended by using one voltage source inverters (VSIs) connected at the point of common coupling (PCC) to a common DC link voltage [20-23].

The set of loads for aircraft system consist of three loads. The first load is a three-phase rectifier in parallel with an inductive load and an

---

***Corresponding author:** Saifullah Khalid, Department of Electrical Engineering, IET Lucknow, India, E-mail: saifullahkhalid@Outlook.com*

unbalanced load connected in a phase with the midpoint (Load 1). The second one is a three phase rectifier connects a pure resistance directly (Load 2). The third one is a three-phase inductive load linked with the ground point (Load 3). Finally, a combination of all three loads connected with system together at a different time interval to study the effectiveness of the control schemes has been used to verify the functionality of the active filter in its ability to compensate for current harmonics. For the case of all three load connected, Load 1 is always connected, Load 2 is initially connected and is disconnected after every 2.5 cycles, Load 3 is connected and disconnected after every half cycle. All the simulations have been done for 15 cycles. The circuit parameters are given in last section - Parameters.

Control Theory

The proposed control of APF depends on Synchronous Reference Frame Strategy, and it has been optimized using artificial intelligent technique like artificial Neural Network and Genetic Algorithm together. Synchronous Reference Frame strategy has been discussed in brief in this section. The following section also deals with the primary application of ANN and GA in the control schemes [19,20].

Synchronous reference frame strategy (SRF)

The reference frame d-q is decided by the angle θ on the d-q frame applied in the p-q theory. Slight modifications are required in the conventional SRF method so that they can be used in aircraft power utility and compensate well the neutral current. For this reasons, the zero-sequence component of current has not been well thought-out. So, the zero sequence subtract block cart from the zero sequence current. The positive sequence and negative sequence component are obtained in output current. After its park transformation, only instantaneous active and reactive current are available in the output current in the

d-q frame. A low-pass filter accomplishes the disunion of the dc and ac factor of the active current for compensation of the reactive and harmonic current. This active current goes through a low-pass filter, and the signal came from dc voltage regulator together though a Park counter-transformation subtracting from the load currents generates the reference current. Reference current is generated (Figure 2).

Application of ANN based control

In this paper, Synchronous reference frame Strategy has been modeled by an artificial neural network (ANN) made up of two hidden layers with 10 neurons each, and one output layer with 3 neurons. The logarithmic activation function is the base of the two hidden layers neurons, and linear activation function for the output layer neurons.

As shown in Figure 3, the ANN has seven inputs (v_a, v_b, v_c, dc voltage error, i_a, i_b, i_c) and three outputs (i_{ra}, i_{rb}, i_{rc}) as observed in the different strategies. The adaptation of the weights (W) and bias (b) in the ANN, is based, first, on the computation of the mean square error (MSE) between the outputs of the PQ technique and those of the ANN, and secondly, on the execution of 'Levenberg-Marquardt back-propagation' algorithm.

The reference current generation unit and dc voltage controller unit has been modeled and their individual and simultaneous effect has been observed.

Application of genetic algorithm

Genetic algorithm (GA) is a search technique used in computing [9] to find an optimal solution for a search problem. It is implemented as a computer simulation in which a population of abstract representations (called chromosomes of the genome) of candidate solutions (individuals) to an optimization problem evolves [11] towards improved solutions. In general, the fittest individuals of any population tend to reproduce and survive to the next generation, thus improving the successive generations [12]. Commonly, the algorithm terminates when either a maximum number of generations has been produced, or a satisfactory fitness level has been reached for the population [12]. The fitness function is defined over the genetic representation and measures the quality of the represented solution. The fitness function is always problem dependent [11]. By starting at several independent points and searching in parallel, the algorithm avoids local minima and also avoids converging to sub optimal solutions. In this way, GAs have been shown to be capable of locating high performance areas in complex domains without experiencing the difficulties associated with high dimensionality, as may occur with gradient decent techniques or methods that rely on derivative information [15].

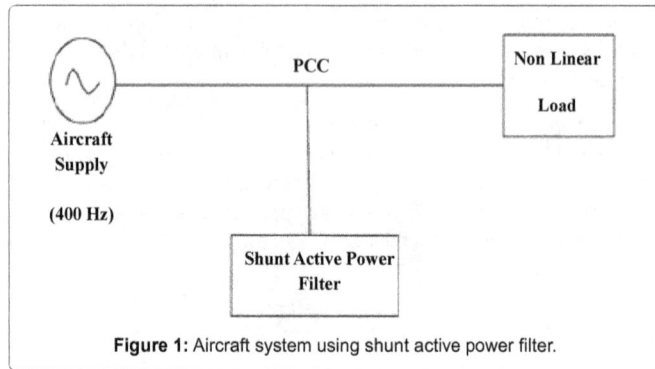

Figure 1: Aircraft system using shunt active power filter.

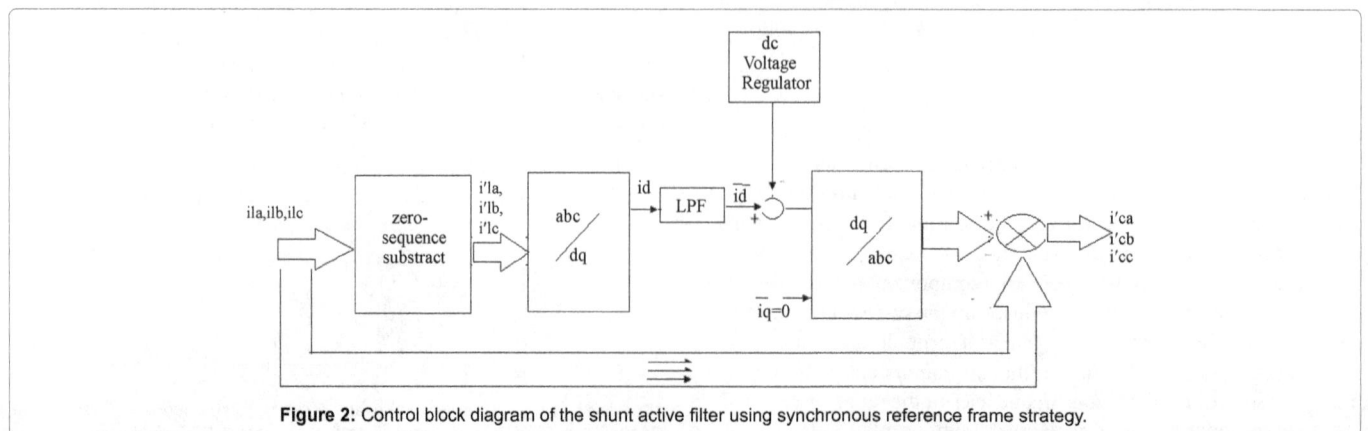

Figure 2: Control block diagram of the shunt active filter using synchronous reference frame strategy.

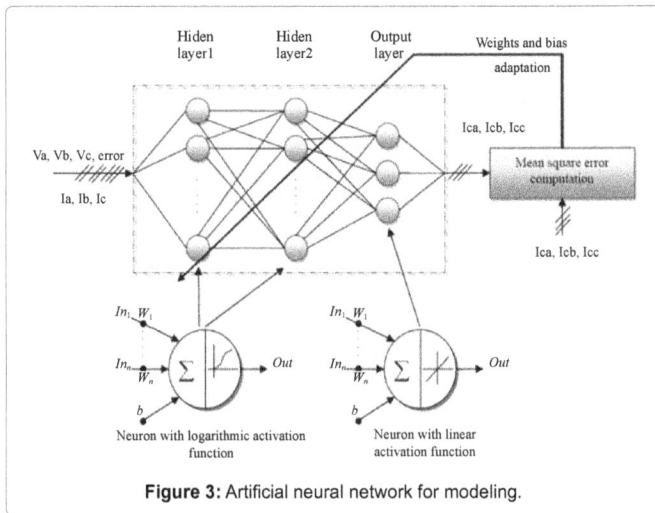

Figure 3: Artificial neural network for modeling.

Most of the published papers use this generalized program to fit their objective function but after the popularity of MATLAB as software for simulation and modeling researchers have started using genetic algorithm optimization tool (GAOT) available. MATLAB provides many built in auxiliary functions useful for function optimization [15]. GAOT is not only easy to implement but is faster in simulating.

In this paper, the GA is applied to determine value of inductor filter (L_f). GA will try to search the best value of the filter inductor. Supply side has been taken as input side for inductor filter. Inductor filter value used in this thesis is 0.25 mH. Offline, computer simulation using MATLAB Simulink has been applied to find out the optimum value for inductor filter.

For the program, the limits, inequality and bounds need to be defined. This paper has attempted to develop a single GA code program for optimizing objective function.

x0=[V_{dc}; V_s; I_c; t; L_f];

lb=[V_{dcmin}; V_{smin}; I_{cmin}; t_{min}; L_{fmin}];

ub=[V_{dcmax}; V_{smax}; I_{cmax}; t_{max}; L_{fmax}];

Aeq=[];

beq=[];

A=[1 -1 1 -1 1; 1 1 -1 1 -1; 0 0 1 1 1; 1 1 -0 0 -1; 1 1 0 1 0];

b=[Values of V_{dc}; V_s; I_c; t; L_f depending upon the equations];

[x, fval, exitflag]=fmincon(@myobj,x0,A,b,Aeq,beq,lb,ub).

The boundary and limits of parameters in the filter has been defined using the data of ANN model. The data has been collected using MATLAB/Simulink. Finally, a program using genetic algorithm has been written to generate the best value of the filter inductor. After the calculation, GA generates the value of 0.187 mH. After using this inductor value, total harmonic distortion of source current and voltage have been reduced so we can say that inductor value calculated is optimum.

Simulation Results and Discussions

The proposed scheme of APF is simulated in MATLAB environment to estimate its performance. Three loads have been applied together at a different time interval to check the affectivity of

the control schemes for the reduction of harmonics. A small amount of inductance is also connected to the terminals of the load to get the most effective compensation. The simulation results clearly reveal that the scheme can successfully reduce the significant amount of THD in source current and voltage within limits.

Uncompensated system

Figure 4 shows the waveforms obtained after the simulation of an uncompensated system. It has been observed that the THD of source current calculated when loads connected with the system is 9.5% and THD of source Voltage were 1.55%. By observing these data, we can easily recognize supply has been polluted when loads have been connected and is obviously not within the limit of the international standard.

Compensated system

The performance of APF under different loads connected, when utilizing ANN Control has been discussed below for the control strategy given below.

For synchronous reference frame strategy: From Figure 5 it has been empiric that that the THDs of source current and source voltage were 2.82% and 1.65% respectively. The compensation time was 0.01 sec. At t=0.01 sec, it is apparent that the waveforms for source voltage and source current have become sinusoidal. Figure 5 shows the waveforms of compensation current, DC capacitor voltage, and load current.

The aberration in dc voltage can be acutely apparent in the waveforms. As per claim for accretion the compensation current for accomplishing the load current demand, it releases the energy, and after that it accuses and tries to achieve its set value. If we carefully observe, we can acquisition out that the compensation current is, in fact, accomplishing the appeal of load current, and afterward the active filtering the source current and voltage is affected to be sinusoidal.

For synchronous reference frame strategy using ANN and genetic algorithm: THDs of source current and source voltage have been found 2.62% and 1.03% respectively after making observations from the simulation results shown in Figure 6. The waveforms for source voltage and source current have become sinusoidal at t=0.0068 sec. Compensation time is 0.0068 sec. The waveforms of compensation current, dc capacitor Voltage, and load current have been shown in

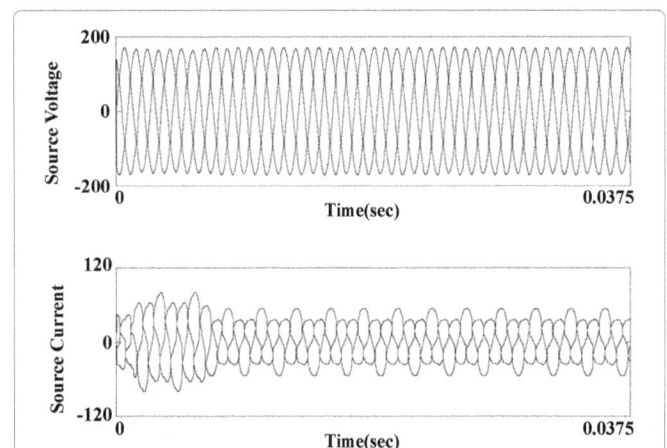

Figure 4: Source voltage and source current waveforms of uncompensated system.

Figure 6. Waveforms show the variations in dc capacitor voltage. Whenever the demand for high load current comes, it releases the energy that in turn increases the compensation current. Later on, it charges and tries to regain its previous set value. By making a simple observation, we can say that compensation current is fulfilling the demand of load current. After the active filtering, the source current and voltage is forced to be sinusoidal.

Comparative analysis of the simulation results

From Table 1, we can easily say that Synchronous Reference Frame using ANN and Genetic Algorithm (SRF-ANN-GA) has been found best for current and voltage harmonic reduction. When these results have been compared based on compensation time, it has been also found that SRF-ANN-GA strategy is the fastest one.

Conclusion

This paper has done an acute analysis of traditional and Artificial Intelligent Technique (ANN and Genetic Algorithm applied together) based controller for shunt APF in aircraft power utility of 400 HZ. Optimum selection of control strategy based on compensation time and THD has been suggested. Overall Synchronous Reference Frame using ANN&GA (SRF-ANN-GA) has been observed as an optimum choice. Synchronous Reference Frame Strategy's performance has been

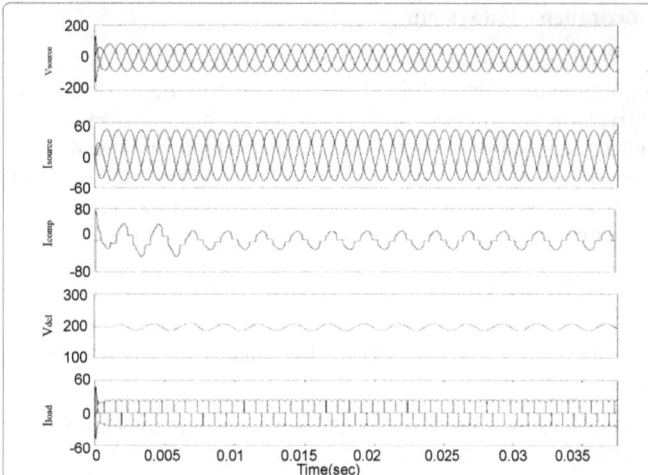

Figure 5: Source voltage, source current, compensation current (phase b), DC link voltage and load current waveforms of active power filter using synchronous reference frame strategy.

Strategy	THD-I (%)	THD-V (%)	Compensation Time (sec)
SRF	2.82	1.65	0.0100
SRF-ANN-GA	2.62	1.03	0.0068

Table 1: Summary of simulation results.

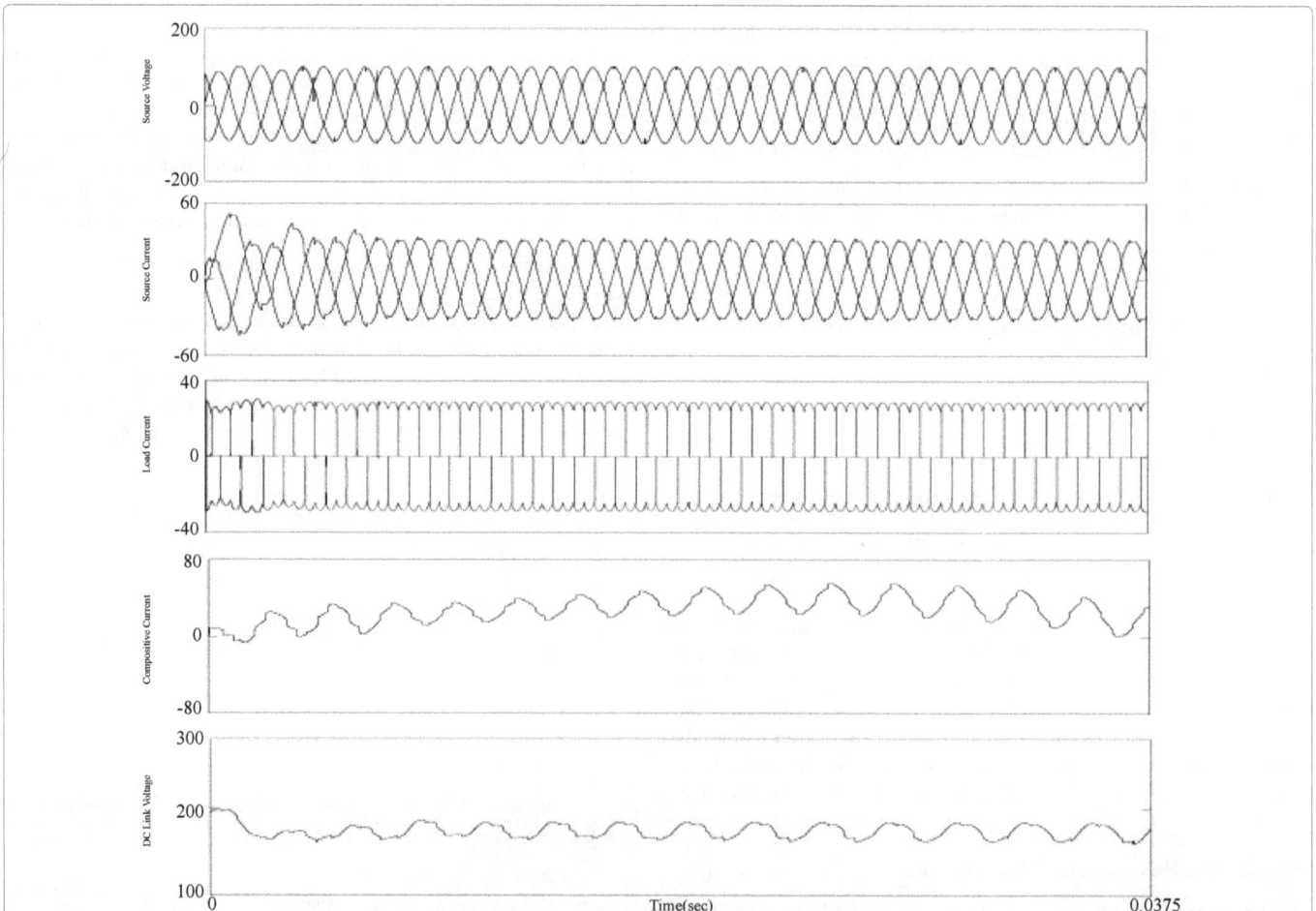

Figure 6: Source voltage, source current, load current and DC link voltage waveforms of active power filter using synchronous reference frame strategy using ANN and genetic algorithm with all three loads connected for aircraft system.

improved, which itself an achievement for the case of optimization in traditional strategies.

Parameters

The aircraft system parameters are [1]:

Three-phase source voltage: 115 V/400 Hz,

Filter capacitor: 5 μF,

Filter inductor=0.25 mH,

DC capacitor: 4700 μF,

DC voltage reference: 400 V.

References

1. Donghua C, Guo T, Xie S, Zhou B (2005) Shunt Active Power Filters Applied in the Aircraft Power Utility. 36th Power Electronics Specialists Conference, PESC 5: 59-63.

2. Saifullah K, Bharti D (2014) Comparative Evaluation of Various Control Strategies for Shunt Active Power Filters in Aircraft Power Utility of 400 Hz. Majlesi Journal of Mechatronic Systems 3: 1-5.

3. Saifullah K, Bharti D (2013) Application of AI techniques in implementing Shunt APF in Aircraft Supply System. Proceeding of SPRINGER - SOCROPROS Conference, IIT-Roorkee 1: 333-341.

4. Guillermin P (1996) Fuzzy logic Applied to Motor Control. IEEE Transactions on Industrial Application 32: 51-56.

5. Hew Wooi AHAP, Hamzah A, Mowed HAF (2002) Fuzzy Logic Control of a three phase Induction Motor using Field Oriented Control Method. Society of Instrument and Control Engineers, SICE Annual Conference 264-267.

6. Jain SK, Agrawal P, Gupta H (2002) Fuzzy logic controlled shunt active power filter for power quality improvement. IEE Proceedings of the Electric Power Applications 149: 317-328.

7. Norman M, Samsul B, Mohd N, Jasronita J, Omar SB (2004) A Fuzzy logic Controller for an Indirect vector Controlled Three Phase Induction Motor. Proceedings Analog And Digital Techniques In Electrical Engineering, TENCON 2004, Chiang Mai, Thailand 4: 1-4.

8. Afonso JL, Fonseca J, Martins JS, Couto CA (1997) Fuzzy Logic Techniques Applied to the Control of a Three-Phase Induction Motor. Proceedings of the UK Mechatronics Forum International Conference, Portugal. pp. 142-146.

9. Chiewchitboon P, Tipsuwanpom P, Soonthomphisaj N, Piyarat W (2003) Speed Control of Three-phase Induction Motor Online Tuning by Genetic Algorithm. Fifth International Conference on Power Electronics and Drive Systems, PEDS 1: 184-188.

10. Kumar P, Mahajan A (2009) Soft Computing Techniques for the Control of an Active Power Filter. IEEE Transactions on Power Delivery 24: 452-461.

11. Ismail KB, Abdeldjebar H, Abdelkrim B, Mazari B, Rahli M (2008) Optimal Fuzzy Self-Tuning of PI Controller Using Genetic Algorithm for Induction Motor Speed Control. Int J of Automation Technology 2: 85-95.

12. Guicheng W, Min Z, Xu X, Changhong J (2006) Optimization of Controller Parameters based on the Improved Genetic Algorithms. IEEE Proceedings of the 6th World Congress on Intelligent Control and Automation, Dalian, China.

13. Radha T, Chelliah TR, Pant M, Ajit A, Grosan C (2010) Optimal gain tuning of PI speed controller in induction motor drives using particle swarm optimization. Logic Journal of IGPL Advance Access.

14. Joao OP, Bimal BK, Eduardo BSL (2001) A Stator-Flux-Oriented Vector-Controlled Induction Motor Drive with Space-Vector PWM and Flux-Vector Synthesis by Neural Networks. IEEE Transaction on Industry Applications 37: 1308-1318.

15. Rajasekaran S, Vijayalakshmi PGA (2005) Neural Networks, Fuzzy Logic and Genetic Algorithm: Synthesis and Applications. Prentice Hall of India, New Delhi, fifth printing.

16. Rojas R (1996) Neural Network - A Systematic Introduction. Spriger-Verlag, Berlin.

17. Zerikat M, Chekroun S (2008) Adaptation Learning Speed Control for a High-Performance Induction Motor using Neural Networks. Proceedings of World Academy of Science, Engineering and Technology 35: 294-299.

18. Seong-Hwan K, Tae-Sik P, YooJi-Yoon, Gwi-Tae P (2001) Speed-Sensorless Vector Control of an Induction Motor Using Neural Network Speed Estimation. IEEE Transaction on Industrial Electronics 48: 609-614.

19. Saifullah K, Bharti D (2011) Power Quality Issues, Problems, Standards & their Effects in Industry with Corrective Means. International Journal of Advances in Engineering & Technology (IJAET) 1: 1-11.

20. Aredes M, Hafner J, Heumann K (1997) Three-Phase Four-Wire Shunt Active Filter Control Strategies. IEEE Transactions on Power Electronics 12: 311-318.

21. Saifullah K, Bharti D (2013) Power quality improvement of constant frequency aircraft electric power system using Fuzzy Logic, Genetic Algorithm and Neural network control based control scheme. International Electrical Engineering Journal (IEEJ) 4: 1098-1104.

22. Saifullah K, Bharti D, Agrawal N, Kumar N (2007) A Review of State of Art Techniques in Active Power Filters and Reactive Power Compensation. National Journal of Technology 3: 10-18.

23. Dugan RC, McGranaghan MF, Beaty HW (1996) Electrical Power Systems Quality. New York: McGraw-Hill.

24. Saifullah K, Bharti D (2010) Power Quality: An Important Aspect. International Journal of Engineering, Science and Technology 2: 6485-6490.

25. IEEE Recommended Practices and Requirements for Harmonic Control in Electrical Power Systems. IEEE Standard 519-1992.

26. Ghosh A, Ledwich G (2002) Power Quality Enhancement Using Custom Power Devices. Boston, MA: Kluwer.

27. Khalid S, Vyas N (2009) Application of Power Electronics to Power System. University Science Press, India.

28. Saifullah K, Bharti D (2013) Comparative Critical Analysis of SAF using Soft Computing and Conventional Control Techniques for High Frequency (400 Hz) Aircraft System. Proceeding of IEEE - CATCON Conference.

14

Handling Electromagnetic Radiation beyond Terahertz using Chromophores to Transition from Visible Light to Petahertz Technology

Langhals H*

LMU University of Munich, Department of Chemistry, Butenandtstr-13, D-81377 Munich, Germany

Abstract

An increase of the operating frequencies of electromagnetic waves leads from the well-estalished terahertz technology to the visual and reaches petahertz radiation. It is shown that electromagnetic radiation close to petahertz is attractive for technology where knowledge about radio waves can be applied. The dimensions of such radiation are still classically macroscopic; however, molecular components such as resonators were used where quantum mechanics rules have to be considered. Constructions of coupled resonators for energy transfer are as well demonstrated as molecular components for optical metamaterials.

Keywords: Terahertz; Petahertz; Electromagnetic radiation; Dyes; Visible light; Resonators; FRET; Energy transfer; Metamaterials; Reflectance

Introduction

The technology of radio waves representing electromagnetic radiation is well-established and dominated by metallic conductors and electronic circuits where both the radiation and the operating devices are macroscopic (large, classical) [1]. There is a tendency for increasing the operating frequency both because of the increase of the number of possible ports, the increase of information recording and the decrease of the size of antennae and other involved components. One may demand if there is a natural technological limit for the increase in frequency.

Materials and Methods

The dyes were prepared according to the literature [2]. Pellets were prepared from 150 mg of pigments in a compacting tool with a polished round steel heading die of 13 mm diameter (compacting tool for KBr pellet in infrared spectroscopy). The compacting tool was evacuated to 0.2 mBar and 10 tons were applied for 5 min; this corresponds to 7500 Bar. The UV/Vis absorption spectra were recorded in chloroform in 1 cm cuvettes with extinctions between 0.7 and 1.0 in the maxima. The fluorescence spectra were recorded in chloroform in 1 cm cuvettes with extinctions between 0.01 and 0.02 in the maxima (Figure 3).

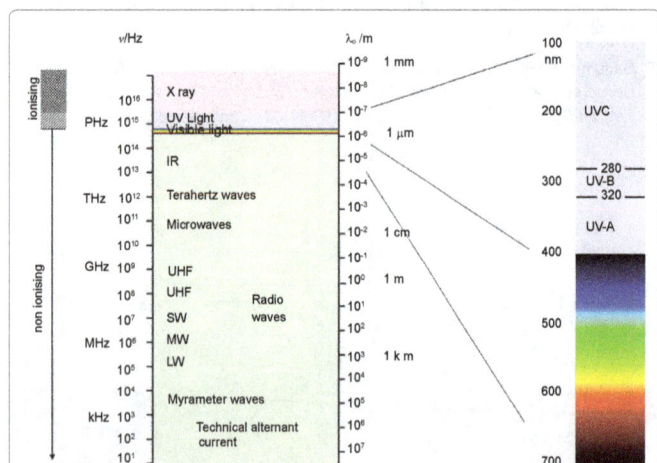

Figure 1: Important ranges of electromagnetic waves. The visible is indicated and enlarged. Ionisation becomes more and more dominating at short wavelengths.

Operating electromagnetic radiation at very high frequencies

Technologically important ranges of electromagnetic radiation are shown in Figure 1. The frequency v of electromagnetic radiation is interlinked with a wavelength λ and the velocity of light c in terms of $\lambda = c/v$. The wavelength in vacuo is reported in Figure 1 according to the velocity c_o of light in vacuo. The index of refraction n_v has to be considered for the propagation of electromagnetic waves [3] in matter where $c = c_o/n_v$ (a complex index of refraction has to be applied for a complete description, however, this is simplified to n_v in this context); correspondingly, the wavelength shrinks to λ/n_v. The index of refraction n_v is equal the square root of the relative permittivity ε_r ($n_v = \sqrt{\varepsilon_r}$) for radio frequencies and non-magnetic materials. More and more deviations are observed for very high frequencies and condensed matter. This is why the rearrangement of the atomic nuclei as a response to an external electric field dominates for low frequencies and polar condensed matter. The comparably slow movement of the heavy nuclei cannot follow the electric field for very high frequencies at about 1 PHz where the response of the movement of the electrons becomes the dominating effect because of their lower mass (this is indicated by the polarizability of matter). However, the behaviour becomes more complicated close to absorption bands by the complex index of refraction (anomalous refraction). As a consequence, the polarizability should be measured by means of the index of refraction well separated from absorption bands. The centre of the visible at about half a micron in vacuum wavelength is a good compromise therefore because it is well situated for the majority of materials between strong absorption band in the UV and in the NIR. The position of the sodium double D spectral line at 589/590 nm was taken as the reference wavelengths for the determination of the indices of refraction at 20°C (n_D^{20}) for historical reasons and is still a good compromise for reference measurements.

$$E = h \cdot v = h \cdot c/\lambda \qquad (1)$$

*Corresponding author: Langhals H, LMU University of Munich, Department of Chemistry, Butenandtstr-13,D-81377 Munich, Germany
E-mail: Langhals@lrz.uni-muenchen.de

Electromagnetic radiation consists of individual quanta where their frequencies and wavelengths, respectively, are related to the energy E according to Einstein's formula (1) where h is Planck's constant. The individual quanta are of minor importance for radio transmitters because of their low energy; sometimes consequences of the quantization are observed in the amplification low-level signals. However, the energy of the individual quanta increases for higher frequencies, finally reaches the energy of the chemical bonds in technical materials, can cleave the bonds and can cause damage of the material. Such radiation is named ionizing radiation. The energy in chemical reactions is commonly reported for a mole of material. As a consequence, equation (1) for individual quanta has to be multiplied with Avogadro's number to obtain equation (2a) and (2b), respectively, for a general purpose comparing with other chemical processes.

$$E = 28591 \text{ kcal·mol}^{-1}/\lambda \qquad (2a)$$

$$E = 119700 \text{ kJ·mol}^{-1}/\lambda \qquad (2b)$$

The energy of quanta reaches the energy of standard chemical bonds at about 1 PHz (1000 THz) in the visible and in the UV; see Figure 1. However, there is no strict limit because very special chemical structures allow radiation-induced chemical processes even in the NIR close to the visible and find applications in photographical infrared films [4]. With increasing frequencies, more and more structures become suitable for radiation-induced processes in the visible and even more in the UV until becoming ubiquitary by X-rays. This causes a damage of materials by ionization and propagates with increasing doses of radiation. As a consequence, the visible becomes attractive for advanced technology of electromagnetic radiation because of the high frequencies of operation and the possibility of controlling the stability of setups by the construction of suitable chemical structures. Moreover, more labile chemical structure may be integrated as an interface to other physical effect such as the construction of sensors [5].

Devices

The very high frequencies of a technology at about 1 PHz require fundamental alterations of the concept for the components of devices. The technology of radio frequencies is dominated by metallic conductors for antennae, wave-propagating lines, resonating circuits and macroscopic (large, classical) switching devices such as transistors [6]. The short wavelength of visible radiation of about half a micron requires $\lambda/4$ antennae with lengths of slightly more than 100 nm and the structure-giving dimensions should be smaller by two powers of ten or even more. Such requirements are beyond the scope of the present nanotechnology by far. Extensions of one nm or less mean the dimensions of molecular structures. Moreover, such small dimensions cannot be altered continuously as in macroscopic, classical structures but only stepwise because of limitations by the extension of chemical bonds such as about 0.15 nm for carbon-carbon bonds.

The molecular approach to petahertz devices would be an alternative where complex chemical structures can be constructed by well-established chemical synthesis (bottom-up approximation instead top-down approximation). Organic molecules are the most attractive structures for such purposes because they can form stable complex three dimensional frameworks with well-defined rigid structures as a σ-skeleton formed by localized covalent centrosymmetric C-C-bonds. Electronic functions can be introduced into the σ-skeleton by additional π-bonds forming double bond structures [7] where the alternation of singles and double bonds (conjugated systems) allows shifting of electrons as is known for metallic conductors and can be applied as antennae for electromagnetic radiation. Generally, the dimensions

of preparatively efficiently accessible and well-defined operating structures of that type extend between 1 and 3 nm where the lengths of such antennae are generally too short with respect to the wavelengths of the interacting radiation and have to be considered for construction and operating. Classical, macroscopic resonating circuits are realized by means of capacitors and coils for radio frequencies or lower [6]. This is difficult to realize for molecular structures because of the restricted variation of structure limited by the arrangements of atoms. Electronic transitions between molecular eigenvalues, molecular energy levels, can be an alternative; these are given by the special chemical structure of the applied molecules. This concept leads to the situation where the electromagnetic radiation of wavelengths of half a micron is still macroscopic (large, classical), whereas both the antennae and the resonators are microscopic (small) in dimensions where physical effects are dominated by quantum mechanic rules.

Chemical structures absorbing visible light are well established in organic dyes [8]. Their chromophores (light-absorbing structures) may be adapted to the special requirements for components of petahertz technology. The chemical and photochemical stability of such materials is of central importance for the realization of long-term operating devices where extended aromatic structures [9] combine both requirements, a solid σ-framework (firm, localized covalent bonds) and π-bonds for operating, both stabilized by aromatism. Strongly fluorescent structures where the absorbed light is re-emitted are preferred as components because there is no leakage of the absorbed energy until re-emission caused by the natural transition probability. Suitable and attractive chromophores of this kind are the *peri*-arylenes [Figure 2] [2,10-12].

Chromophores: *peri*-Arylenes

Figure 2 indicates the general chemical structure of *peri*-arylenes [2]. The light absorption is strongly influenced by the length of the chromophore where the absorption for $n=1$ is in the UV, for $n=2$ in the middle of the visible, for $n=3$ in the region of the visible at long wavelengths and extends for $n=4$ until the NIR; see Figure 3.

The thermal, chemical and photochemical stability of the *peri*-arylenes is extraordinarily high and means a good prerequisite for the construction and application of complex functional structures. The purely organic components render recycling uncritical and make such materials suitable for industrial mass production. Moreover, strong fluorescence is observed for homologous with $n>1$ and means that the energy of the absorbed radiation remains on the chromophores until natural decay with a lifetime of about 4 ns proceeds by emission of fluorescence light. The substituents R in Figure 2 are decoupled

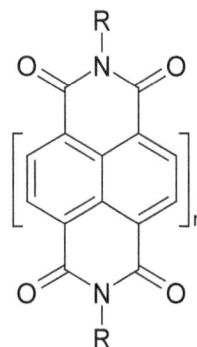

Figure 2: *peri*-arylenes **1**.

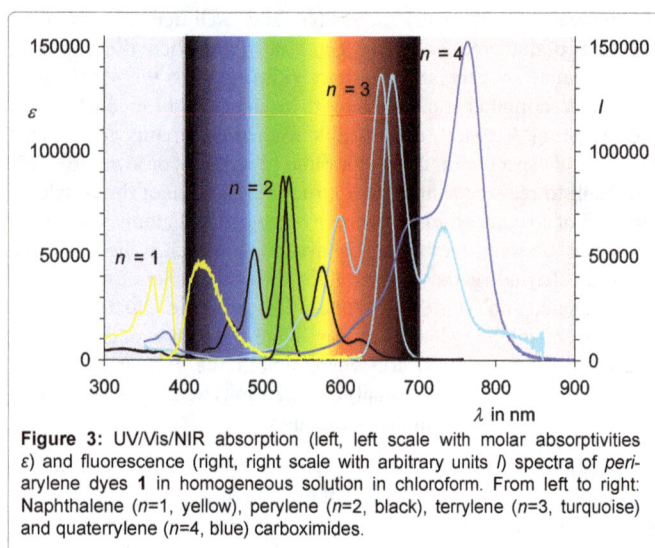

Figure 3: UV/Vis/NIR absorption (left, left scale with molar absorptivities ε) and fluorescence (right, right scale with arbitrary units *I*) spectra of *peri*-arylene dyes **1** in homogeneous solution in chloroform. From left to right: Naphthalene (*n*=1, yellow), perylene (*n*=2, black), terrylene (*n*=3, turquoise) and quaterrylene (*n*=4, blue) carboximides.

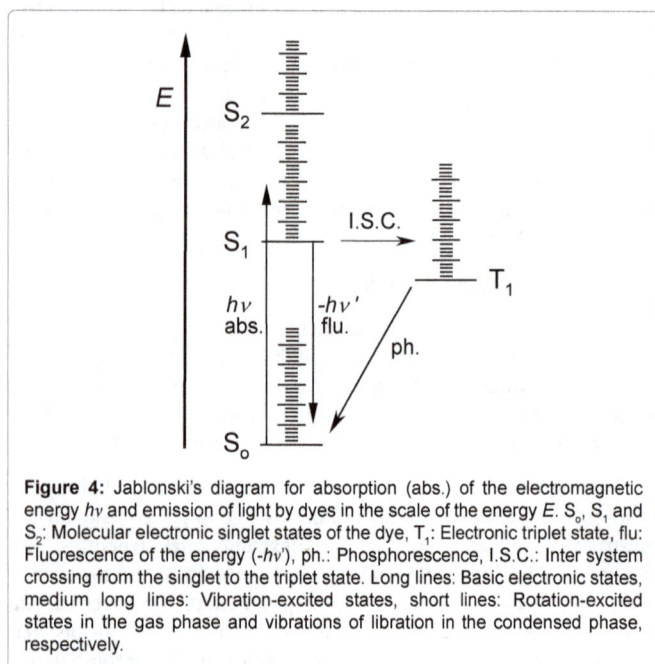

Figure 4: Jablonski's diagram for absorption (abs.) of the electromagnetic energy $h\nu$ and emission of light by dyes in the scale of the energy *E*. S_o, S_1 and S_2: Molecular electronic singlet states of the dye, T_1: Electronic triplet state, flu: Fluorescence of the energy (-$h\nu'$), ph.: Phosphorescence, I.S.C.: Inter system crossing from the singlet to the triplet state. Long lines: Basic electronic states, medium long lines: Vibration-excited states, short lines: Rotation-excited states in the gas phase and vibrations of libration in the condensed phase, respectively.

from the light-absorbing process by nodes in the orbitals (segment for the population by electrons) responsible for light absorption. As a consequence, these positions can be used for the linking to other chromophores or to operating structures without interfering the optical properties of **1**. Long-chain *sec*-alkyl substituent R (swallow-tail substituents such as **1**-hexylheptyl) renders such materials soluble [13]. The absorption spectra of **1** can be tuned by means of substituents at the aromatic nuclei and thus, can be adapted to special requirements of the operating structure [2].

Absorption, fluorescence and phosphorescence

The electronic systems of chromophores such as **1** form distinct allowed molecular energy levels such as S_o, S_1, S_2 and so on and are known as eigenvalues determined by quantum mechanics; see Figure 4 (these terms may be differently used in other subjects such as solid-state physics). The energy of radiation according to formula (1) can be resonantly transferred to chromophores inducing electronic transitions

between eigenvalues such as between S_o and S_1. At room temperature, the dye molecules remain in the electronic ground state S_o because the upper molecular electronically excited states such as S_1 cannot be thermally reached. There are also vibration levels of the chromophores (medium long lines in Figure 4) forming more dense ladders of states beginning at the individual electronic states. Finally, there are even more densely packed states of rotational levels in the gas phase and less well defined and overlapping vibrations of libration in the liquid phase (small lines in Figure 4). The latter form quasi continua rather than individual lines being responsible for the continuous spectra of absorption and fluorescence in Figure 3.

Electronic transitions with absorption of electromagnetic waves at room temperature ($h\nu$ abs. in Figure 4) start at the basic electronic S_o state and reach electronically excited states with possibly excitation of vibration and vibration of libration where there is a limit at low energy (low frequency and short wavelengths, respectively) determined by the energetic difference between the S_o and S_1 ground states; this corresponds to the band gap in semiconductors. The degrees of freedom both of vibration and vibration of libration and even of further electronic excitation are strongly coupled with the electronic excitation; as a consequence, a fast relaxation with the dissipation of the excess energy proceeds within femtoseconds to the S_1 ground state; the latter exhibits a longer lifetime of commonly several nanoseconds. The S_1 ground state is the starting point for further processes such as photochemistry and fluorescence (-$h\nu'$ flu. in Figure 4). Electrons are Fermi-type particles where each energetic level can be occupied by maximal two electrons with antiparallel spin resulting in the singlet S_o ground state reported in Figure 4. The electrons of the highest occupied orbital were re-located in two different orbitals as a consequence of the electronic excitation so that two realizations, the S_1 and the T_1 state, become possible where the electron spins of the latter are parallel. An allowed photo-induced transition requires the retaining of the spin so that the S_1 state will be reached. Relaxation processes (I.S.C., inter system crossing in Figure 4) allow the formation of the energetically lower lying and longer living T_1 state [14] where phosphorescence (ph. in Figure 4) may occur as light-emission. The similar energetic ladders of vibronic states in S_o and S_1 causes a mirror-type structuring of the absorption and fluorescence spectra; see Figure 3. However, this mirror-type is not perfectly realized where the intensities of the higher vibronic bands are relatively more intense in absorption than in fluorescence. This is a consequence of the mismatch between the small molecular dimensions of about 1 nm as antennae with the dimensions of the wavelengths of light of about 500 nm. This is even more pronounced in fluorescence than in the absorption because of longer wavelengths. A quantitative relation between lengths and efficiency of short antennae is given by Rüdenberg's equation [15-17] where the efficiency given by the radiation resistance is proportional to the square of quotient of the effective lengths of a radio antenna over the wavelengths and finds in counterpart in the similar Ross' equation [18] in quantum chemistry.

Line shape of UV/Vis spectras

A single relaxation process would lead to an exponential damping of the propagating electromagnetic wave in the domain of time; this would result in a Lorentzian line shape in the frequency domain (see Fourier transformation); such line shape is known for simple resonating circuits. However, multi relaxation processes are possible concerning the interaction of chromophores with UV/Vis radiation. As a consequence, a Gaussian line shape is expected such as is given in equation (3) where E_{max} is the absorpivity at the maximum of the band and $E_{\nu,\lambda}$ at the individual frequency ν and the wavelength λ, respectively. ν_{max} and λ_{max}, respectively, are the positions of maximum of the band. σ

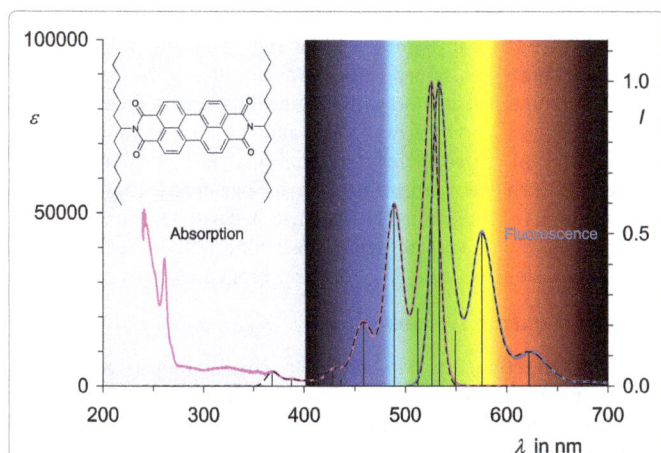

Figure 5: Absorption (magenta curve, ε on the left side) and fluorescence (blue curve, normalized intensity I on the right side) spectra of **1** ($n=2$, R=1-hexylheptyl) in chloroform. Black-dashed lines: Simulated spectra on the basis of a Gaussian analysis according to equation (3); Bars: Positions and intensities of the individual Gaussian bands for $\lambda > 350$ nm.

is the standard deviation and characterizes the extension of the bands. The compatibility of equation (3) with experimental data was tested [19,20] where a transformation according the right part of equation (3) is recommended [21] because UV/Vis spectra are commonly linearly recorded in wavelengths. The factor 100 in the exponent is useful because the wavelength is commonly reported in nm and extensions of bands such as σ in kK (10000 cm^{-1}) (Figure 5).

$$E_{v,\lambda} = E_{max}.e^{\frac{(v - v_{max})^2}{2\sigma^2}} = E_{max}.e^{\frac{100.(\frac{1}{\lambda} - \frac{1}{\lambda_{max}})^2}{2\sigma^2}} \qquad (3)$$

Typical structured absorption and fluorescence spectra were found for **1** ($n=2$, R=1-hexylheptyl) in homogeneous solution in chloroform; see Figure 5. The spectra could be split into individual Gaussian bands according to equation (3) as is shown in Figure 5 ($E_{v,\lambda}$ and E_{max} have to be replaced by $I_{v,\lambda}$ and I_{max} for fluorescence). The re-calculated spectra on the basis of these bands perfectly fit the experimental spectra (see dashed black curves in Figure 5). The Gaussian line-shape characterizes more generally the absorption and fluorescence bands of dyes as is shown with many examples [21].

Absorption and reflectance

The common linear optical behaviour of matter as a function of time is described by equation (4) [2], where E is the electrical field and P the resulting polarization.

$$\mu.P^{..} + f.P^{.} + D.P = N.q^2.E \qquad (4)$$

μ is a measure for inertia, f for friction, and D for elasticity concerning the applied field, where as $N \cdot q^2$ characterizes the interacting charge. Equation (4) can be simplified for the response of a single light quantum when $E=0$ and by setting the quotient f/μ to $1/\tau$, the reciprocal of a characteristic lifetime and D/μ to ω^2, the square of a characteristic frequency, the plasma frequency; thus, equation (5) results.

$$P^{..} + \frac{P^{.}}{\tau} + \omega^2.P = 0 \qquad (5)$$

The solution of equation (5) is given by equation (6) where A_1 and A_2 are the two constants of integration and t is the time.

$$P = e^{-\frac{t}{2\tau}}.(A^1.e^{\sqrt{\frac{1}{4\tau^2} - \omega^2}.t} + A_2.e^{-\sqrt{\frac{1}{4\tau^2} - \omega^2}.t}) \qquad (6)$$

The radicands of the square roots in equation (5) are negative for the majority of materials in the visible because of the very high frequency ω of about 1 PHz (1 000 THz). Thus, the exponents become imaginary and represent a propagating wave for the underdamped solution (oscillatory solution) of equation (4). The first exponential factor in equation (6) represents the damping of the wave and corresponds to the damping according to Lambert Beer's law where the light flux ϕ is proportional to the square of the electric field respectively.

$$E = \varepsilon \cdot c \cdot d \qquad (7)$$

An increase of absorptivity and a concomitant increase in $1/\tau$ and $1/\tau^2$, respectively, lead finally to the critically damped solution of equation (5) because the square roots in equation (6) become zero or even positive for a further increase. As a consequence, there is no propagating wave in the material and the reflection of the wave becomes dominant; compare ref. [22-24]. For this limiting behaviour, the pre-exponential damping factor can be compared with absorptivity E (a confusion between E and the electrical field E should be avoided) as the damping according to Lambert-Beer's law (7) where ε is the molar absorptivity, c the chemical concentration of chromophores and d the path lengths of the propagating wave corresponding to t in (6). A critical molar absorptivity ε for a pure dye is defined by equation (8) where n_v is the real component of the index of refraction, M_n the molecular weight, λ_o is the vacuum wavelength and ρ_d the density of the dye.

$$\varepsilon = \frac{4\pi \cdot n_v \cdot M_n}{\lambda_o \cdot \rho_d \cdot \ln 10} \qquad (8)$$

This limit might be reached by highly absorbing organic dyes [25,26]. The perylene tetracarboxylic bisimides (**1**, $n=2$) [11,12] and the anhydride **2** are well-known for their high molar absorptivities and light-fastness and are therefore suitable for testing the application of equation (6) and (8), respectively (Figure 6).

The anhydride **2** exhibits a density of 1.84 g·cm^{-3} as a solid and a molecular weight of 674. The critical absorptivity according to equation (6) is obviously not reached for the solid powder and thus, **2** is generally employed as a red pigment selectively absorbing light; see Figure 7a (Figure 7).

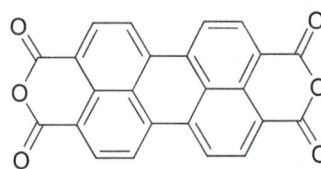

Figure 6: perylene tetracarboxylic bisimides anhydride **2**.

Figure 7: Left (a) pigment **2**: Red powder in the left and a golden shiny plate in the right after the application of pressure (7500 bar with evacuation) by means of a polished piston. Right (b) pigment **1** ($n=2$, R=H): black powder in the left and metallic plate after the application of pressure.

However, the density and packing of **2** can be increased by the application of pressure with 7500 bar (110000 psi) with evacuation to obtain a metallic golden shiny material. A golden shiny mirror-like plate was obtained with a polished press die; see Figure 7a. Obviously, the critical damping according to equation (4) could be exceeded with the higher density and packing. The increase of density by the application of pressure could be verified with other dyes such as for the perylene bisimide (**1**,n=2, R=H, pigment red 179), see Figure 7b for the pigment powder and the compressed material with a more dark bluish metallic lustre, and the DPP dye irgazin red 254. Nearly no effects are observed for malachite green and for crystal violet because the chromophores are highly light-absorbing and the pure crystals seem to be already so densely packed that they form metallic shining solids.

The opposite effect can be obtained with solid metallic gold forming a mirror-like reflecting material because of the high absorptivity of gold for visible light. A dilution of gold by silicate forms the well-known gold ruby glass selectively absorbing light with no reflection.

Interacting of chromophores

A single chromophore acts as a single resonator for electromagnetic waves. A complex behaviour is observed for interacting chromophores where two identical chromophores cause a splitting of the initial band (black curves in Figure 8, upper left) into two novel bands at shorter and longer wavelengths known as the Davydov splitting [27,28] by exciton interactions. The intensities of the absorption bands depend on the orientation of the transition dipoles schematically indicated as thick, black lines in Figure 8.

Absorption at long wavelengths become favoured for collinear arrangements of dipoles and the transition at shorter wavelengths suppressed (red line in Figure 8 upper right). On the other hand, a coplanar arrangement suppresses the transition at long wavelengths and favours the transition at short wavelengths (blue line in Figure 8 lower left). Finally, a skew-type arrangement allows both transitions (Figure 8 right, bottom) where the relative intensities and positions depend on the exact geometry. An orthogonal arrangement extinguishes the exciton interactions; this could be demonstrated with a trichromophoric dye [29] where a Cartesian-named backbone [30] places the three chromophores orthogonally into the three room directions.

Resonance energy transfer

The interaction of two different chromophores allows the establishment of an energetic gradient where the chromophore operating at shorter wavelengths (the donor D) allows the absorption of electromagnetic radiation with higher energy. The subsequent resonance energy transfer as a consequence of dipole dipole interactions transfers the absorbed energy between the components. The fast thermal dissipation of excess vibronic energy causes that the absorbed energy finally is set on the chromophore with the lower energy between the S_1 and S_o named the acceptor (A). The energy transfer was firstly [31] described as a consequence of a resonant interaction of the dipoles of the energy donor D and the energy acceptor A by Perrin [32]. Förster was able to develop Formula (9) for the quantitative description [33-36] of such a process where k_{FRET} is the rate constant for the process; such processes are generally named FRET (Förster Resonant Energy Transfer).

$$k_{FRET} = \frac{1000.(ln10).\kappa^2.J_{DA}.\phi_D}{128.\pi^5.N_A.\tau_D.|R_{DA}|^6} \qquad (9)$$

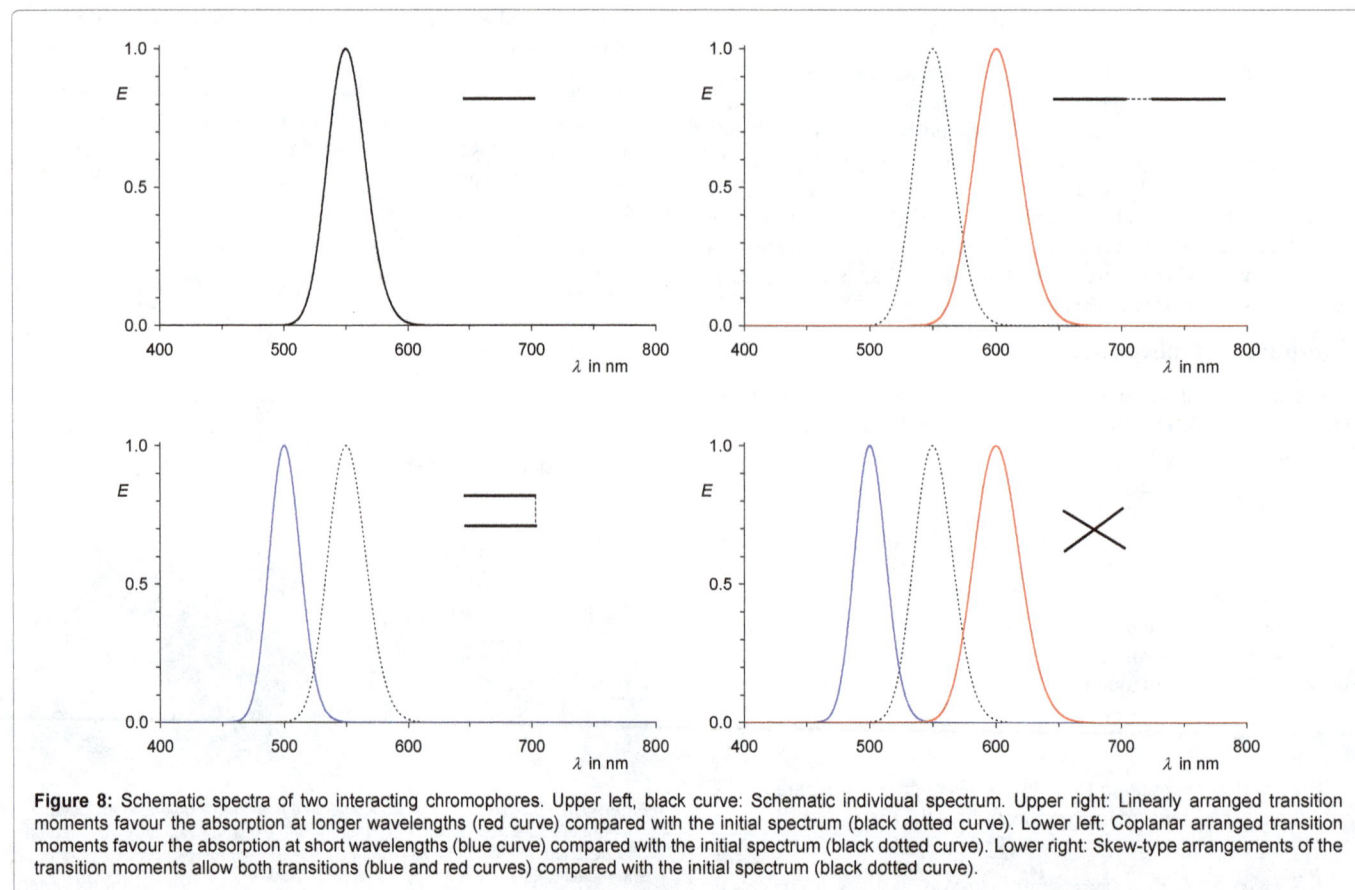

Figure 8: Schematic spectra of two interacting chromophores. Upper left, black curve: Schematic individual spectrum. Upper right: Linearly arranged transition moments favour the absorption at longer wavelengths (red curve) compared with the initial spectrum (black dotted curve). Lower left: Coplanar arranged transition moments favour the absorption at short wavelengths (blue curve) compared with the initial spectrum (black dotted curve). Lower right: Skew-type arrangements of the transition moments allow both transitions (blue and red curves) compared with the initial spectrum (black dotted curve).

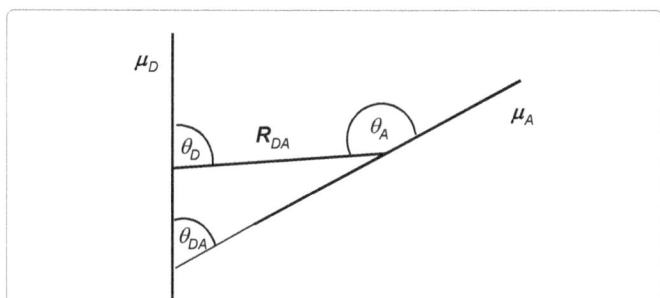

Figure 9: Orientation of the transition dipoles of the energy donor μ_D and acceptor μ_A and the interconnecting vector R_{DA} (projection into the plane).

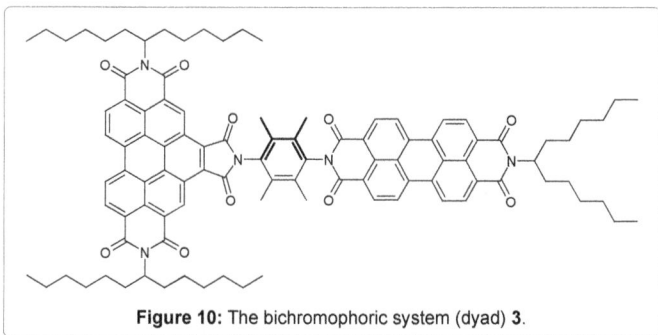

Figure 10: The bichromophoric system (dyad) **3**.

There are mathematical terms and natural constants such as Avogadro's number (N_A) in equation (9) as well as the overlap integral J_{DAv} between the fluorescence spectrum of the energy donor (D) and the absorption spectrum of the acceptor (A). There are two chemically important physical variables, the distance R between the middle points of the transition dipoles (see Figure 9) and the orientation factor κ. The dependence of the rate constant with the sixth power of the distance allows as well an efficient controlling of the process as the determination of distances and changes of distances in molecular dimensions such as with a molecular ruler (Figure 9).

$$\kappa = \left(\hat{\mu}_D . \hat{\mu}_A\right) - 3\left(\hat{\mu}_D . \hat{R}_{DA}\right).\left(\hat{R}_{DA} . \hat{\mu}_A\right) \qquad (10)$$

$$\kappa = \cos\left(\theta_{DA}\right) - 3\left(\cos\left(\theta_D\right).\cos\left(\theta_A\right)\right) \qquad (11)$$

The orientation factor κ is formed by a sum of scalar products of the transition dipole moments and the interconnecting vector according to equation (10) where $\kappa^2 = 2/3$ is obtained for statistical orientation. Equation (10) can be simplified to the arrangement of cos functions in equation (11); for the angles in (11) see Figure 7. The orientation factor κ should become zero for orthogonally arranged transition dipoles if the interconnecting vector of the middle points of the dipoles is also orthogonal to one of the vectors where the energy transfer is expected to be extinguished (Figure 10).

The bichromophoric system (dyad) **3** was synthesized [37] for a test where the orthogonal arrangement according to equation (11) was fulfilled because the transition moment are parallel to an N-N-connection line both for the left chromophore as the energy donor (D) and the right chromophore as the energy acceptor (A). An extinguishing of the energy transfer is expected according to equation (11), however, a very fast (9.4 ps) and efficient ($\approx$100%) energy transfer was observed for **3** and similar molecules [38,39]. More recent results indicate that the dependence of the rate constant k_{FRET} on the distance R is better described with an R^{-3} dependence for proximate chromophores and alters to an R^{-6} dependence for more distant chromophores [40].

The dipole dipole interactions seem to be important for the energy transfer, however, they are not the exclusive process and molecular dynamics seem to be equally important. This has to be considered for the construction of complex interacting chromophoric systems. On the other hand, the energy transfer allows the construction of controlled molecular pathways such as in conventional electronics.

Molecular metamaterials

The electric component μ_e is dominating in molecular optical transitions. The introduction of a magnetic component μ_m may introduce non-conventional optical effects. An orthogonal arrangement of μ_e and μ_m is of special interest because molecular components for optical metamaterials are obtained where negative indices of refraction can be established [41]. Such metamaterials are popular because of the invisible cloak (magic cap, magic cloak), however, also may be important for the construction of very thin lenses and other advanced optical components. Metamaterials were constructed for microwaves with split ring resonators [42,43], however, a molecular realization of such a component is difficult. On the other hand, metamaterials can be established by means of two parallel resonators for the electric component μ_e where the magnetic component μ_m will be formed by the dielectric displacement current in between; see Figure 11, (Figure 12).

An arrangement according to Figure 11a was realized on a molecular scale in **4** [44] where the terminal chromophores represent the electric conductors because of their pure electric transition moments polarized

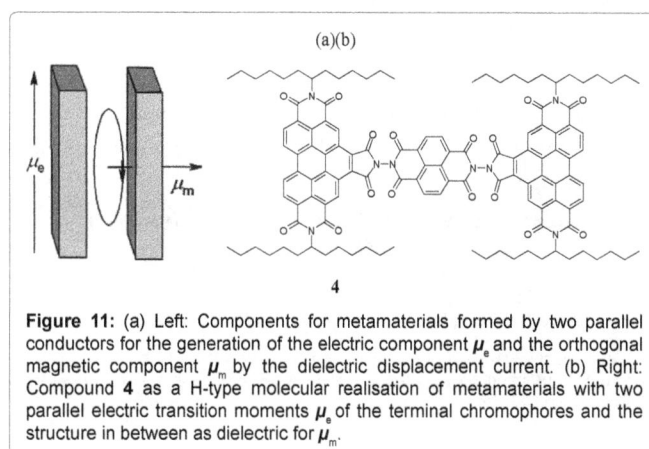

Figure 11: (a) Left: Components for metamaterials formed by two parallel conductors for the generation of the electric component μ_e and the orthogonal magnetic component μ_m by the dielectric displacement current. (b) Right: Compound **4** as a H-type molecular realisation of metamaterials with two parallel electric transition moments μ_e of the terminal chromophores and the structure in between as dielectric for μ_m.

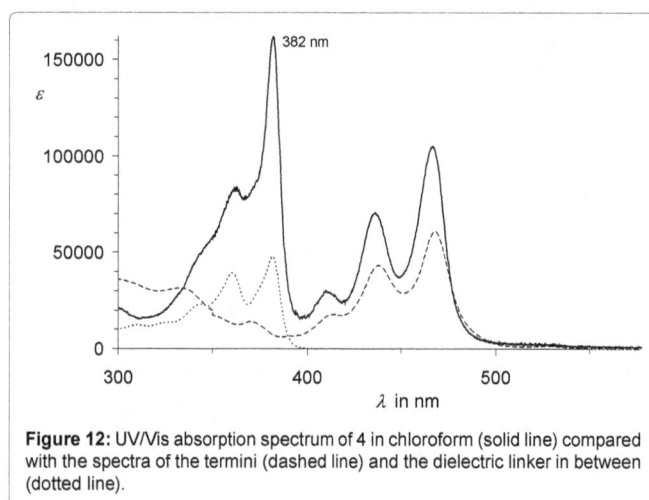

Figure 12: UV/Vis absorption spectrum of 4 in chloroform (solid line) compared with the spectra of the termini (dashed line) and the dielectric linker in between (dotted line).

along their long axis. The interlinking structure in the middle of the H forms the dielectric (it absorbs in the UV). The UV/Vis spectrum of **4** mainly consists of an addition of the spectra of the components except one sharp, intense additional absorption at 382 nm; this is attributed to the special H-type meta arrangement of the chromophores in **4**. As a consequence, molecular components for metamaterials can be constructed by special arrangements of chromophores. A special pattern of these molecular components may form bulk optical metamaterials; the application of liquid crystals is an attractive way therefore.

Conclusions

A continuation of terahertz electromagnetic radiation to higher frequencies and shorter wavelengths, respectively, passes the visible and reaches 1 petahertz in the UVB where the region of 0.1 to 10 PHz may be named petahertz radiation in analogy to the comparable region of terahertz [45]. This may extend the well-established technologies of electromagnetic waves. There, the visible light and with some constraints extended to the UVA is of special technological interest because the very high frequencies just below the problematic ionizing radiation allow comparably hazard-free [46] handling and unproblematic interaction with materials. Many feature of radio waves were found in the region of petahertz radiation because the dimensions of half a micron in the visible are still macroscopic (large and classical), however, chromophores as molecular resonators are microscopic (small) where quantum mechanics have to be considered. Special features can be established with the interaction of two or more chromophores where absorption spectra may be altered by exciton interactions, pathways for energy transport opened by FRET and special optical properties established by means of molecular metamaterials. As a consequence, although many aspects such as carrying optical signals have to be solved, the handling of petahertz radiation with interacting components can be considered as the target for devices of the next-generation technology.

Acknowledgements

This work was supported by the *Fonds der Chemischen Industrie*.

References

1. Lange K, Löcherer KH (1992) Meinke Gundlach. Taschenbuch der Hochfrequenztechnik (5thedn.). Springer Verlag, Berlin, Germany.

2. Langhals H (2013) Chromophores for picoscale optical computers:Sattler K (edn.), Fundamentals of picoscience, Taylor & Francis Inc. CRC Press Inc., Bosa Roca US.

3. Bergmann L, Schaefer Cl, Gobrecht H, Matossi F (1966) Lehrbuch der Experimentalphysik, Band III, Optik. (4thedn.).Walter de Gruyter& Co., Berlin,Germany.

4. Dürr H, Bouas-Laurent H (1990) Photochromism. Molecules and systems. Series: Studies in Organic Chemistry, Elsevier, Amsterdam,Netherlands.

5. Wang X-D, Wolfbeis OS (2013) Fiber-Optic Chemical Sensors and Biosensors (2008–2012). Anal Chem 85: 487-508.

6. American Radio Relay League (2013) The ARRL handbook for radio communications 2013. (90th edn.).The American Radio Relay League, Inc., Newington, CT, 06111-1494,USA.

7. Heeger AJ, SariciftciNS, Namdas EB (2010) Semiconducting and metallic polymers. Oxford University Press, Oxford,England.

8. Zollinger H (1993) Color chemistry. Synthesis, properties and applications of organic dyes and pigments.(3rd edn.).Wiley, Weinheim,Germany.

9. Clar E (1964) Polycyclic Hydrocarbons. Academic Press, London, UK.

10. Langhals H (2008) Molecular devices. Chiral, bichromophoric silicones: Ordering principles in complex molecules. In F. Ganachaud, S. Boileau, B. Boury (edns.). Silicon based polymers,, Springer, Berlin,Germany.

11. Langhals H (2005) Control of the interactions in multichromophores: Novel

12. Langhals H (1995) cyclic carboxylic imide structures as structure elements of high stability. Novel developments in perylene dye chemistry. Heterocycles 40: 477-500.

13. Langhals H, Demmig S, Potrawa T (1991) the relation between packing effects and solid state fluorescence of dyes. J Prakt Chem 333: 733-748.

14. Hunt GR, McCoy EF ,Ross IG (1962) Excited states of aromatic hydrocarbons: Pathways of internal conversion. Aust J Chem 15: 591-604.

15. Rothammel K (1976) Antennenbuch. (5th edn.). Telekosmos Verlag, Stuttgart, Germany.

16. Rüdenberg R (1908) Der Empfang elektrischer Wellen in der drahtlosen Telegraphie. Ann D Phys 330: 446-466.

17. Rüdenberg R (1908) Ann d Phys Leipzig 25: 466-500.

18. McCoy EF, Ross IG (1962) Electronic states of aromatic hydrocarbons: The Franck-Condon principle and geometries in excited states. Aust J Chem 15: 573-590.

19. Langhals H (2000) A re-examination of the line-shape of the electronic spectra of complex molecules in solution. Log-normal function versus Gaussian. Spectrochim Acta 56: 2207-2210.

20. Langhals H (2002) UV-visible spectroscopy and the potential of fluorescent probes. In F. H. Frimmel, Refractory Organic Substances in the Environment, p. 200-214, Wiley-VCH, Weinheim,Germany.

21. Langhals H (2002) The rapid identification of organic colorants by UV/Vis-spectroscopy. AnalBioanal Chem 374: 573-578.

22. Weiser G, Fuhs W, Hesse HJ (1980) Study of polariton resonances in a cyanine dye crystal. Chem Phys 52: 183-191.

23. Penelly RR, Eckhardt CJ (1976) Quasi-metallic reflection spectra of TCNQ single crystals. Chem Phys 12: 89-105.

24. Philpott MR (1973) Advances in chemical physics.(I. Prigogine, S. A. Rice, edns.) 23:227, Academic Press, New York, USA.

25. Schmelzer H (1976) Origin of bronzing in glossy black printing inks. Deutsche Farben-Zeitschrift 30: 277-278; ChemAbstr 85: 162017.

26. Schmelzer H (1976) Cause of bronzing in tinted black printing inks. FATIPEC Congress 1976, 13: 572-574; ChemAbstr 86: 56822.

27. Davydov, AS (1848) Theory of absorption spectra of molecular crystals. Zhur. Eksptl. i Teoret. Fiz. 18: 210-218; ChemAbstr 43: 24604

28. Davydow AS (1962) Theory of Molecular Excitations. Transl. Kasha H.Oppenheimer, Jr., M. McGraw-Hill, New York, USA.

29. Langhals H, Rauscher M, Strübe J, Kuck D (2008) Three orthogonal chromophores operating independently within the same molecule. J Org Chem 73: 1113-1116.

30. Kuck D (2006) Functionalized aromatics aligned with the three Cartesian axes: Extension of centropolyindane Pure Appl Chem 78: 749-775.

31. Clegg RM (2006) The history of FRET: From conceptions to the labours of birth. In Reviews in Fluorescence 2006. (C. D. Geddes, J. R. Lakowicz, edns.).Springer US, New York, USA.

32. Perrin JB (1925) Fluorescence et radiochimie Conseil de Chemie. (2nd edn.). Solvay, Paris.

33. Förster T (1946) Energy migration and fluorescence. Naturwiss.33: 166-175.

34. Förster T (1948) Intermolecular energy transference and fluorescence. AnnPhys2: 55-75.

35. Förster T (1949) Experiments on intermolecular transition of electron excitation energy. Z. Elektrochem. 53: 93-99; ChemAbstr43: 33629.

36. Förster T (1949) Experimental and theoretical investigation of intermolecular transfer of electron activation energy. Zeitschr. Naturforsch. 4a: 321-327; ChemAbstr44: 43074.

37. Langhals H, Poxleitner S, Krotz O, Pust T, Walter A (2008) FRET in orthogonally arranged chromophores. Eur. J Org Chem 73: 4559-4562.

38. Langhals H, Esterbauer AJ, Walter A, Riedle E, Pugliesi I (2010) Förster

concepts. Perylene bisimides as components for larger functional units.Helv Chim Acta 88: 1309-1343.

resonant energy transfer in orthogonally arranged chromophores. J Am Chem Soc 132: 16777-16782.

39. Pugliesi I, Walter A, Langhals H, Riedle E (2011) Highly efficient energy transfer in a dyad with orthogonally arranged transition dipole moments: Beyond the Limits of Förster. In M. Chergui, D. Jonas, E. Riedle, R. W. Schoenlein, A. Taylor (edn.). Ultrafast Phenomena XVII, Oxford University Press, Inc., New York, USA.

40. Nalbach P, Pugliesi I, Langhals H, Thorwart M (2012) Noise-induced Förster resonant energy transfer between orthogonal dipoles in photoexcited molecules. Physical ReviewLett108: 218302(1)-218302(5).

41. Liu N, Giessen H (2010) Coupling effects in optical metamaterials. AngewChemInt Ed 49: 9838-9852.

42. Pendry JB, Holden AJ, Robbins DJ, Stewart WJ (1999) IEEE Trans. Microwave Theory Tech. 47: 2075-2084.

43. Linden S, Enkrich C, Wegener M, Zhou JF, Koschny T et al.(2004) Magnetic Response of Metamaterials at 100 Terahertz. Sci 306: 1351-1353.

44. Langhals H, Hofer A (2013) Chromophores arranged as magnetic meta atoms: Building blocks for molecular metamaterials. J Org Chem 78: 5889-5897.

45. Perenzoni M, Paul DJ, eds. (2014) Physics and applications of terahertz radiation.Springer Series in Optical Sciences 173, Heidelberg, Germany.

46. Fitzpatrick, TB (1974) Sunlight and man. Normal and abnormal photobiologic responses.University Press, Tokyo, Japan.

15

Design of Efficient Linear Feedback Shift Register for BCH Encoder

Aiswarya S[1]*, Manikandan SK[2] and Aravinth S[1]

[1]Embedded System and Technologies Velalar College of Engineering and Technology Erode, Tamilnadu, India
[2]Department of Electrical and Electronics Engineering Velalar College of Engineering and Technology, Erode, Tamilnadu, India

Abstract

The sequential circuit designed was Look-Ahead Transformation based LFSR in which a hardware complexity was present and it may limit their employ in many applications. The design of efficient LFSR for BCH encoder using TePLAT (Term Preserving Look-Ahead Transformation) overcame this limitation by opening the employ of minimizing the iteration bound and hardware complexity in wide range of applications. A TePLAT convert LFSR formulation behaves in the same way to achieve much higher throughput than those of a native implementation and a Look-Ahead Transformation-based.

Keywords: Linear Feedback Shift Register (LFSR); Term preserving look-ahead transformation iteration (TePLAT) bound; Loop unrolling; Look-Ahead transformation

Introduction

Linear Feedback Shift Register is used to generate test vectors. It uses feedback and modifies itself on every rising edge of clock. LFSR algorithms have found wide applications in wireless communication, including scrambling, error correction coding, encryption, testing, and random number generation. A LFSR is specified by its generator polynomial over the Galois Field GF (2). Some generator polynomials used on modern wireless communication applications are summarized in Tables 1 and 2 [1,2] respectively.

Traditionally LFSR can be implemented in hardware. But due to complexity in hardware LFSR can be implemented in software defined radios [3]. Due to the mismatch of data types between the bit-serial operations of the LFSR and the word-based data path, it has been reported that 33 percent of CPU cycles of those for implementing an OFDM transmitter are dedicated to the scrambler operations. The software-implementation of the LFSR algorithm is also too slow to support real-time implementation of the 802.11 standard. This work will focus on efficient implementation of LFSR for BCH Encoder. The first approach aims at increasing execution.

Speed at the expense of additional special purpose hardware [4]. These hardware units may interface with the host microprocessor via instruction set extensions or interrupt. The second approach seeks to reformulate LFSR algorithm so that inherent bit-level parallelism afforded by a word-based micro architecture may be fully exploited [5]. Since a word may be regarded as a vector of binary bits, traditional vector zed compilation techniques such as loop unrolling [6] may be applied. The iteration bound is the inverse of theoretical maximum throughput rate an algorithm may achieve. Many LFSR polynomials such as those listed in Tables 1 and 2 have rather large loop bounds and hence cannot take full advantage of the benefit of unrolling. Fortunately, a look-ahead transformation (LAT) [6] promises to resolve this difficulty. However, LAT comes with a price: it often introduces additional operations. For LFSR, this implies LAT-transformed LFSR formulation may contain many more terms [7] than the original LFSR. These overhead may offset the potential benefit of applying LAT.

The main contribution of this work is on exploiting the low overhead property of term-preserving look-ahead transformation (TePLAT) which guarantees the number of terms of the transformed generator polynomial will remain unchanged. This term preserving property makes it feasible to apply TePLAT aggressively to achieve maximum throughput rate with respect to a particular micro architecture discussed in the context of parallel recurrent equations. This work provides critical implementation details such as initial conditions, experimental outcomes as well as applications to specific SDR platforms. The speedup factor varies from 1.5 to 18 depending on the structure of the generator polynomials.

LFSR Index	Generator Polynomial
1	$1+X+x^3$
2	$1+X+x^4$
3	$1+x^2+x^5$
4	$1+X+x^6$
5	$1+x^4+x^5+x^6+x^7$
6	$1+X+x^5+x^6+x^8$
7	$1+x^4+x^9$
8	$1+x^3+x^{10}$
9	$1+x^3+x^4+x^7+x^{12}$
10	$1+x^3+x^{16}$
11	$1+x^5+x^{12}+x^{16}$
12	$1+x^5+x^{23}$
13	$1+x^2+x^3+x^7+x^{32}$
14	$1+X+x^2+x^4+x^5+x^7+x^8+x^{10}+x^{12}+x^{16}+x^{22}+x^{23}+x^{25}+x^3$
15	$1+x^{10}+x^{33}$
16	$1+x7+x^{42}$
17	$1+x^{35}+x^{42}$
18	$1+x^9+x^{49}$
19	$1+x^{49}+x^{52}$
20	$1+x^{35}+x^{74}$
21	$1+x^{18}+x^{29}+x^{42}+x^{57}+x^{67}+x^{80}$

Table 1: Common LFSR-Polynomials in [7] and [8].

*Corresponding author: Aiswarya S, M.E Student, Embedded System and Technologies Velalar College of Engineering and Technology Erode, Tamilnadu, India, E-mail: aishu1309@gmail.com

Wireless Standards	Generator Polynomial
Wi -Fi	$1+x^4+x^7$
Wimax	$1+x^{14}+x^{15}$
LTE (Gold Code)	$1+x^{28}$
	$1+x^{28}+x^{29}+x^{30}+x^{31}$

Table 2: LFSR as Scrambler in SDR.

Background and Definitions

Linear feedback shift register

LFSR is a shift register whose input bit is a linear function of its previous state. The XOR gate is used to provide feedback to the register that shifts bits from left to right. The maximum sequence contains all possible state except the "0000" state. Normally XOR gate is preferred for linear function of single bits. Thus, an LFSR is often a shift register whose input bit is driven by the exclusive-or (XOR) of some bits of the overall shift register value.

An LFSR can be specified by its generator polynomial over a Galois field GF (2)

$$P(x) = 1 + \sum_{k=1}^{K} p_k \chi^k \tag{1}$$

Where both x and $pk \in \{0,1\}$, and K is the order of P(x). Each generator polynomial uniquely characterizes a linear difference equation in GF (2).

$$Y[n] + \sum_{k=1}^{K} p_k.y[n-k] \tag{2}$$

Where $y[n] \in \{0, 1\}$, "+" is the logical XOR (exclusive-OR) operator, "." (Multiplication) is the logical AND operator.

The beginning value of the LFSR is called the seed. Since the operation of the register is deterministic, the large number of values occurring from the register is completely determined by its present (or previous) state. Similarly, the register has a fixed number of possible states; it must finally enter a repeating cycle. However, an LFSR with a better feedback function can generate a sequence of bits which appears without any definite purpose and which has a larger cycle. The existing state may be obtained by right shifting the current state by 1 bit and filling in y[n-K-1]. Substituting n by n -1 into (2), and XO Ring y[n-K-1] on both sides, one has (note that p0 = pk= 1).

$$y[n-K-1] = \sum_{k=0}^{K=1} p_k.y[n-k-1] \tag{3}$$

The series of numbers created by a LFSR or its Exclusive-NOR counter section can be treated as binary just as Gray code or the natural binary code. The settlement of taps for feedback in an LFSR can be declared in finite field arithmetic with a polynomial of mod 2. The coefficients of the polynomial should be 1's or 0's. This in terms known as the reciprocal characteristic polynomial.

In the Galois configuration, when the system is clocked, the bits which are not tapped are right shifted one position unchanged. Before they are stored in the next position the taps, are XORed with the output bit. The new output bit is the next input bit. All the bits in the register shift to the right unchanged, and the input bit becomes zero only when the output bit is zero. When the output bit is one, the bits in the tap positions all flip (if they are 0, they become 1, and vice versa, finally the input bit becomes 1 only when the entire register is shifted to the right.

To generate the large number of same output, the tap order is the similar function (see above) of the order for the normal LFSR; otherwise the stream will be in reverse. It is not necessary that the LFSR's internal should be same. The Fibonacci register and Galois register has the large number of same output as the in the initial section.

The large number of output of LFSR is based on determinism. You can predict the next state only when you know the current state and the arrangement of the XOR gates in the LFSR. It is not possible when random events occur. It is much easier to calculate the next state, with minimal-length LFSRs, as there are only an easily limited number of them for each length. The stream of output is reversible; an LFSR with similar taps will occur through the output sequence in reverse order.

A block diagram of this LFSR is depicted in Figure 1. Applying (4), one has (with K=16).

$$y[n-17] = y[n-1] \oplus y[n-4] \tag{4}$$

Iteration bound

Recursive and adaptive digital filters are belongs to DSP algorithms which contain feedback loops, that impose an inherent essential lower bound on the achievable iterative steps or sample period. Iteration bound is the maximum of loop bound, a fundamental limit for recursive algorithms. Loop bound is the computation time divided by delay element in a loop.

Iteration bound is the inverse of theoretical maximum throughput rate an algorithm may achieve. If no delay element in the loop, then iteration bound is infinite. Clock period is lower bounded by the critical path computation time. Critical path of a DFG is the path with the longest computation time among all paths that contain zero delays.

The data dependence imposes an upper bound on how many times a loop can be unrolled to explore the inherent Inter operation parallelism. Theoretically, this kind of inter operation dependence relation is Characterized by a notion called iteration bound. Roughly, the iteration bound equals to the inverse of the number of iterations that can be unrolled into the same iteration. To increase throughput, the iteration bound must be minimized.

When the DFG is recursive, the iteration bound is the fundamental limit on the minimum sample period of a hardware implementation of the DSP program. Two algorithms to compute iteration bound are Longest Path Matrix Algorithm (LPM) and Minimum Cycle Mean (MCM). In Longest Path Matrix Algorithm (LPM) a series of matrix is constructed and the iteration bound is found by examining the diagonal elements of the matrices.

An arbitrary reference node is chosen in Gd (called this node s). The initial vector f (0) is formed by setting f(0)(s)=0 and setting the remaining nodes off(0) to infinity and find the min average length of the edge in the loop in orderto compute iteration Bound by using Minimum Cycle Mean (MCM) Method.

Loop unrolling

Loop unrolling (loop unfolding) is a well-known compiler

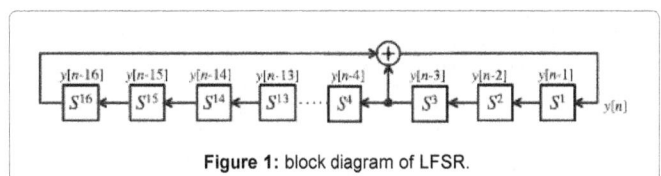

Figure 1: block diagram of LFSR.

optimization technique [3]. It consolidates loop bodies of consecutive iterations into a single iteration to expose inherent parallelism. For example, the LFSR-10 depicted in Figure 2 can be represented as a loop (^: bitwise XOR).

However, loop unrolling cannot achieve arbitrary level of parallelism. Using LFSR-10 as an example, if one wants to unroll the loop three times, instead of two times, the following equation will need to be added into the unrolled loop body.

$$y[n+3] = y[n]^\wedge y[n-13] \tag{5}$$

However, this statement cannot be executed in the same iteration with the statement.

$$y[n] = y[n-3]^\wedge y[n-16] \tag{6}$$

Since y[n] needs to be evaluated first before it can be used to evaluate y [n+3]. This data dependence imposes an upper bound on how many times a loop can be unrolled to explore the inherent interoperation parallelism.

The iteration bound equals to the inverse of the number of iterations that can be unrolled into the same iteration. To increase throughput, the iteration bound must be minimized. In Figure 2, the LFSR-10 has an iteration bound of 1/3. Hence, three successive iterations can be unrolled into the same iteration. Any path in the original DFG containing J or more delays leads to J paths with 1 or more delay in each path. Therefore, it cannot create a critical path in the J-unfolded DFG. Unfolding a DFG with iteration bound T∞ results in a J-folded DFG with iteration bound JT∞. J-unfolding of a loop with wl delays in the original DFG leads to gcd(wl,J) loops in the unfolded DFG, and each of these gcd(wl,J) loops contains wl/ gcd(wl , J) delays and J/ gcd(wl,J) copies of each node that appears in the loop.

Unfolding a DFG with iteration bound T results in a J unfolded DFG with iteration bound JT. Unfolding preserves the number of delays in a DFG. This can be stated as follows:

$$[w / J] + [(w+1) / J] + ... + [(w+J-1)] = w \tag{7}$$

Look ahead transformation

Look Ahead transformation is a kind of block transformation and has the properties of block processing. In look-ahead transformation, the linear recursion is first iterated a few times to create additional concurrency. The iteration bound of this recursion is same as the original version, because the amount of computation and the number of logical delays inside the recursive loop have both doubled Look ahead Approach is also applied for Sequential Nature Decoder Algorithms. Look Ahead technique can enhance its parallel processing or block processing implementations.

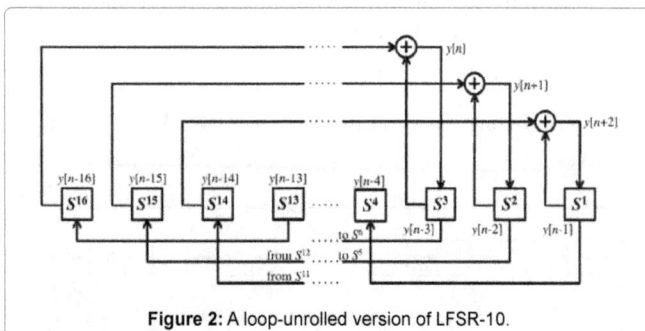

Figure 2: A loop-unrolled version of LFSR-10.

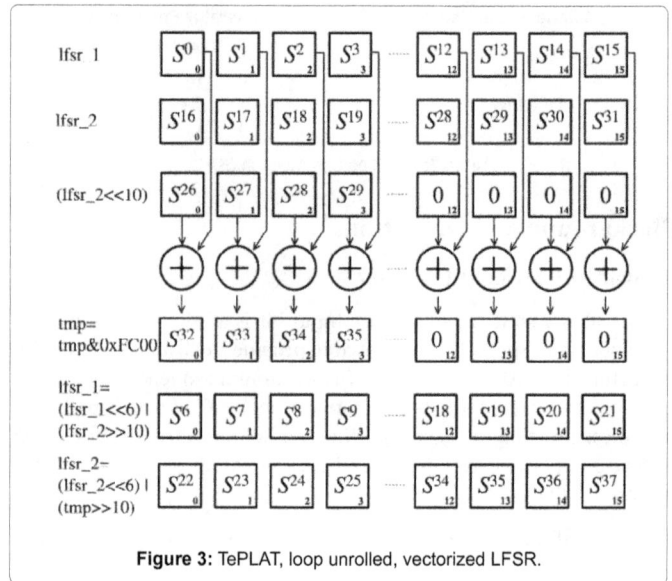

Figure 3: TePLAT, loop unrolled, vectorized LFSR.

Higher-order IIR digital filters can be pipelined by using clustered look-ahead or scattered look-ahead techniques. (For 1st-order IIR filters, these two look-ahead techniques reduce to the same form).In Clustered look-ahead Pipelined realizations require a linear complexity in the number of loop pipelined stages and are not always guaranteed to be stable. Scattered look-ahead can be used to derive stable pipelined IIR filters.

In this the poles of the systems will approach origin. This implies the system is more stable and limit cycle effects are reduced. The data dependency relation will be reduced. A generalized look-ahead transform in GF (2) is

$$Q(x) = P(x)G(x) = P(x) + \sum_{m=1}^{M} g_m x^m P(x) \tag{8}$$

Term Preserving Look Ahead Transformation

TePLAT of a LFSR with a generator polynomial P(x) is a LFSR with a generator polynomial Q(x)=[P(x)] ^2. Although Q(x) is a polynomial of twice the order of P(x), both of them have the same number of terms. Since this property is always true for power of two but may not necessarily hold for other exponents, we only consider power of two in TePLAT.

If the iteration bound of an LFSR is T, then the iteration bound after a TePLAT=T/2. While TePLAT promises full exploitation of bit-level parallelism, one should not lose sight on the fact that the purpose of LFSR implementation is to generate the maximal length pseudo-random bit stream.

Each time the TePLAT is applied, the order of the transformed generator polynomial doubles. As such, twice as many bits of on-chip storage space will be needed to store the increased number of states. To fully exploit inherent bit-level parallelism, auxiliary data format conversion operations such as bit-vector packing and unpacking, and word boundary alignment will be needed (Figure 3).

If the number of states after TePLAT transformation approaches the length of the original maximal length sequence, then it may be more cost effective to simply cache the entire maximal length sequence and save all the computation. Assume that the TePLAT is repeatedly applied m times, termed as mth-level TePLAT thereafter, and then the number of bits to store the states will be 2^m*K bits. After TePLAT parallelize the LFSR to the full length of word, the number of consumed registers doubles each time we apply the technique.

LFSRs are designed for efficient implementation such as Grain stream cipher. They are generally very long and the iteration bounds are small. TePLAT result in high-order generator polynomials, more registers will be required to hold the additional bits. Hence, the memory and register-footprint of executing the LFSR algorithm should be treated as a cost function In this case; the parallelism can be achieved by simply applying loop unrolling.

In terms of cost, since LAT and TePLAT both result in high-order generator polynomials, more registers will be required to hold the additional bits. Hence, the memory and register-footprint of executing the LFSR algorithm should be treated as a cost function. The improvement does not induce any hardware overhead. However, after TePLAT parallelize the LFSR to the full length of word, the number of consumed.

The conventional LFSRs are similar to the applied loop unrolling (LU) technique. TePLAT may also be applied, with careful tradeoffs between area and throughput, to hardware-based LFSR implementation. Assume that the TePLAT is repeatedly applied m times, termed as mth-level TePLAT thereafter, and then the number of bits to store the states will be $2^m K$ bits. Using above argument, one must limit m such that $2m.K \leq 2k$. After simplification, one has m ≤ [K -log2Kc].

Simulation Result

The cycle-accurate simulator can profile convincible outcome for demonstrating this algorithm. Therefore, Texas instruments Inc., Code Composer Studio (CCS) and advanced RISC Machines Ltd. Instruction Set Simulator (ARMulator) is used.

In this work, an in-house source-to-source compiler that generates LFSR codes with TePLAT factors ranging from 2^0 to 2^8 is built. The generated codes on the corresponding simulators are simulated and determined the best TePLAT factor for the LFSR.

The best performance look-ahead transformed LFSR found based on the experiments is termed as "best" in the following results. Popular and representative processor for mobile devices such as TI-C6416 digital signal processor is used. The best look-ahead transformed LFSR's improvement depends on the LFSR generator polynomial and the Processor architecture. A comparison of the optimization technique is provided in Figure 4.

Throughput numbers are given for all the LFSRs. The conventional LFSRs are similar to that applied loop unrolling (LU) technique.

The best look-ahead transformed LFSR's improvement depends on the LFSR generator polynomial and the processor architecture. Our experimental results show that the best LFSR can usually be found by TePLAT factor ranging from 2^0 to 2^8. The best look-ahead transformed LFSRs can perform at most 18 * (LFSR-8) to 50 percent (LTE) faster.

In [8], bit manipulation unit (BMU) hardware was proposed to accelerate a communication DSP and was implemented on XILINX VirtexII. The throughput of Wifi scrambler in [8] is 0.6 bit/cycle. Our method can achieve 0.7 bit/cycle on ARM and 2.9 bit/cycle on TI.

If the throughput is measured using bit/sec (bps), the 180 nm DSP in [8] achieves throughput of 168 Mbps and our proposed framework is 280 Mbps and 1.74 Gbps on 130 nm ARM926 and 130 nm TI C6416, respectively. They are generally very long and the iteration bounds are small. In this case, the parallelism can be achieved by simply applying loop unrolling, and thus our TePLAT methods have negligible improvement.

Conclusion

The Design of Efficient Linear Feedback shift Register is to minimize iteration bound without introducing any additional operations. A term preserving look-ahead transformation (TePLAT) is used for efficient parallel implementation of LFSR in several applications. Compared to existing approaches there will be significant speedup performance and also the hardware utilization will be minimized. Compared to existing approaches, significant speedup has been observed in numerous simulations. TePLAT may also be applied, with careful tradeoffs between area and throughput, to hardware-based LFSR implementation.

References

1. IEEE Std. 802.16e-2005, Part 16: "Air Interface for Fixed Broadband Wireless Access Systems," IEEE Std. 802.16, 2009.

2. Hell M, Johansson T, Meier W, "Grain-A Stream Cipher for Constrained Environment," Int'l J. Wireless and Mobile Computing, 2: 86-93.Mitola J (2011) Cognitive Radio Architecture, John Wiley & Sons.

3. Mitola J (2011) Cognitive Radio Architecture, John Wiley & Sons.

4. Glossner J (2009) "A Software-Defined Communications Baseband Design," Proc IEEE Comm. Magazine, 41: 120-128.Parhi KK (2005) "VLSI Digital Signal Processing Systems Design and Implementation. "John Wiley & Sons, Inc.

5. Tang Y, Qian L, Wang Y (2009) "Optimized Software Implementation of a Full-Rate IEEE 802.11a Complaint Digital Baseband Transmitter on a Digital Signal Processor," Proc. IEEE Global Comm. Conf, 4: 2194-2198.

6. Parhi KK (2005) "VLSI Digital Signal ProcessingSystems Design and Implementation."John Wiley & Sons, Inc.

7. Sriram S, Sundararajan V (2011) "Efficient Pseudo-Noise Sequence Generation forSpread-Spectrum Applications," Proc. IEEE Workshop Signal Processing Systems (SIPS '02) 2: 80-86.

8. Linetal J (2010) "Cycle Efficient Scrambler Implementation for Software Defined Radio," Proc. Int'l Conf. Acoustics, Speech, and Signal Processing (ICASSP '10) 10: 1586-1589.

16

Ultraweak Electromagnetic Wavelength Radiation as Biophotonic Signals to Regulate Life Processes

Hugo J. Niggli*

BioFoton AG, Rte. D'Essert 27, CH-1733 Treyvaux, Switzerland

Abstract

In recent years the low level analysis of ultra-weak photon emission in human cells is achieved using sophisticated Photomultiplier Technique (PMT). The basis of photonic measurements goes back to the theoretical finding of Einstein that a photon, which hits a metal plate, causes an electrical impulse. This current can be detected by single photon detection device as mentioned before. As shown in a variety of analytical laboratories worldwide using this sensitive workhorse it is evident that all cells from plants over animals up to humans emit a low level biophotonic emission. The measured electromagnetic wavelengths of this miniscule 0.01 Femto Watt (10^{-17} W) radiation are ranging from ultraviolet light over the visible up to the infrared region. In order to visualize the size of this very weak light source: the luminous power of a candle in a Lunar Distance (LD) (1 LD equal to 384'400 km) still can be measured using the photomultiplier system mentioned above. From biophotonics investigations so far, the origin of ultra-weak photon emission is the DNA as well as proteins coupled with radical reactions. In order to determine this radiation in human cells, a fibroblastic differentiation system was developed using dermal fibroblasts of skin. Since normal cells store efficiently ultra-weak photons, it has been shown that older cells as well as cancer tissue tend to lose this retention capacity. From all these results it seems evident, that this low level radiation serve as biophotonic signals in order to transfer information in biological systems. Further intense basic research is needed in order to show evidence that ultraweak electromagnetic radiation plays the key role in life.

Keywords: Biophotonic signals; Photomultiplier technique; Photon emission; Photon detection; Electromagnetic wavelengths; Radiation; Luminous power

Introduction

Photons participate in most atomic and molecular interactions and changes in life regulation processes. In this respect, the development of photomultiplier tubes (PMT's) in the early fifties of the last century was a key event in order to detect photons in biological tissue as reported by Colli et al. [1]. This practical discovery of light emission in cells by a highly sophisticated physical method was based on the theoretical finding of Einstein in 1905, the so-called photoelectric effect. For this breakthrough, Einstein received the Nobel Prize in Physics for the year 1921. The discovery of ultraweak photon emission by a physical device confirmed the so-called mitogenetic radiation determined by onion roots as biological detectors from the Russian biologist Alexander Gurwitsch as I reviewed 22 years ago [2]. The discovery measured latter on by PMT, that plant cells emit a low level radiation, was a coincidence. In the early nineteen fifties, PMT's were the breakthrough analytical technique in Physics, in order to detect single photons as side products of nuclear collisions. They allowed in the late fifties of the last century to provide evidence for the immense zoo of subatomic particles in modern physics. Colli as physical scientific researcher tried to test this new, highly sensitive device. At this time, Colli's wife bought some fresh vegetables on the street market and visited the physical laboratory of his husband, thereafter. She placed her shopping bag with the fresh bought plants in it, beside the new PMT. Shortly after this event, Colli checked his new device and realized an unexplained counting rate of the PMT. He concluded that the PMT was defect and returned it immediately to the factory in order to receive shortly after a new apparatus on guarantee. He tested then the substituted PMT again with fresh vegetables. To his surprise, the new, replaced PMT was again showing ultraweak photons (Emilio del Giudice, personal communication on Summer School 2010 in Neuss (Germany)). Therefore, Colli and his co-workers performed experiments with plant cells and published thereafter the observation of ultraweak photon emission in plants as mentioned above [1]. But it needed another twenty years, until Popp in Germany, Quickenden in Australia and Inaba in Japan confirmed in the middle 1970s independently this ultraweak radiation in plant cells as measured already beginning of the 1950s in Italy [1,2]. In the meantime, it is now scientifically accepted that plants, animals and humans emit a weak, so called - biophotonic radiation which can be readily detected with an appropriate photomultiplier system [3,4]. Although the emission is extremely low in mammalian cells, it can be efficiently induced in human cell cultures by ultraviolet light as shown almost ten years ago [5]. This process is most probably coupled with a very weak delayed luminescence (DL). The scientific researchers, the physicist Fritz-Albert Popp and the biologist Yu Yan described this phenomenon twelve years ago as coherent states in biological tissues [6]. In the last years, we developed a cell culture model for biophotonic measurements using fibroblastic differentiation [7-15]. An *in vivo* application of ultraweak radiation on human skin was developed by Cohen and Popp [16] and most recently confirmed by Musumeci and co-workers [17]. Schrader et al. used this method in order to test antioxidant stress *in vivo* on human skin [18]. All these results open the highly interesting question if ultraweak photons serve as electromagnetic signals in biological systems in order to regulate life processes in plant, animal and human cells [19]. This review presented here, will show new insights on this highly interesting hypothesis and further, intense biophotonic research has to be undertaken, as initiated already by the German biophotonic research program [20] in order to show clear evidence of ultraweak photons as

*Corresponding author:** Hugo J. Niggli, BioFoton AG, Rte. D'Essert 27, CH-1733 Treyvaux, Switzerland, E-mail:biofoton@hispeed.ch

inducers of cellular activation processes shown for example more than 30 years ago in the rhodopsin protein molecule of the eye [21].

Information Transfer by Electromagnetic Waves by Structured Water

Rarely is it thought about the admirable achievement to transfer sound over large distances. Such a transmission is difficult because acoustic waves disappear quickly. Additionally their transfer of 343 m/sec in air at 20°C is slow compared to the velocity of electromagnetic waves. They propagate considerably faster with 299792458 m/sec. Clerk Maxwell predicted electromagnetic waves based on mathematical correlation of his Maxwell-equation as summarized elsewhere [19]. Heinrich Rudolf Hertz proofed the existence of such waves in the 1880s more than 130 years ago. In the year 1896 Guglielmo Marconi succeeded to transfer electromagnetic waves on short distances. Alexander Gurwitch, a brilliant biologist from Russia developed between 1920 and 1935 a biological hypothesis with mitotic onion roots in order to prove that light transfer information in cells [2,22]. In the early 1950s, Herbert Fröhlich confirmed theoretically the transfer of information by electromagnetic waves in biological systems as summarized recently by his last pupil Gérard Hyland [23]. The biophysicist Fritz-Albert Popp proposed together with the biologist Walter Nagl in the year 1983 an electromagnetic model of cell differentiation [24]. This was the first genius concept in biology of modern natural science based on ultraweak photon research that electromagnetic waves may be important in order to regulate life processes. In cellular systems, this low level radiation has been measured by a various of scientific researchers from Australia [25], Austria [26], Belgium [27], [Brazil [28-30], Czech Republic [31-32], China [6,33-35], Germany [3,4,6,16,18-20,24,34-37], Great Britain [23,37-38], Holland [33,39-42], India [43,44], Italy [1,5,14-15,17,33-34,45-47], Japan [48-54], New Zealand [25], Poland [55-57], Russia [22,58-62], South-Korea [63], Switzerland [2,5,11-15,19,45,64-68], Ukraine [69-71] and USA [72-76].

Water is structured energy. A typical example is a whirlpool showing excessive power. As reported by Wernet et al. [77] in the year 2004, triangular chains, 3-dimensional rings and even triangulary pyramids are found in the molecular arrangement of water. The nuclear physicist Emilio del Giudice from Milan (Italy) found coherent regions in water [78]. Light is composed normally from photons of different wavelengths compared to the colours of a rainbow. As del Giudice emphasizes, coherent photons are very likely to a single, intense colour. According to his research, water is able to pick up information of other molecules similar to a DVD writer and player. This is even possible if the originated molecule is disappeared. This important observation is further corroborated by the Japanese physicist Kunio Yasue and his colleagues [79]. They found that water molecules are capable to transform disordered energy in coherent photons, a process called supra-radiation according to the physicist Dicke [80]. In summary, water, the natural medium of all cells, is most probably the crucial controller of the transmission of information from cellular molecules as DNA and proteins. Water molecules may perform this process through arrangement of highly organized patterns, as proposed by Wernet and co-workers mentioned above, which then can memorize wave frequencies. 70% of the human cell is not else than structured water. In the brain the amount is elevated to 90%. Only blood (92%) and saliva (98%) contain more water. Most interestingly the high level of water in the brain decreases with age. The American biochemist Albert-Szent Györgyi, originated from Hungary, received in 1937 the Nobel Prize for Medicine. He claimed that water is the mother for vital processes in the cell [81].

DNA Damage and Repair: Photochemical Processes in Order to Induce Intercellular Communication by Electromagnetic Waves

We have 15 years ago shown that *in vivo* induction of pyrimidine dimers in human skin by UVA radiation is not only a sign of DNA damage, but may initiate intercellular communication [12]. As reviewed most recently [19], DNA changes are mostly induced by interaction of UV radiation with molecules of biological significance. Absorption of UV-light by nucleic acids are inducing the most fatal effects in cells. As summarized by Niggli [82], pyrimidine photodimers are the major photoproducts induced by UV light with crucial biological effects. It was James Cleaver in San Franciso in the year 1968 [83] who found the cause of skin cancer after intensive exposure to the sun in humans with the illness Xeroderma Pigmentosum. Missing DNA excision repair was the biochemical reason for high levels of skin cancer in young people suffering on this well investigated genetic disease. Beside enzymatic cutting of photoinduced DNA changes, in order to regain hereditary stability of the damaged genome induced by UV-light, the photo repair system as studied by Richard Setlow and William Carrier in the year 1964 [84], is another highly relevant repair process of UV damaged DNA. Also a highly important molecule of life is oxygen. A few minutes without oxygen may lead to the fatal event of death. Most interestingly, nerve tissue and cells of the brain, composed of only to 2% of the body weight; consume more than 20 % of the oxygen taken by breath. The Russian chemist Vladimir Voeikov established the hypothesis that the photons stored in oxygen gain access to the cells by radical reactions [85]. Radicals are chemically, highly active substances [86]. They arise in plant, animal or human cells by irradiation with ultraviolet light or ionizing electromagnetic waves like X-rays. Vitamin E and superoxide dismutase, for example, influence significantly radical reaction processes which happen in aqueous solutions. According to Voeikov radical processes are indispensable to life in order to gain the stored light energy in oxygen. Together with Emilio del Giudice he showed that water respiration is the basis of the living state [87]. As mentioned above, the Italian physicist Del Giudice showed that excited atoms and molecules are interacting with the zero point vacuum fields as summarized by Lynne Mc Taggart [88]. In these processes arise small areal units of high order state. These so-called Coherent Domains (CD) have in water a diameter of about 100 nm fitting to about 10^7 water molecules. The lifespan of these coherent domains is estimated to months or even long-lasting periods over years. A CD is surrounded by normal water, so-called bulk water. As summarized by Voeikov and Del Giudice [87] water hydrating hydrophilic surface areas is significantly different from bulk water. These diversities in density, freezing temperature, viscosity and relative permittivity are so striking that Zheng and co-workers [89] considered it as a fourth aggregate state. They reported that the thickness of this layer is probably hundreds of microns. Pollack [90] defined water adjacent to hydrophilic surfaces as so-called "Exclusion Zone Water" (EZ-water). As outlined by Voeikov and Del Giudice [87] the most important features of EZ-water is the prominent peak of light absorption in the UVC range of 270 nm. Most interestingly EZ-water excited by this wavelength emits low level photon emission. In addition, the thickness of the EZ-water layer increases with illumination by visible light. IR radiation enhances the size of EZ-water significantly stronger, confirming the results by Albrecht-Buehler showing IR radiation effects in cells [73,75]. Based on these above-mentioned qualities, Voeikov and Del Giudice conclude that EZ-water can store energy. Their most important conclusion is that a CD in water is able to transform low grade energy with high entropy in high grade energy with low entropy. In this process oxygen is needed, in short: Cells, which contain at least 70% H_2O, perform water respiration.

Patrick and Rahn [91] investigated in the seventies of the last century the yield of thymine dimerization upon UVB-irradiation in DNA as a function of photosensitizers in the presence and absence of oxygen. They observed that that the pyrimidine dimer yields are about three- to six fold lower in the presence of oxygen confirming the report of Greenstock et al. [92] of reduced dimer yields of UV-Irradiated thymine solution. These quenching results from the transfer of the excitation from the sensitizer to the oxygen. This reaction is very fast and accounts for almost all quenching of triplets to oxygen as reviewed by Foote in the year 1976 [93]. I have shown in the year 1983 [82], that after UVC and UVB irradiation superoxide dismutase in the irradiation medium of the UV exposed cells enhance the pyrimidine dimer yield for both UV ranges by 25%, while catalase lower the dimerization in the size of 15%. It was discussed in this doctoral thesis [82], that the decrease in dimer yields observed in fibroblasts treated with the enzyme catalase is probably due to the quenching of molecules in excited states by the catalase produced molecular oxygen which is confirming the report of Patrick and Rahn [91]. Surprisingly, superoxide dismutase which is producing during its catalytic reaction oxygen too, shows higher dimer yields. It is well known that the uncatalyzed reaction is producing singlet oxygen [94]. These authors also show, that superoxide dismutase suppresses singlet oxygen production which has been shown to be involved in cell damage and carcinogenesis [86]. A report of Goda et al. [95] demonstrates that in the catalytic reaction the resultant O_2 is formed in the triplet state in contrast to the non-enzymatic reaction. The difference may be accounted for by the presence of catalytically active Cu^{2+} ions in the dismutase molecule [96]. However, triplet oxygen may be capable to introduce pyrimidine dimers. In this respect we have also shown that UVA radiation which produces pyrimidine photodimers mostly via photosensitization [97] are introduced on human skin *in vivo* [12] as detected by antibody staining of dimers. Lamola has shown chemical induction of dimers in the dark [98]. In this respect, Giuseppe Cilento has created a photobiochemistry without light as reviewed in the year 1988 [29]. As Voeikov and Del Giudice emphasize, it has now become clear that life is possible without sunlight [87]. Fantastic ecosystems have been found at the bottom of the ocean. There is no light and no oxygen from the atmosphere at 10 kilometer below the ocean surface. The temperature is rarely more than 2-4°C. Most surprisingly, highly active aerobic animals of various, different species, live in this cold darkness. It is possible that they use as energy source the thermal heat emerging from the hot, inner core of our Earth. It is well know from marine research that the basic physiology and biochemistry of these living beings in the cold darkness of the deep ocean is not significant different from those animals living on the land under the sun. The continuous flow of energy in their bodies required for multiplication, growth and life formation is preserved by their efficient capacity to utilize energy from their scanty surrounding area. But it has to be noted that this ability is not the specific property of deep-sea creatures. It is the general strategy of all life. Another astonishing example of light in the cells is the desert mice Monodelphis domestica. As Ley published almost 30 years ago, these animals mostly living in the dark have an efficient DNA repair system based on photo reactivation [99]. As Lynne Mc Taggert reviewed [88], Popp found in 1975 that carcinogens absorb ultraviolet-A light (UVA) in the range of 380 nm. He tested first benzo[a]pyrene by irradiation in the UVA range and found that this carcinogen, causing for example scrotal cancer in chimney sweepers, absorb the UV radiation and re-emit then photons at a complete different frequency. Most surprisingly, the harmless benzo[e]pyrene allowed the UV-light to pass through it unchanged. Popp was astonished by this observation and performed his test on 37 different chemicals. Several tested substances are cancer causing, others

in this testing system are unoffending chemicals. Using his test he was able to predict the carcinogenic potential of his tested substance as reviewed in the year 1979 [100]. In the meantime it is clear that the carcinogenic substances predicted by the UVA-test of Popp can interchelate with the hereditary DNA source. Benzo[a]pyrene for example is a so-called procarcinogen which is activated by detoxification enzymes like cytochromes P 450 in the mitochondria to the highly carcinogenic Benzo[a]pyrene-7,8-dihydrodiol-9,10-epoxide interchelating with the guanine residue of the DNA [86]. This activation process is similar to that of aflatoxin B_1 found for example in *Aspergillus flavus* and *Aspergillus parascitus*. *Aspergillus flavus* show accelerated growth in decomposed peanuts, but also in rotten hay or grains in a decay action. The epoxide is here induced by the cytochrome P 450 pathway in hepatocytes. Most interestingly for this enzymatic processes water and oxygen is needed. As pyrimidine dimers, these highly toxic chemicals change the DNA source and may induce mutations, leading finally to carcinogenesis [82,86]. We have in the year 1999 reported that *in vivo* induction of pyrimidine dimers in human skin by UVA radiation is not only initiation of cell damage, but may be a common biochemical pathway of intercellular communication [12]. As reviewed elsewhere [19,66], ultra weak photon emission is most probably involved in this communication process. It is well known, that biophotonic emission has been detected both in the ultraviolet as well as in the visible region of the electromagnetic spectrum [2,19]. There is substantial evidence that DNA is an important source of ultra-weak photon emission [2,11,19,50,66]. As published more than 20 years ago, normal fibroblasts tend to store UV-light efficiently , while excision repair deficient Xeroderma Pigmentosum cells [83] loose the capacity to store ultraweak photons. This observation was confirmed then years ago in report showing temperature dependence of ultraweak photon emission in fibroblastic differentiation after artificial UV-exposure [101,102]. In several reports [12,66,102], we have proposed that pyrimidine dimers are inducers of photonic activation of biochemical pathways via excision repair of pyrimidine photodimers in DNA similar to the more than forty years ago published light-driven activation process of chemical reactions by bacteriorhodopsin [103]. As reported by Albrecht-Buehler [73,75], cells use infrared radiation in order to perform intercellular communication. In this respect, it is of high interest to observe that EZ-water emit in the UVC region as reported by Voeikov and Del Giudice recently [87]. Theoretical consideration of Fröhlich, propose intercellular communication by coherent states of cells. As discussed by John Swain [72], there are at least two types of quantum electromagnetic communication systems associated with the living state. One is in the microwave frequency range as hypothesized by Fröhlich [23], the other is from the UV, over the visible to the infrared region: UV and visible is suggested by Popp [3,4,34,58], the infrared data is based on the scientific observations of Albrecht Buehler [73,75]. According to Swain [72] there exists a coupling by resonance between these two regions. As Swain [72] emphasizes, a biological system can easily store low microwave energy and transfer it to high energy photons as found in IR, visible and UV regions. In John Swain's view [72], there exists, as cited, "a natural framework not just for a biological molecule to experience long-range forces pulling it to where it should go, but also for the appropriate amount of energy to be transferred between them. The use of single photons as part of cell-to-cell signaling is also fascinating and the sort of system here could allow for a high degree of selectivity with little cross-talk by choosing slightly different optical frequencies for different communication". Based on Fröhlich's hypothesis of quantum, coherent behavior in cells, microtubules as light conductor are important as proposed 25 years ago by the scientist Bornens from Belgium [27]. In this concept, structured water and their respiration processes, as brilliant implicated by Voeikov and Del Giudice, play an highly important role as infinite energy source in the dark [87]. Further intense, bio photonic research has to be realized in order to show further evidence, that the electromagnetic waves are the essential key in intra- and intercellular communication processes.

Acknowledgements

I would like to thank Professor Lee Ann Laurent-Applegate (CHUV, Lausanne, Switzerland) as well as Dr. Max Bracher (BioFoton AG, Treyvaux, Switzerland) for critical reading of the manuscript and many helpful discussions. This work is dedicated to Emilio Del Giudice who died on the 31st of January 2014. He is the Italian pioneer of ultraweak photonic research. He ingeniously introduced the basis of quantum physics into high sophisticated, modern cell research.

References

1. Colli L, Facchini U, Guidotti G, Dugnani Lonati R, Arsenigo, M et al. (1955) Further measurements on the bioluminescence of the seedlings. Experientia 11: 479-481.

2. Niggli HJ (1992) Ultraweak photons emitted by cells: biophotons J Photochem Photobiol B: Biol 14: 144-146.

3. Popp FA, Li KH, Gu Q (1992) Recent Advances in Biophoton Research and its Application, World Scientific, Singapore.

4. Chang JJ, Fisch J, Popp FA (1998) Biophotons Kluwer Academic Publishers Dordrecht, Netherlands, Europe.

5. Niggli HJ, Tudisco S, Privitera G, Applegate LA, Scordino A, et al. (2005) Laser-Ultraviolet-A induced ultraweak photon emission in mammalian cells. J Biomed Opt 10: 024006.

6. Popp FA, Yan Y (2002) Delayed luminescence of biological systems in terms of coherent states. Physics letters A 290-293.

7. Bayreuther K, Rodemann HP, Hommel R, Dittman K, Albiez M, et al. (1988) Human skin fibroblasts in vitro differentiate along a terminal cell lineage, Proc natl Acad Sci USA 85: 5112-1516.

8. Niggli HJ, Francz PI (1992) May ultraviolet light-induced ornithine decarboxylase response in mitotic and postmitotic human skin fibroblasts serve as a marker of aging and differentiation? Age 15: 55-60

9. Niggli HJ (1993) Aphidicolin inhibits excision repair of UV-induced pyrimidine photodimers in low serum cultures of mitotic and mitomycin C-induced postmitotic human skin fibroblasts. Mut Res 295: 125-133.

10. Niggli HJ , Applegate LA (1997) Glutathione response after UVA irradiation in mitotic and postmitotic human skin fibroblasts and keratinocytes. Photochem Photobiol 65: 680-684.

11. Niggli HJ (1996) the cell nucleus of cultured melanoma cells as a source of ultraweak photon emission. Naturwissenschaften 83: 41-44.

12. Applegate LA, Scaletta C, Panizzon R, Niggli HJ, Frenk E (1999) In vivo induction of pyrimidine dimers in human skin by UVA radiation: Initiation of cell damage and/or intercellular communication? Int J of Mol Med 3: 467-472.

13. Niggli HJ, Scaletta C, Yu Y, Popp FA, Applegate LA (2001) Ultraweak photon emission in assessing bone growth factor efficiency using fibroblastic differentiation. J Photochem Photobiol B: Biol. 64: 62-68.

14. Niggli HJ, Tudisco S, Privitera G, Applegate LA, Scordino A, et al. (2005) Laser-ultraviolet-A-induced biophotonic emission in cultured mammalian cells. In: Biophotonics: Optical Science and Engineering for the 21st century (X. Shen and R. Van Wijk,edn.) Springer New York USA

15. Niggli HJ, Tudisco S, Lanzanò L, Applegate LA, Scordino A, et al. (2008) Laser-ultraviolet-A-induced ultraweak photon emission in human skin cells: A biophotonic comparision between keratinocytes and fibroblasts. J Exp Biol 46: 358-363.

16. Cohen S, Popp FA (1997) Biophoton emission of the human body, J Photochem Photobiol B: Biology 40: 187-189.

17. Musumeci F (2007)Spectral analysis of photoinduced delayed luminescence from human skin in vivo. Conference Paper of European Conference on Biomedical Optics. Munich, Germany.

18. Jain, Rieger I, Rohr M, Schrader A (2010), Antioxidant efficiency on human skin in vivo, Skin Pharmacol Physiol 23: 266-272.

19. Niggli HJ (2014) Biophotons: Ultraweak light impulses regulate life processes in aging Journal of Gerontology & Geriatric Research in Press.

20. Liedtke S, Popp J, Laser, Licht und Leben (2006) Visions for better Health Care. WILEY-VCH Verlag GmbH&Co. KGaA, Weinheim, Germany.

21. Stryer L (1987) the molecule of visual excitation. Scientific American 257: 42-50.

22. Beloussov L (1997) Life of Alexander G. Gurwitsch and his relevant contribution to the theory of morphogenetic fields. Int Dev Biol 41: 771-779.

23. Hyland GJ (2009) Fröhlich's physical theory of cancer- Fröhlich's path from theoretical physics to biology and the cancer problem. Neural Network World 19: 337-354.

24. Nagl W, Popp FA (1983) A physical (electromagnetic) model of differentiation: basic considerations, Cytobios 37: 45-62.

25. Tilbury RN, Quickenden TI (1987) The effect of cosmic-ray shielding on the ultraweak bioluminescence emitted by cultures of Escherichia Coli. Radiat Res 112: 398-402.

26. Schwabl H, Klima H (2005) Spontaneous ultraweak photon emission from biological systems and the endgenous light field. Forsch Komplementermed Klass Naturheilkunde 12: 84-89.

27. Bornens M (1979) The centriole as a gyroscopic oscillator. Implications for cellorganization and some other consequences. Biologie Cellulaire 35: 115-132.

28. Cilento G (1984) Generation of electronically excited triplet species in biochemical systems. Pure Appl Chem 56: 1179-1190.

29. Cilento G (1988) Photochemistry without light. Experientia 44: 572-576.

30. Gallep CM, Moraes TA, Santos SRD, Barlow PW (2013) Coincidence of biophoton emission by wheat seedlings during simultaneous. Transcontinental germination tests Protoplasma 250: 793-796.

31. Cifra M, Fields JM, Farhadi A (2011) Electromagnetic cellular interactions. Progress in Bioph Mol Bio 105: 223-246.

32. Pokorny J, Vedruccio C, Cifra M, Kucera O (2011) Cancer physics, diagnostics based on damped cellular elastoelectrical vibrations in microtubules. Eur Biophys J 40: 747-759.

33. Shen X, Wijk RV (2005) Biophotonics, Springer Science Buisness media. New York, USA.

34. Chang JJ, Fisch J, Popp FA (1998) Biophotons, Kluwer Academic Publisher Dordrecht, Netherland.

35. Popp FA, Chang JJ, Herzog A, Yan Z, Yan Y (2002) Evidence of non-classical (squeezed) light in biological sysems, Phys Lett A 293: 98-102.

36. Rattemeyer M, Popp FA, Nagl W (1981) Evidence of photon emission from DNA in living systems, Naturwissenschaften.68: 572-573.

37. Ho MW, Popp FA, Warnke U (1994) Bioelectrodynamics and Biocommunication. World Scientific Singapore.

38. Ho MW (1995) Bioenergetics and the coherence of organisms. Neuronetwork World 5: 733-750.

39. Hyland GJ (2000) Physics and biology of mobile telephony. Lancet 356: 1833-1836.

40. Wijk EV, Kobayashi M, Wijk RV, Greef JV (2013) Imaging of ultra-weak photon emission in mammalian cells. PLos One 8: e84579.

41. Wijk RV, van Aken JM (1992) Photon emission in tumor biology. Experientia 48: 1092-1102.

42. Wijk RV, Aken HV. (1991) Spontaneous and light-induced photon emission by rat and by hepatoma cells. Cell Biophys. 18: 15-29.

43. Bajpai RP (2003) Symposium in Print on Biophoton, National Institute of Science Communication and Information Resources, CSIR, New Delhi, India, Indian J Exp Biol 41: 5.

44. Bajpai RP (2008) Biophotons and Alternative Therapies, National Institute of Science Communication and Information Resources, CSIR, New Delhi (India) Indian Journal of Exp. Biol. 46: 5

45. Scordino, Baran I, Gulino M, Ganea C, Grasso R, et al. (2014) Ultra-weak delayed luminescence in cancer research: a review of the results by the ARETUSA equipment submitted.

46. Giudice ED, Elia V, Tedeschi A (2009) Role of water in the living organism. Neural Network World 19: 355-360.

47. Giudice ED, Preparata G , Vitello G (1988) Water as a free electric dipole laser. Phys Rev Lett 61: 1085-1088.

48. Inaba H (1988) Super-high sensitivity systems for detection and analysis of ultraweak photon emission from biological cells and tissues. Experientia 44: 550-559.

49. Scott RQ, Mashiko S, Kobayashi M, Hishinuma K, Ichimura T (1989) Two-dimensional detection of ultraweak bioluminescence using a single-photon image acquisition system, J Opt Soc AM 4: 183-185.

50. Devaraj B, Scott RQ, Roschger P, Inaba H (1991) Ultraweak photon emission from rat liver nuclei. Photochem. Photobiol. 54: 289-293.

51. Nakamura K, Hiramatsu M (2005) Ultra-weak photon emission from human hand: Influence of temperature and oxygen concentration on emission. J Photochem Photobiol B. Bio 80: 156-160.

52. Ichimura T, Hiramatsu M, Hirai N, Hayakawa T (1989) Two-dimensional imaging of ultra-weak emission from intact soybean roots. Photochem. Photobiol 50: 283-286.

53. Makino T, Kato K, Iyozumi H, Honzawa H, Tachiiri Y (1996) Ultra-weak luminescence generated by sweet potato and fusarium oxysporum interactions associated with a defense response. Photochem. Photobiol 64: 953-956.

54. Inaba H, Shimizu Y, Tsuji Y, Yamagishi A (1979) Photon counting spectral analyzing system of extraweak chemi and bioluminescence for biochemical applications. Photochem Photobiol 30: 169-175.

55. Slawinska D, Slawinski J (1983) Biological chemiluminescence. Photochem Photobiol 37: 709-715.

56. Chwirot WB, Dygdala RS, Chwirot S (1986) Quasimonochromatic-light-induced photon emissionfrom microsporocytes of larch showing oscillations decay behavior predicted by an electromagnetic model of differentiation. Cytobios 47: 137-146.

57. Chwirot WB (1988) Ultraweak photon emission and anther meiotic cycle in laryx europea. Experientia 44: 594-599.

58. Popp FA, Beloussov LV (2003) Integrative Biophysics: Biophotons, Kluwer Academic Publishers Boston,USA.

59. Voeikov V (2001) Reactive oxygen species, water, photons and life, Riv Biol Biol Forum 94: 237-258.

60. Gurwitsch AG (1947) Une théorie du champ biologique cellulaire. Bibliotheca Biotheoretica Series D.V. Leiden, Netherland.

61. Popov GA, Tarusow BN (1963) Nature of spontaneous luminescence of animal tissues. Biofizika 8: 317-320.

62. Beloussov LV, Voeikov VL, Martynyuk VS (2007) Biophotonics and Coherent Systems in Biology, Springer Science-Buisness Media New York, USA.

63. Choi C, Woo WM, Lee MB, Yang JS, Soh KS, et al. (2002) Biophoton emission from the hands, J Korean Phys Soc 41: 275-278.

64. Bischof M (1995) Biophotonen das Licht in unseren Zellen, Verlag Zweitausend und Eins Leipzig. Deutschland.

65. Niggli HJ (1998) Biophotons: Our body produces light. Network 65: 16-17.

66. Niggli HJ, Scaletta C, Yan Y, Popp FA, Applegate LA (2001) UV-induced DNA damage and ultraweak photon emission in human fibroblastic skin cells: parameters to trigger intra- and extra-cellular photobiostimulation, Trends in Photochem Photobiol 8: 53-65.

67. Fels D (2009) Cellular communication through light. PLoS One 4: E5086.

68. Weilenmann U (2010) Mögliche wissenschaftliche Ansätze als Erklärungsgrundlage komplementärmedizinischer Therapien, Swiss J Integrative Med 22: 171-178.

69. Brizhik L, Musumeci F, Scordino A, Trigla A (2000) The soliton mechanism of the delayed luminescence of biological systems. Europhys Lett 52 :238

70. Brizhik L, Scordino A, Trigla A, Musumeci F (2001) Delayed luminescence of biological systems arising from correlated many soliton-states. Phys Rev 64: 031902.

71. Brizhik L (2003) Dynamical properties of Davydov solitons. Ukr J Phys 48 : 611-622

72. Swain J, on the possibility of large upconversions and mde coupling between Fröhlich states and visible photons in biological systems.

73. Albrecht-Buehler G (1992) Rudimentary form of cellular vision. Proc Natl Acad Sci USA 89: 8288-8292.

74. Eller MS, Yaar M, Gilchrest BA (1994) DNA damage and melanogenesis. Nature 372: 413.

75. Albrecht-Buehler G (2000) Reversible excitation light-induced enhancement of fluorescence of live mammalian mitochondria. FASEBJ 14: 1864-1866.

76. Swain J (2006) on the possibility of large upconversions and mode coupling between Fröhlich states and visible photons in biological systems.

77. Wernet P, Nordlund D, Bergmann U, Cavalleri M, Odelius M, et al. (2004) The structure of the first coordination shell in liquid water. Science 304: 995-999.

78. Giudice ED, Preparata G (1988) Water as a free electric dipole laser. Phys Rev Lett 61: 1085-1088.

79. Jibu M, Yasue K, Hagan S (1997) Evanescent (tunneling) photon and cellular vision. Bio systems 42: 65-73.

80. Dicke RH (1954) Coherence in spontaneous radiation processes. Phys Rev 93: 99-110.

81. Szent-Györgyi A (1937) Oxidation, energy transfer and vitamins Nobel Prize Lecture of December 11th.

82. Niggli H (1983) Formation and excision of cyclobutane-type pyrimidine dimers in human skin fibroblasts after irradiation with ultraviolet light. Doctoral thesis in Sciences ,University of Lausanne,Switzerland 22:1390-1395.

83. Cleaver JE (1968) Defective repair replication of DNA in xenoderma pigmentosum. Nature, London, 218: 652-656.

84. Setlow R RB, Carrier WL (1964) The disappearance of thymine dimers from DNA: an erro-correcting mechanism. Proc Natl Acad Sci USA, 51: 226-231.

85. Voeikov VL (2005) Biophotonic analysis of spontaneous self-organizing oxidative processes in aqueous systems, In: Biophotonics: Optical Science and engineering for the21st Century, Springer Verlag ,New York, Xun Shen and Roeland Van Wijk, edn. 141-145.

86. Cerutti PA (1985) Prooxidant states and promotion Science. 227: 375-381.

87. Voeikov VL, Giudice ED (2009) Water respiration – The basis of the living state. Water 1: 52-75.

88. Mc Taggert L (2003) The Field: The Quest for the Secret Force of the Universe. Harper Collins Publisher, New York.

89. Zheng JM, Chin WC, Khijniak E, Jr Khijniak E, Pollack GH (2006) Surfaces and interfacial water: evidence that hydrophilic surfaces have long-range impact. Adv Colloid Sci 23: 19-27.

90. Pollack GH (2001) Cells, Gels and Engines of Life. Ebner and Sons, Seattle.

91. Patrick MH, Rahn RO (1976) Photochemistry and Photobiology of Nucleic Acids. Academic Press, New York, 35-95.

92. Greenstock CL, Brown IH, Hunt JW, Johns HE (1967) Photodimerization of pyrimdine nucleic acid derivatives in aqueous solution and the effect of oxygen. Biochem Biophys Res Commun 27: 431-436.

93. Foote CS (1976) Photosensitized Oxidation and Singlet Oxygen: Consequences in Biological Systems. Free radicals in Biology II Academic Press, New York, 2: 85-133.

94. Mayeda EA, Bard AJ (1974) Singlet oxygen. Suppression of its production in dismutation of superoxide ion by superoxide dismutase. J A ChemSoc 96: 4023-4024.

95. Goda K, Kimura T, Thayer AL, Kees K, Schaap AP (1974) Singlet molecular oxygen in biological systems: Non quenching of singlet oxygen-mediated chemiluminescence by superoxide dismutase. Biochem Biophys Res Commun 58: 660-666.

96. Agro F, Rinaldi A, Floris G, Rotilio G (1984) A free-radical intermediate in the reduction of plant Cu-amine oxidase. FEBS Lett 176: 378-380.

97. Rochette PJ, Therrien JP, Drouin R, Perdiz D, Bastien N, et al. (2003) UVA-induced cyclobutane pyrimidine dimers form predominantly at thymine-thymine dipyrimidines and correlate with the mutation spectrum in rodent cells. Nucleic Acids Res 31: 2786-2794.

98. Lamola AA (1968) Excited state precursors of thymine photodimers. Photochem Photobiol 7: 619-632.

99. Ley RD (1985) Photo reactivation of UV-induced pyrimidine dimers and erythema in the marsupial Monodelphis domestica. Proc Natl Acad Sci USA 82: 2409-2411.

100. Popp FA, Becker G, König HL, Peschka W (1979) Electromagnetic Bio-Information Urban & Schwarzenberg. Baltimore.

101. Niggli HJ (1993) artificial sunlight irradiation induces ultraweak photon emission in human skin fibrolbasts. J Photochem Photobiol B: Biol 18: 281-285.

102. Niggli HJ (2003) Temperature dependence of ultra-weak photon emission in fibroblastic differentiation after irradiation with artificial sunlight. Indian J Exp Bio 41: 419-423.

103. Oesterhelt D, Stoeckenius W (1971) Rhodopsin-like protein from the purple membrane of Halo bacterium halobium. Nature,London 233: 149-152.

17

Optimal Choosing of Hybrid Renewable Source for Varying Condition for Effective Load Management

Anand BB* and Ramesh V

Department of Electrical Engineering, VIT University, Vellore, Tamil Nadu, India

Abstract

The goal of this thesis is to evaluate the performance of a hybrid solar wind, fuel cell and battery energy system through Mat lab simulation studies. The main objective of this study is the development of dynamic models of an off grid (standalone) solar wind turbine, fuel cell and battery system which is used for charging and discharging. Maximum power can be extracted from PV module by using Maximum Power Point Tracking (MPPT) by controlling the duty cycle of a switch. The developed model of the wind energy conversion system consists of dynamic models for a wind turbine as well as induction motor or Permanent Magnet Synchronous Machine. In this thesis we have to control the sources depending upon the load condition. The solar installation cost is less than the remaining sources. So solar supplies the load continuously. If solar cell generated power is not sufficient, then the wind supplies. And if solar and wind generated power are not sufficient, then fuel cell will be operated along with solar and wind. A battery which is used for backup purpose will supply if solar, wind, fuel cell generated power is not sufficient.

Keywords: Solar cell; Wind turbine; Fuel cell; Battery; Converter

Introduction

The usage of non renewable energy sources (conventional energy sources) causes a hike in generation cost. In the world India is the 5th largest in generation and usage of electricity. By 2017 Electricity demand is approximately expected to be 1400 billion kwh. Now it is around 900 billion kwh. There is a deficit of 12% electricity by using non renewable energy. So to meet this demand we have to use other sources like renewable energy. Conventional energy sources also increase fuel consumption and pollution. Solar is the only entirely renewable alternative energy source to satisfy the energy needs of India.

Wind power is unreliable due to irregular nature of wind; the combination of solar photo voltaic (PV) and wind energy conversion system is preferred to improve the reliability of hybrid power generation systems in remote locations. Off grid, i.e., stand alone Hybrid Energy Systems (HES) based on renewable energy resources can provide supply of electrical energy in remote locations. Hybrid Energy Systems (HES) [1-4] can provide higher reliability and power quality. Standalone (off grid) system has been generated by single source systems using solar PV, wind turbine, fuel cell, biomass, diesel generators and battery for back up purpose or by the combining two or more types of these electricity generating sources (Hybrid Energy Systems). These systems include energy source like Lead Acid batteries. A hybrid system can supply power to AC and DC loads. So it requires AC, DC or both buses (Figure 1).

Solar Cell

A semiconductor device convert's sunlight into electrical energy is called Photovoltaic cell. PV module means group of PV cell [5,6] to produce high voltage. PV array means a group of PV module. Electricity is produced when a photon light energy is greater than the band gap of the semiconductor. PV modules are of two types. They are Current input PV module and voltage input PV module. The current input PV module is suitable when PV modules are connected in series and have same current. Voltage input PV module is suitable when PV modules are connected in parallel and have the same voltage.

Photovoltaic cell modelling

Photovoltaic cell contains a diode and current source. PV cell is shown in Figure 2 and equations given below.

Using KCL

$$I_{sc} - I_D - \frac{V_D}{R_P} - I_{pv} = 0 \tag{1}$$

Diode Characteristic equation

$$I_D = I_0(e^{V_D/V_T} - 1) \tag{2}$$

Using KVL $\tag{3}$

Where V_D is voltage across diode, I_0 is reverse saturation current, $V_T = (K * T / q)$

K=Boltzmann constant = 1.38×10^{-19} Joule/k, T is junction temperature, q= electron charge

PV Panel Current-voltage and power-voltage equations are shown in Figure 3.

Generally solar cell efficiency is very low and it is around 20%. So we have to improve the efficiency of solar cell by maximum power point Tracking (MPPT). Maximum power [7] can be obtained by varying source but it is very difficult to vary source because Current – voltage curve of PV cell is non linear (Figure 4). So that by varying the duty cycle of switch by using various MPPT algorithms [5,6,8] like Perturb and Observe, Incremental Conductance, Fuzzy logic Control, artificial neural networks etc.

*Corresponding author: Anand BB, Department of Electrical Engineering, VIT University, Vellore, Tamil Nadu, India
E-mail: bandreddyanandbabu@gmail.com

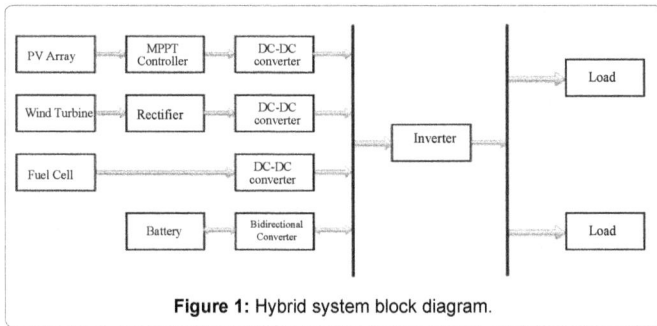

Figure 1: Hybrid system block diagram.

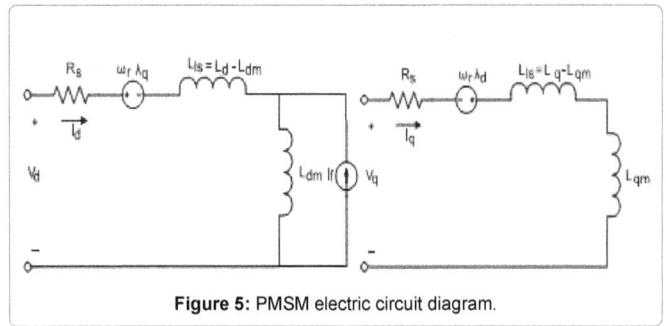

Figure 2: PV cell circuit.

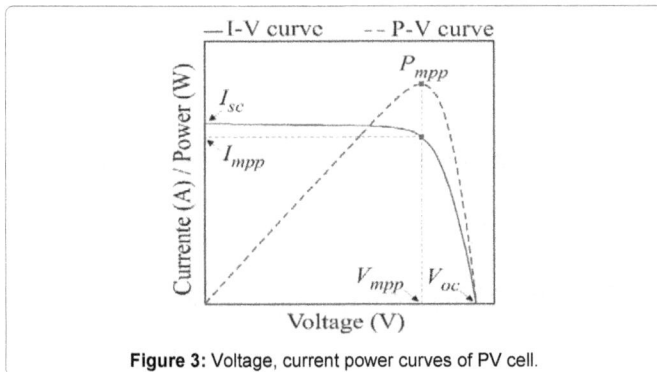

Figure 3: Voltage, current power curves of PV cell.

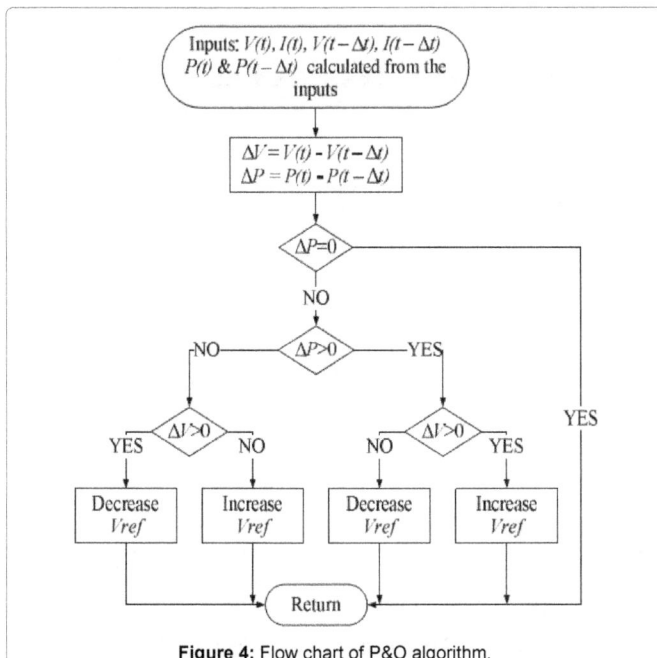

Figure 4: Flow chart of P&O algorithm.

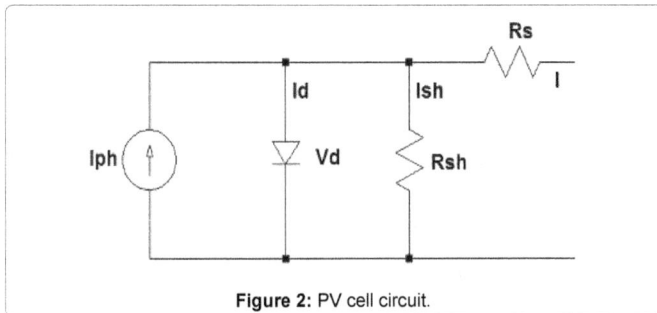

Figure 5: PMSM electric circuit diagram.

Perturbation	Change in power	Next perturbation
Positive	Positive	Positive
Positive	Negative	Negative
Negative	Positive	Negative
Negative	Negative	Positive

Table 1: Summary of the hill-climbing algorithm.

Flow chart of P&O MPPT Algorithm is shown in Figure 4 and Table 1.

Wind Turbine

One of the energy sources is wind energy which is the renewable energy source. The main drawback of wind energy is output power of wind turbine mainly depends upon nature of wind. So we have to control the speed of wind turbine. Wind turbine is coupled to Permanent magnet synchronous Machine. It depends upon nature of torque, Permanent magnet synchronous machine acts as generator when the torque is negative and acts as motor when torque is positive.

Wind turbine mechanical power [9] is

$$P_m = \frac{1}{2}\pi\rho R^2 C p v^3_{wind} \tag{4}$$

Where

$$C_p = (0.44 - 0.0167\beta)\sin\frac{\pi(\lambda-2)}{13-0.3\beta} - 0.00184(\lambda-2)\beta \tag{5}$$

$\rho = Air\ density = 1.226 kg/m^3$

R=Radius of rotor blade,

λ=Tip speed ratio,

V_{wind}= wind speed in m/sec,

β=blade pitch angle

Let us assume C_p=0.4, λ=8.5, β=0, P_m=850 KW

Substitute above values in equation (1) and find R value

Tip Speed Ratio (TPR) is

$$\lambda = \frac{wR}{v} \tag{6}$$

Substitute λ=8.5, wind speed v=12 m/sec, R value obtained by using equation (1) in (2) and then find w value, w value is around 2.81 rad/sec. Increase β value, v value and fine C_p such that power should be maintain constant.

Permanent magnet synchronous machine

Wind turbine is connected to shaft of Permanent Magnet Synchronous Machine (PMSM) [6,7]. Output power of PMSM is not varies with change of wind speed and also PMSM has high torque permanent Magnet Synchronous machine acts as generator if torque is negative and it acts as motor if torque is positive. Three phase input can be converted into two quadrant by using parks transformation [8] and two quadrants d, q axis be rotor reference frame (Figure 5).

$$\begin{matrix} v_q \\ v_d \\ V_o \end{matrix} = \frac{2}{3} \begin{pmatrix} \cos\theta_r & \cos(\theta_r - 120) & \cos(\theta_r + 120) \\ \sin\theta_r & \sin(\theta_r - 120) & \sin(\theta_r + 120) \\ \frac{1}{2} & \frac{1}{2} & \frac{1}{2} \end{pmatrix} \begin{matrix} V_a \\ V_b \\ V_c \end{matrix} \qquad (7)$$

$$V_q = \frac{2}{3}((V_a * \cos\theta_r + V_b * \cos(\theta_r - 120) + V_c * \cos(\theta_r + 120)) \quad (8)$$

$$\lambda = \frac{wR}{v} \quad V_d = \frac{2}{3}(V_a * \sin\theta_r + V_b * \sin(\theta_r - 120) + V_c * \sin(\theta_r + 120)) \quad (9)$$

Flux linkages along d axis $[\lambda_d]$ q axis $[\lambda_q]$ equations are give below

$$\lambda_d = L_d i_d + \lambda_f \qquad (10)$$

$$\lambda_q = L_q i_i \qquad (11)$$

Voltage equations are

$$V_q = R_s i_q + w_r \lambda_d + \rho \lambda_q \qquad (12)$$

$$V_d = R_s i_d - w_r \lambda_q + \rho \lambda_d \qquad (13)$$

Where ρ is $\frac{d}{dt}$

Substitute λ_d, λ_q in above equations

$$V_q = R_s i_q - w_r (L_d i_d + \lambda_f) + \rho L_q i_q \qquad (14)$$

$$V_q = R_s i_q - w_r L_d i_d - w_r \lambda_f = \rho L_q i_q \qquad (15)$$

$$i_q = \frac{\int [V_q - R_s i_q - w_r L_d i_d - w_r \lambda_f]}{L_q} \qquad (16)$$

$$V_d = R_s i_d - w_r (L_q i_q) + \rho (L_d i_d + \lambda_f) \qquad (17)$$

$$V_d = R_s i_d + w_r (L_q i_q) - \rho \lambda_f = \rho L_d i_d \qquad (18)$$

$$i_d = \frac{\int [V_d - R_s i_d + w_r (L_q i_q)]}{L_d} \qquad (19)$$

Torque T_e is

$$T_e = \frac{3}{2}\frac{P}{2}(\lambda_d i_d - \lambda_q i_d) \qquad (20)$$

$$T_e = \frac{3}{2}\frac{P}{2}(i_q (L_d i_d + \lambda_f) - i_d (L_q i_q)) \qquad (21)$$

$$T_e = \frac{3}{2}\frac{P}{2}(Flux * i_q + (L_d - L_q) * i_q i_d) \qquad (22)$$

$$T_e = T_L + Bw_m + J\frac{dw_m}{dt} \qquad (23)$$

Rotor mechanical speed

$$w_m = \int [\frac{T_e - T_L - Bw_m}{J}] dt \qquad (24)$$

$$w_m = w_r * (2/P) \qquad (25)$$

Where w_r is rotor electrical speed.

Fuel Cell

Fuel cell defines as a device that converts chemical energy into electrical energy through a chemical reaction. Generally fuel cell uses hydrogen, methanol, and natural gas as fuel and produces protons, electrons, heat and water. Now a day's various types' fuel cells are available and all those have anode, cathode, electrolytes but different electrolytes and fuel is different (Figure 6). One of fuel cell generally used is Proton Exchange Membrane Fuel cell [10]. Chemical reaction is done at anode when hydrogen is reacts with anode then hydrogen is separated into two parts one is protons (H+ ions) and other one is electrons (e-). Electricity means flow of electrons (Figure 6).

Chemical reaction at anode is

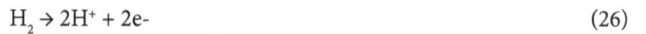

$$H_2 \rightarrow 2H^+ + 2e- \qquad (26)$$

Chemical reaction at cathode is

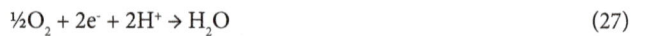

$$\frac{1}{2}O_2 + 2e^- + 2H^+ \rightarrow H_2O \qquad (27)$$

Overall reaction:

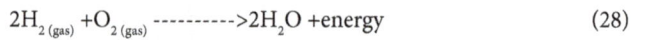

$$2H_{2\,(gas)} + O_{2\,(gas)} ----------> 2H_2O + energy \qquad (28)$$

Controlling Algorithm for generation sources

```
function y= fnc (u)
if ((u>2.5) && (u<3.5))
y= [1; 1; 1; 0];
elseif (u<1.2)
y = [1; 0; 0; 0];
```

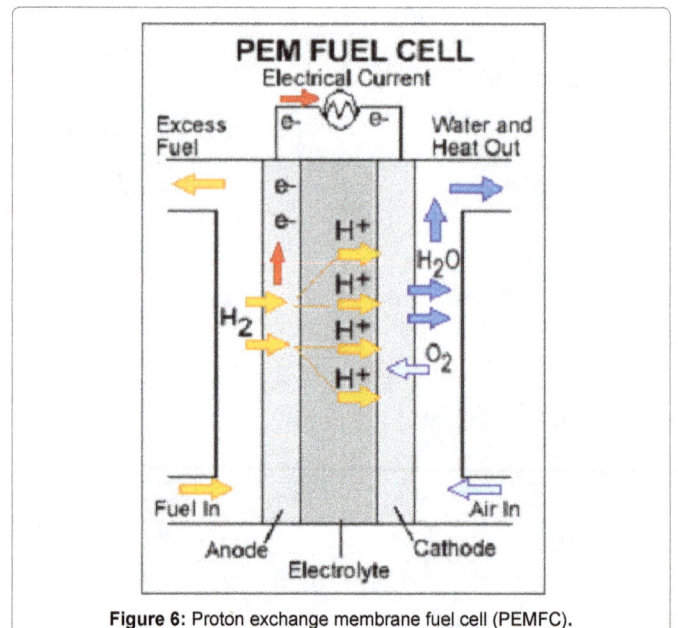

Figure 6: Proton exchange membrane fuel cell (PEMFC).

```
elseif ((u>1.21) && (u<2.5))
y= [1; 1; 0; 0];
else
y= [1; 1; 1; 1];
end
```

Matlab Simulation

MATLAB simulation of the system which has power generation solar, wind, fuel cell, battery. DC bus voltage is nearly 1 KV and the loads are of AC load type and 5 loads and power rating of loads are 1 KW, 2.5 KW, 1 KW, 1 KW, 1 KW respectively. So DC should be converted into AC by using inverter. In each source one breaker is connected to control the source depending upon the load connected to the system [11,12]. The input condition of breaker is output of control algorithm code and code is given above.

Simulation Results and Discussion

When input of circuit breaker of load (1) is 1 that means circuit breaker closes then only solar supplies generated power up to current be below 1.2 A. When the circuit breaker of load 2 is ON and the load is between 1.21 A to 2.5 A wind supplies generated power along with solar. Before wind supplies power fuel cell is used because wind turbine blades have to rotate and have to settle [13]. After wind supplies generated power then fuel cell turns off. When circuit breaker of 3rd load turns on and the load reaches between 2.5 A to 3.5 A fuel cell also supplies generated power to loads along with solar, wind. When the load is beyond 3.5 A then all the sources supply generated power to load. Battery has to be charge when the input of battery is 0 and at this time bi directional converter will acts as buck converter , the source to buck converter is DC bus. Battery input is 1 that means battery discharges and at this time bi directional converter works as boost converter, battery supplies generated power to loads (Figures 7-11).

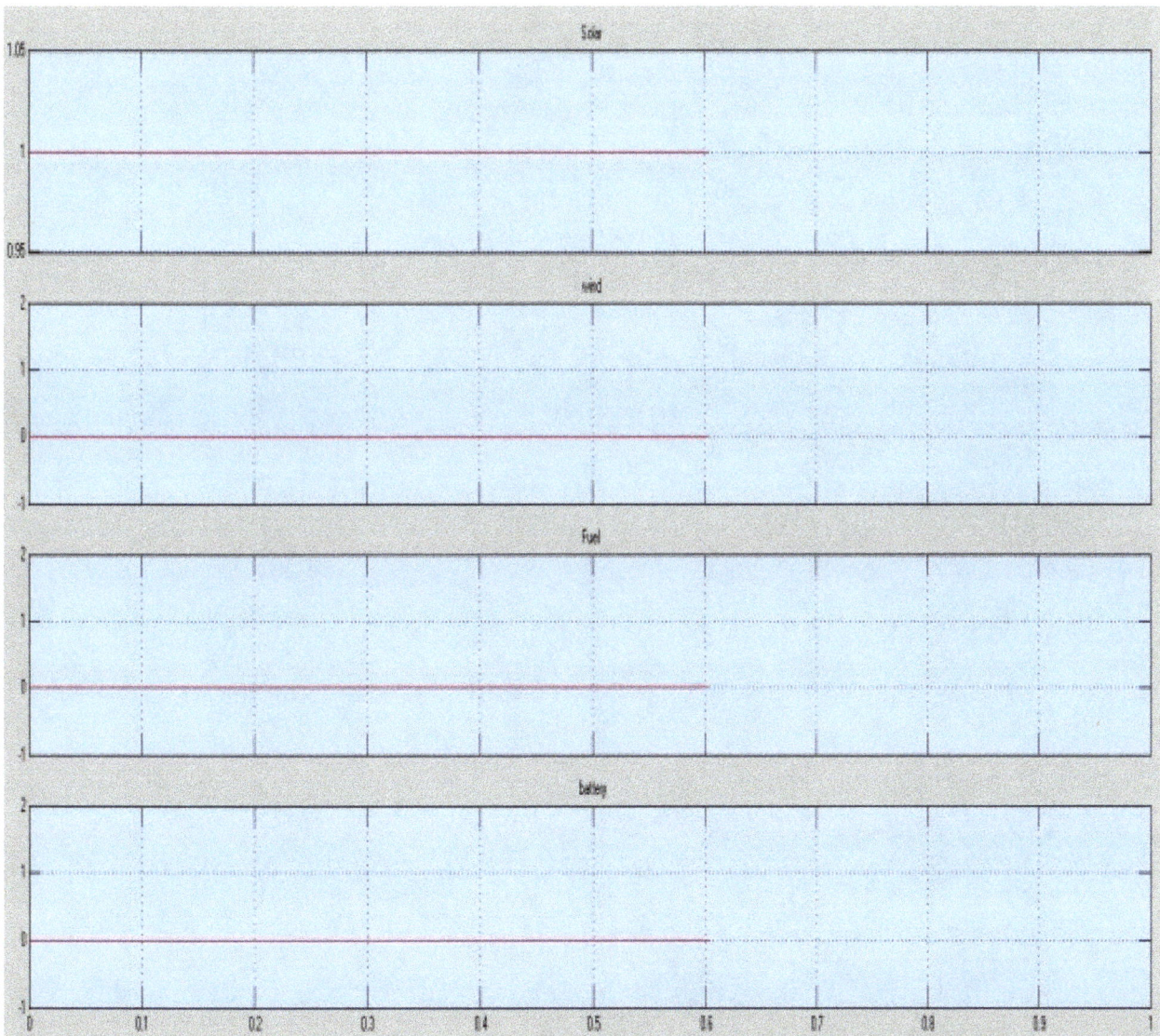

Figure 7: Solar ON state when load is below 1.2 A.

Figure 8: Solar and wind ON state when load is in between 1.2 to 2.5 A.

Figure 9: Solar and wind ON state when load is in between 1.2 to 2.5 A.

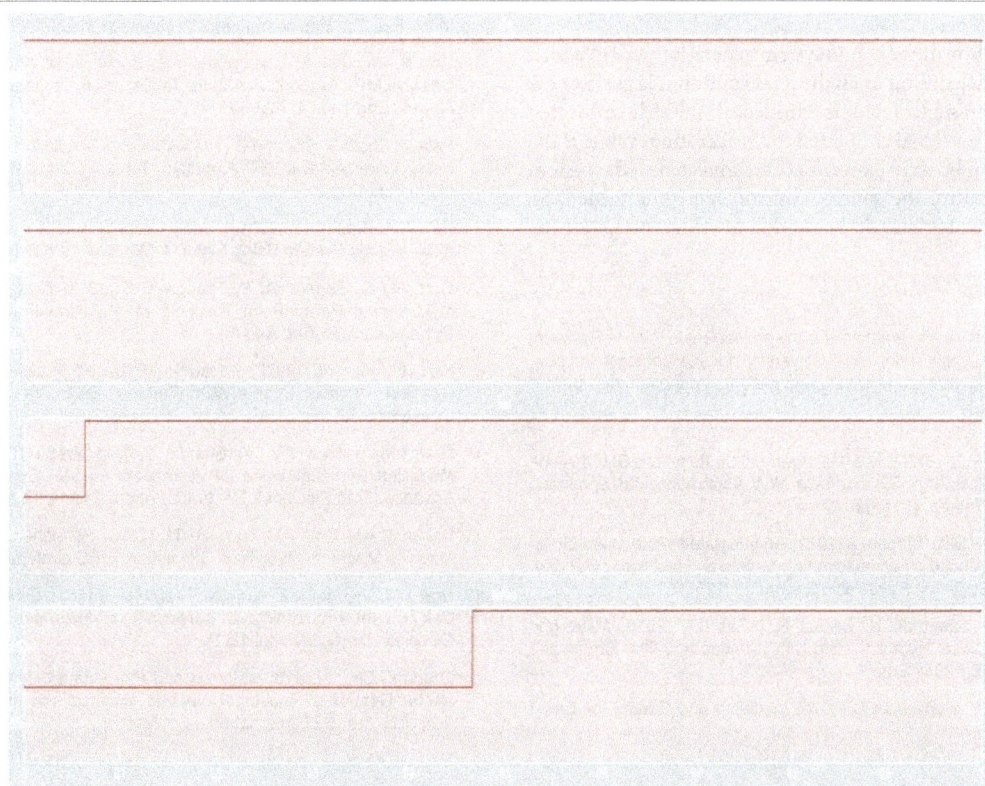

Figure 10: All sources is ON state when load is in beyond 3.5 A.

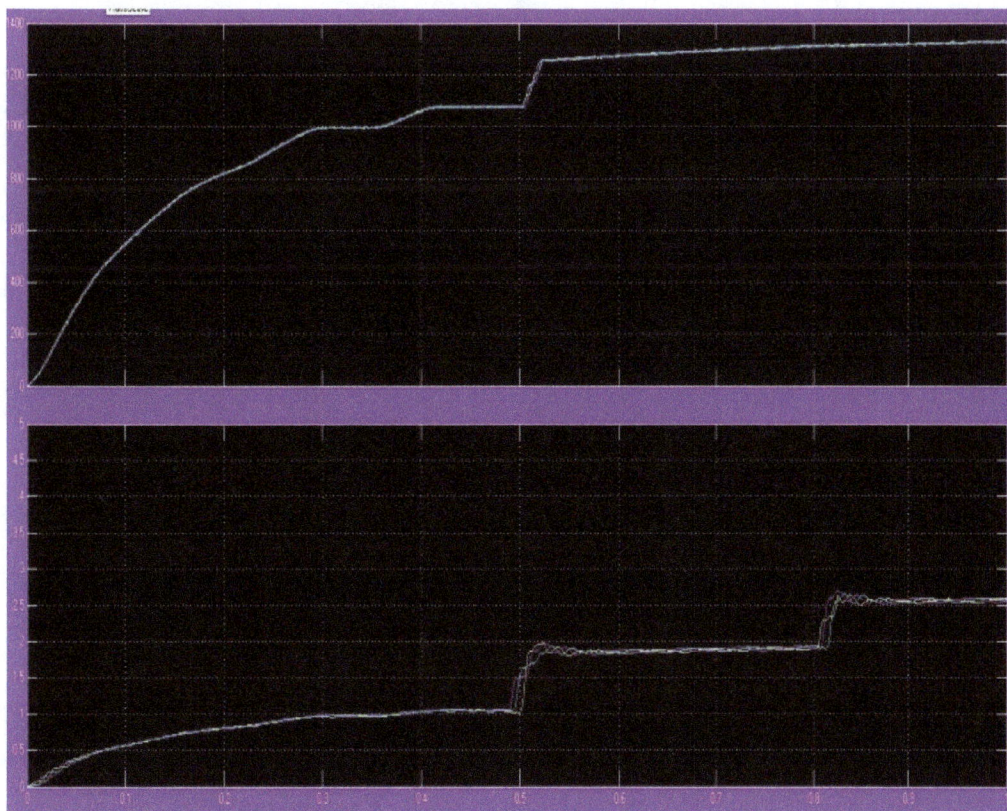

Figure 11: RMS values of load voltage and current.

Conclusion

Finally the simulation of the hybrid system generation system which controls the sources depending upon the load condition is performed. Here 5 loads are present. Load is connected and disconnected depending upon control signal connected to breaker input value. If the control signal is 1 then breaker closes, load is connected to the system and vice versa. To control the sources control algorithm needed. So in each source one breaker should be needed to switch on or off the sources.

Acknowledgment

I would like to express my sincere thanks to my project Guide Prof. V. Ramesh, Department of Electrical Engineering, VIT University, for his constant support, timely help, guidance, sincere co-operation during the entire period of my work.

References

1. Pinto PJR, Rangel CM (2010) A Power Management Strategy for a Stand-Alone Photovoltaic/Fuel Cell Energy System for a 1kW Application, 3rd Seminario Internacional Torres Vedras, Portugal.

2. Ghoddami H, Delghavi MB, Yazdani A (2012) An integrated wind-photovoltaic-battery system with reduced power-electronic interface and fast control for grid-tie d and off-grid applications, Renewable Energy 45: 128e137.

3. Arlampalam A, Mithulananthan N, Bansal RS, Sab TK (2010) Micro-grid Control of PV-Wind-Diesel Hybrid System with Islanded and Grid Connected Operations, EEE ICSET 2010 6-9.

4. Kumar SS, Swapna B, Nagarajan C (2014) Modeling and Control for Smart Grid Integration with MPPT of Solar/Wind Energy Conversion System, Int J Innov Res Sci, Engineering and Technology 3: 920-929.

5. Das B, Jamatia A, Chakraborti AP, Bhowmik M (2012) New Perturb And Observe MPPT Algorithm And Its Validation Using Data From PV Module, Int J Advanc Eng Tech 4: 579-591.

6. Patel J, Sheth V, Sharma G (2013) Design &Simulation of Photovoltaic System Using Incremental MPPT Algorithm, Int J Advance Res Electrical, Electron Instrument Eng 2.

7. Patil SN, Prasad RC (2014) Design And Simulation of MPPT Algorithm For Solar Energy System Using Simulink Model, Int J Res Eng Appl Sci 02: 37-40.

8. Gomathy S, Saravanan S, Thangavel S (2012) Design and Implementation of Maximum Power Point Tracking (MPPT) Algorithm for a Standalone PV System, Int J Sci Eng Res 3.

9. Kim SK, Kim ES (2007) Pscad/Emtdc-Based Modelling and Analysis of A Gearless Variable Speed Wind Turbine, IEEE Transac Tions On Energy Conversion 22.

10. Gangadhara Rao SV, Vattikonda S, Mandarapu S (2013) Mathematical Modelling And Simulation Of Permanent Magnet Synchronous Motor, Int J Advanced Res Electrical, Electronics and Instrument Eng 2: 3720-3726.

11. Dehkordi AB, Gole AM, Maguire TL (2005) Permanent Magnet Synchronous Machine Model for Real-Time Simulation, IEEE conference.

12. Ural Z, Gençoğlu MT, Gümüş B (2007) Dynamic Simulation of a PEM Fuel Cell System, Proceedings 2nd International Hydrogen Energy Congress and Exhibition IHEC, Istanbul 13-15.

13. Zhang Z, Shu J (2013) Matlab-based Permanent Magnet Synchronous Motor Vector Control Simulation, Computer Science and Information Technology (ICCSIT), 3rd IEEE International Conference.

18

QoS Performance Evaluation of Voice over LTE Network

Ahmed J Jameel[1]* and Maryam M Shafiei[2]

[1]Department of Telecommunication Engineering, Ahlia University, Manama, Bahrain
[2]Department of Information Technology, Ahlia University, Manama, Bahrain

Abstract

This paper describes the QoS performance evaluation of voice over LTE network using OMNeT++; an open-source system-level simulator and SimuLTE. OMNet++ is a well-known, widely-used modular simulation framework, which offers a high degree of experiment support. As a result, it can be integrated with all the network oriented modules such as INET. We describe the voice over LTE, and show performance evaluation results obtained using the simulator.

Keywords: OMNeT++; INET; VoLTE; LTE; Voice over LTE simulation

Introduction

In the past few decades, the mobile communication industries have evolved very fast to shift between each generation ranging from the 1G to LTE. This evolution from 1G to 4G was not as easy as it took a lot of work to make 4G technology the fastest network rollout. Long Term Evolution (LTE) was designed for the data transfer and also as a packet switched all-IP system. It does not contain any circuit switched domain for the purpose of providing with the regular voice and SMS services. The increase of the data traffic raised the issue of mobile broadband services by the consumers. The latest of these developments is the voice over LTE (VoLTE) it is devised scheme for standardized system between the mobile operators to carry out voice over Long Term Evolution (LTE) technology by replacing voice over the old technologies. The idea of the voice over LTE based on simply adapt to a completely new infrastructure based on internet protocol (IP) to replace the old legacy (2G-3G). The VoLTE specifications are based on air-interface, which is based on orthogonal frequency division multiplexing (OFDM) [1].

The new technology VoLTE can provide a combined system for transfer the voice traffic over the long term evolution air network access and employ the voice-over-IP (voice over internet protocol) technology, which is based on the (IP)- multimedia and IMS sub system to provide an appropriate service and video calling. The setup protocol for connection control is the session initiated protocol (SIP) which built to work with generic open IP network. The LTE is acclimatized with the current networks (3GPP, GSM, WCDMA, HSPA) and support for full forward and backward compatibility, until the LTE network voice service is fully implemented so the voice calls will automatic shift and full back to the best old bearer available (2G and 3G) [2].

This paper is organized as follows: section 2 presents the LTE network architecture, in section 3, the simulation results are presented in section 4, and section 5 concludes the paper.

LTE Network Architecture

The high level of the LTE network architecture as shown in Figure 1 is mainly composed of three main components:

UE (the User Equipment)

This component mainly consists of few of the functionalities of Mobile Terminal (MT) that is held responsible for all kinds of functioning of the call. On the other hand, the Terminal Equipment, which is also considered as one of the major devices of UE serves the function of data streaming and Universal Subscriber Identity Module (USIM). The USIM stores the network identification and user information. In this simulation of the LTE network, the User Equipment such as mobile, tablet, laptop etc., has been used [1].

E-UTRAN

The Evolved UTMS Terrestrial Radio Access Network used to handle the radio communication between the user equipment (UE) and the EPC. E-UTRAN composed of one or more base station called eNB or eNodeB. One of the parts of E-UTRAN is also termed as The Radio Access Network (RAN). The eNB or eNodeB serves the function of providing E-UTRA user plane and also it controls the plane protocol terminators along the user equipment [3].

EPC (Evolved Packet Core)

Evolved Packet Core is composed of two main elements: The Service Gateway (S-GW) which allows the user to communicate with other users of LTE network and PDN Gateway (P-GW) which is responsible to provide the connectivity between UE and external network like Internet. It serves the function of controlling the network access, management of mobility, and the other functions of network management. The Home Subscriber Server (HSS) present in the EPC stores all the information related with the subscriber. The entity of management of mobility controls the release and set-up of connections existing between the packet data network and user. It also accomplishes its activity through the registration of UE authentication location and using valuable information from the HSS. The Packet Data Gateway (P-GW) does the function of GGSN and SGSN, which also signifies the connectivity to the IP network. This system is assigned with the varied tasks of assignment of IP address, DHCP functions, user authentication, Quality of Service (QoS), charging data creation Deep Packet Inspection (DPI) [3].

Simulator Overview

Simulation software accomplishes a major role in the analysis of complex automation system and non-linear control system. Few of

---

*Corresponding author: Ahmed J Jameel, Department of Telecommunication Engineering, Ahlia University, Manama, Bahrain
E-mail: adulaimi@ahlia.edu.bh

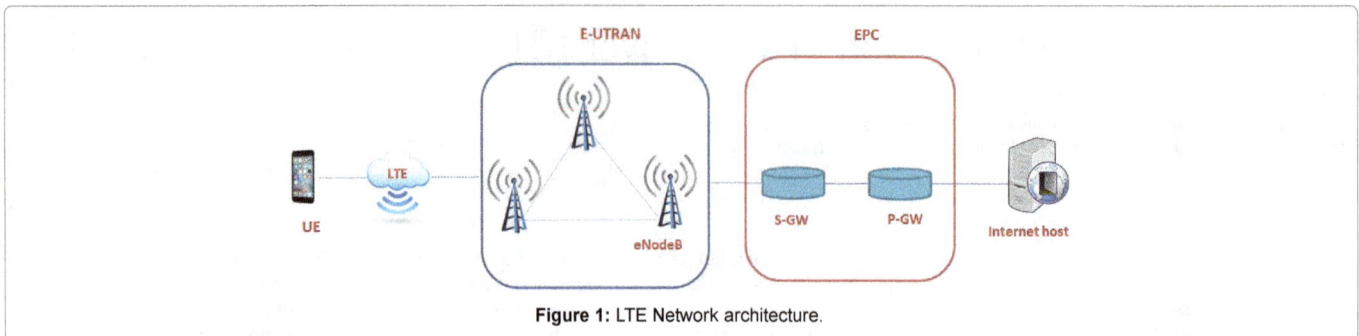

Figure 1: LTE Network architecture.

the software of computer that are designed for the dynamic system simulation at higher level than that of programming languages can be named as simulation languages, simulation software, simulation system, simulation environment and the simulators. Basically, simulation is explained as a particular method which is used for the solving of a problem in the dynamical systems, and which also finds out the model of the system rather than the real system. Simulation process follows few of the steps in sequence, which can be listed as formulation of problem, collection of data, mathematical modelling, identification of the model, and experiments with the model, representation of the result and interpretation of the result. Simulation software is usually used for designing, studying and analyzing the network communications. There are various software's available in the market that can serve this purpose. Most of the simulation software's are commercial but some of them are free for non-commercial use such as OMNET++.

OMNeT framework

OMNET++ is and extensible open-source library and framework primarily used to simulate networks. It can be used in various problem domains such as modelling of wired and wireless communications networks, evaluating performance aspects of complex networks. OMNET++ is widely used by academic institutions and educational environments for teaching purpose. It's also used by students and researchers to study and analyze the performance of the networks.

The basic building block of OMNeT++ is modules, either simple module or compound module. These modules communicate through messages that are sent and received through connection linking the gates of the modules. OMNeT++ facilitates the user to keep the implementation, description and parameter values of the model separate. C++ is used as the coding of the implementation. The files written in Network Description (NED) language is used for expressing the description. Theses NEDs also allow for writing of the parametric topologies. The major reasons for selecting OMNeT++ as the major tool for the simulation is that it is one of the most mature, stable and enriched with features framework [4].

INET framework

OMNET++ has some of external extensions that can be used to design and simulate the wireless network such as INET Framework. The INET framework is an open-source model that should be installed on top of OMNET++. In addition to the wireless network, it can be used to simulate wired and mobile networks. It contains IPv4, IPv6, TCP, SCTP, UDP protocol implementations and some of the other application models. As that of OMNET++, INET framework also uses the similar modules that communicate through the passing of message [5].

SimuLTE

SimuLTE for OMNeT++ can be used to analyze and evaluate the performance of LTE and LTE Advanced networks. It is an open source project developed by group of researchers to evaluate the complex network environments. It should be installed on top of OMNET++ and INET Framework. It simulates the data plane of the LTE Radio Access Network and Evolved Packet Core. SimuLTE has a special feature that it contains around 40000 lines of codes that helps in the extra functionalities such as the applications, mobility, event queues, ID/UDP, and so on. However, in this particular study of the voice over LTE, this distinguished feature of SimuLTE has been extracted from the OMNeT++ and INET frameworks. [5].

Simulation Results

This section describes the implemented simulation topology in OMNET++, GUI and explains the simulation parameters used in the experiment.

Quality of Service (QoS) criteria

The performance of Voice over LTE can be measured with the help of various criteria. In this experiment the major focus will lie on the following four of the major criteria.

Mean Opinion Score (MOS): MOS is the grading system that is used for the measuring of the quality of a voice call. It is usually graded by the user with the scale of 1 to 5, which means bad to excellent. This particular score is determined by few of the factors such as end to end delay, jitter and packet loss. One of the empirical formula that can be used for the calculation of MOS score from the packet loss in terms of percentage in milliseconds is as follows [6].

$$MOS = \ln(\text{loss}) - 0.1\ln(\text{size}) \qquad (1)$$

The following Table 1 shows the standard and the ideal quality values for the Mean Opinion Score (MOS).

End to end delay: End to End Delay is the time taken for a voice packet to be transmitted from the source UE to the destination UE across the LTE network. In simple words, it can be explained as the difference in the time between the sending and receiving of the packet. It basically takes place due to the performance of the network and the distance that exists between two of the nodes. This parameter is crucial so as to receive more information on the voice of a real time. There would be difficulty in having the effective communication in case of too much delay.

The following Table 2 shows the average and the ideal quality values for the VoLTE End to End Delay.

Packet loss: Packet Loss can be defined as the number of the transmitted packets that are failed to reach its destination. It can also be described as the particular rate in which the packets that are being sent do not reach at the receiving end. The real time communications

MOS	Quality
5	Excellent
4	Good
3	Fair
2	Poor
1	Bad

Table 1: MOS standard.

End to end delay	Quality
<50 ms	Ideal
<150 ms	Average

Table 2: End to end delay standard.

Packet loss rate	Quality
<1%	Ideal
<5%	Average

Table 3: Packet loss rate standard.

Jitter	Quality
<20 ms	Ideal
<50 ms	Average

Table 4: Jitter standard.

are based on the UD protocols. This protocol is usually without any connections and it cannot be send again if the packet is lost. The loss of the packages can also take place by removing all those packets that do not arrive to the end of the receiver on time. It becomes problematic whenever the loss of packet takes place in a bulk. The highest rate of packet loss so the voice can be heard with enough quality must be 1%.

The following Table 3 shows the average and the ideal quality values for the Packet Loss during Voice over LTE session.

Jitter: Jitter is the variation in the latency of the voice packets sent from the source to the destination. This basically occurs due to the congestion in the network. These similar cases can be solved with the addition of jitters buffers. This is an important parameter to be considered while measuring the quality of service since the high jitter can lead to poor quality of voice. The high jitter usually leads to the weaker quality of call as the information of the voice will not be received within the timely manner and thus, the information will not make any sense. In the technical terms, jitter is the measure of the variability of the latency over the time and also across the network [7].

The jitter that exists between the starting and final point of the communication must always be less than 100 ms. If the value of the jitter becomes smaller than 100 ms, it can be adjusted with the addition of jitter buffers [7]. The following Table 4 shows the average and the ideal quality values for the Jitter:

Simulation configuration

This section presents all the general parameters used in the conduction of the simulation.

- Ethernet Link Data Rate: 10 Mbps
- Simulation Time: 20 sec
- Packet Size: 40 byte
- Queue Size: 1 MB

Voice Over LTE Scenarios

Scenario (1 and 2) Voice over LTE network: OMNET++

architecture for the first and second scenarios is illustrated in Figure 2. The high level VoLTE network is composed by the following elements:

- Two User Equipment Support Voice Over LTE
- Two eNodeB
- Four Routers
- Two S-GW
- One P-GW
- One Internet Host

Scenario (1 and 2) Voice over LTE network: The third scenario of the VoLTE network (Figure 3) is composed with the help of below mentioned elements:

- Six User Equipment
- Two eNodeB
- Four Routers
- Two S-GW
- One P-GW
- One Internet Host.

Simulation analysis and result

This section presents the simulation analysis and result for the conducted experiment. The quality of service can be measured by several of the factors. In this experiment, the quality of the network for each scenario has been compared in terms of MOS, End to End Delay, Packet Loss Rate and Jitter.

Scenario 1: This scenario has been implemented to conduct an evaluation analysis of the performance of VoLTE between two UEs. The speed of the sender and receiver of the voice is 0 m/s. The following Figures have been obtained after running the simulation of the first scenario.

MOS: The MOS of the first scenario stayed above 4 during the simulation which falls under the category ranging between the scale of good and excellent. The average of MOS we obtained is 4.36, which is the normal value of any VoLTE service.

The following Figure 4 shows the MOS obtained after running the simulation of Scenario 1

End to end delay: As we can see in Figure 5, we have delay for about 8 sec between the period 7 s-11 s. The average end to end delay is 1.77 ms which meets the standard since it's below 50 ms.

Packet loss rate: According to the explanation and description of the packet loss, the average value obtained from the simulation is 0.21%. This rate is very small as compared to that of the ideal value (1%) (Figure 6).

Jitter: The jitter as seen in Figure 7 remained static and didn't change over the time. It stayed at the same value (8 ms) till the end of the call. Since the result is less than 20 ms this means that the quality of voice in this scenario was excellent.

Scenario 2: In this scenario, we studied the evaluation of the performance of VoLTE between the sender (UE1) and receiver (UE2). The simulation was conducted for about 20 s and the speed of the UEs

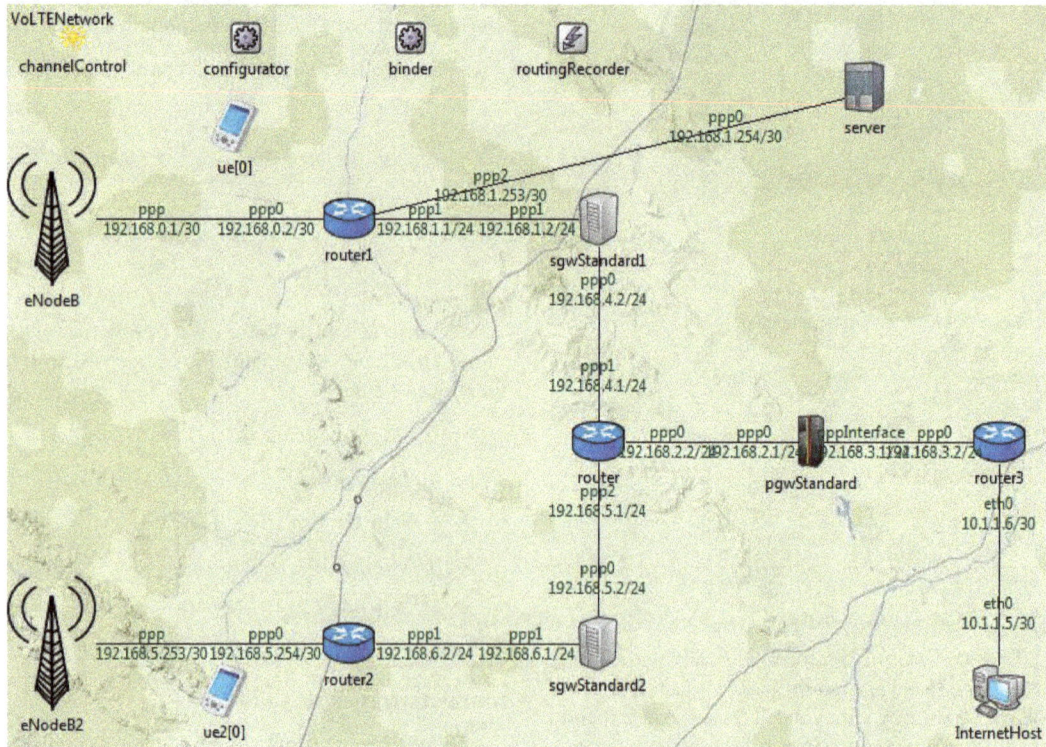

Figure 2: OMNeT++ LTE Network topology (Scenarios 1 and 2).

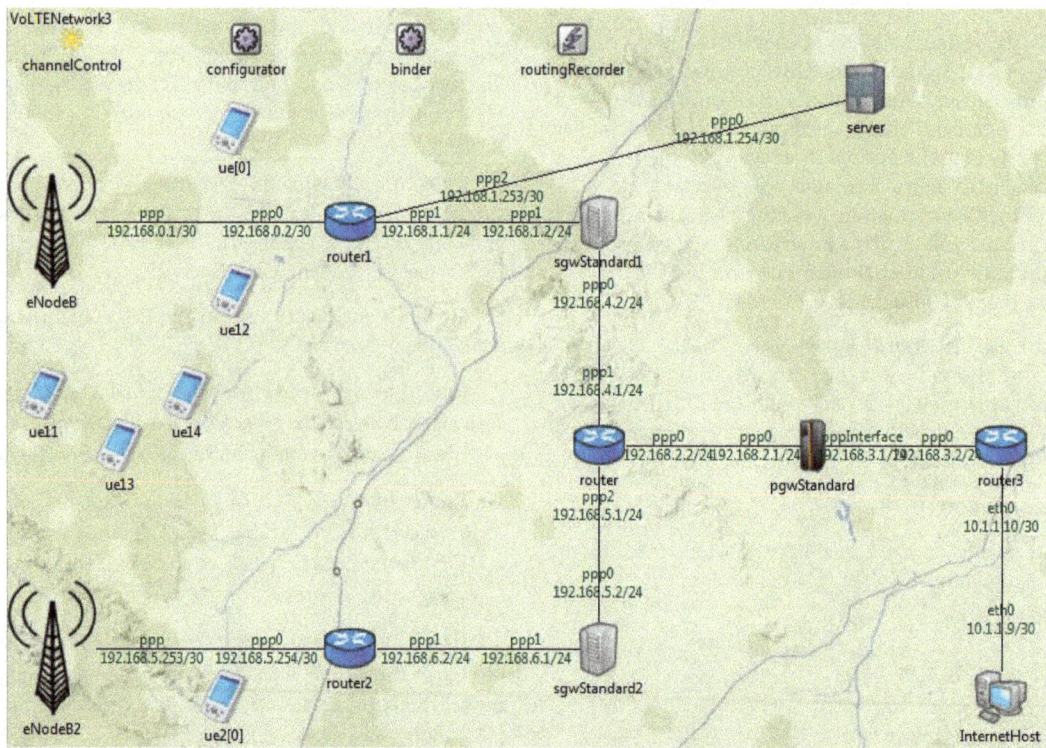

Figure 3: OMNET++ LTE Network topology (scenario 3).

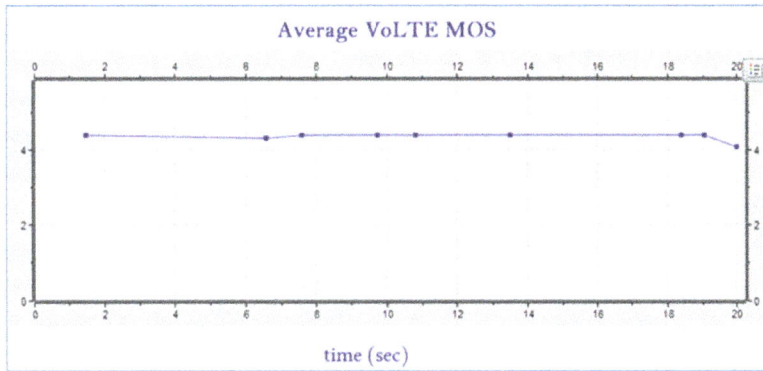

Figure 4: Scenario 1 MOS.

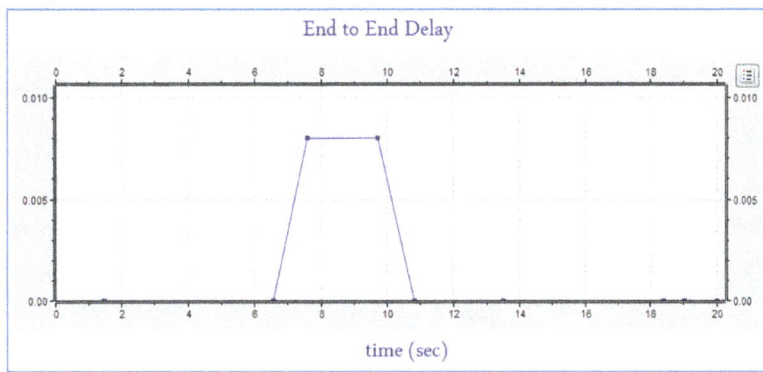

Figure 5: Scenario 1 end to end delay.

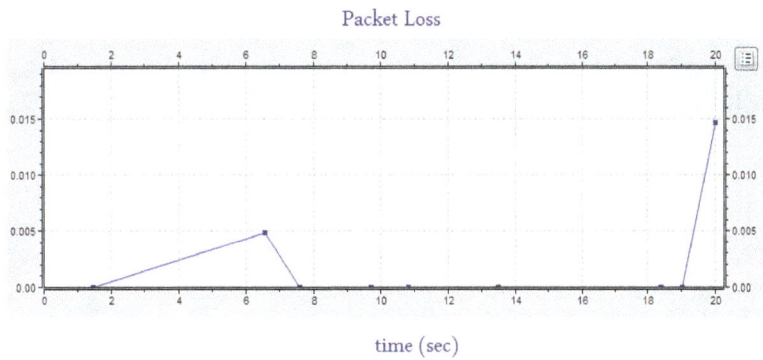

Figure 6: Scenario 1 packet loss rate.

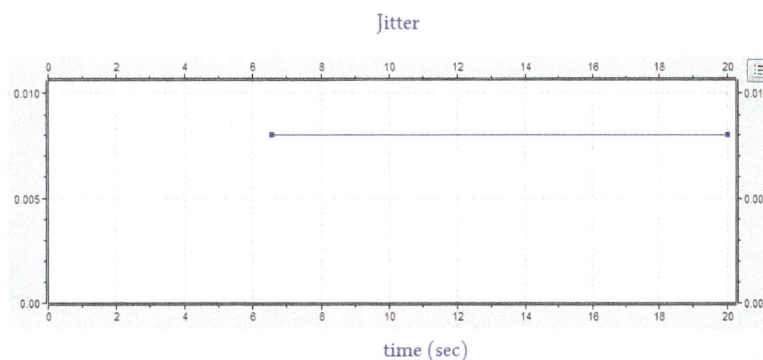

Figure 7: Scenario 1 jitter.

during the voice conversation was 100 km/h (28 m/s).

The following Figures are obtained after running the simulation of the second scenario:

MOS: As the following Figure 8 shows, the MOS for the second scenarios is varying over the time. It started with 4.4, and then dropped to 1.7, then again increased to 4.4 and finally ended up with 1.7. The average value is 3.35 and based on the standard rating, it can be determined that the quality of the voice is ranging between fair and good.

End to end delay: As we can see in the Figure 9, the delay started with 0 ms then after 4 seconds from the beginning of the conversation

it reached 8 ms then again it decreased to 0 ms. The average delay we got in this term is 2.6 ms, which is acceptable since it is less than 50 ms.

Packet loss rate: The average percentage of the packets loss for scenario 2 is 7.86%. This value is more than the ideal (1%) and the average (5%) percentage of the acceptable loss in VoLTE service. This variance in the percentage may affect the quality of the voice between the sender and the receiver.

From this result we can conclude that the speed of UEs during the VoLTE session possess the capacity of directly affecting the performance of the call (Figure 10).

Jitter: The following figure shows the jitter result for the second

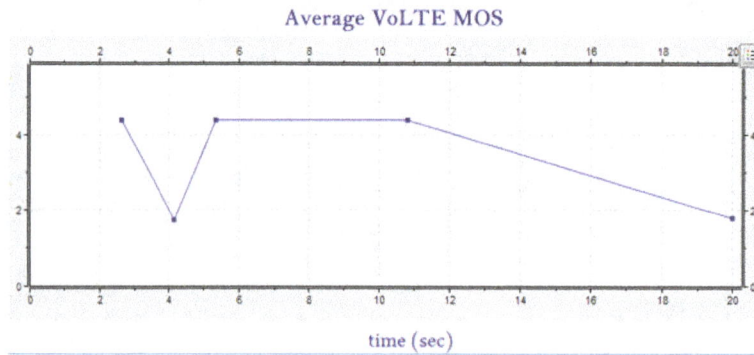

Figure 8: Scenario 2 MOS.

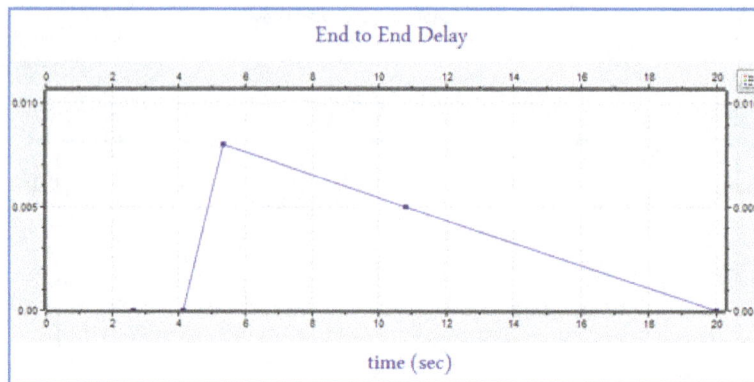

Figure 9: Scenario 2 end to end delay.

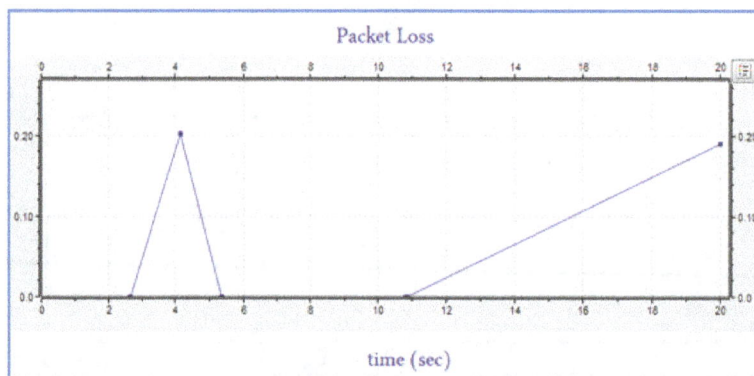

Figure 10: Scenario 2 packet loss rate.

scenario. As we can see from the chart, the jitter started to increase after 4 seconds from the start of the call. It increased from 8 ms to 23 ms and then dropped to 3 ms. The average jitter we got is 7.7 ms, which is acceptable as per the standard grading.

From the result we can conclude that making VoLTE call while driving may affect the quality of voice depending on the speed of the UEs (Figure 11).

Scenario 3: In this scenario, the sender (UE1) is calling the receiver (UE2) via VoLTE service while other four UEs are downloading a video of size 200 mb from the server. The other four UEs are connected to the same eNodeB as UE1.

MOS: The MOS of scenario 3 does not have a big difference when it is compared to the first Scenario (Figure 12). The MOS average value

we obtained after conducting the simulation is 4.192, which can be considered to be close to the value of scenario 1 (4.3619).

End to end delay: The average delay value obtained from the simulation of scenario 3 is 2.5 ms which is much less than the ideal value (50 ms) and the average value (150 ms). This means that having a congested eNodeB while making a call in LTE network should not lead to a big delay in the packets sent (Figure 13).

Packet loss rate: The line chart in Figure 14 shows the packet loss during the simulation of scenario 3. As we can see the average percentage of the packet loss are less than 1% of the total packets sent which is acceptable (Figure 14).

Jitter: The below mentioned diagram represent the jitter result from the third scenario. It can be seen that the jitter started to increase

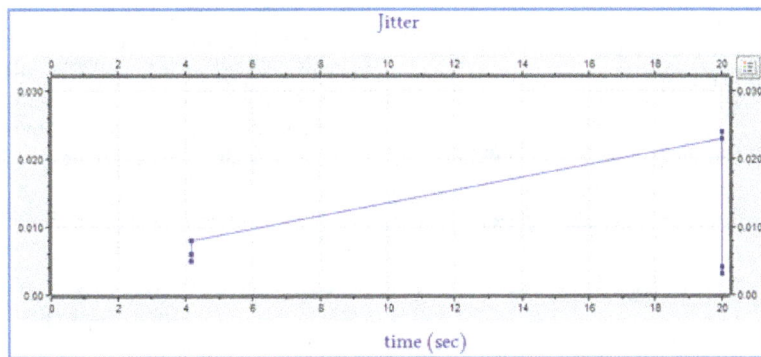

Figure 11: Scenario 2 jitter.

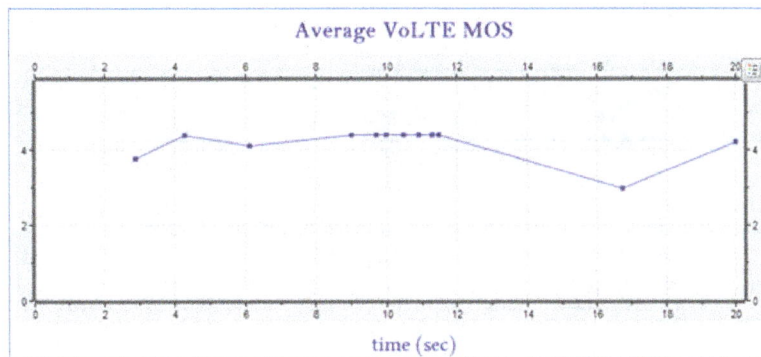

Figure 12: Scenario 3 MOS.

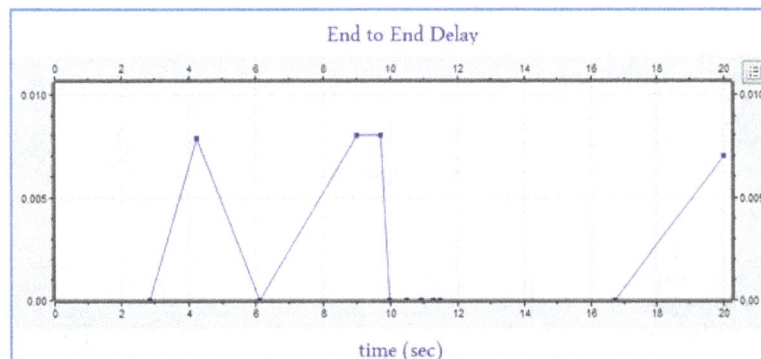

Figure 13: Scenario 3 end to end delay.

after 3 sec from the start of the call and again decreased to 1 ms after 17 sec. By 17 sec, it again increased to 7 ms and by 20 sec, it again declined to 1 ms Figure 15.

Scenarios result side by side: In this section, a brief comparison between the three scenarios based on the average QoS parameters values has been introduced.

The following Table 5 and charts compare the performance of the three conducted scenarios.

MOS: In the following diagram, the comparison of the voice call over LTE has been compared by keeping into consideration the three of the scenarios. Three of the different scenarios are displaying the irregular frequency of the VoLTE MOS vector.

End to end delay: The average delay value of the three scenarios has been compared in the above diagram. It can be observed that the first scenario is increasing at 7 sec and then again declining to o ms at 11 sec. The second scenario shows that at 4 sec, it is increasing and again it declines at slow rate. The final third scenario is showing a fluctuating rate with both increase and decline (Figures 16 and 17).

Packet loss: In the above diagram, the comparison of the packet loss during the simulation of three of the scenarios is demonstrated. The diagram shows a fluctuating rate in the different scenarios. Thus, if there is a fluctuation in the rate of packet loss, it might have adequate effect on the quality of voice between the sender and receiver [8,9].

Jitter: Similar to that of the above diagrams, in this, the comparison of the jitter results of all the three scenarios has been mentioned. The second scenario's jitter result show that it is increasing from 4 sec and again at 20 sec, it declined. In the first scenario, the jitter result is seen to be stable throughout the call. Finally, the third scenario declines initially and then rises to certain point and again declines drastically (Figures 18 and 19).

Conclusion

In this paper, performance analysis of Voice over LTE network is presented by studying the quality of service based on four of the major factors such as MOS, End to End Delay, Packet Loss Rate and Jitter. The simulation is designed and implemented with major simulation tools of OMNeT++ 4.6, INET Framework 2.6 and SimuLTE. Based on the simulated scenarios, we found that the speed of the sender and Receivers (UEs) are the crucial motivators that possess the capacity of seriously affecting the quality of the call. Once, the speed of the UEs

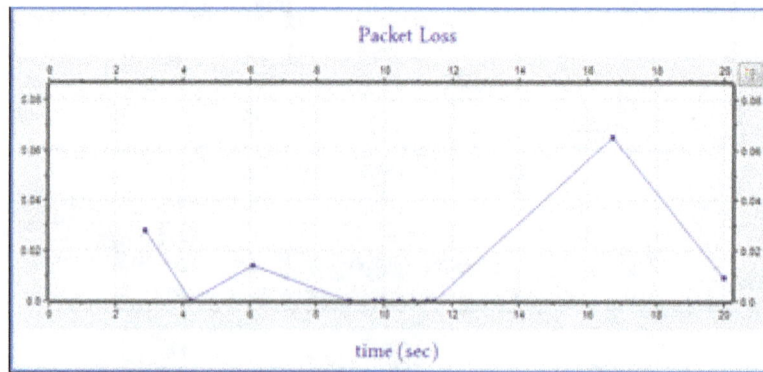

Figure 14: Scenario 3 packet loss.

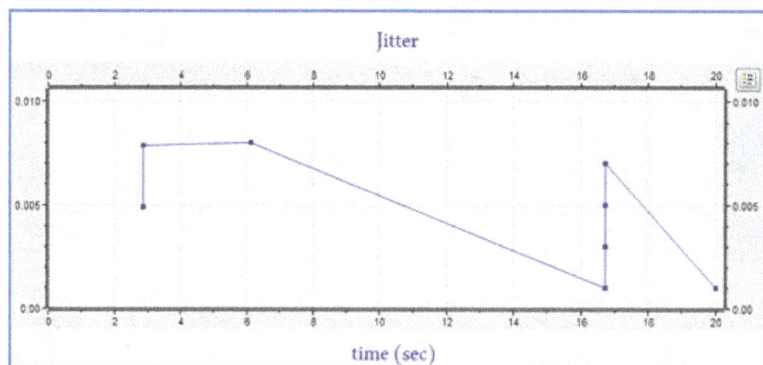

Figure 15: Scenario 3 jitter.

Scenarios	QoS			
No.	MOS	End to end delay	Packet loss	Jitter
Scenario 1	4.3619	1.7 ms	0.0021	8 ms
Scenario 2	3.3489	2.6 ms	0.0786	7.7 ms
Scenario 3	4.1942	2.5 ms	0.0096	2.8 ms

Table 5: Scenarios comparison.

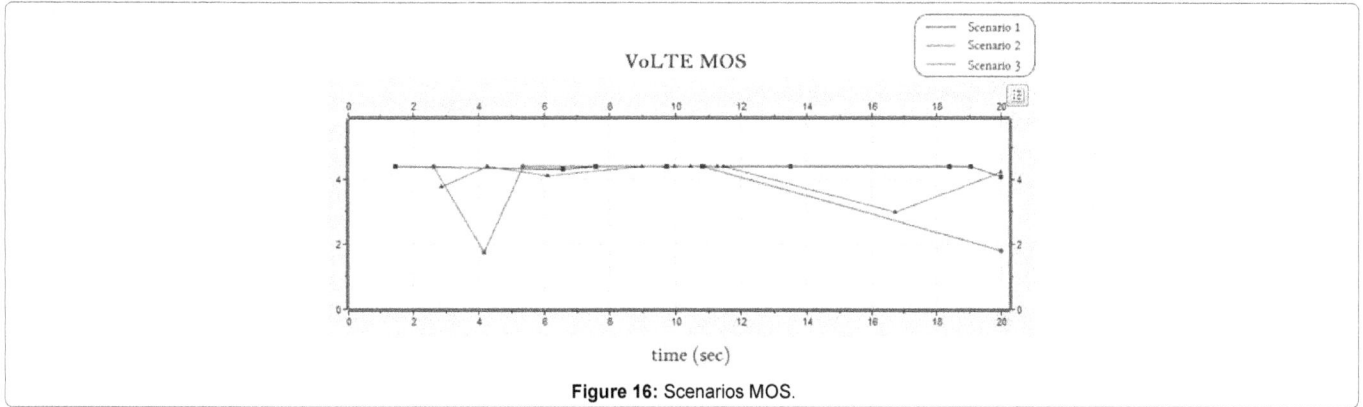

Figure 16: Scenarios MOS.

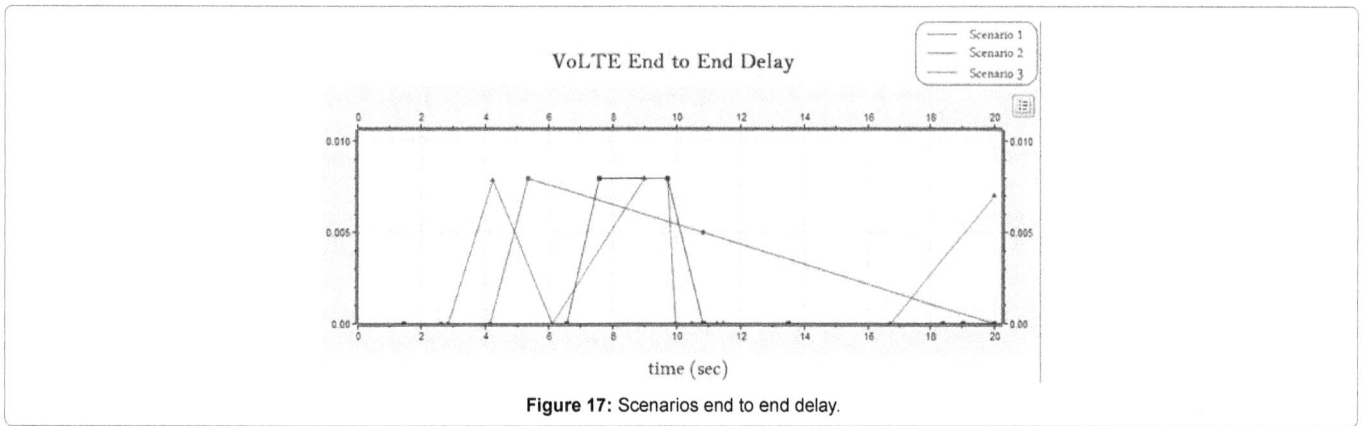

Figure 17: Scenarios end to end delay.

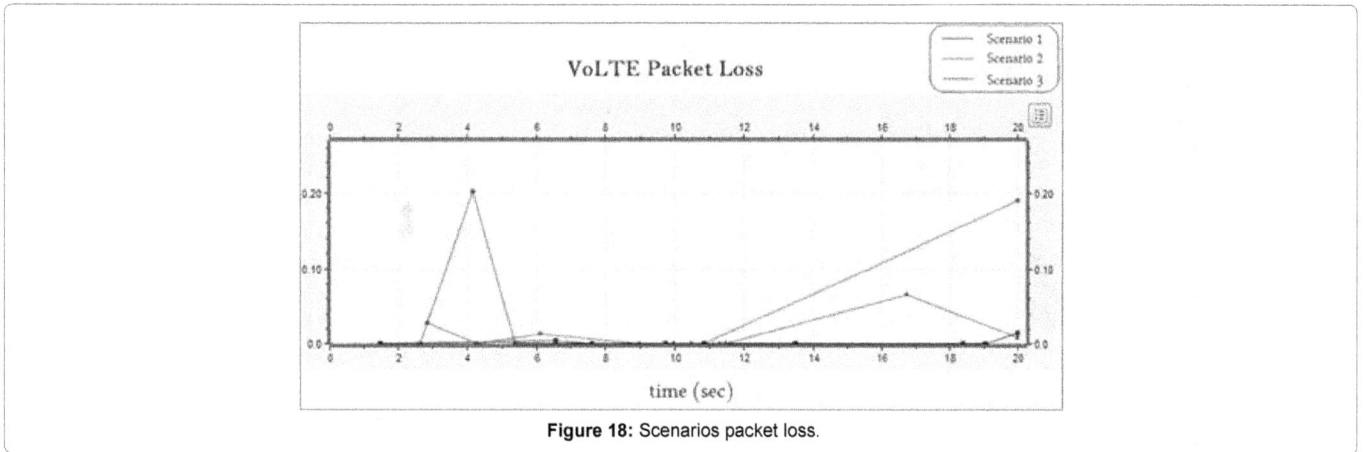

Figure 18: Scenarios packet loss.

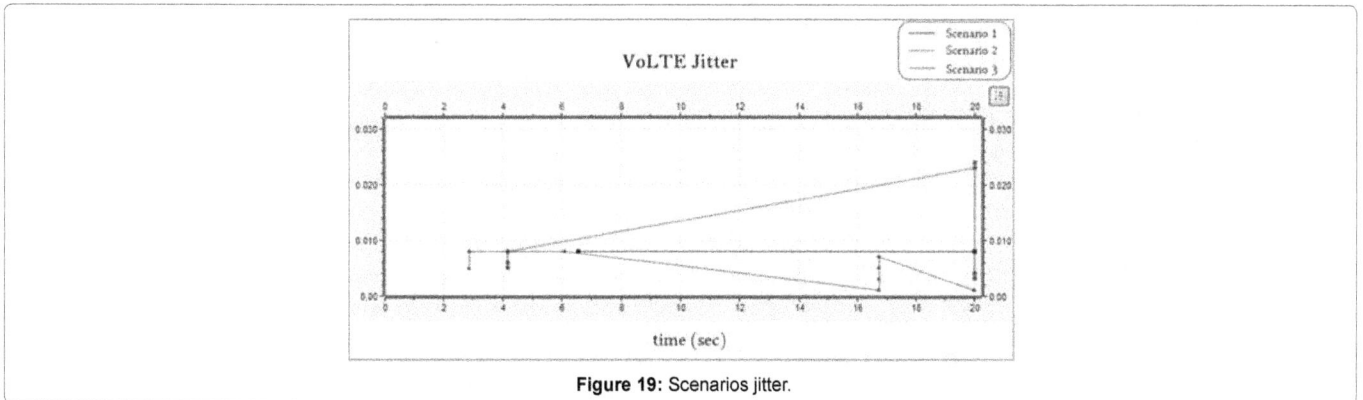

Figure 19: Scenarios jitter.

is changed from 0 to 100 km/s, the average value of MOS has been dropped from 4.3619 to 3.3489.

The jitter and the packet loss percentage were also affected by the speed of UEs. Therefore, taking into consideration, all these facts and other measures, the simulation module of the LTE network with the help of OMNeT++, INET Framework and SimuLTE has been successfully done for this particular study. Apart from the above mentioned information, this particular section explains the operation of the LTE network under the variety of scenarios. Each of the scenarios explains the performance of data and voice under the different configurations. In all the scenarios, the description of the performance of voice has been explained according to the increase in the times of the general response due to the increase in the demand for traffic along with maximum bit rate from the different users and the maximum throughput.

References

1. Poikselka M, Holma H, Hongisto J, Kallio J, Toskala A (2012) Voice over LTE (VoLTE). John Wiley & Sons.

2. Roessel S, Faerber M, Raaf B, Hausner J (2014) Radio Network Evolution towards LTE-Advanced and Beyond. Intel Technology Journal 18: 204-227.

3. Korowajczuk L (2011) LTE, WiMAX and WLAN Network Design, Optimization and Performance Analysis. John Wiley & Sons.

4. OpenSim Ltd (2015) OMNeT++. OMNeT++ 5.0rc released. Accessed April 02, 2016.

5. Virdis A, Stea G, Nardini G (2014) SimuLTE- A Modular System-level Simulator for LTE/LTE-A Networks based on OMNeT++. International Conference on Simultech.

6. Paessler (2016) How does PRTG calculate the MOS score for QoS sensors? Knowledge Base.

7. Think V (2015) QoS - Quality of Service. Jitter.

8. Virdis A, Nardini G (2015) SimuLTE. What is SimuLTE: simulator for LTE networks?

9. Breitenecker F, Troch I (2015) Simulation Software-Development and Trends. Control Systems, Robotics and Automation 4: 1-13.

19

Chaotic Pulse Generation Induced by a Specific Class of Autonomous Oscillator

Ndombou GB[1], Marquié P[2], Fomethe A[3], Yemélé D[3], Jeutho MG[3] and Kenmogne F[4]*

[1]*Laboratory of Electronics and Signal Processing, Faculty of Science, University of Dschang, Dschang, Cameroon*
[2]*Laboratory of Electronic, Informatics and Image (LE2I), University of Burgundy, Dijon Cedex, France*
[3]*Laboratoire de Mécanique et de Modélisation des Systèmes Physiques L2MSP, Faculté des Sciences, Université de Dschang, Cameroon*
[4]*Laboratory of Modelling and Simulation in Engineering, Biomimetics and Prototype, Faculty of Science, University of Yaoundé I, Yaoundé, Cameroon*

Abstract

The nonlinear dynamics of an autonomous chaotic oscillator, using two different stages operational amplifier coupled by mean of diode employed as the nonlinear device, recently introduced by Giannakopoulos and Deliyannis is considered with some particular modifications. These modifications are necessary for generating new type of oscillations, the regular and chaotic pulse oscillations according to the nature of operational amplifiers. Based on the nonlinear diode equation, the transfer voltage function of operational amplifiers in open loop configuration, and an appropriate selection of the state variables, a mathematical model is derived for a better description of the dynamics of the system. The complexness of oscillations is characterized using the bifurcation diagrams and the phase portraits. Some PSPICE simulations of the nonlinear dynamics of the oscillator are presented in order to confirm the ability of the oscillator to generate both the regular and chaotic pulse oscillations, according to the appropriate choice of its components.

Keywords: Pulse signals; Chaotic pulse signals; Transient chaotic pulse signal; Autonomous oscillator

Introduction

In recent years, communications via open networks such as satellites and internet occur more and more frequently. In most communication society, the transmitted messages are not to the public destination and must be protected against the pirate's access. In digital communications, only periodic pulse are used as carrier wave to modulate signals before their transmission [1], which remain a very promising practical application since the modulated signal by periodic pulse will propagate over very long distance without a significant attenuation, but remains nevertheless unsatisfactory since these information can easily be accessible.

Recent works on chaotic communication have revealed that information secured by chaos can be transferred from one place to another with higher security [2,3]. Since the chaotic signal is non-periodic, it cannot be stored in the receiver as a reference in order to achieve coherent detection of the transmitted signal [4]. This is why the interest in chaotic oscillators [5-7] and their possible applications in secure communication remain increasing. The use of fully digital chaos communication, necessary for the confidentiality of information requires a chaotic pulse train, which is not easy to generate [1]. The chaotic pulse train means the inter-spike interval has a broad distribution and is uncorrelated. Chaotic pulses have been already found in the literature [1,8] but the circuits are not simple and the pulse generation need a given chaotic time series and a pulse converter, using for a deep amplitude modulation of stationary chaotic signal at the output of chaotic source and which require permanent operation of chaotic oscillator. Chaotic radio-pulses are used as an information carrier in wideband and ultra-wideband communication systems [9,10].

More recently [11], the autonomous chaotic oscillator, consisting of the Deliyannis single amplifier biquad [11,12] and a LC resonant circuit coupled by means of a diode has been considered. In this chaotic oscillator, the negative resistance has been introduced in order to preserve the oscillations and has been implemented by using a negative impedance converter. It has been proved that this oscillator is described by a set of four differential equations, which exhibits a chaotic-like behavior, according to the resistance of the negative resistor.

In this paper, we reconsider the chaotic oscillator previously introduced in [11], in which some particular modifications have been carried out in order to introduce new effects on its dynamics, and we study the effects due to operational amplifiers on the dynamics of the system. To this end, the paper is organized as follows: In Section 2 we present the circuit under consideration, derive the equation of state and study the fixed point stability. In section 2, Numerical simulations will be carried out first, based on the derived state equations, in order to study the dynamic bifurcation and the pulse train generation. Next Pspice simulations are used to check numerical investigations and to justify the ability of the system to generate chaotic pulse train-like signal. Finally concluding remarks are devoted to section 3.

Circuit Description and State Equation

Circuit description

In this section, we describe the physical structure of the autonomous chaotic oscillator as depicted in Figure 1. This oscillator consists of two different stages operational amplifiers connected by means of a nonlinear diode. The first stage contains the linear inductor L, the linear capacitor, the resistors R_0, R_1 and R_2 in the positive feedback, and also the resistors R_0 and R_3 in the negative feedback. In this stage, the operational amplifier associate to both two identical resistors, R_0 and R_3 act as negative resistance (if the voltages at the positive and negative

*Corresponding author: Kenmogne F, Laboratory of Modelling and Simulation in Engineering, Biomimetics and Prototype, Faculty of Science, University of Yaoundé I, Po Box 812, Yaoundé, Cameroon, E-mail: kenfabien@yahoo.fr

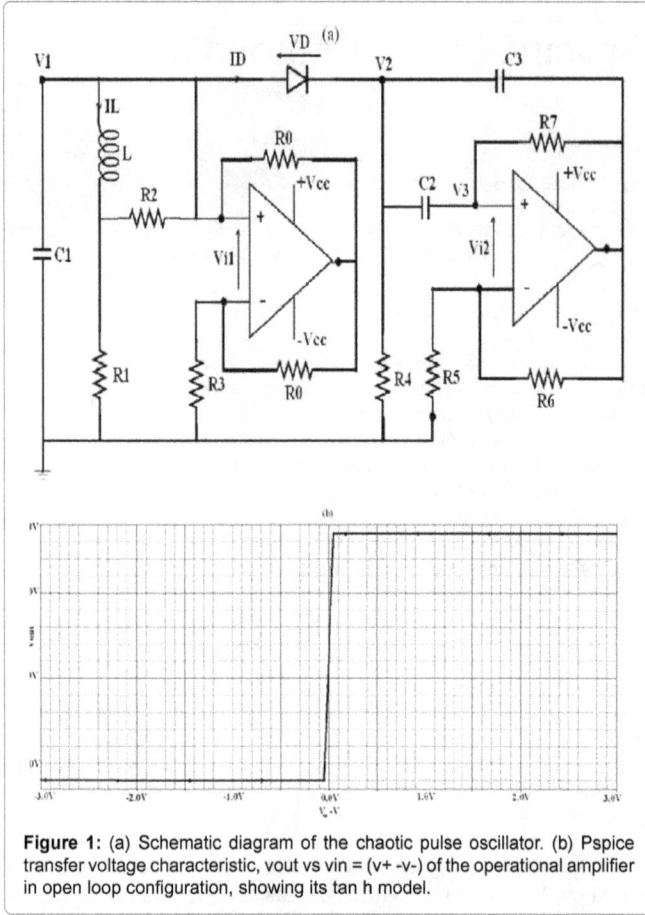

Figure 1: (a) Schematic diagram of the chaotic pulse oscillator. (b) Pspice transfer voltage characteristic, vout vs vin = (v+ -v-) of the operational amplifier in open loop configuration, showing its tan h model.

terminals are supposed to be identical) and can reduce to a negative resistance which $-R_3$ is introduced to preserve oscillations. The second stage contains both the linear capacitors c_2 and c_3, the resistors R_4 and R_7 in the positive feedback, and also the resistors R_5 and R_6 in the negative feedback. We notice that in [11], the resistors R_1 and R_2 had been not introduced and the effects due to operational amplifiers were neglected. In present work and in order to approximate our studies to the results which can be obtained with physical realistic operational amplifiers with the best accuracy, we suppose that the voltage difference between the positive and negative terminals $V_{in} = V_+ - V_-$ operational amplifiers are nonzero, that is the operational amplifiers are supposed to be non ideal. As proved in Figure 1a, the transfer voltage function characteristic of the operational amplifier in the open loop configuration, from input to output is nonlinear and is expressed as

$$v_{out} = V_{cc} tanh\left(\beta \frac{v_{in}}{V_{cc}}\right),\tag{1}$$

Where β is the voltage gain of operational amplifiers and the supply voltage used.

Equation of state

Denoting by i_L the current flowing through the inductor L, v_1 the voltage across the capacitor c_1, v_2 and v_3 he voltages at the left and the right of the capacitor c_1, respectively, and applying Kirchhoff's laws to the circuit of Figure 1, we obtain the following set of ordinary differential equations governing the dynamics of the system:

$$c_1 \frac{dv_1}{dt} = -\left(\frac{1}{R_1 + R_2} - \frac{1}{R_3}\right)v_1 - \left(\frac{1}{R_0} + \frac{1}{R_3}\right)v_{i1} - \frac{R_2}{R_1 + R_2}i_L - i_d ,$$

$$L \frac{di_L}{dt} = \frac{R_2}{R_1 + R_2}\left(v_1 - R_1 i_L\right),$$

$$c_3 \frac{dv_2}{dt} = \frac{1}{1-\left(1+\frac{R_5}{R_6}\right)\frac{dv_{i2}}{dv_3}}\left[\frac{R_5}{R_6}\left(\frac{v_2}{R_4} - i_d\right) + \frac{1}{R_7}\right.$$
$$\left.\left(\left(1+\frac{R_5}{R_6}\right)v_{i2} - v_3\right)\left[1 + \frac{c_3}{c_2}\left(1+\frac{R_6}{R_5}\left(1-\left(1+\frac{R_5}{R_6}\right)\frac{dv_{i2}}{dv_3}\right)\right)\right]\right],$$

where i_d is the current in the diode, which is related to the diode voltage as

$$V_d = V_1 - V_2 \tag{2}$$

$$i_d = I_s\left(exp\left(\frac{v_d}{\eta V_T}\right) - 1\right), \tag{3}$$

where I_S is the saturation current of the junction, V_t is the thermal voltage [13]. This thermal voltage is proportiona l to absolute temperature and takes the value 26 mV at room temperature, that is at 293 K, while η, with $1 < \eta < 2$ is the ideality factor of the diode. The input voltages v_{i1} and v_{i2}, are related to v_1 and v_2 by the following equations:

$$v_1 = v_{i1} + \frac{R_3 v_{cc}}{R_0 + R_3} tanh\left(\frac{\beta v_{i1}}{v_{cc}}\right),$$

$$v_3 = v_{i2} + \frac{R_5 v_{cc}}{R_5 + R_6} tanh\left(\frac{\beta v_{i1}}{v_{cc}}\right), \tag{4}$$

while dv_{i1}/dv_2 is the derivative of v_{i2} in term of v_3 When the operational amplifiers are supposed to be ideal, that is when $v_{i1} = v_{i2} = 0$, the set of Eq.(2) reduces to the following set of differential equations:

$$c_1 \frac{dv_1}{dt} = -\left(\frac{1}{R_1 + R_2} - \frac{1}{R_3}\right)v_1 - \frac{R_2}{R_1 + R_2}i_L - i_d ,$$

$$L \frac{di_L}{dt} = \frac{R_2}{R_1 + R_2}\left(v_1 - R_1 i_L\right),$$

$$c_3 \frac{dv_2}{dt} = \frac{R_5}{R_6}\left(\frac{v_2}{R_4} - i_d\right) - \frac{v_3}{R_7}\left(1 + \frac{c_3}{c_2}\left(1 + \frac{R_6}{R_5}\right)\right),$$

$$c_3 \frac{dv_3}{dt} = \frac{R_5}{R_6}\left(\frac{v_2}{R_4} - i_d\right) - \frac{v_3}{R_7}\left(1 + \frac{c_3}{c_2}\right), \tag{5}$$

In order to modify the above set of differential equations in the dimensionless form in a way convenient for analytical and numerical analysis, we introduce the following change of variables:

$$x_1 = v_1/\eta V_T, x_2 = \rho i_L/\eta V_T, x_3 = v_2/\eta V_T,$$
$$x_4 = v_3/\eta V_T, y_1 = v_{i1}/\eta V_T, y_2 = v_{i2}/\eta V_T , \tag{6}$$
$$t = \tau\sqrt{Lc_1}\rho = \sqrt{L/c_1}.$$

which lead the set of Eq. (2) to the following set of four ordinary differential equations governing signal voltage in the system

$$\begin{cases} \dot{x}_1 = \sigma(x_1 - y_1) - a_1 y_1 - \frac{1}{1+a_0 b_0}(x_2 + a_0 x_1) - \gamma g(x_1, x_3), \\ \dot{x}_2 = \frac{1}{1+a_0 b_0}(x_1 - b_0 x_2), \\ \dot{x}_3 = -\frac{1}{1-(\varepsilon+1)\frac{dy_2}{dx_4}}\left[\varepsilon_1 \varepsilon(b_1 x_3 - \gamma g(x_1, x_3)) + b_2\left(\varepsilon_1 + \varepsilon_2\left(1+\frac{1}{\varepsilon}\right)\left(1-\frac{dy_2}{dx_4}\right)\right)((\varepsilon+1)y_2 - x_4)\right], \tag{7} \\ \dot{x}_4 = \frac{1}{1-(\varepsilon+1)\frac{dy_2}{dx_4}}\left[\varepsilon_1 \varepsilon(b_1 x_3 - \gamma g(x_1, x_3)) + b_2(\varepsilon_1 + \varepsilon_2)((\varepsilon+1)y_2 - x_4)\right], \end{cases}$$

with

$$x_1 = y_1 + \frac{a_1}{\gamma_1(a_1 + \sigma)} tanh(\beta\gamma_1 y_1), x_4 = y_2 + \frac{\varepsilon}{\gamma_1(1+\varepsilon)} tanh(\beta\gamma_1 y_2), \quad (8)$$

and

$$\frac{dy_2}{dx_4} = \frac{cosh^2(\beta\gamma_1 y_2)}{\frac{\varepsilon\beta}{(1+\varepsilon)} + cosh^2(\beta\gamma_1 y_2)}, g(x_1, x_3) = exp(x_1 - x_3) - 1. \quad (9)$$

where by the dots we mean the derivative with respect to τ and where the following parameters are introduced

$$\sigma = \frac{\rho}{R_3}, a_0 = \frac{\rho}{R_2}, a_1 = \frac{\rho}{R_0}, b_0 = \frac{R_1}{\rho}, b_1 = \frac{\rho}{R_4}, b_2 = \frac{\rho}{R_7},$$

$$\varepsilon = \frac{R_5}{R_6}, \varepsilon_1 = \frac{c_1}{c_3}, \varepsilon_2 = \frac{c_1}{c_2}, \gamma_1 = \frac{\eta V_T}{v_{cc}}, \gamma = \frac{\rho I_s}{\eta V_T}. \quad (10)$$

For simplifying our investigations, the following values of parameters have been kept constant as follows:

L = 188.8 mH, R_1 = 2 kΩ, R_2 = 20 Ω, R_0 = 5.6 kΩ, R_4 = 200 kΩ, R_5 = 4 kΩ

R_6 = 100.1 kΩ, R_7 = 12.99 kΩ, c_1 = 62 nF, c_2 = 60 nF, c_3 = 70 nF (11)

The nonlinear diode used is the D1N4148 model with the characteristics I_s = 2.682 × 10^{-9} A and η = 1.9 leading to the following parameters values:

$$a_0 = 0.873, a_1 = 0.312, b_0 = 5307.045, \varepsilon = 3.996 \times 10^{-2},$$

$$\varepsilon_2 = \frac{31}{30}, \gamma = 9.474 \times 10^{-5}, \gamma_1 = 3.293 \times 10^{-3}, \varepsilon_1 = \frac{31}{35}, b_1 = 8.725 \times 10^{-3}, b_2 = 0.134. \quad (12)$$

while σ (that is, R_3 since σ = 1745.039051/R_3) and β are the tuning parameter

Fixed point analysis

It is easy to show that system [7] has one equilibrium point. E_0 (0,0,0,0) Physically, a steady state solution corresponds to an equilibrium state of the system and the behavior of the system may depend on its stability. To test this stability, let us consider the state E = E_0 + δ E vector, where δ E is the perturbation of the equilibrium solution E_0 (0,0,0,0). The stability of this equilibrium state against the perturbation δ E depends on the properties of the eigenvalues of the Jacobian matrix $J(E_0)$, which can be easily computed as follows:

$$\begin{bmatrix} -\left(\gamma + \frac{a_0}{1+a_0b_0} + \frac{a_1+\sigma(1-\beta)}{1+\beta+\sigma/a_1}\right) & -\frac{1}{1+a_0b_0} & \gamma & 0 \\ \frac{1}{1+a_0b_0} & -\frac{b_0}{1+a_0b_0} & 0 & 0 \\ \frac{\varepsilon_1\gamma(1+\varepsilon(1+\beta))}{1+\varepsilon-\beta} & 0 & \frac{\varepsilon_1(b_1+\gamma)(1+\varepsilon(1+\beta))}{\beta-1-\varepsilon} & -b_2\left(\varepsilon_1 + \frac{\beta\varepsilon_2(1+\varepsilon)}{1+\beta(1+\varepsilon)}\right) \\ \frac{\varepsilon_1\gamma(1+\varepsilon(1+\beta))}{1+\varepsilon-\beta} & 0 & \frac{\varepsilon_1(b_1+\gamma)(1+\varepsilon(1+\beta))}{\beta-1-\varepsilon} & -b_2(\varepsilon_1+\varepsilon_2) \end{bmatrix} \quad (13)$$

This Jacobean matrix evaluated at the equilibrium point E_0 satisfies the following characteristic equation:

$$\lambda^4 + j_3\lambda^3 + j_2\lambda^2 + j_1\lambda + j_0 = 0, \quad (14)$$

with

$$j_0 = \frac{\varepsilon_1\varepsilon_2 b_2}{1+a_0b_0}\left[(\gamma+b_1)\left(1+b_0\left(\gamma + \frac{a_1+\sigma(1-\beta)}{1+\beta+\sigma/a_1}\right)\right) - \gamma^2 b_0\right],$$

$$j_1 = \frac{j_0(1+\varepsilon(\beta+1))}{\varepsilon_2 b_2\varepsilon(\beta-1-\varepsilon)} + b_2\left(\gamma + \frac{a_0}{1+a_0b_0} + \frac{a_1+\sigma(1-\beta)}{1+\beta+\sigma/a_1}\right)$$

$$\left[\varepsilon_1\varepsilon_2(\gamma+b_1) + \frac{(\varepsilon_1+\varepsilon_2)b_0}{1+a_0b_0}\right] + \frac{b_2(\varepsilon_1+\varepsilon_2)}{(1+a_0b_0)^2} + b_0b_2\varepsilon_1\varepsilon_2\left(\frac{(\gamma+b_1)b_0}{1+a_0b_0} - \gamma^2\right),$$

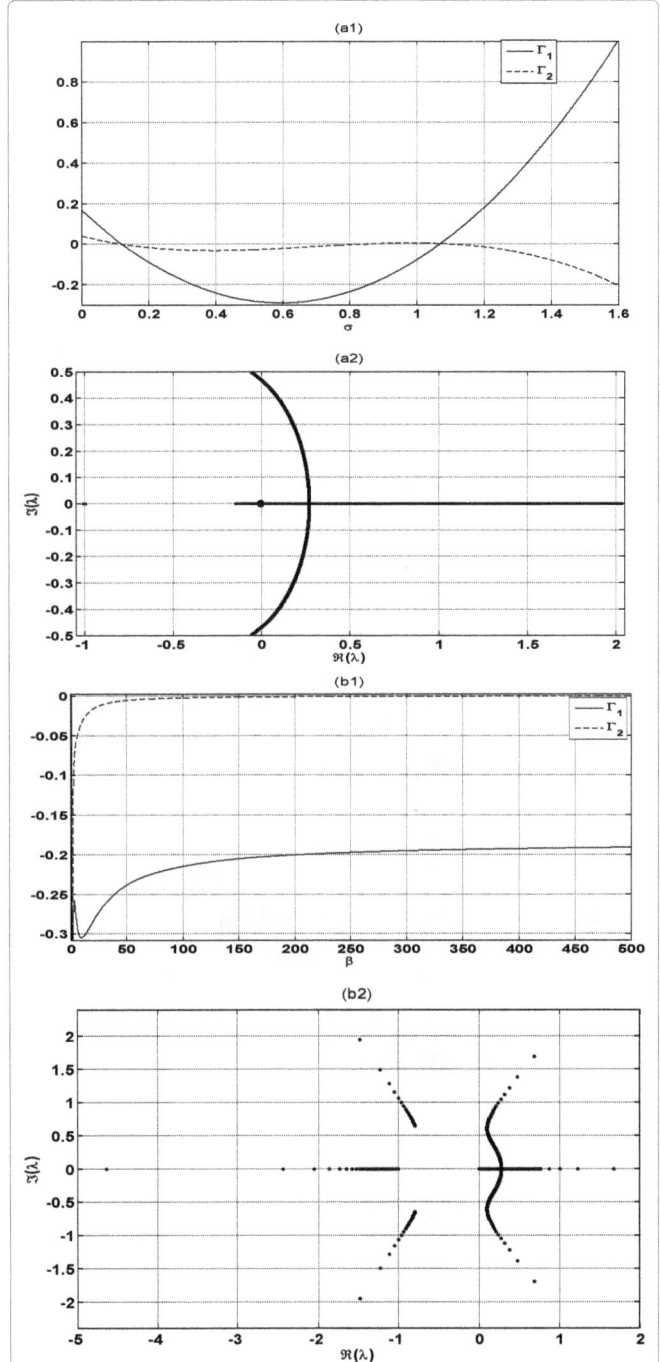

Figure 2: (a1) Parameters Γ_1 and Γ_2, and (a2) Representation in the complex plane of the eighteen values of the Jacobian matrix solutions of Eq. (14). The parameters of the system are listed in (12) (b1) β = 1500 and the parameter σ is used as the control parameter of the system. (b2) σ = 0.88 and β chosen as the control parameter.

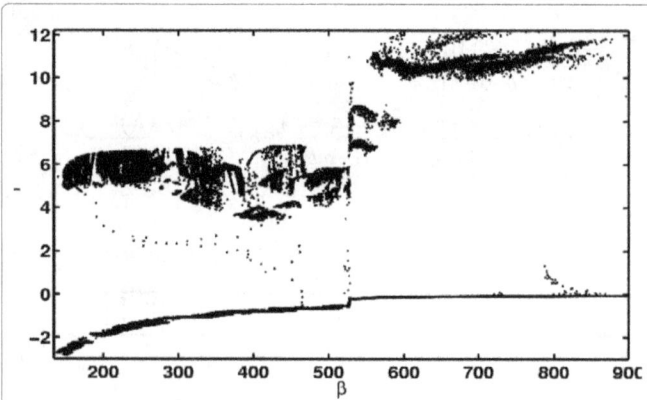

Figure 3: Bifurcation diagram obtained for σ = 0.88and the parameters given in Eq. 12, while is β chosen as a control parameter.

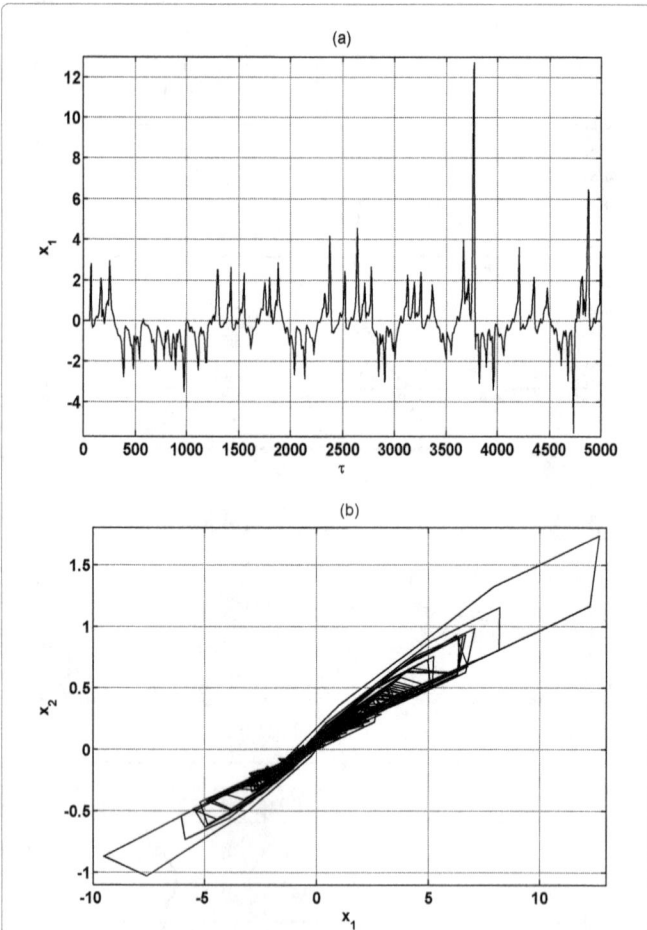

Figure 4: (a) Time waveform of variable $x_1(t)$, (b) Phase space plot x_2 against x_1. The parameters of the system are taken as in Figure 2, but with σ = 0.88 and β = 1500.

$$j_2 = \frac{\varepsilon_1\left(1+\varepsilon(\beta+1)\right)}{1+\varepsilon-\beta}\left[(\gamma+b_1)\left(\gamma+\frac{b_0+a_0}{1+a_0b_0}+\frac{a\_1+\sigma(1-\beta)}{1+\beta+\sigma/a_1}\right)-\gamma^2\right]$$

$$+\left(\gamma+\frac{a_0}{1+a_0b_0}+\frac{a_1+\sigma(1-\beta)}{1+\beta+\frac{\sigma}{a_1}}\right)\left(\frac{b_0}{1+a_0b_0}+b_2(\varepsilon_1+\varepsilon_2)\right)b_2\left[\frac{b_0(\varepsilon_1+\varepsilon_2)}{1+a_0b_0}+\varepsilon_1\varepsilon_2(\gamma+b_1)\right],$$

$$j_3 = \gamma+\frac{b_0+a_0}{1+a_0b_0}+\frac{a_1+\sigma(1-\beta)}{1+\beta+\sigma/a_1}-\frac{\varepsilon_1(\gamma+b_1)\left(1+\varepsilon(\beta+1)\right)}{\beta-1-\varepsilon}+b_2(\varepsilon_1+\varepsilon_2). \quad (15)$$

A set of necessary and sufficient conditions for all the roots of (14) to have negative real parts is given by the well-known Routh-Hurwitz criterion expressed in the form:

$$j_i > 0, \left(i=0,1,2,3\right), \quad \Gamma_1 = j_2 j_3 - j_1 > 0,$$
$$\Gamma_2 = j_3\left(j_1 j_2 - j_0 j_3\right) - j_1^2 > 0. \quad (16)$$

The plot of parameters Γ_1 and Γ_2 are shown in Figure 2 for certain choice of the system parameters, from where it appears that when β = 1500 and σ chosen as a control parameter as shown in Figure 2a in one hand, the corresponding equilibrium set E_0 is unstable for σ > σ_0 = 0.121, which implies that the orbit of system [7] starting from the equilibrium set is unstable and asymptotically tend to limit cycle or attractor or infinite. Otherwise for σ <σ_0 the equilibrium set E_0 is attractive, implying that regular oscillations may appear, but the system will not be chaotic. In the other hand when σ = 0.88 and β chosen as a control parameter as shown in Figure 2b, the equilibrium point E_0 is always unstable, provided that Γ_1 and Γ_2 are negatives.

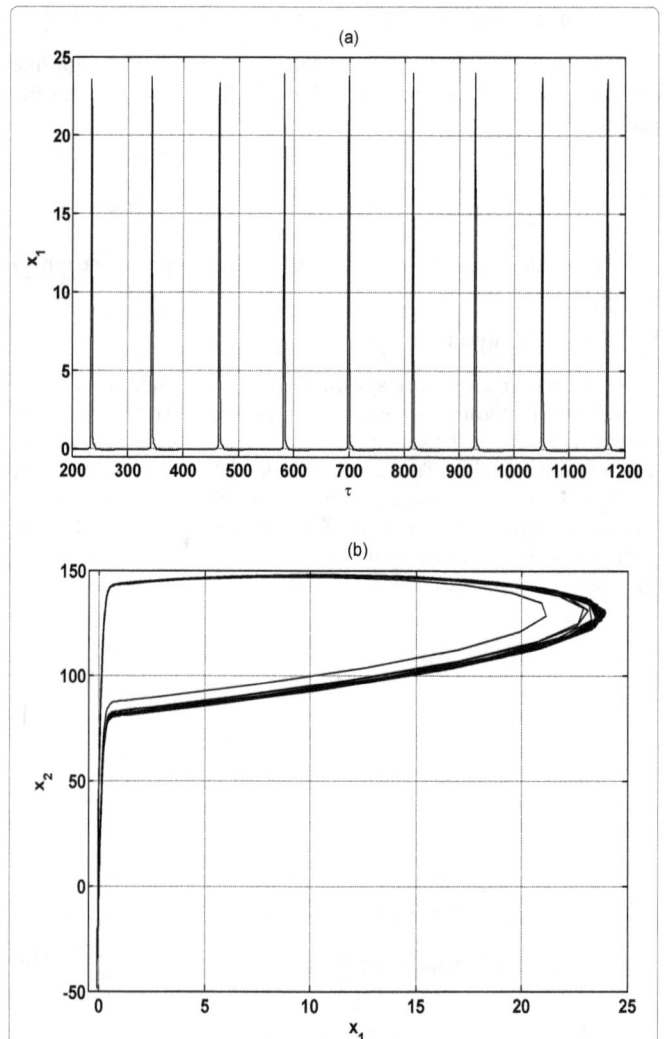

Figure 5: (a) Time waveform of $x_1(t)$ (b) Phase space plot x_2 against x_1. The parameters of the system are taken as in Figure 2, but with σ = 0.88 and β = 1500.

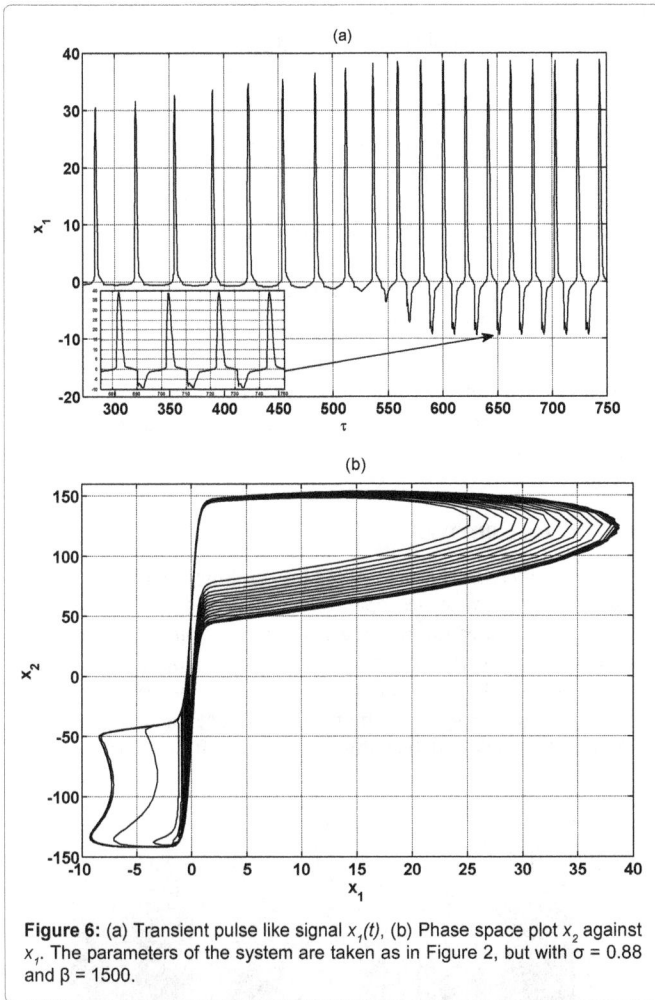

Figure 6: (a) Transient pulse like signal $x_1(t)$, (b) Phase space plot x_2 against x_1. The parameters of the system are taken as in Figure 2, but with $\sigma = 0.88$ and $\beta = 1500$.

Results of Numerical and Pspice Simulations

Numerical investigations

In order to investigate the dependency of the dynamics of the system to the choice of the operational amplifier, system [7-9] is integrated numerically using the classical fourth-order Runge-Kutta integration scheme, with the integration time step $\Delta\tau = 0.002$ and the computations are performed out using variables and constants parameters in extended mode given in [12]. For $\sigma = 0.88$, system [7-9] is integrated for a sufficiently long time to discard the transient, leading to the bifurcation diagram plotted as a function of the amplifier gain to prove the sensitivity of the system to the choice of the operational amplifier. This bifurcation diagram is obtained by plotting the local maxima of state variables in terms of the bifurcation control parameter β. As illustrated in Figure 3, when $\beta < 140$, there is no oscillation in the system. Otherwise the dynamics of the system varies for increasing value of β and rise to periodic 1 limit cycle for $\beta > 875$. To confirm the ability of the system to generate regular and chaotic pulse-like signals, sample time traces with the corresponding phase portraits are computed for some discrete values of the control parameters σ and β. The chaotic pulse train is obtained for $\beta = 1500$ and $\sigma = 0.98$ as depicted in Figure 4, while in Figure 5, only regular pulses are obtained. In Figure 6 it shows two aspects of the behavior of the system, the transient regular pulses, and the incoherent pulse train obtained after the transient.

Pspice simulations and bifurcation of phase portrait

Based on the theoretical analysis presented above, realistic Pspice simulations of the system shown in Figure 1 are simulated, in order to validate the mathematical model proposed in this work, and to evaluate the effects of ideal diode model and operational amplifiers on the dynamics of the system. For this end, the parameters used for the circuit simulation are chosen as indicated in Eq. (11), with varying values of R_3, first in order to confirm the sensitivity of the system with respect to

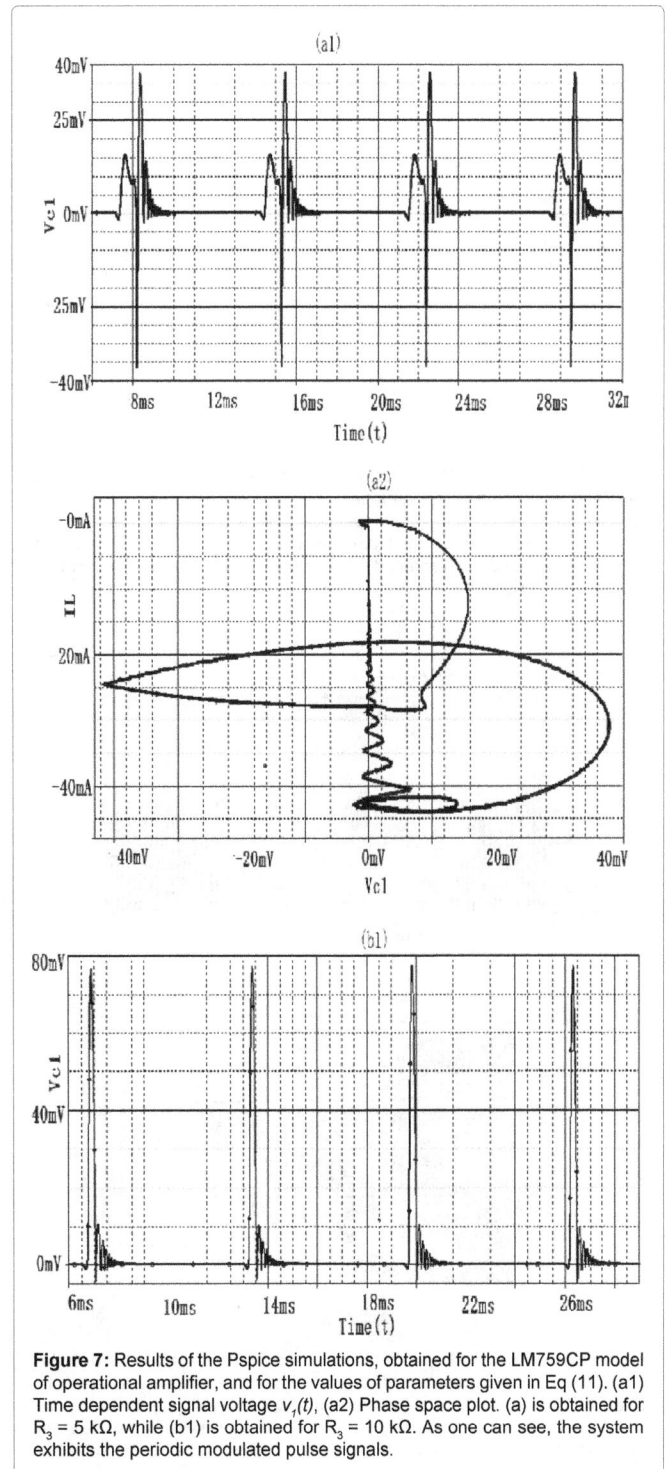

Figure 7: Results of the Pspice simulations, obtained for the LM759CP model of operational amplifier, and for the values of parameters given in Eq (11). (a1) Time dependent signal voltage $v_1(t)$, (a2) Phase space plot. (a) is obtained for $R_3 = 5$ kΩ, while (b1) is obtained for $R_3 = 10$ kΩ. As one can see, the system exhibits the periodic modulated pulse signals.

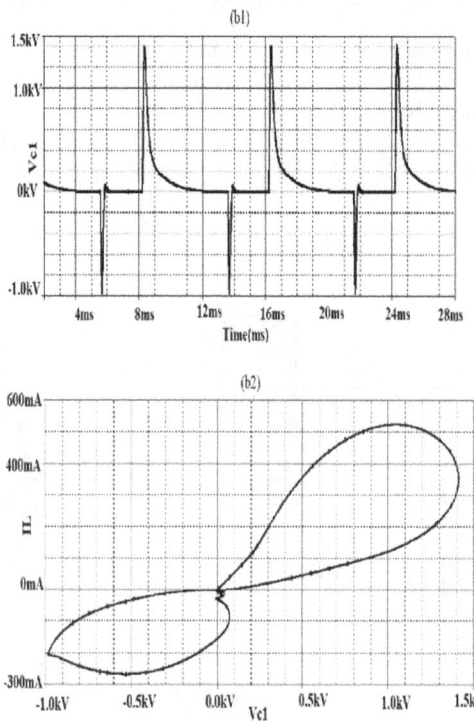

Figure 8: Results of the Pspice simulations, obtained for R_3 = 5 kΩ and for the LM675 model of operational amplifier. The values of the others parameters are given in Eq.(11). (b1): Time dependent signal voltage v_1(t), (b2): Phase space plot. The phase portrait is the homoclinic orbit, that is the orbit which begins on one fixed point and end on the same fixed point, corresponding to pulse train-like signals.

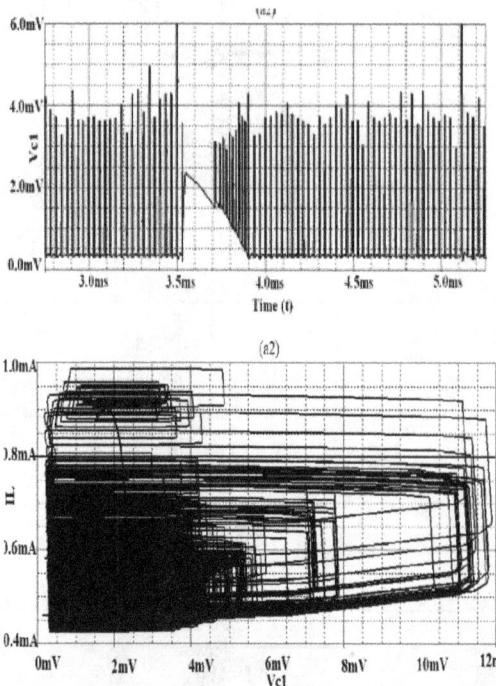

Figure 9: Results of the Pspice simulations, obtained for R_3 = 7 kΩ and for the LM675 model of operational amplifier. The values of the others parameters are given in Eq(11). (a2): Time dependent signal voltage v_1(t) (a2): Phase space plot. Showing the chaotic pulse signals generated by the system, provided that the behaviour of the system is sensitive to the choice of operational amplifier.

parameter as σ outlined in Section 2. Secondly for two different choices of the operational amplifiers, that is to show that the dynamics of the system is also sensitive to the nature of operational amplifiers. And the obtained results are given as follows:

When the LM759CP model of the operational amplifier is used, as shown in Figure 7, and for different choice of R_3, it appears that the dynamics of the system is not more rich and the system exhibits only regular periodic pulse-like behavior, but R_3 must be chosen in intervals *4.68 kΩ ≤ R_3 ≤ 15.6 kΩ* and, *19.5 kΩ ≤ R_3 ≤ 50 kΩ* agreeing with results of numerical simulations. That is the ability of the system to generate regular pulses signals.

When the LM675 model of operational amplifier is used, as depicted in Figures 8-11, the dynamics of the system is more complex and the system exhibits a regular pulse like behavior (Figure 8), chaotic pulse like behavior (Figures 9 and 10) and transient chaotic pulse like behavior (Figure 11).

Concluding Remarks

In this paper, we have presented an autonomous chaotic pulse oscillator, using two stages operational amplifier coupled by mean of diode employed as the nonlinear device element, as well as a new mathematical model for a better description of its nonlinear dynamics. Using the bifurcation diagram, the dynamics of the system has been characterized with respect to the transfer voltage gain of operational amplifiers. It was found that this system exhibits regular pulse and

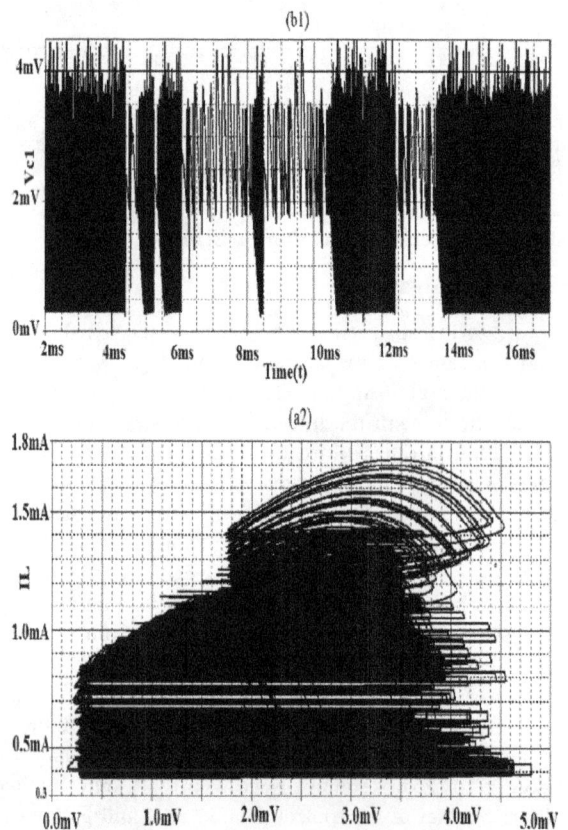

Figure 10: Results of the Pspice simulations, obtained for R_3 = 10 kΩ and for the LM675 model of operational amplifier. The values of the others parameters are given in Eq.(11). (b1) Time dependent signal voltage v_1(t), (a2) Phase space plot showing the chaotic behavior of the system.

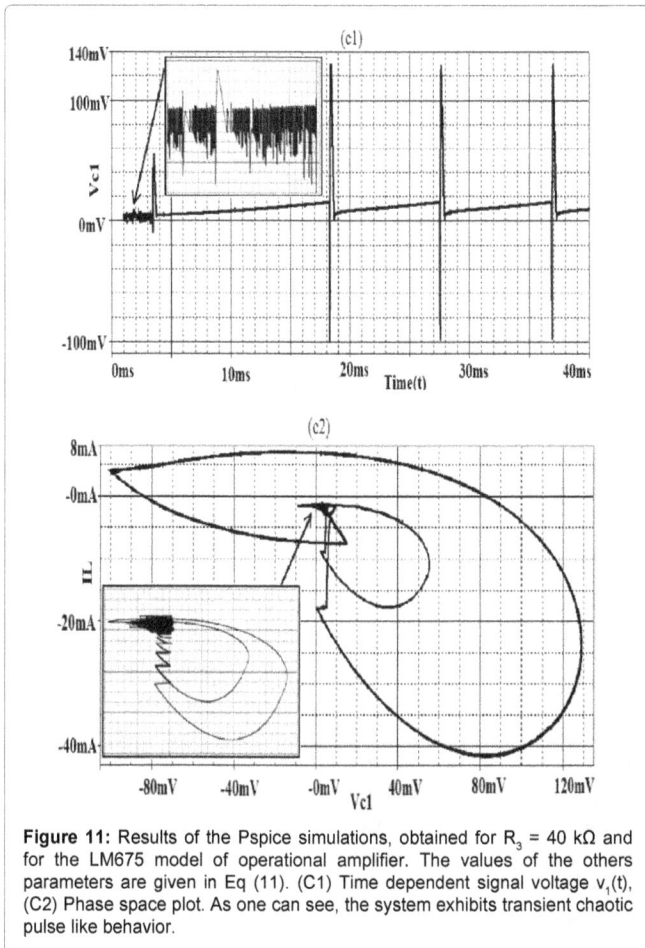

Figure 11: Results of the Pspice simulations, obtained for $R_3 = 40$ kΩ and for the LM675 model of operational amplifier. The values of the others parameters are given in Eq (11). (C1) Time dependent signal voltage $v_1(t)$, (C2) Phase space plot. As one can see, the system exhibits transient chaotic pulse like behavior.

pulse, not predicted by our analytical investigations, but which can be attributed to the layout of the components of operational amplifiers, which are not considered analytically. In future work, we will furthermore explore the influence of the layout of operational amplifier components, establish the sufficient conditions for the occurring of pulse and chaotic pulse signal, and try to find the exact pulse solution of the obtained dynamical equations.

chaotic pulse oscillations for certain range of the control parameters that are the negative resistance and the transfer voltage gain. Justifying then the dependency of the dynamics of the system on the nature of operational amplifiers. These obvious results were agreed by Pspice investigations, which confirm the dependency of oscillations on the nature of operational amplifiers and which for certain classes of operational amplifiers show new behavior, like transient chaotic

References

1. Bhowmick SK (2014) How to generate chaotic pulse. International Journal of Nonlinear Science 17: 67-70.

2. Francis CM, Lau, Chi K, Tse (2003) Chaos-based digital communication systems: operating principles, analysismethods, and performance evaluation. Springer.

3. Argyris A, Syvridis D, Larger L, Lodi VA, Colet P, et al. (2005). Chaos-based communications at high bit rate using commercial fiber-optic links. Nature 438: 343-346.

4. Rulkov NF, Sushchik MM, Tsimring LS, Volkovskii AR (2001) Digital Communication Using Chaotic-Pulse-Position Modulation. IEEE Transactions On Circuits And Systems-I: Fundamental Theory And Applications 48: 12.

5. Kengne J, Kenmogne F (2014) On the modeling and nonlinear dynamics of autonomous Silva-Young typechaotic oscillators with flat power spectrum. Chaos.

6. Poliashenko M, McKay SR, Smith CW (1991) Hysteresis of synchronous-synchronous regimes in a system of two coupled oscillators. Physical Review A 43: 5638-5641.

7. Kwuimy CAK, Woafo P (2008) Dynamics, chaos and synchronization of self-sustained electromechanical systems with clamped-free flexible arm. Springer 53: 201-213.

8. Torikai H, Saito T, Schwarz V A (1997).Chaotic Pulse Generator and Sawtooth Control for Information processing. IEEE International Symposium on Circuits and Systems 1: 9-12.

9. Dmitriev AS, Kyarginsky, Ye B, Panas AI, Starkov SO (2003) Experiments on ultra wideband direct chaotic information transmission in microwave band. Int. J. Bifurcation and Chaos 13: 1495-1507.

10. Dmitriev A, Efremova E, Kuzmin L, Atanov N (2007) Forming pulses in non-autonomous chaoticoscillator. Int. J. Biffurcation and Chaos 17: 3443-3448.

11. Giannakopoulos K, Deliyannis K (2009) Autonomous 4-D hyperchaotic oscillator and synchronization. IEEE Press Piscataway.

12. Deliyannis T, Sun Y, Fidler JK (1999) Continuous-Time Active Filter Design. CRC Press LLC.

13. Hanias MP, Giannaris G, Spyridakis A, Rigas A (2006) Time series analysis in chaotic diode resonator circuit. Chaos, Solitons and Fractals 27: 569-573.

20

A Novel Type of Wireless V2H System with a Bidirectional Single-Ended Inverter Drive Resonant IPT

Hideki Omori[1]*, Shinya Ohara[1], Masahito Tsuno[2], Noriyuki Kimura[1], Toshimitsu Morizane[1] and Mutuo Nakaoka[3]

[1]*Department of Electrical and Electronics Systems Engineering, Osaka Institute of Technology, Japan*
[2]*Nichicon Co.Ltd., Kyoto Japan*
[3]*University of Malaya, Kuala Lumpur, Malaysia*

Abstract

Electric vehicles (EV) offer promise as an effective solution to environmental problems. One of the keys to their successful diffusion is the provision of adequate battery charging infrastructure. In order to create a charging infrastructure by installing equipment in such as locations as carports in private homes, the wireless battery charging system is very suitable. EVs can be used in smart house systems to supplement the energy storage. This vehicle to home (V2H) system essentially requires a bidirectional power transfer feature between the EV and home. This paper presents a new bidirectional inductive power transfer (IPT) system for wireless V2H with simplest components and low cost aiming at wide diffusion for home use. Proposed is a novel type of bidirectional wireless EV charging system with an efficient and compact type single-ended quasi-resonant high-frequency inverter for V2H.

Keywords: Wireless; Charger; Bidirectional; Single-ended inverter; Home use; Power transfer; EV

Introduction

In recent years, Electric Vehicles (EV) which are highly efficient as well as do not create air pollution, and offer promise as an effective solution to environmental problems with great advances of power electronic technology.

One of the key issues to their successful and wide diffusion is the provision of adequate battery charging infrastructure. In order to create a battery charging infrastructure by installing equipment in such as locations as carports in private homes, the inductive power transfer (IPT)-wireless battery charging system is indispensable for spread. The wireless battery charging system eliminates the use of power cables with plug. Merely by parking the car in a designated spot, the battery can be charged. It is a promising system for wider diffusion because it is easy and safe to use for a broad range of users including the elderly.

Usually as wireless EV charging power supply topologies have been half-bridge, push-pull, full bridge, boost half-bridge and boost full-bridge circuit configuration which have a plurality of power switching devices [1-4].

The authors have previously put into practice the cost-effective and high-efficiency single-ended quasi-resonant soft-switching inverter [5]. Although EVs are primarily consider as a method of clean transport, they can also be used in smart house systems to supplement the energy storage. This vehicle to home (V2H) system essentially requires a bidirectional power transfer feature between the EV and home.

This paper presents a new system with the simplest components and low cost aiming at wide diffusion for home use. From a practical point of view, simple high-frequency inverter circuit topologies have to be effectively selected in accordance with specific cost effective applications. Proposed is a novel type of wireless EV charging system based on IPT technology with an efficient and compact type single-ended quasi-resonant high-frequency inverter. The single-ended inverter, which can operate in the frequency range from 20-30 kHz under a self-excited ZVS control and its zero voltage crossing detector of resonant capacitor voltage is evaluated from an experimental point of view. Furthermore transfer power and efficiency have been successfully improved by pick-up circuit with resonant component. The output power of a proposed system is successfully improved

by a resonant IPT circuit. EVs can be used in smart house systems to supplement the energy storage. This vehicle to home (V2H) system essentially requires a bidirectional power transfer feature between the EV and home. This paper presents a new bidirectional inductive power transfer (IPT) system for wireless V2H with simplest components and low cost aiming at wide diffusion for home use also. Proposed is a novel type of bidirectional wireless EV charging system with an efficient and compact type single-ended quasi- resonant high-frequency inverter for V2H. And a result of feasibility study by simulation and experiment is indicated for V2H [6].

Single-Ended EV Charging System Descriptions

System configuration

An IPT-based wireless EV charging system is schematically illustrated in Figure 1. Figure 2 shows a proposed total system with a single-ended quasi-resonant high-frequency inverter. The system is mainly composed of a single-phase diode D_1 rectifier with a L_3-C_3 filter, a single-ended quasi-resonant high-frequency inverter operating with a ZVS-PFM power regulation scheme in the frequency range of 20-30kHz, resonant capacitor C_1 with the primary coil L_1 which is loosely coupled to the pickup coil L_2 as load side, a single-phase diode D_2 rectifier with a L_4-C_4 filter connected with the battery bank of EV, and a specific power regulation control circuit due to the self-excited timing signal processing.

Operating principle of a proposed wireless EV charger

The periodic steady-state voltage and current operating waveforms of the single-ended quasi-resonant high-frequency inverter- fed DC-

*Corresponding author: Hideki Omori, Department of Electrical and Electronics Systems Engineering, Osaka Institute of Technology, Japan
E-mail: hideki.omori@oit.ac.jp

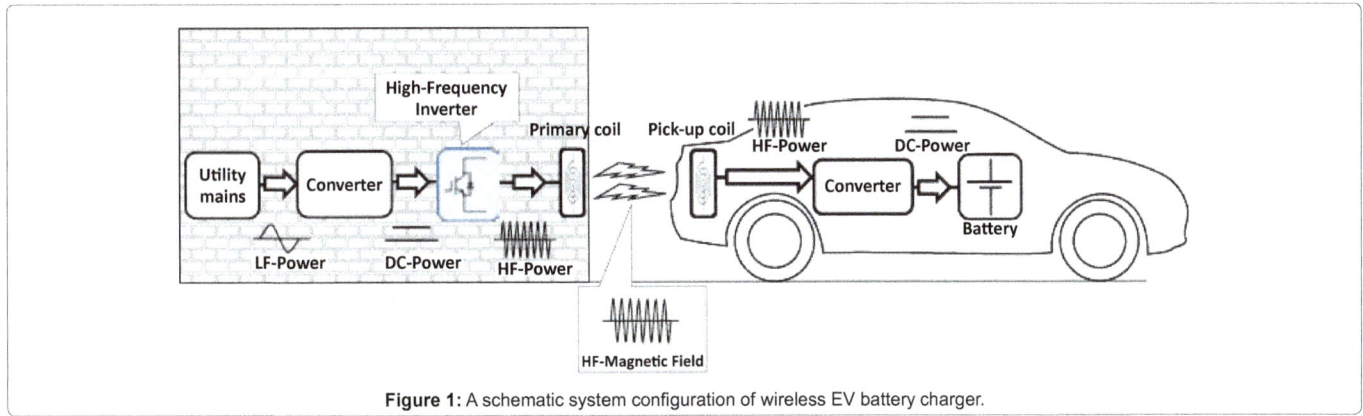

Figure 1: A schematic system configuration of wireless EV battery charger.

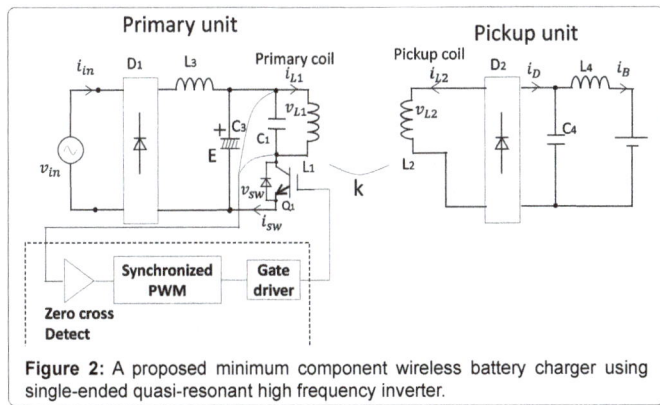

Figure 2: A proposed minimum component wireless battery charger using single-ended quasi-resonant high frequency inverter.

DC converter in Figure 2 are illustrated in Figure 3, which include Mode I, II, III, IV, during one switching cycle.

Figure 4 illustrates switching-mode equivalent circuits in accordance with on-off operating mode due to the single active switch Q_1 and passive switches in secondary-side diode bridge D_2. The circuit state in which the active switch Q_1 in the primary-side is cut off and passive switch D_2 in the secondary-side is conducting is defined as Mode I and II (Figure 4a and 4b).

In Mode I (Figure 4a), when the active switch Q_1 is turned off at t_0, LC resonant tank circuit can operate. The inductor current and capacitor voltage become resonant state. Because of resonant operation, the voltage across the switch Q_1 begins to increase sinusoidally from zero voltage. The active switch Q_1 is turned off with ZVS (Zero Volt Switching) transition. The inductor current iL_1, iL_2 through the primary coil L_1, the secondary coil L_2 are decreasing. As soon as the inductor current iL_2 reaches zero at t_1, the state of D_2 moves to Mode II shown in Figure 4b, the inductor voltage v_{L2} across the secondary coil L_2 becomes $-V_{C4}$ from V_{C4}.

As soon as the resonant capacitor voltage reaches the supply DC voltage V_{C3} at t_2, the voltage across Q_1 becomes zero. At this point, the antiparallel diode of Q_1 turns on naturally. The operating mode becomes Mode III shown in Figure 4c. The circuit state in which the active switch Q_1 in the primary-side and passive switch D_2 in the secondary-side are both conducting is defined as Mode III, IV (Figure 4c and 4d). The inductor current iL_1 through the primary coil L_1 and the active switch current iQ_1 increase in a function time as illustrated in Mode III, IV of Figure 3 during Ton period.

When the secondly current iL_2 reaches to zero at t_3, the state of

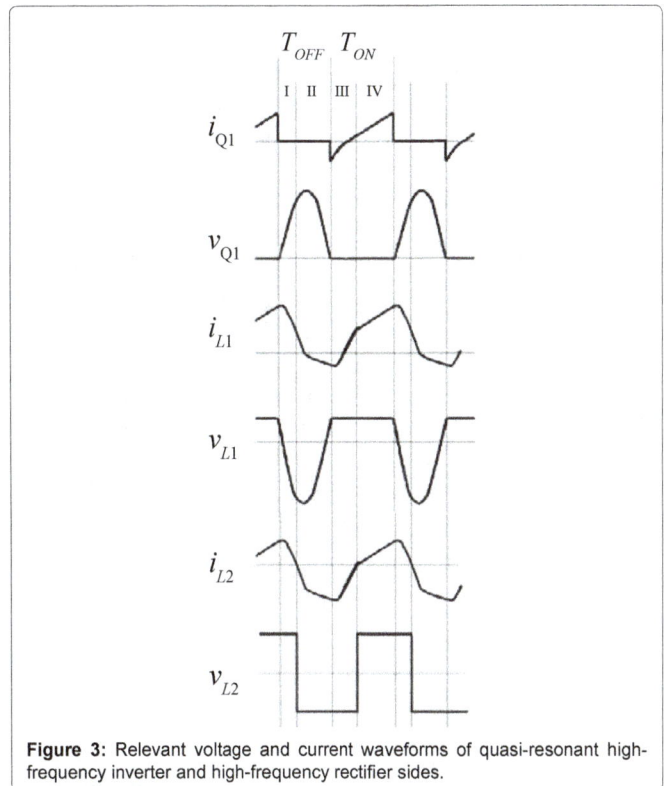

Figure 3: Relevant voltage and current waveforms of quasi-resonant high-frequency inverter and high-frequency rectifier sides.

D_2 moves to Mode IV shown in Figure 4d, and the secondary voltage v_{L2} changes to V_{C4} from $-V_{C4}$. The active switch Q_1 is turned off in accordance with its gate pulse duration time Ton

Judging from the operating voltage v_{Q1} and current i_{Q1} of the active switch, it is understood that the switch Q_1 can achieve a soft-switching turn-on transition with ZVS. Note that the diode D_2 in the secondary side can operate under a principle of ZCS (Zero Current Switching). As a result, the recovery current of D_2 is considerably small, and switching power loss of D_2 as well as switching noise can be minimized effectively. The active switch Q_1 can achieve the complete soft switching transitions with ZVS at turn-on and turn-off. The synchronized PWM oscillator generates pulses synchronized with v_{L1} to achieve ZVS turn-on.

This system can supply an approximately constant current to battery owing to leakage inductance of wireless coupling planar coils without sensing any signals from the secondary-side pick-up coil circuit in EV.

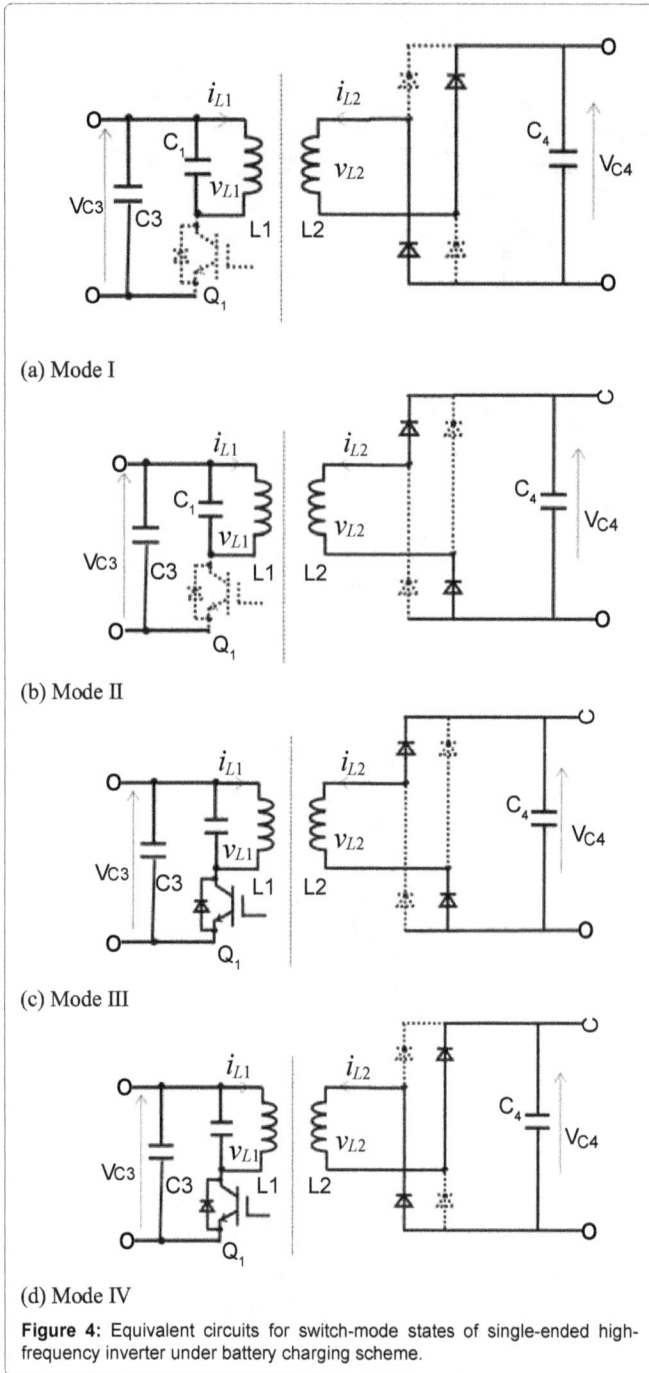

(a) Mode I

(b) Mode II

(c) Mode III

(d) Mode IV

Figure 4: Equivalent circuits for switch-mode states of single-ended high-frequency inverter under battery charging scheme.

Transfer-power improvement by a resonant pick-up Circuit equations of two planar spiral coils are given by

$$v_{L1} = r_1 i_1 + L_1 \frac{di_{L1}}{dt} + M \frac{di_{L2}}{dt} \left.\right\}$$
$$v_{L2} = r_2 i_2 + M \frac{di_{L1}}{dt} + L_2 \frac{di_{L2}}{dt} \left.\right\}$$

(1)

Where,

L_1, L_2: self-inductance of the primary-side power feeding coil and secondary side power receiving coil.

r_1, r_2: internal resistances of each planar coil.

K: electromagnetic coupling coefficient of two planar spiral coils.

M: mutual inductance between L_1 and L_2, which is influenced upon gap distance variable.

$$M = k\sqrt{L_1 L_2}$$

Output power of the system shown in Figure 2 is approximately represented by the following Equation:

$$P_{nr} = \frac{|V_{L2}|^2}{R_0} = \frac{k^2}{R_0 + \omega^2 L_2^2 (1-k^2)^2 \frac{1}{R_0}} \frac{L_2}{L_1} V_{L1}^2$$

(2)

Where,

V_{L1}, V_{L2}: effective voltages of the power feeding coil and the power receiving coil.

R_0: equivalent resistance of the output circuit.

ω: operating frequency of the inverter.

On the other hand, output power P_r in the case of a power receiving coil L_2 with a parallel connected resonant capacitor C_2 is approximately represented as following equation:

$$P_{nr} = \frac{|V_{L2}|^2}{R_0} = \frac{k^2}{R_0 \left\{1 - \omega^2 L_2 C_2 (1-k^2)\right\}^2 + \omega^2 L_2^2 \frac{(1-k^2)^2}{R_0}} \frac{L_2}{L_1} V_{L1}^2$$

(3)

P_r can be compared to P_{nr} as follows:

$$\frac{P_{nr}}{P_r} = \frac{R_0 \left\{1 - \left(\frac{\omega}{\omega_0}\right)^2 (1-k^2)\right\}^2 + \omega^2 L_2^2 \frac{(1-k^2)^2}{R_0}}{R_0 + \omega^2 L_2^2 \frac{(1-k^2)^2}{R_0}}$$

$$= 1 - \frac{R_0 \left(\frac{\omega}{\omega_0}\right)^2 (1-k^2) \left\{2 - \left(\frac{\omega}{\omega_0}\right)^2 (1-k^2)\right\}}{R_0 + \omega^2 L_2^2 \frac{(1-k^2)^2}{R_0}}$$

(4)

With $\omega_0 = \frac{1}{\sqrt{L_2 C_2}}$

The resonant transfer power P_r is higher than the non-resonant transfer power P_{nr} under the condition of, then proposed is a resonant wireless EV charger with the pick-up side resonant capacitor in Figure 2 system [7].

Experimental Results

Wireless coupling coil units

It is noted that the primary- side power feeding unit L_1 is loosely coupled to the pick-up coil L_2 of the secondary side power receiving unit for the battery charging power supply. In actual, these circular coils in the primary and secondary-side are assembled by the power litz wires in order to reduce power losses due to the skin effect. Two contactless planar and circular coil units with ferrite core sheets; power sending coil and power receiving coil are depicted in Figure 5.

We propose a power transfer configuration through a rear glass, because it is easy to install the pick-up unit. Then the diameter of coils is

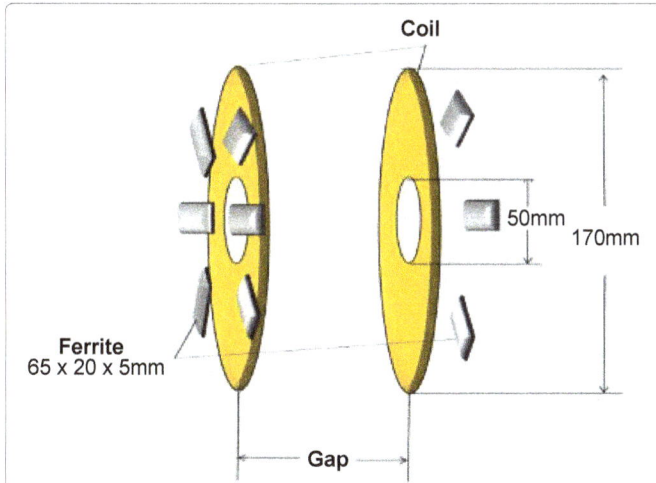

Figure 5: A wireless electro-magnetic structure with two planar and circular type coil units. (Winding 22 turns, Diameter Φ 170 mm, Thickness 2 mm, Inductance 60 μH, Resistance 20 mΩ, Gap distance typ. 30 mm).

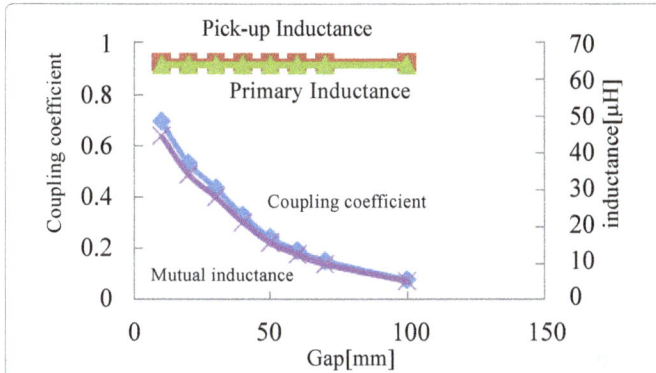

Figure 6: Relationship of circuit parameters of two planar coils for the gap distance variable.

170 mm, and the typical gap distance is 30 mm which include thickness of the rear glass and clearance.

As a matter of fact, some circuit parameters; mutual inductance M and magnetic coupling coefficient k of two planar spiral coils can depend upon the gap distance variable. Primary inductance L_1 and secondary inductance L_2 is not depending on gap (Figure 6).

Operating characteristics

The measured waveforms in a proposed EV charging system are shown in Figure 7. The active switch Q_1 turn-on and turn-off by ZVS, and the self-excited ZVS control scheme well functions. In particular, in single-ended quasi-resonant inverter, note that the maximum voltage applied to the active switch Q_1 becomes relatively high because of quasi-resonant operation of primary side.

The measured value of output power in non-resonant case and resonant case vs. load characteristics for gap distance from 30 to 50 mm is shown in Figure 8. The output power with resonant pick-up circuit is 1.5-2 times higher than that with non-resonant one. Measured power transfer efficiency is 90% under 30 mm gap in the resonant pick-up case.

A prototype of EV equipped with pick-up coil unit of the resonant IPT-wireless charging system and power feeding coil unit is shown in

Figure 9a. The wireless charging system can charge 4.5 kWh in 5 hours as a feasibility study result as shown in Figure 9b.

A Wireless V2H System With Bidirectional Resonant Single-Ended Inverter

A smart house with a vehicle to home (V2H) system is shown in Figure 10. Connected EV plays a role to supplement a storage battery system. As it is easy to connect EV the house by a wireless vehicle to home system, the EV is efficiently used for the smart house. The vehicle to home (V2H) system essentially requires a bidirectional power transfer feature between the EV and home. A new system configuration by a single-ended inverter with a resonant IPT circuit for bidirectional power transfer is shown in Figure 11. The proposed system is with simplest components and low cost.

In the EV charging mode, a home-side single-ended inverter produces high frequency power which is transferred to vehicle side circuit by resonant IPT as shown in Figure 12 left side. Capacitance C_1 in home-side circuit operates as resonant capacitance for ZVS transition and capacitance C_2 in vehicle-side circuit operates as resonant capacitance for resonant IPT. Input current I_{L3} smoothed by DC choke L_3 is constant without ripple. Also output current I_{L4} smoothed by DC choke L_4 is constant without ripple (Figure 12).

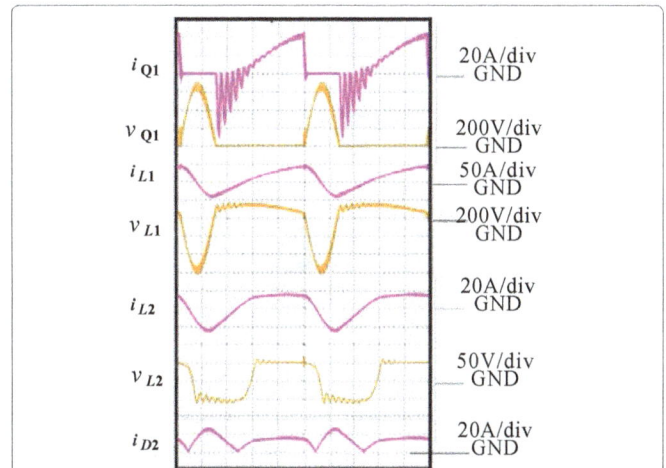

Figure 7: Measured voltage and current waveforms of quasi-resonant high-frequency inverter. (10 μs/div).

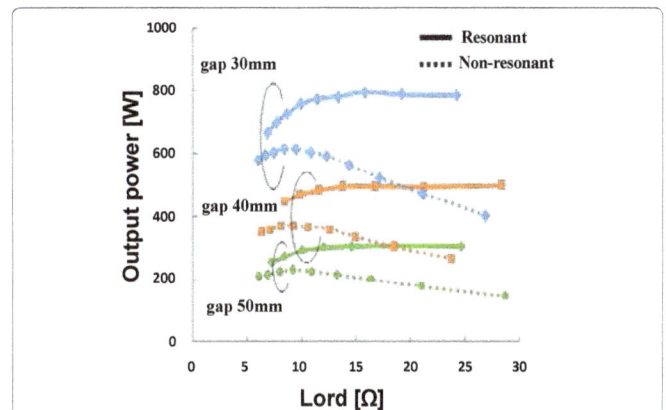

Figure 8: Observed output power vs. lord characteristics. (Diameter 170 mm, C1 0.3 μF, C2 0.3 μF, C3 2200 μF, C4 6 μF, L1 60 μH, L2 60 μH, L3 90 μH, L4 90 μH, VB 55 V, Vac 100 V 60 Hz).

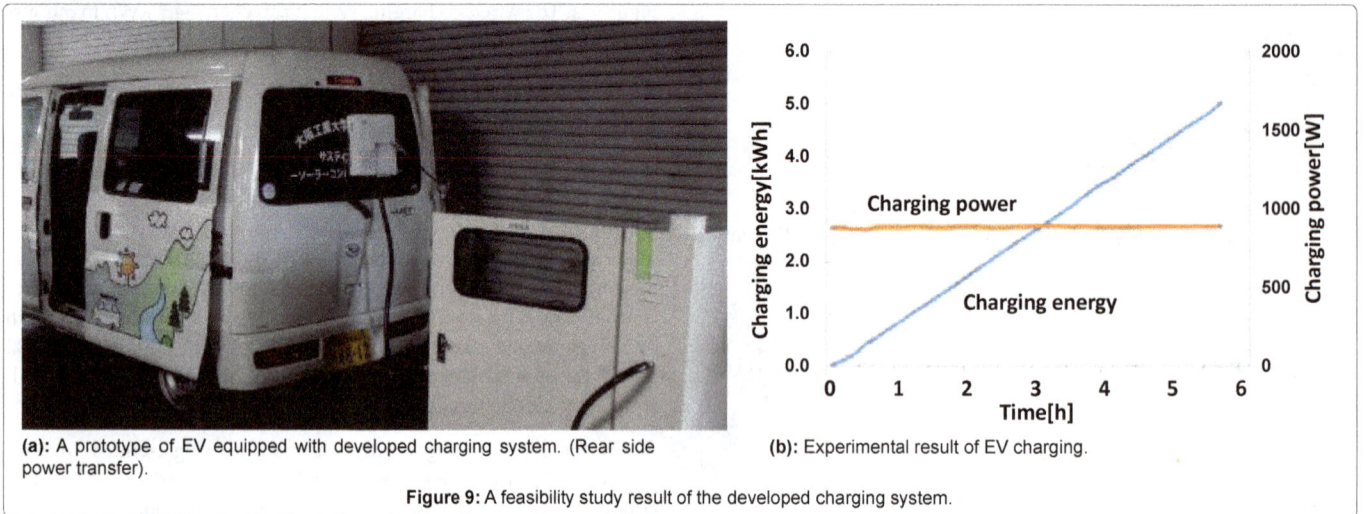

(a): A prototype of EV equipped with developed charging system. (Rear side power transfer).

(b): Experimental result of EV charging.

Figure 9: A feasibility study result of the developed charging system.

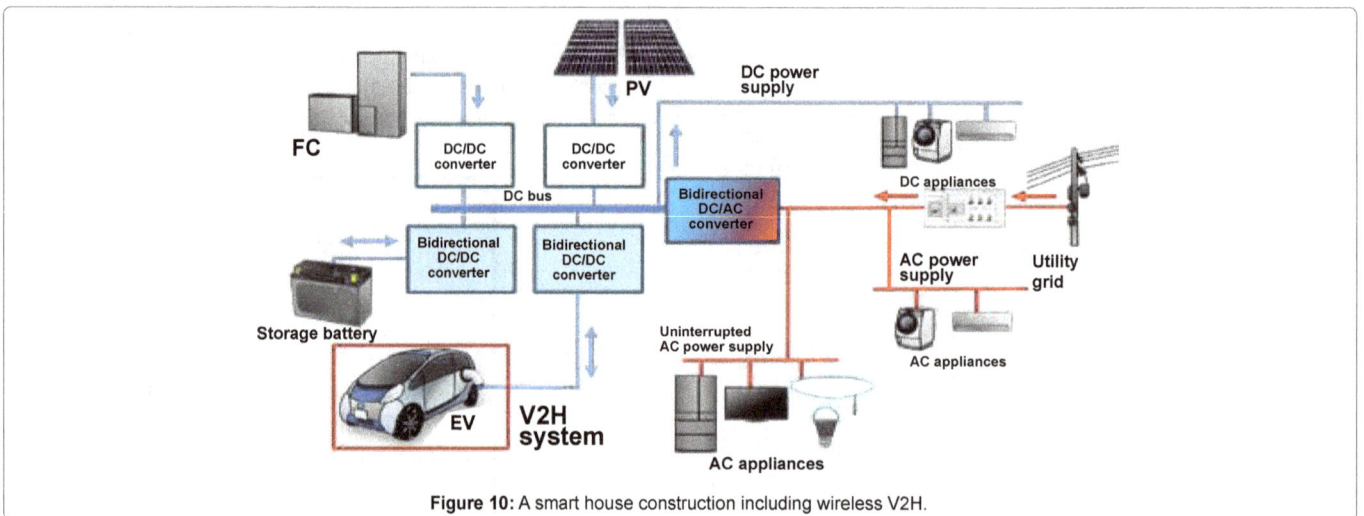

Figure 10: A smart house construction including wireless V2H.

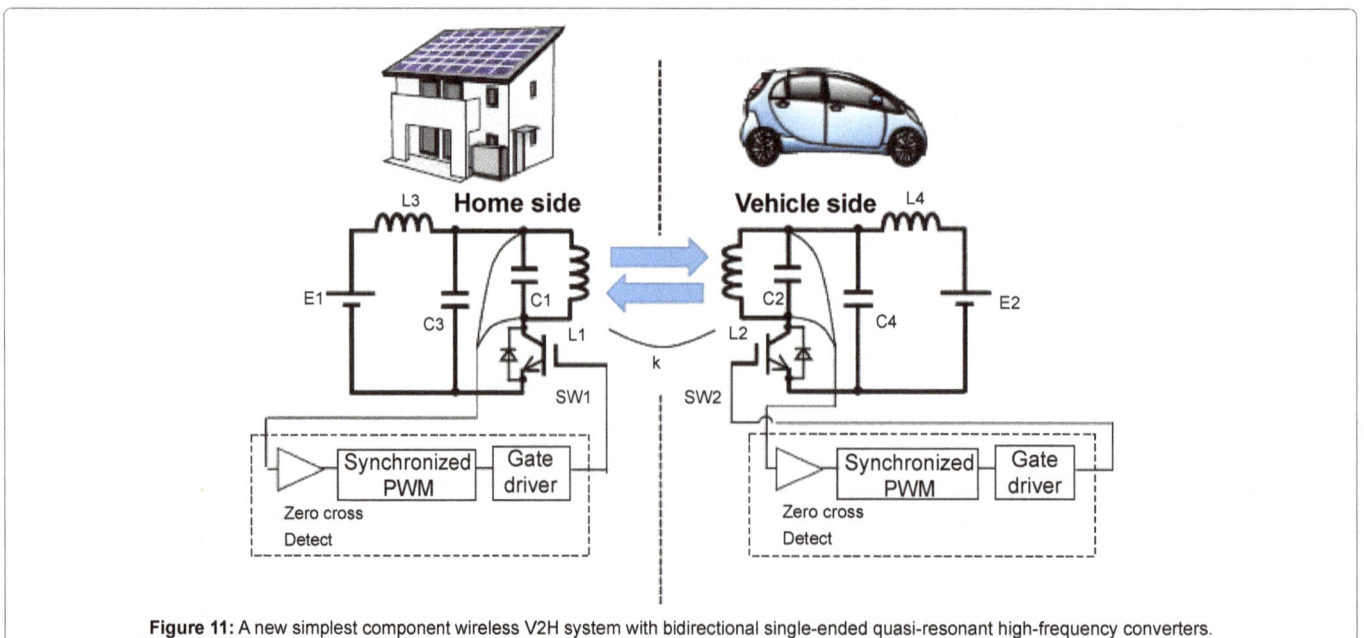

Figure 11: A new simplest component wireless V2H system with bidirectional single-ended quasi-resonant high-frequency converters.

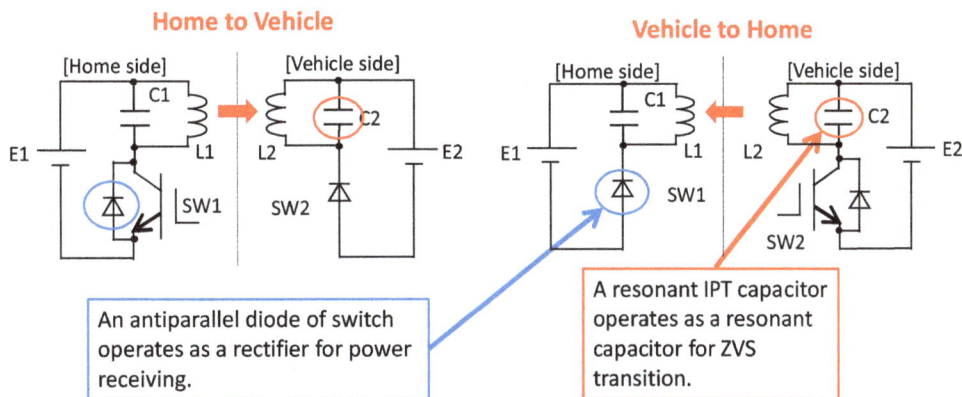

Figure 12: Operation principle of the proposed wireless V2H system with bidirectional single-ended converters.

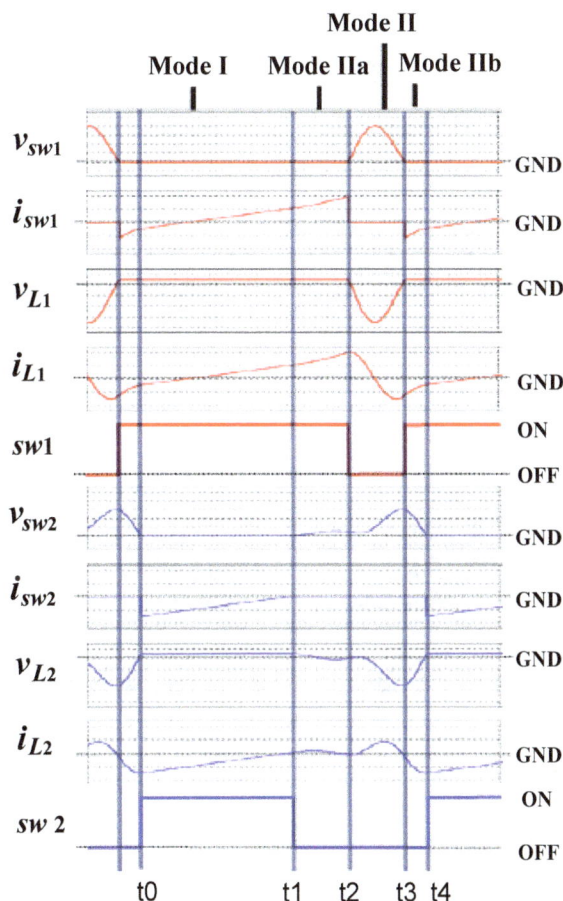

Figure 13: Simulated waveforms of the new bidirectional circuit using quasi-resonant high-frequency inverters (Home to Vehicle).

On the other hand, in the EV battery energy using mode, a vehicle-side single-ended inverter produces high frequency power which is transferred to home side circuit by resonant IPT indicated in Figure 12 right side. Capacitance C_2 in vehicle-side circuit operates as resonant capacitance for ZVS transition and capacitance C_1 in home-side circuit operates as resonant capacitance in power receiving for resonant IPT.

Figure 13 shows simulated operating waveforms of a proposed bidirectional circuit using a quasi-resonant high-frequency inverter in home to vehicle mode. The active switch SW1 is turned off and turned on with ZVS transition. And the active switch SW2 is kept off. A vehicle side circuit operates as a rectifier with a pick-up resonant circuit. This system operates as a resonant forward converter. Figure 14 shows operating modes of the bidirectional single-ended converter with resonant IPT in home to vehicle mode.

Figure 15 shows measured operating waveforms of proposed wireless V2H system. These waveforms are same as above mentioned simulation waveforms. The ZVS control scheme works well. Measured

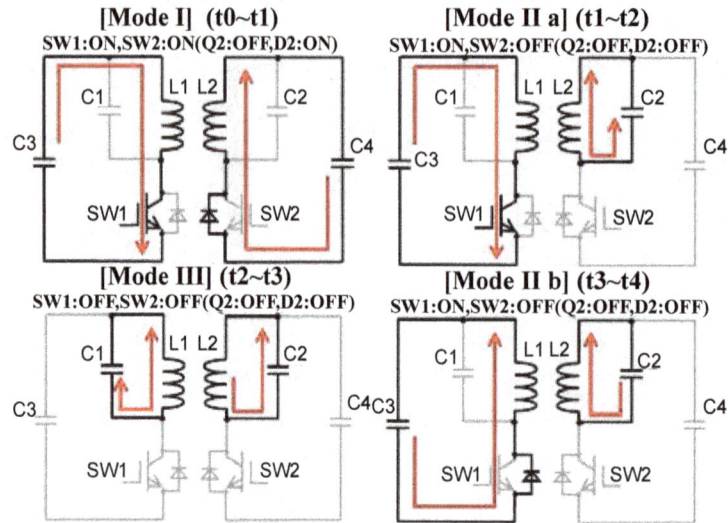

Figure 14: Operating modes of the new bidirectional circuit using quasi-resonant high-frequency inverters (Home to Vehicle)

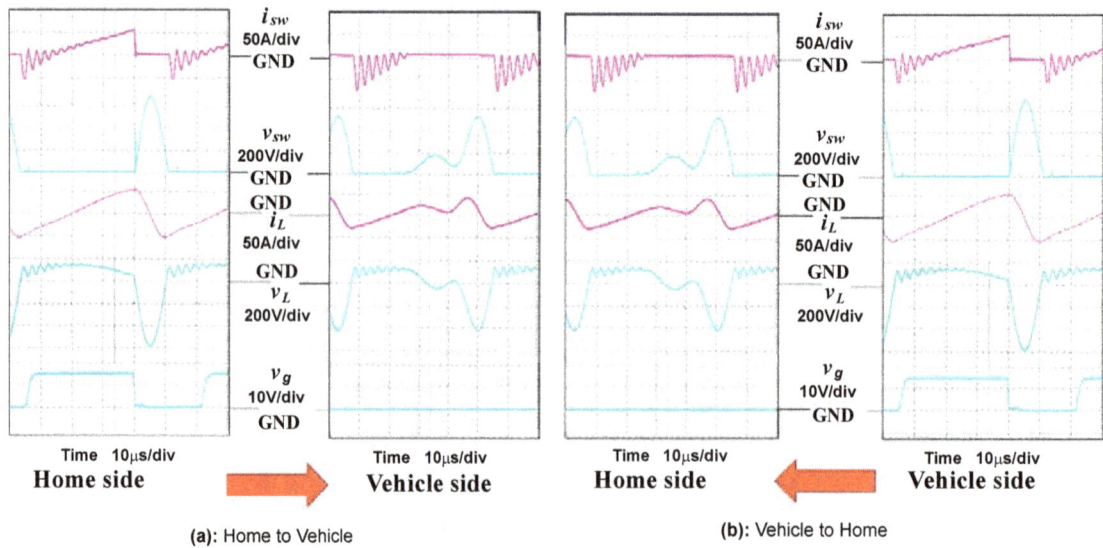

(a): Home to Vehicle   **(b):** Vehicle to Home

Figure 15: Measured waveforms of the bidirectional circuit using quasi-resonant high-frequency converters.

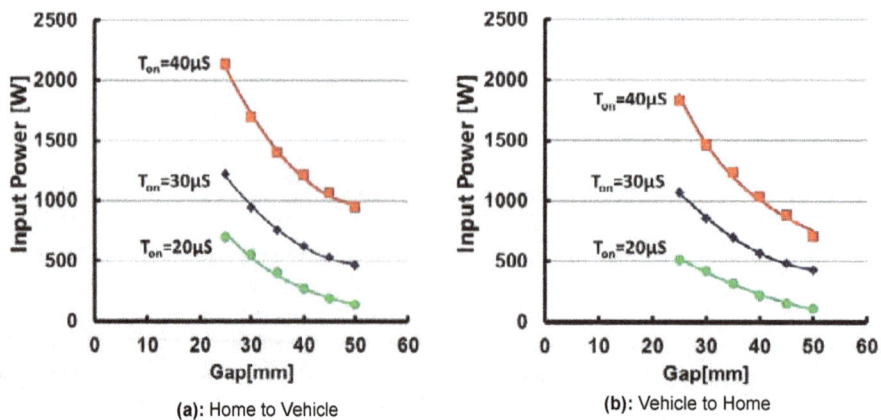

(a): Home to Vehicle   **(b):** Vehicle to Home

Figure 16: Observed output power vs. gap distances of the bidirectional circuit using quasi-resonant high-frequency inverters. (Diameter 170 mm, C1 0.3 μF, C2 0.3 μF, C3 6 μF, C4 6 μF, L1 35 μH, L2 35 μH, L3 90 μH, L4 90 μH, E1 100 V, E2 100 V).

waveforms of vehicle to home mode also show the reverse operation scheme well function.

Figure 16a shows output power characteristics of home to vehicle mode. The output power decreases depending on gap distance, and can be controlled by conduction time TON. Figure 16b shows the output power characteristics in home to vehicle power transfer mode which same as those in vehicle to home mode.

Conclusion

Presented has been a new wireless EV charging system based on IPT technology with minimum components and low cost aiming at wide diffusion for home use. From a practical point of view, an efficient and compact type single-ended quasi-resonant high-frequency inverter was selected. The single-ended inverter which can operate in the frequency range from 20-30 kHz under a synchronized self-excited ZVS control and its zero voltage crossing detector of resonant capacitor voltage is evaluated from an experimental point of view. Output power of a proposed system has been successfully improved by a resonant IPT circuit. The output power with resonant pick-up circuit was 1.5-2 times higher than that with non-resonant one. Furthermore proposed has been a new resonant IPT wireless V2H system with a simplest and low cost bidirectional single-ended converter aiming at wide diffusion for home use also. And a result of feasibility study for V2H by simulation and experiment was indicated.

References

1. Thrimawithana DJ, Madawala UK, Shi Y (2010) Design of a Bi-Directional Inverter for a Wireless V2G System. Proc. IEEE International Conf. on Sustainable Energy Technologies, ICSET, Kandy, Sri Lanka.

2. Huh J, Lee W, Cho GH, Lee B, Rim CT (2011) Characterization of Novel Inductive Power Transfer Systems for On-Line Electric Vehicles. Proc Annual IEEE Applied Power Electronics Conference and Exposition, APEC, Texas, USA.

3. Machino S, Kozako M, Harada K, Hikita M, Hotta K, et al. (2011) Construction and Characteristics of Wireless Resonance Type Inductive Power Supply. Proc Japan Industry Applications Society Conf Okinawa, Japan, CD-ROM 2-13.

4. Kai T, Kraisorn T, Minagawa Y (2011) A Study on Receiver Circuit Topology of Non-contact Charger for Electric Vehicle. Proc Japan Industry Applications Society Conf. Okinawa, Japan, CD-ROM 2-16.

5. Omori H, Nakaoka M (1997) Generic Circuit Topologies and Their Performance Evaluations of Single-Ended Resonant High-Frequency Inverters for Induction-Heated Cooking Appliances. Trans on IEE Japan 117: 150-159.

6. Omori H, Iga Y, Morizane T, Kimura N, Nakagawa K, et al. (2012) A Novel Wireless EV Charger using SiC Single-Ended Quasi-Resonant Inverter for Home Use. International Power Electronics and Motion Control Conference and Exposition, Novisad, Serbia.

7. Omori H, Iga Y, Fukuoka H, Morizane T, Kimura N, et al. (2013) A New Bidirectional Resonant IPT EV Charging System with Single-Ended Inverter. Electrical Drives and Power Electronics, Dublovnik, CROATIA.

21

Ac to Ac Frequency Changer THD Reduction Based on Selective Harmonic Elimination

Ibraheem Mohammed Khaleel*

Department of Computer Communications Engineering, Al-Rafidain University College, Baghdad, Iraq

Abstract

AC to AC frequency converters had a challenge to be proposed as a reduction of THD. This challenge due to eliminate the most effective harmonics in frequency spectrum with low cost and no difficulty. However, the challenge is represented by producing new technique for AC to AC frequency changer with lower total harmonic distortion compared with regular AC to AC frequency changer. This proposed work introduces reduction based on eliminate the 3rd predominant harmonic by imposing a waveform with same frequency, amplitude and poly phase of presented harmful harmonic the elimination for the most effective harmonic will cause reduction in THD. The proposed technique gave a significant reduction in THD (about 52%) with respect to regular AC to AC frequency changer with low cost implementation requirements.

Keywords: Power electronics; Cycloconverters; Frequency changers; Harmonic reduction techniques; SPWM

Introduction

Variable frequency converters became important now a day with every growing demand of industrial applications [1], in most direct AC to AC frequency changers (step down) high THD has been appeared [2], the increasing in THD occurs due to higher magnitude of predominant harmonics compared with amplitude of fundamental frequency after dividing the frequency of supply source by (M) where M=2, 3, 4,… etc. [3], i.e., for frequency changer with M=2 and main supply frequency of 50 Hz the output frequency of frequency changer will be 25 Hz, the frequency spectrum of the output shown only the odd harmonics with different amplitudes and different phases [4] while the even harmonics had zero magnitude due to even wave symmetry, and harmonic distortion result by the odd harmonics only [5]. One of the important parameters that used to measure the performance in power electronic converters is nth harmonic distortion [6] caused by the each harmonic which can be estimated using $(\frac{A_n}{A_1} \times 100)$ where A_n represent the amplitude of corresponding harmonic while A_1 represent the magnitude of fundamental, for divide by two frequency changer described above the nth harmonic distortion resulted by apparent harmonics are subjected in Table 1.

According to the Table 1, the 3rd harmonic has most effective distortion with respect to other harmonics appeared in frequency spectrum of output waveform for regular frequency changer which is shown in Figure 1a, which frequency spectrum shown in Figure 1b, and the overall distortion result by each apparent harmonics gives THD about 62% [5].

Harmonics generated by AC to AC frequency changers, increase power system heat losses and power bills of end users. These harmonics related losses reduce system efficiency, cause apparatus overheating, and increase power and air conditioning coasts [7], according that harmonic currents can have a significant impact on power electronics

devises [8]. However many techniques have been investigated in order to reduce THD of frequency changers such as (Comb Filtering, Multi-Phase Comparative Commutation, Half Cycle Omission, PWM Switching) [9-12], the techniques comb filtering, multi-phase comparative commutation depend on estimating the magnitude of predominant harmonic which caused high THD, and trying to reduce the magnitude of these predominant harmonics in order to reduce THD, however the pervious methods depend on magnitude of harmonics and forgetting the phase shift of predominant harmonic effect, also to implement single phase undistorted output waveform, it is necessary to provide at least two balanced distorted output waveforms, or for the PWM technique the output voltage control can be obtained without any external components and PWM minimizes the lower order harmonics, while the higher order harmonics with high n^{th} harmonic distortion has been appeared, which can be eliminated using a filtering techniques, also the value of root mean square voltage of fundamental frequency reduced to significant amount and caused increasing in current that subjected to the load, the previous problem has been appeared in half cycle omission technique, however the problems that described in each technique gives a bad performance for power electronics devices generally and frequency changers specially, according that a new technique to reduce THD of AC to AC frequency changers based on the phase shift of predominant harmonic has been proposed.

Proposed Technique

The suggested method represents the effective predominate harmonic of regular AC to AC frequency changer with M=2 in complex polar form as a magnitude and phase shift the elimination

N	F (Hz)	n^{th} Harmonic distortion 100%
3	75	60.28
5	125	15.12
7	175	6.44
9	225	4.07

Table 1: n^{th} Harmonic distortion for AC to AC frequency changer with M=2.

*Corresponding author: Ibraheem Mohammed Khaleel, Department of Computer Communications Engineering, Al- Rafidain University College, Baghdad, Iraq, E-mail: ibraheemmohammed1978@gmail.com

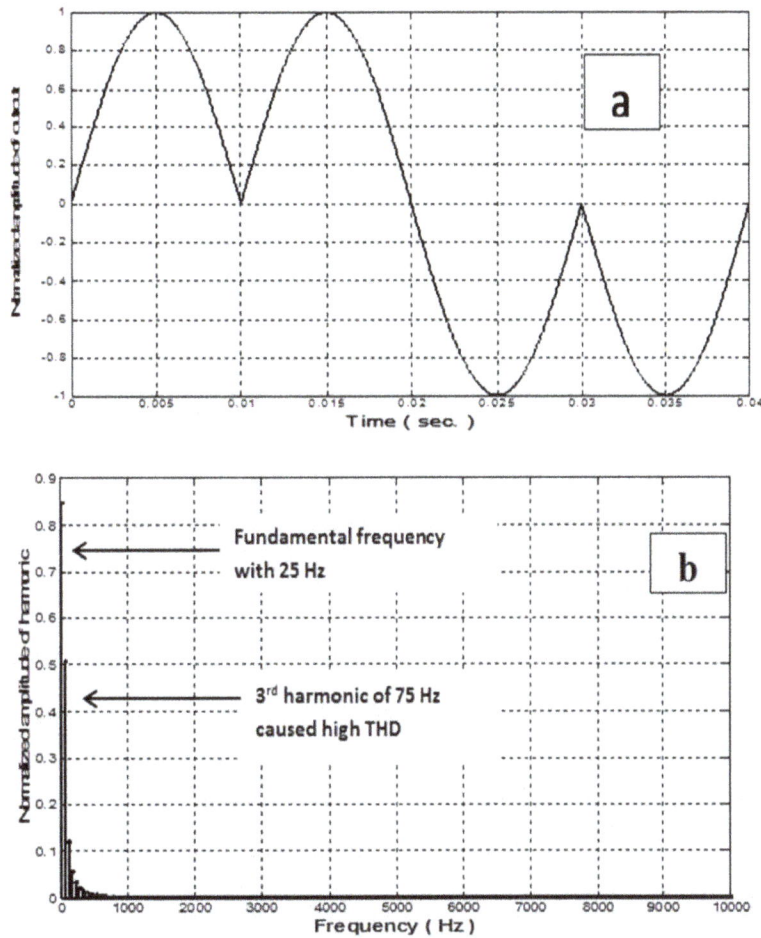

Figure 1: (a) Output waveform of regular AC to AC frequency converter; **(b)** Frequency spectrum of regular AC to AC frequency converter.

process depend on imposing a poly-phase shifted signal with amplitude and frequency equal to magnitude and frequency of harmful third harmonic which can be described in complex form as $A_3 \angle (\theta_3 + \pi)$, the complex addition can be determined as shown in phasor diagram of in Figure 2.

Which is shows two complex components one of them represent harmful 3rd harmonic and the other represent the imposed poly-phase signal, the resultant of adding these two complex components yields in an elimination of harmful 3rd harmonic, which can mathematically represented:

A₃<θ3+A3 < (θ3+Π)=A3 cos (θ3)+jA3 sin (θθ3)+A3 cos (θ3+Π)+jA3 sin (θ3+Π)=A3 cos (θ3)+jA3 sin (θ3)+A3 cos (θ3) cos (Π)+A3 sin (θ3) sin (Π)+j[A3 cos (θ3) sin (Π)+j+A3 sin (θ3) cos (Π)]=A3 cos (θ3)+jA3 sin (θ3) - [A3 cos (θ3)+jA3 sin (θ3)]=0

However the proposed system depends on generating the poly phase of 3rd harmonic using simple H bridge inverter, and generating distorted waveform using regular AC to AC single phase frequency changer and adding the two singles to-gather as shown in Figure 3.

Mathematical Model and Calculation

The proposed technique depend on adding two waveforms of different converters which are differs in frequency, phase, and amplitude one of them represents output of regular AC to AC converter operates

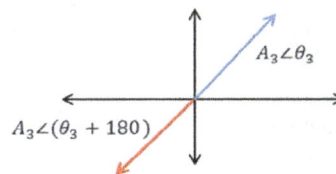

Figure 2: Phasor diagram of proposed technique to eliminate 3rd harmful harmonic.

with M=2 with frequency can be represented according $f_c = \dfrac{f_{supply}}{M}$ which achieve higher THD and the other represents the imposed signal that eliminate the harmful harmonic which is has frequency of $f_p = \dfrac{3 \times f_{supply}}{M}$. The output of the proposed converter can be represented by the following equation:

$$V_{(with\ low\ THD)} = V_{(with\ high\ THD)} + V_{(3rd\ harmonic\ imposed\ wave)} \qquad (1)$$

The Fourier transform for both sides according to linearity theorem can be represented as shown in equation (2):

$$F(jw)_{(with\ low\ THD)} = A_1\ F(jw)_{(with\ high\ THD)} + A_3\ F(jw)_{(3rd\ harmonic\ imposed\ wave)} \qquad (2)$$

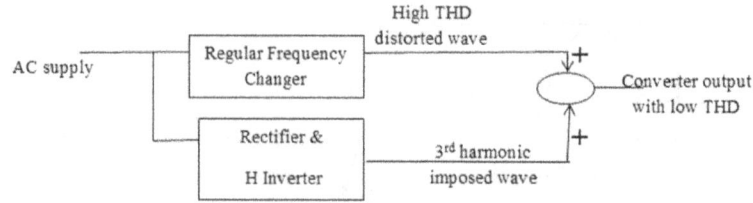

Figure 3: Low THD proposed converter.

Calculation for regular AC to AC converter			Calculation of SPWM inverter		
Order of harmonics for AC to AC converter	Normalized amplitude of harmonic 100%	Phase shift of harmonic	Order of harmonics inverter	Normalized amplitude of harmonic 100%	Phase shift of harmonic
N=1 F_N= 25 Hz	100	0.0°	N=1 F_N= 75 Hz	100	180°
N=3 F_N= 75 Hz	60	0.0°	N=3 F_N= 225Hz	30	-90°
N=5 F_N=125 Hz	15	180°	N=5 F_N=375 Hz	5	-20°
N=7 F_N=175 Hz	6	-2.0°	N=7 F_N=525 Hz	4	-20°
N=9 F_N=225 Hz	4	180°	N=9 F_N=675 Hz	2	180°

Figure 4: Amplitude and phase calculations for fundamental and nth harmonic for both converters.

The polar format of Fourier transform estimated, and manipulated at amplitude and phase of each harmonic appears in the frequency spectrum as shown in equation below:

$$F((T_P, T_C)Nw) \angle \theta_{((T_P, T_C)Nw)} = F(T_C Nw) \angle \theta_{T_C Nw} \tag{3}$$

Where (T_C) represent the period of distorted wave result by regular AC to AC frequency step down changer, while (T_p) represents the period of imposed wave converter, and the frequency spectrum can determined as shown in equation (4):

$$F((T_P, T_C)Nw) = \begin{cases} either F(T_C Nw) \, or F(T_P Nw) \, when (T_P Nw) \neq (T_C Nw) \\ F(T_C Nw) + F(T_P Nw) \, when (T_P Nw) = (T_C Nw) \end{cases} \tag{4}$$

However the output of AC to AC frequency with $\frac{f_s}{f_c} = M = 2$,

where f_c represent the output frequency of regular AC to AC frequency changer and f_s represent main supply frequency, can be represented using equation (5):

$$V_{(with\ high\ THD)} = (-1)^z \, abs(A \sin wt) where \begin{cases} z = 0 \quad 0 \leq t \angle (\frac{T_P}{2}) \\ z = 1(\frac{T_P}{2}) \leq t \angle (T_P) \end{cases} \tag{5}$$

The general form for Fourier series expansion to a periodic waveform can be established using equation (6).

$$F(Nw) = a_0 + \sum_{N=1}^{\infty} a_n \cos Nwt + \sum_{N=1}^{\infty} b_n \sin Nwt \tag{6}$$

Where,

$$a_N = T \int_{-\frac{T_P}{2}}^{\frac{T_P}{2}} V_{(with\ high\ THD)} \cos Ntdt \tag{7}$$

However due to even wave symmetry, the Fourier coefficients (a_N) and (a_0) are zeros, so that we need evaluate coefficients (b_N), which can be determined according (b_N) integral, which can be established according to equation (8):

$$b_N = T \int_{-\frac{T_P}{2}}^{\frac{T_P}{2}} V_{(with\ high\ THD)} \sin Ntdt \tag{8}$$

After estimating the above integrals which is outcome the sine Fourier coefficients of frequency spectrum which represented using equation (9) [13].

$$b_N = \frac{8 \sin(\frac{N\pi}{2})}{\pi(4 - N^2)}$$

And these coefficients exist for odd integer of N, so that the Fourier series expansion can be established as shown in equation (10):

$$F(Nw) = \sum_{N=1,3,5....}^{\infty} \frac{8 \sin(\frac{N\pi}{2})}{\pi(4 - N^2)} \sin Nw_C t \tag{10}$$

As well as the Fourier series expansion of the imposed signal which generated using unipolar SPWM inverter can be represented in equation (11) [14].

$$F(Nw_p t) = \sum_{N=1,3,5....}^{\infty} \frac{4}{N\pi} [\cos(N\alpha_1) - \cos(N\alpha_2) + \cos(N\alpha_3)] \sin Nw_p t \tag{11}$$

Where α1, α2, α3 represent the switching angles of unipolar PWM.

The amplitude and phase of fundamental and nth harmonics appeared in spectrums according to equations (10 and 11) were denoted in Figure 4, assuming that the imposed waveform with reference signal has poly phase of 3rd harmful harmonic of regular AC to AC frequency changer waveform.

The THD of each waveform was estimated and gives 62% for waveform that represent regular AC to AC frequency converter, while for imposed waveform gives THD of 48%, finally resultant spectrums according to equations (10) and (11) separately were determined as shown in Figure 5.

And the Fourier series of resultant waveform that established according the proposed technique which is described in equation (1), with low THD can be determined using equation (12):

$$F((T_P, T_C)Nw) \angle \theta_{((T_P, T_C)Nw)} = \sum_{N=1,3,5....}^{\infty} \frac{8\sin(\frac{N\pi}{2})}{\pi(4-N^2)} \sin Nw_C t \angle \theta_{Nw_C}$$
$$+ \sum_{N=1,3,5....}^{\infty} \frac{-4}{N\pi} [\cos(N\alpha_1) - \cos(N\alpha_2) + \cos(N\alpha_3)] \sin Nw_P t \angle \theta_{Nw_P} \quad (12)$$

Finally the output of the proposed AC to AC frequency changer is estimated according to equation (1) and it is denoted in Figure 6.

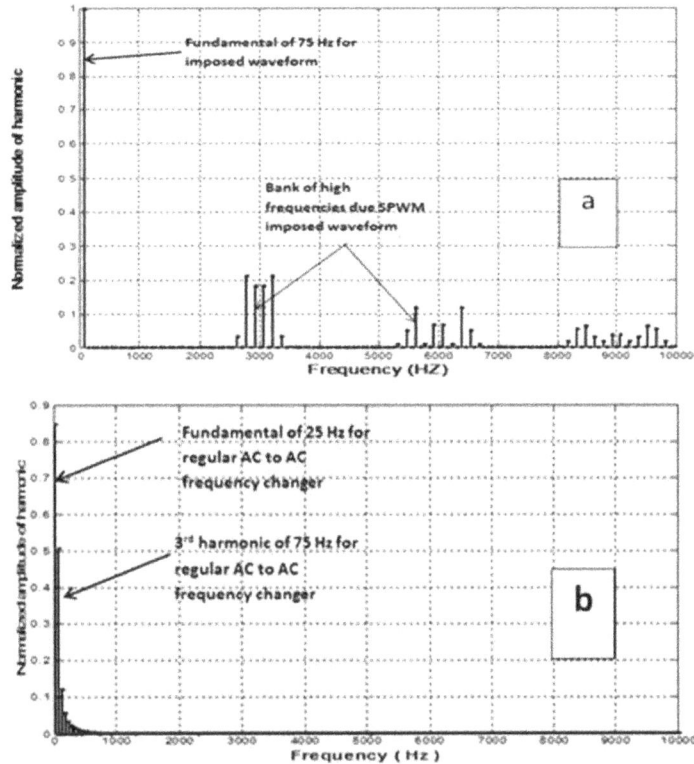

Figure 5: (a) Frequency spectrum of SPWM imposed waveform; **(b)** Frequency spectrum of regular AC to AC frequency converter.

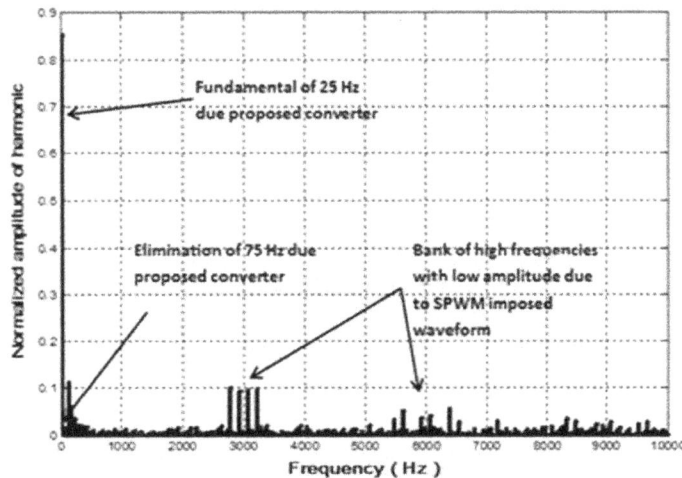

Figure 6: Output waveform of proposed AC to AC frequency converter.

The spectrum of output waveform that is denoted in Figure 6, can be estimated using equation (12), the estimated spectrum is shown in Figure 7, which is a combination of the two spectrums described according to equations (10, 11), that is mean the spectrum of regular AC to AC frequency converter and the spectrum of the unipolar SPWM imposed waveform are merged together and the apparent harmonics in the new spectrum have frequencies of (25*N and 75*N) Hz respectively where (N=1, 3, 5...). The first harmonic (N=1 fundamental) has a frequency of 25 Hz with suitable normalized amplitude while 3rd harmonic which has a frequency of 75 Hz was completely eliminated, finally a bank of high frequencies with low amplitude which have n^{th} harmonic distortion does not exceed 12% have been appeared, and the overall THD for proposed converter reduced to 30%.

Simulation and Result

The proposed system has been implemented using simulink matlab package, the implementation of this system is divided into two levels: the first one represents a bidirectional bridge converter used to implement regular AC to AC frequency changer which is designed to give 25 Hz high distorted output waveform, while the second level represents rectifier and H bridge circuit used to implement unipolar SPWM inverter which is designed to investigate 75 Hz imposed voltage with phase shift=180° and amplitude equal to amplitude of 3rd harmonic related by regular AC to AC frequency changer , the addition process of the two voltages resulted from the two converters has been implemented according to superposition theorem across unity linear transformer. The circuit configuration that represents the proposed technique is described in Figure 8.

The output of each level denoted in the proposed Ac to AC frequency converter circuit is described in Figure 9.

The spectrum of resultant simulated output voltage can be represented using FFT algorithm simulator as shown in Figure 10, which is showing a predominant fundamental frequency of 25 Hz with

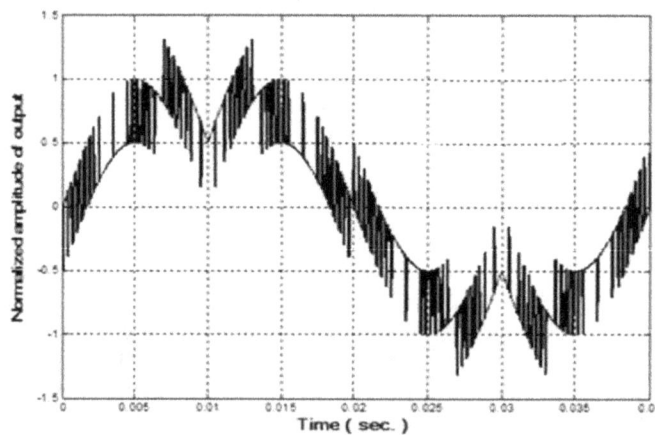

Figure 7: Frequency spectrum of proposed AC to AC frequency converter.

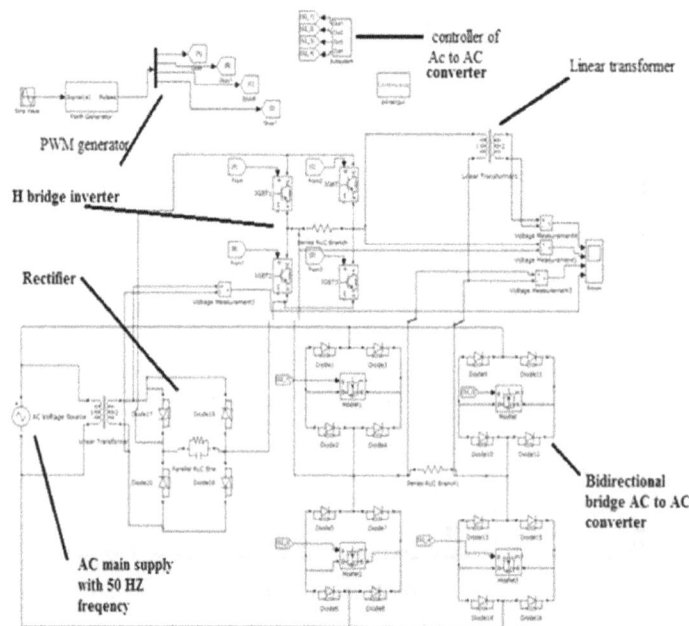

Figure 8: Proposed converter simulated circuit.

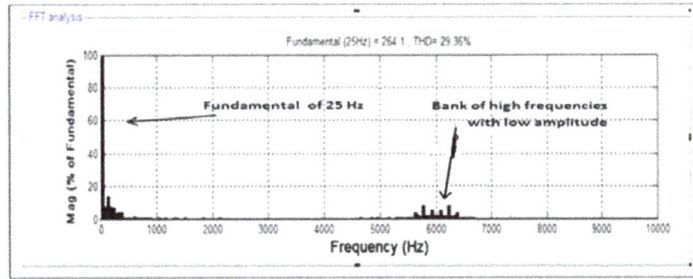

Figure 9: Proposed AC to AC frequency changer stages output waveforms.

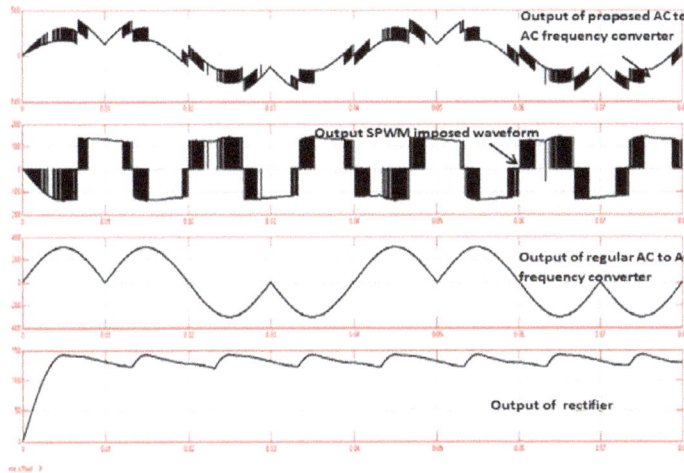

Figure 10: Simulated frequency spectrum for proposed AC to AC frequency changer.

acceptable normalized amplitude, also elimination of harmful third harmonic that appears in frequency spectrum of regular AC to AC frequency changer, the apparent harmonics have frequencies of (25*N) and (N*75) respectively and N=1, 3, 5, 7…, with very low normalized voltages. Finally a bank of high order harmonics with low normalized amplitudes has been appeared without any effect on THD value of the proposed converter.

Due to elimination of 3rd harmonic, the proposed technique gives output waveform with THD about 30% and this value considered to be significant reduction if it is compared with regular AC to AC frequency changer, the reduction in THD enhanced the performance parameters of converter like (efficiency, system cooling requirement cost, overheating current reduction) according to IEEE standard and recommended values [15,16].

Conclusion

The proposed system was investigated to give enhanced single phase output waveforms frequency changer if it is compared to the regular AC to AC frequency changer output due to its effective reduction in THD% (about 56% reduction). The proposed system considers being simple and with low cost of implementation, however H bridge inverter used to impose waveform designed to be poly phase and same frequency of 3rd harmful and adding the imposed waveform to output of regular AC to AC frequency converter, the addition process will eliminate 3rd harmful third harmonic, and result a high frequency harmonic components but with low amplitude and doesn't have significant effect on THD.

References

1. Kim S, Seung-Ki Sul, TA Lipo (2000) AC/AC Power Conversion Based on Matrix Converter Topology with Unidirectional Switches. IEEE Transactions on Industry Applications 36: 139-145.

2. Ahmed I, Nomani KMS (2013) Designing And Analysis Of Cycloconverter To Run Variable Frequency Drive Motor. International Journal of Technology Enhancements and Emerging Engineering Research 1: 149-153.

3. James JG, Saha TK (2003) An Investigation of Harmonic Content in a Remote mine Site. The Sixth International Power Engineering Conference (IPEC2003), Singapore.

4. Lazim MT (2013) Power Frequency Converter Based on Integral-cycle Triggering Mode of Thyristors with Multi-control Periods. Electric Power Components and Systems 41: 1173-1187.

5. Ali WAN, Ali A, Qadeer S (2016) Simulation Model and Harmonic Analysis of Step Down Bidirectional Controlled AC-AC Converter. International Journal of Computer Science and Information Security 14: 642-647.

6. Chattopadw AK (1997) A Review Cycloconverter and cycloconverters Fed-drives. J Indian Inst Sci 77: 397-419.

7. Shmilovitz D (2005) On the definition of total harmonic distortion and its effect on measurement interpretation. IEEE Transactions on Power Delivery 20: 526-528.

8. Al-duaij OS (2015) Harmonics Effects in Power System. Int Journal of Engineering Research and Applications 5: 01-19.

9. Al-Khesbak MS, Lazim MT (2013) Envelppe Cycloconverter Based on Integral Half Cycle Selection and Half Cycle Omission Technique. Journal of Electrical Engineering (JEE) 13: 62-69.

10. Khesbak MSMA, Khaleel IM, Ali RM (2015) Enhanced ac-to-ac Frequency Changer Based on Multi-Phase Smart Comparative Commutation. International Journal of Scientific & Engineering Research, Volume 6: 31-44.

11. Khesbak MSMA, Lazim MT (2009) Harmonic Reduction in Envelope Cycloconverters Using comb Filter Technique. 6th International Multi-Conference on Systems, IEEE Conference Publications Signals and Devices.

12. Govil VK, Chaurasia Y (2012) Modeling & Simulation of PWM Controlled Cycloconverter FED Split Phase Induction Motor. International Journal of Advanced Research in Electrical, Electronics and Instrumentation Engineering 1: 126-133.

13. Ashraf N, Hanif A, Farooq U, Asad MU, Rafiq F (2013) Half Cycle Pairs Method for Harmonic Analysis of Cycloconverter Voltage Waveform. International

Conference on Open Source Systems and Technologies, pp: 97-102.

14. Chiasson J, Tolbert LM, McKenzie K, Du Z (2004) A Complete Solution to the Elimination Harmonic Problem. IEEE Journals & Magazines 19: 491-499.

15. IEEE Power and Energy Society (2014) IEEE Recommended Practice and Requirement for Harmonic Control in Electrical Power system. IEEE Std 519TM- 2014 (Revision of IEEE Std 519- 1992).

16. Haplin (2006) Revisions to IEEE slandered 519-1992. Transmission and Distribution Conference and Exhibition, 2005/2006 IEEE PES.

22

Closed Loop Control of Zero Voltage Switching DC-DC Converter to Generate Three Outputs

Prasad JS[1]*, Obulesh YP[2] and Babu CS[3]

[1]*LBR College of Engineering, Mylavaram, Andhra Pradesh, India*
[2]*KL University, Vijayawada, Andhra Pradesh, India*
[3]*JN Tech. University, Kakinada, Andhra Pradesh, India*

Abstract

Hard switching specifies the stressful switching behavior of the controlled switches. During the turn-off and turn-on processes, the power electronic device has to withstand high current and voltage simultaneously, resulting in high stress and switching losses. The switching loss is directly proportional to the switch frequency, thus reducing the maximum switch frequency of the power electronic converter. The concept was to incorporate resonant tanks in the converters to create oscillatory (usually sinusoidal) voltage and/or current waveforms, so the zero current switching (ZCS) or zero voltage switching (ZVS) conditions can be achieved for the power control switches. The Soft-switched power converters are generally utilizing the resonance condition. Resonance condition is generally occurred just during the turn-off and turn-on processes, so as to create ZCS and ZVS across each switch. The Regulated three and five multiple-output dc-dc converter under zero-voltage switching (ZVS) condition is proposed. The converter is consists of three outputs altogether. With the help of two asymmetric half bridge converters, the first and second outputs are controlled. Based on the phase shift between two asymmetric half bridge converters, the third output is controlled. At high switching frequency, these multiple-output dc–dc converters can give higher efficiency. The various stages of operation, soft switching condition and controlling schemes are also explained. A closed loop and open loop control techniques of the three multiple output converter is explained.

Keywords: DC-DC converter; Zero Voltage Switching (ZVS); Zero Current Switching (ZCS); Switching losses; Full bridge; Inverter; Duty cycle; Switching period; Pulse Width Modulation (PWM)

Introduction

A DC-DC converter is an electronic circuit which converts a source of direct current (DC) from one voltage level to another. It is a class of power converter. The DC-DC converters are widely used for battery power supply in different electronic devices like mobile phones, MP3 players and laptops. There is a scope for developing DC-DC converters to generate multiple dc output voltage from single dc power supply. These multiple output voltages are feed to the different dc load applications. This scheme of developing multiple dc voltage levels from a single dc supply source can reduce the overall device area. The dc voltage provided by rectifier or battery contains more ripples and it is not a constant value and it is not suitable for many electronic devices. To overcome this problem, the dc-dc voltage regulators are used to control the ripples even when change in the input voltage or load current.

The switching mode type dc-dc converters power supply is widely used because it uses a switch in the form of transistor type and less loss components such as transformers, inductors and capacitors for controlling the output voltage. The switched mode power supply contains two different parts: control part and power part. The majority of the work is carried out by the control part for getting better control of output voltage. Generally the MOSFET is used as a control switch in Switched mode power supply for stabilizing the required output voltage. The MOSFET switches are not to be conducted continuously and they operate only under specific frequency interval, hence these switches are useful for a long future and also provide less power loss the converter circuit. The basic structure of Switched mode power supply is used for stepping up or stepping down of input DC voltage. The SMPS circuit is basically consisting of a filter at the output side for removing the ripples due to switching [1].

The main objective of the project is to regulate three multiple output voltages with dc-dc zero-voltage switching (ZVS) converter. The converter is consisting of three multiple outputs voltages. With the help of two asymmetric half bridge converters, the first and second outputs are controlled. Based on the phase shift between two asymmetric half bridge converters, the third output is controlled. ZVS is realized for all the main switches. At high switching frequency, these multiple-output dc-dc converters can give higher efficiency. The various stages of operation, soft switching condition and controlling schemes are also proposed. A closed loop and open loop control techniques of the three multiple output converter is explained.

Closed-loop System Transfer Function. The Transfer Function of any electrical or electronic control system is the mathematical relationship between the systems input and its output, and hence describes the behavior of the system.

Proposed Three Multiple Output Zvs Dc-Dc Converter

The Figure 1 shows the main diagram of three multiple outputs converter consists of the two asymmetrical half bridge

ZVS DC-DC converter. Figure 2 shows the control signal of the converter sand it is connected to the three single phase

---

*Corresponding author: Prasad JS, LBR College of Engineering, Mylavaram, Andhra Pradesh, India, E-mail: janapatisivavaraprasad@gmail.com

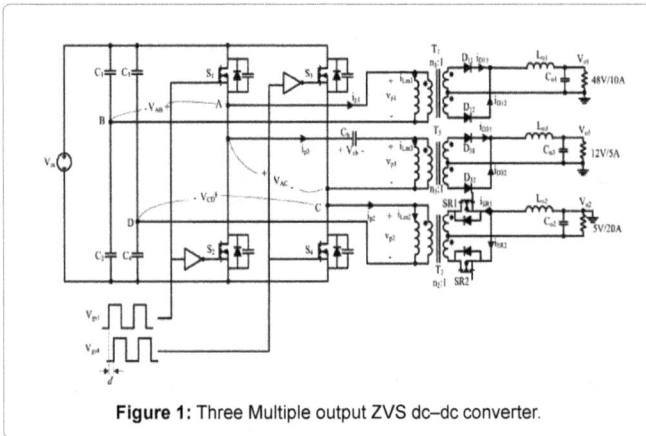

Figure 1: Three Multiple output ZVS dc–dc converter.

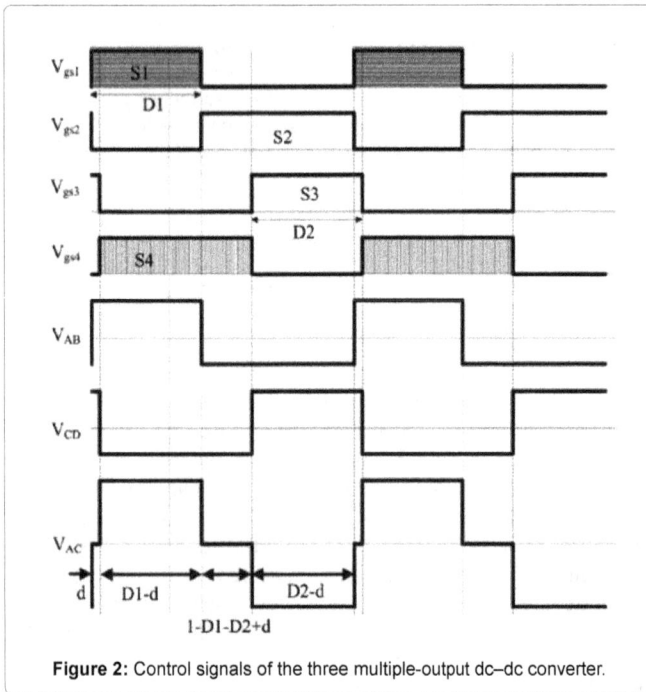

Figure 2: Control signals of the three multiple-output dc–dc converter.

Multiple output converter. Multiple outputs ZVSDC-DC transformers [2].

Characteristic of the three multiple output converters

The multiple-output ZVS dc–dc converter is shown in Figure 1 Since the second output $_{02}$ is low voltage and high current output, we use self-driven synchronous rectification to reduce rectification loss and improve efficiency. All three outputs are regulated through the switches in the primary side. The control signals are shown in Figure 3, where and are the voltage across transformer T1 and T2, respectively, while is the voltage applied to blocking capacitor and transformer T3.

The first output $_{01}$ is regulated through the duty cycle of the first asymmetrical half bridge converter which is composed of switches S1–S2 and capacitors C1–C2. Based on the phase shift between two asymmetric half bridge converters, the third output is controlled which is composed of switches S3–S4 and capacitors C3–C4 (Figures 4-9). Based on the phase shift between two asymmetric half bridge converters, the third output is controlled. Since ZVS can be realized for

all the main switches, this converter operates at high efficiency under high switching frequency [3].

According to the volt-second balance of the output inductors, we can derive the output voltages of the converter. The first output voltage 01 is

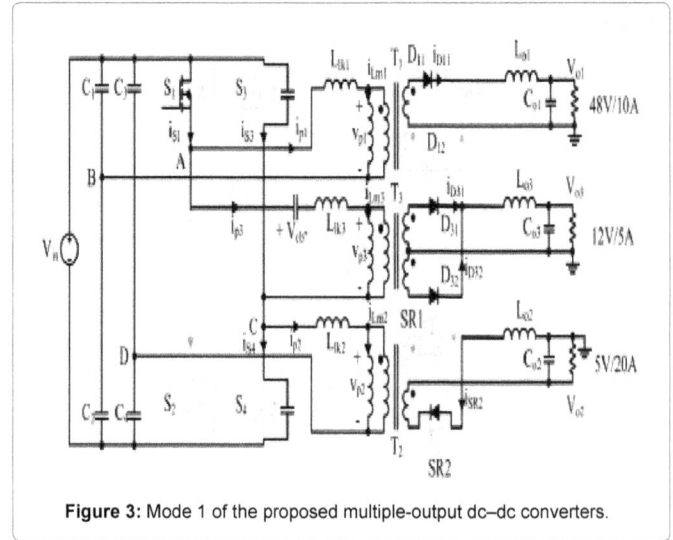

Figure 3: Mode 1 of the proposed multiple-output dc–dc converters.

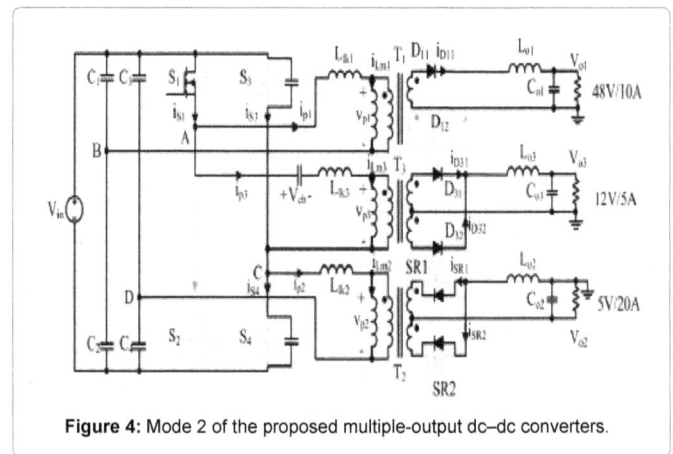

Figure 4: Mode 2 of the proposed multiple-output dc–dc converters.

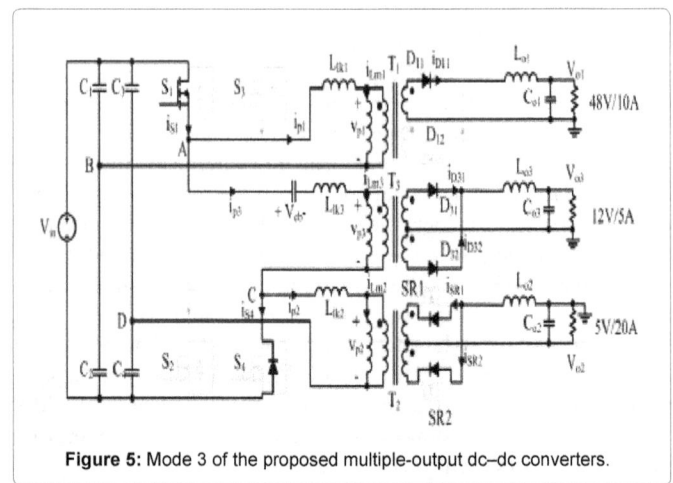

Figure 5: Mode 3 of the proposed multiple-output dc–dc converters.

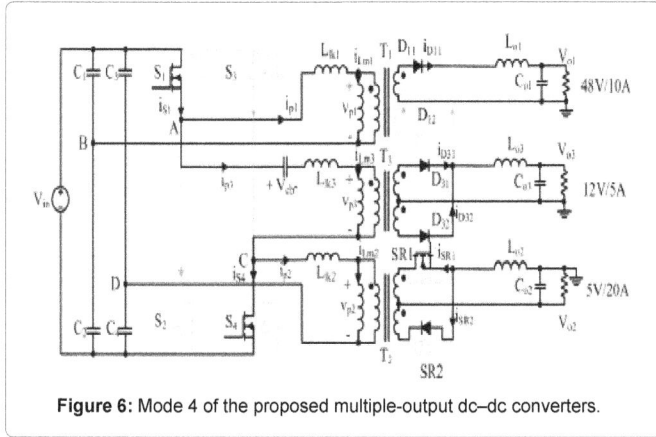

Figure 6: Mode 4 of the proposed multiple-output dc–dc converters.

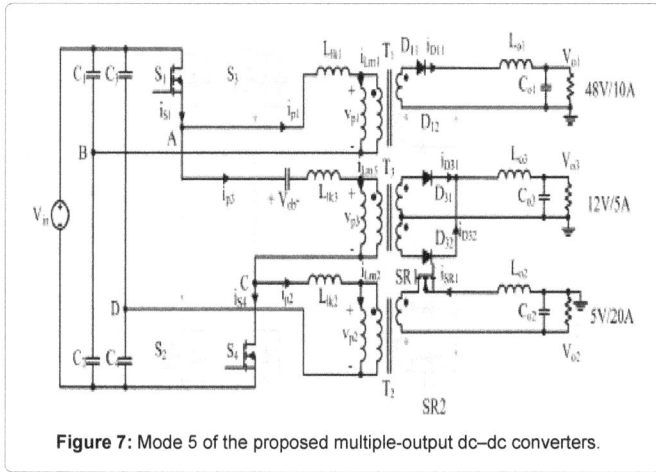

Figure 7: Mode 5 of the proposed multiple-output dc–dc converters.

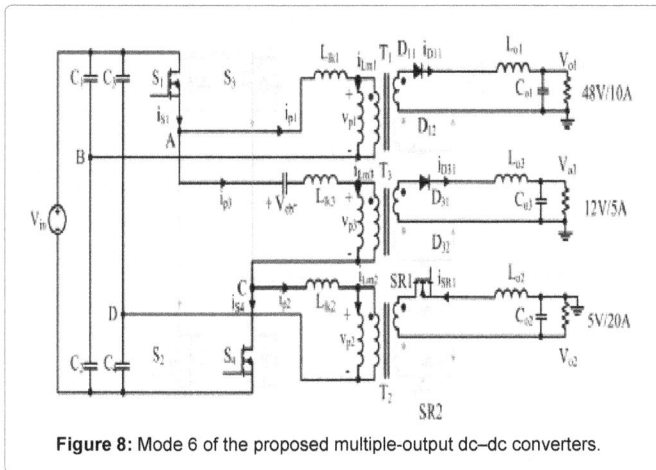

Figure 8: Mode 6 of the proposed multiple-output dc–dc converters.

$$V_{01} = 2.V_{in}.D_1(1 - D_1).\frac{1}{n_1} \qquad (1)$$

the first asymmetrical half bridge converter, equation1 is the turn ratio of transformer T1. The second output voltage02 is

$$V_{02} = 2.V_{in}.D_1(1 - D_1).\frac{1}{n_1} \qquad (2)$$

Where is the input dc voltage, equation 2 is the duty cycle of the second asymmetrical half bridge converter, equation 2 is the turn ratio

of transformer T2.

The expression of the third output voltage 03 is different and depends on both duty cycles 1 and 2. When duty cycle 1 is larger than duty cycle 2, the third output voltage 03 is

$$V_{03} = V_{in}.(D_1 + D_2 - 2.d).\frac{1}{n_3} + V_{in}.(D_1 - D_2).(1 - 2.D1 + 2.d1n3) \qquad (3)$$

When duty cycle 1 is smaller than duty cycle 2 , the third output voltage 03 is

$$V_{03} = V_{in}.(D_1 + D_2 - 2.d).\frac{1}{n_3} + V_{in}.(D_1 - D_2).(1 - 2.D2 + 2.d1n3) \qquad (4)$$

Where is the input dc voltage, $_1$ and $_2$ are the duty cycles of the first and second asymmetrical half bridge converters respectively, is the phase shift of the switch S4 to the switch S1, and 3 is the turn ratio of transformer T3.

For simplifying the expression of the third output voltage 03, we compare the two equations. Therefore, the third output voltage 03 can be simplified as

$$V_{03} = V_{in}.(D_1 + D_2 - 2.d).\frac{1}{n_3} \qquad (5)$$

Operation stages of the proposed three multiple output Dc-Dc converter

Before the analysis, we first make the following assumptions.

1. The duty cycles 1 and 2 are both near 0.5 and almost same, so the voltage across capacitor, which is (1 - 2). (When duty cycle 1 is larger than duty cycle 2) or (2 - 1). (When duty cycle 1 is smaller than duty cycle 2), and is very small compared to the input voltage, and therefore can be ignored.

2. The ZVS of the switches S1–S2 is realized through the energy stored in the output inductor of the third output and the ZVS of the switches S3–S4 is realized through the energy stored in the leakage inductor of transformer T2 and T3.

3. The capacitors C1–C4 and are so large that the voltages across them are considered to be constant.

4. The output inductors are so large that they are considered as current sources.

Stage 1 (t0–t1): Before the starting of this stage, the controlled switches S1 and S3 are on, and the first and second outputs are transferring energy to the load while the third output is freewheeling. At time t0, the switch S3 is turned off. The sum of primary current 1 and 2 begins to charge the parasitic capacitance of the switch S3, and discharge the parasitic capacitance of the switch S4. The voltage turns positive, so the current through the diode D31 begins to increase from zero. The diodes D31 and D32 conduct simultaneously and the voltage across the transformer T3 3 is clamped to zero [4].

Stage 2 (t1–t2): At the time of t1, the parasitic capacitance voltage of switch S4 is discharged to 2. Since the voltage across the capacitor C4 is 2, the voltage is zero. Therefore, the current through the synchronous rectifier SR 2 begins to decrease and the current through the synchronous rectifier SR1 begin to increase from zero. Since the synchronous rectifiers SR1 and SR2 conduct simultaneously, the voltage across the transformer T2 2 is clamped to zero.

Stage 3 (t2–t3): At the time of t2, the parasitic capacitance voltage of switch S4 is discharged to zero. Then the sum of current 2 and 3 begins to flow through the body diode of the switch S4, creating ZVS condition for the switch S4.

Stage 4 (t3–t4): At time t3, the gate signal of the switch S4 is applied, and the switch S4 is ZVS turned on. Then the sum of current $_2$ and $_3$ begins to flow through the switch S4.

Stage 5 (t4–t5):At time t4, the current through the synchronous rectifier SR2 decreases to zero and the current through the synchronous rectifier SR1 increases to the second output current [5].

Stage 6 (t5–t6): At time t5, the current flowing through the secondary side diode D32 starts decreasing to zero and the current through the diode D31 increases to the third output current [6].

Stage 7 (t6–t7): At time t6, the switch S1 is turned off, and the sum of primary current 1 and 3 begins to charge the parasitic capacitance of the switch S1 and discharge the parasitic capacitance of the switch S2 (Figures 9-13).

Stage 8 (t7–t8): At the time of t7, the parasitic capacitance voltage of switch S2 is discharged to 1. Since the voltage across the capacitor C2 is 2, the voltage is zero. The current through the diode D11 begins to decrease and the current through the diode D12 begins to increase from zero. As the diode D11 and D12 conduct simultaneously, the voltage across the transformer T1 1 is clamped to zero [7].

Stage 9 (t8–t9): At the time of t8, the parasitic capacitance voltage of the switch S2 is discharged to zero. Then the sum of primary $_1$ current $_3$ and begins to flow through the body diode of the switch S2, creating ZVS condition for the switches S2.

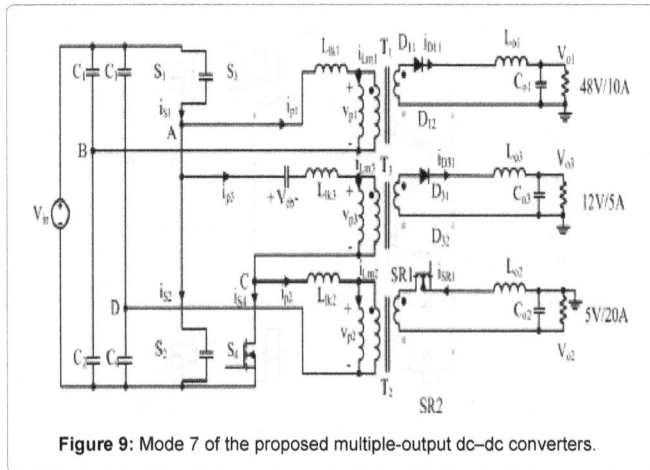

Figure 9: Mode 7 of the proposed multiple-output dc–dc converters.

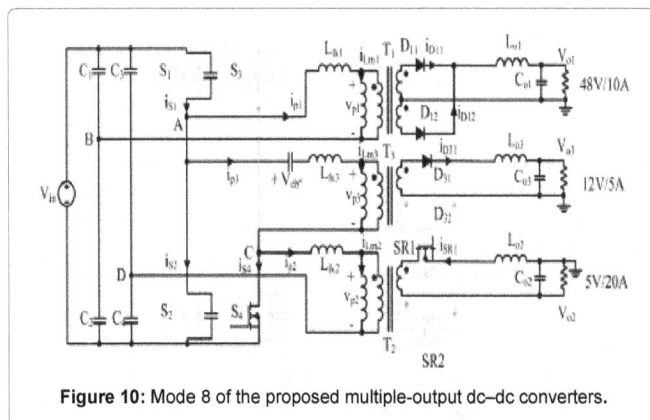

Figure 10: Mode 8 of the proposed multiple-output dc–dc converters.

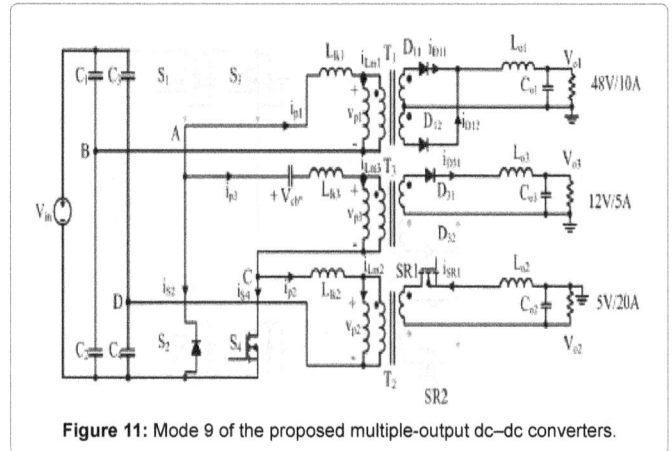

Figure 11: Mode 9 of the proposed multiple-output dc–dc converters.

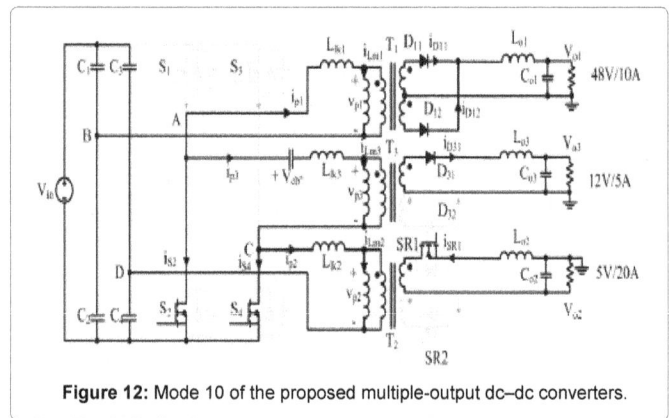

Figure 12: Mode 10 of the proposed multiple-output dc–dc converters.

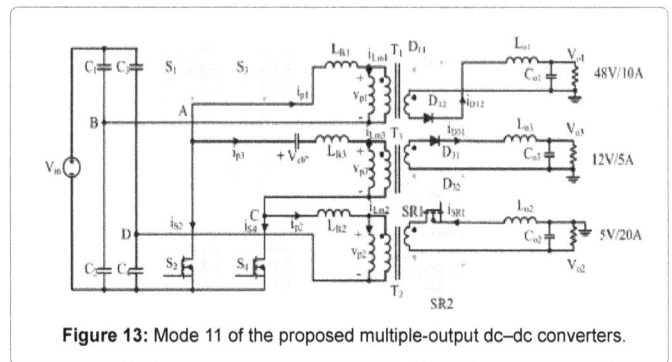

Figure 13: Mode 11 of the proposed multiple-output dc–dc converters.

Stage 10 (t9–t10): At time t9, the gate signal of the switch S2 is applied, and the switch S2 is ZVS turned on. The sum of primary current 1 and 3 begins to flow through the switch S2.

Stage 11 (t10–t11): At time t10, the current through the diode D11 decreases to zero and the current through the diode D12 increases to the first output current.

Simulink Model of the Three Multiple Outputs Dc-Dc Converter

Simulink model of the three outputs converter is shown in the Figure 14. In this model we have taken 400 V input to the full bridge inverter. The full bridge inverter which is converted direct current

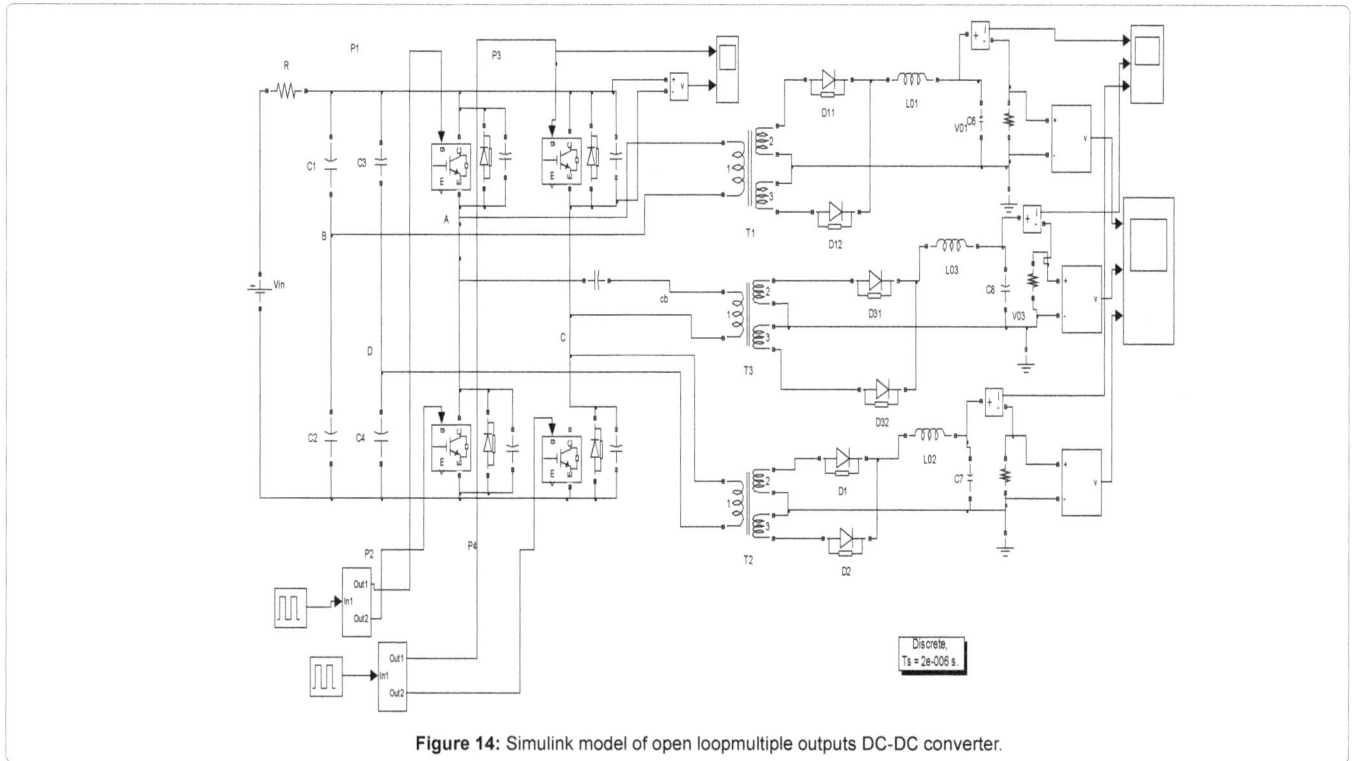

Figure 14: Simulink model of open loopmultiple outputs DC-DC converter.

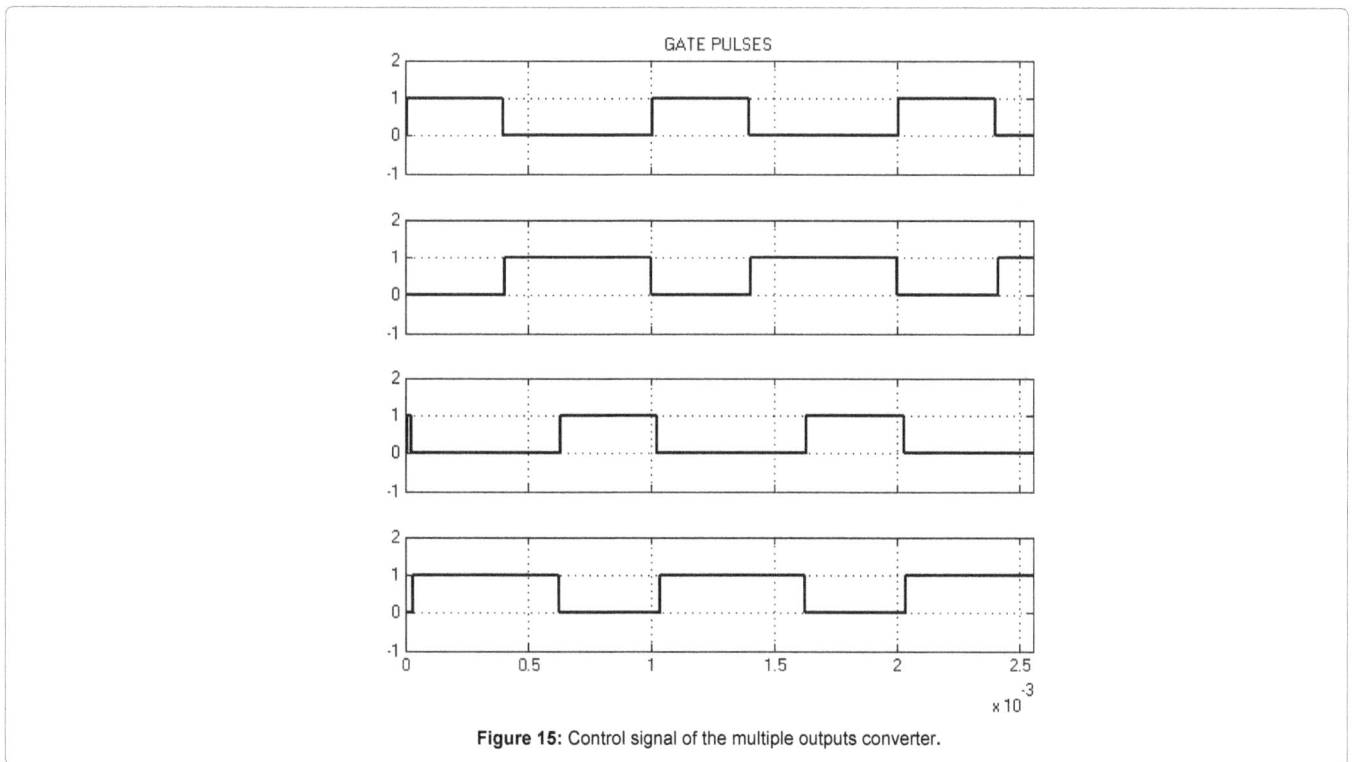

Figure 15: Control signal of the multiple outputs converter.

into alternative current. The output of the inverter is fed to the linear transformer, and connected to diode bridge rectifier which is converted alternative current into direct current [8].

The first output $_{01}$ is regulated through the duty cycle control of the asymmetrical half bridge converter which is composed of switches S1-S2 and capacitors C1-C2. The second output voltage outputs $_{02}$ is regulated through the duty cycle control of the asymmetrical half bridge converter which is composed of switches S3-S4 and capacitors C3-C4. The third output voltage outputs $_{03}$ is regulated through the phase shift of two asymmetrical half bridge converters. The control signal of the proposed three output converter as shown in the Figure 15.

Zvs conditions of the three outputs Dc-Dc converter

The Figures 16 and 17 shows the pulse and across the switch voltage. Two stages of ZVS conditions of switch as shown, when pulse is ON the voltage across the switch is zero and when pulse is OFF the voltage across the switch is zero. Figure 18 (a) shows ON state at ZVS condition of the switch 1 pulse and switch across the capacitor voltage and (b) shows OFF state of ZVS condition of the switch 1 pulse and switch across the capacitor voltage [9].

The Figures 19 and 20 shows the pulse and across the switch voltage. Two stages of ZVS conditions of switch as shown, when pulse is ON the voltage across the switch is zero and when pulse is OFF the voltage across the switch is zero. Figure 21a shows ON state at ZVS condition of the switch 2 pulse and switch across the capacitor voltage and (b) shows OFF state at ZVS condition of the switch 2 pulses and switch across the capacitor voltage.

Closed Loop Control of the Three Multilpe Outputs Dc-Dc Converter

In this closed loop mode we are also taken 400V as an input voltage. The output voltage of the inverter is connected to the three single phase transformers and to diode bridge rectifier to convert alternative voltage onto direct voltage. Three different output voltages 48V/10A, 12V/5A and 5V/20A are obtained. The block diagram of proposed closed loop control of ZVS DC -DC converter is shown in Figure 21. The Simulink model of the closed loop proposed three output voltage converter as shown in the Figure 22 [10].

The detailed control block of the closed loop proposed three output converters as shown in the Figure 22. In this model we are taken three outputs as a feedback control. Compare the reference voltage and feedback voltage by using error amplifier. The error signal is given to PWM comparator to compare the ramp signal which is having 1000 Hz frequency and error signals and gives pulse to the switch S4 and

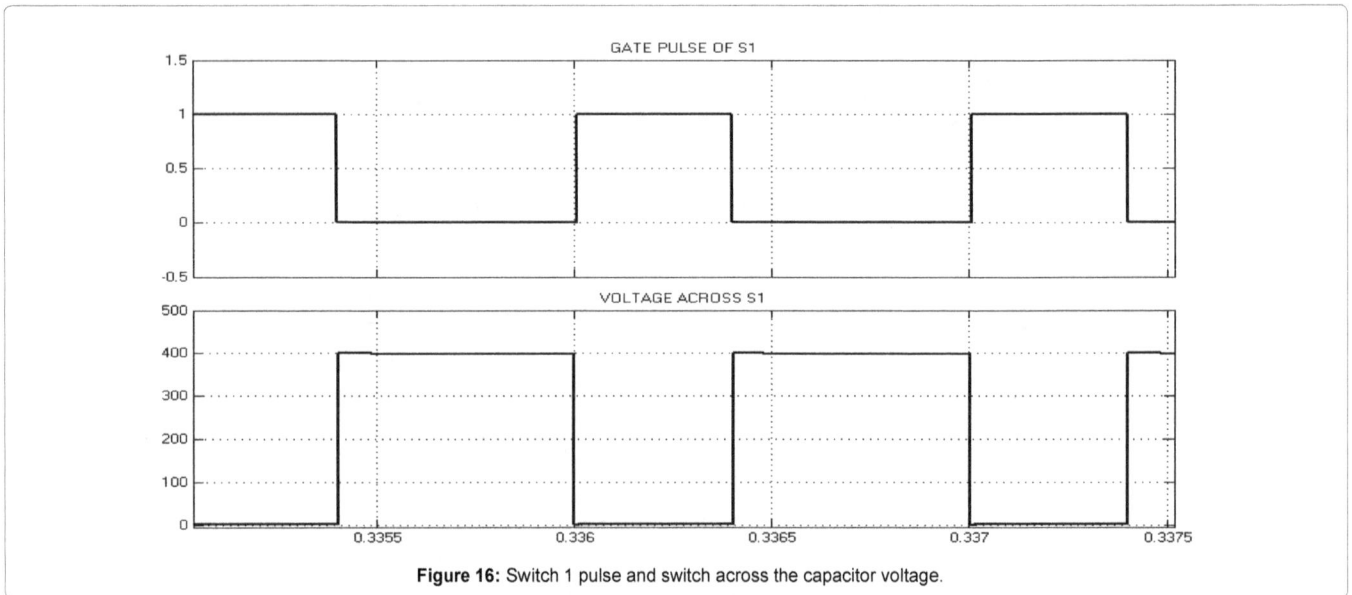

Figure 16: Switch 1 pulse and switch across the capacitor voltage.

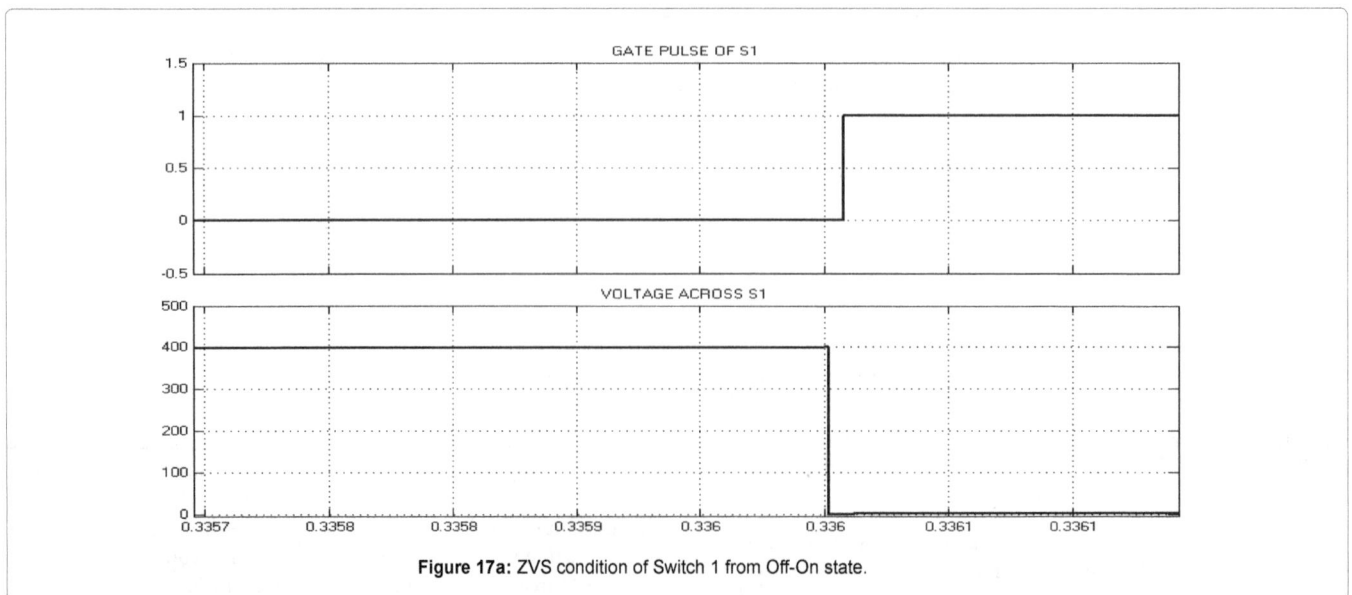

Figure 17a: ZVS condition of Switch 1 from Off-On state.

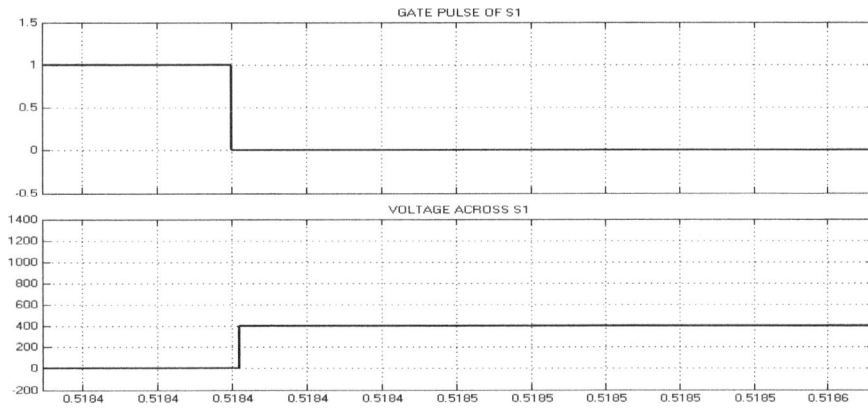

Figure 17b: ZVS condition of Switch 1 from On-Off state.

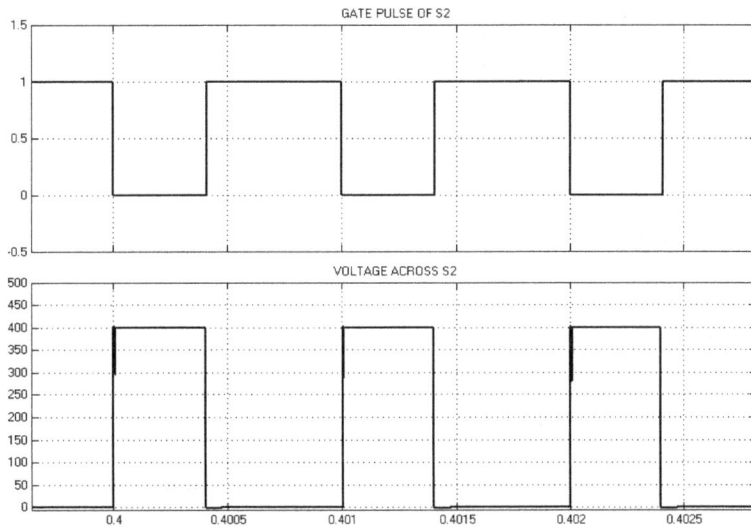

Figure 18: Switch 2 pulse and switch across the capacitor voltage.

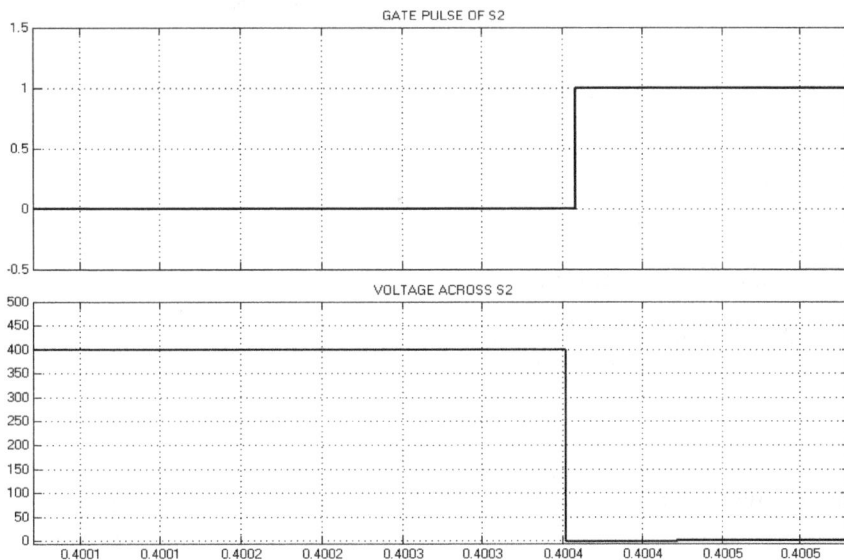

Figure 19: Voltage across S2.

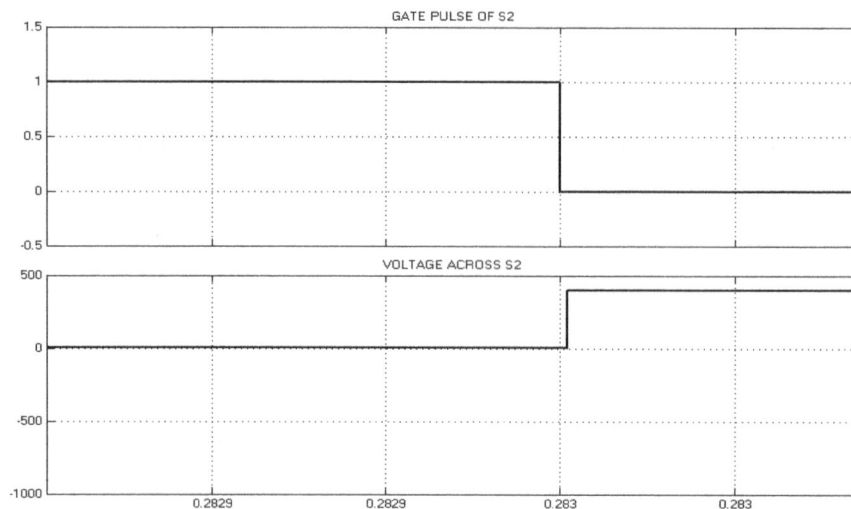

Figure 20: output voltage wave forms of the proposed three outputs converter.

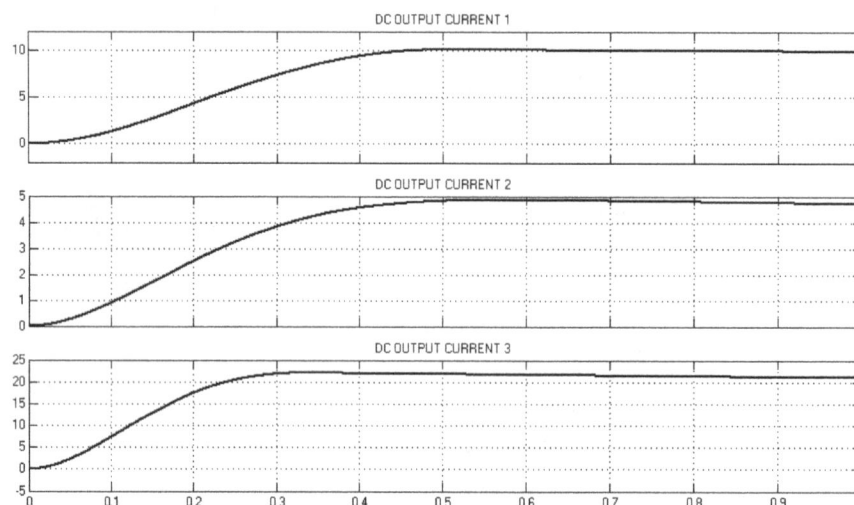

Figure 21: Output current wave forms of the proposed three outputs converter.

inverted pulse is given to the switch S3 of the proposed three outputs converter. By using remaining two outputs we generate a pulse and it is given to the switch S1 and inverted pulse is given to the switch S2 of the proposed three outputs converter [5].

During the time of turn-off and turn-on conditions, the power electronic devices have to withstand large currents and voltages, thus resulting in the high switching stresses and switching losses. To reduce the switching losses and stress we are using zero voltage switching (ZVS). Zero voltage switching (ZVS) conditions of the converter are pulse and switch across capacitor voltages as shown in figures. The zero voltage switching (ZVS) conditions of the four switches at ON time and OFF time period of the pulses and switch across the capacitor voltages are as shown in the below Figure 23.

Zvs conditions of the three multiple outputs Dc-Dc converter under closed loop

The Figures 24 and 25 shows the pulse and across the switch voltage

of S1. Two stages of ZVS conditions of switch S1 as shown, when pulse is ON the voltage across the switch is zero and when pulse is OFF the voltage across the switch is zero.

The Figures 26 and 27 shows the pulse and across the switch voltage of S1. Two stages of ZVS conditions of switch S1 as shown, when pulse is ON the voltage across the switch is zero and when pulse is off the voltage across the switch is zero [4].

By using the closed loop control signal we can regulate the three outputs with changing load value. The output voltage and current wave forms of closed loop control of the proposed three outputs converter as shown in the Figure 28 [7].

Conclusions

Outputs were regulated through primary side switches when proportional sharing was directly used in Multiple-output ZVS dc-dc converter. All the main switches can realize ZVS, therefore the converter can work with higher switching frequency and higher

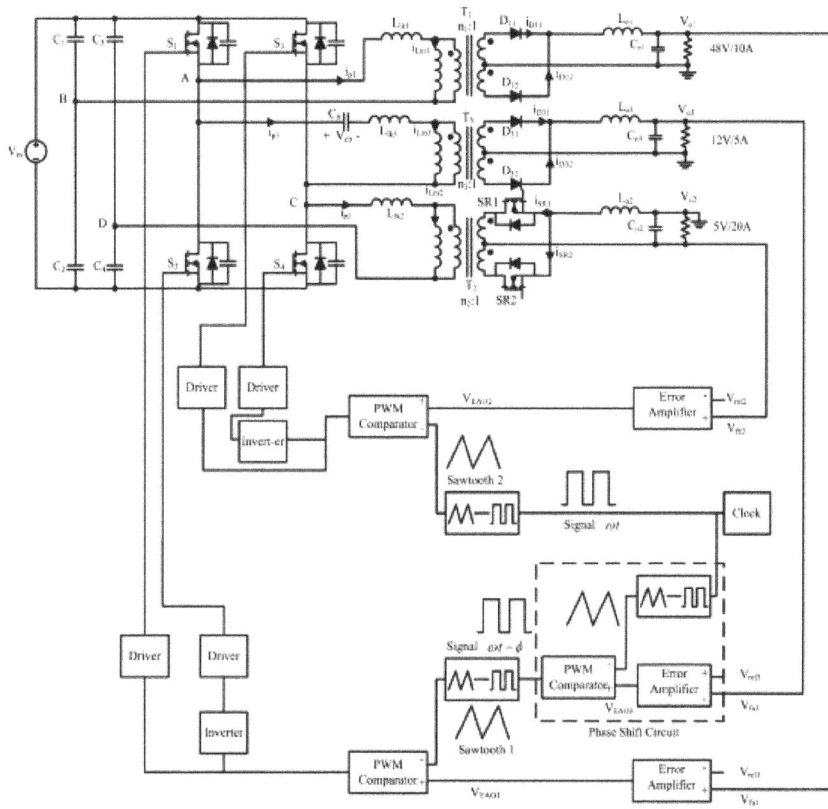

Figure 22: Proposed closed loop control circuit of ZVS DC-DC converter.

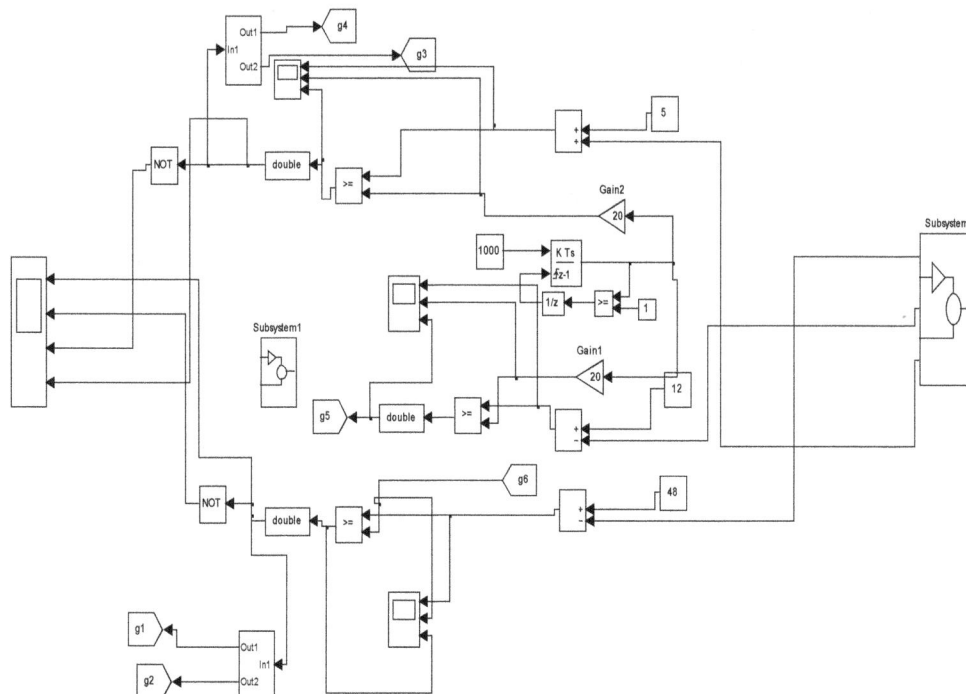

Figure 23: Simulink model of closed loop control of proposed three multiple outputs dc-dc converter.

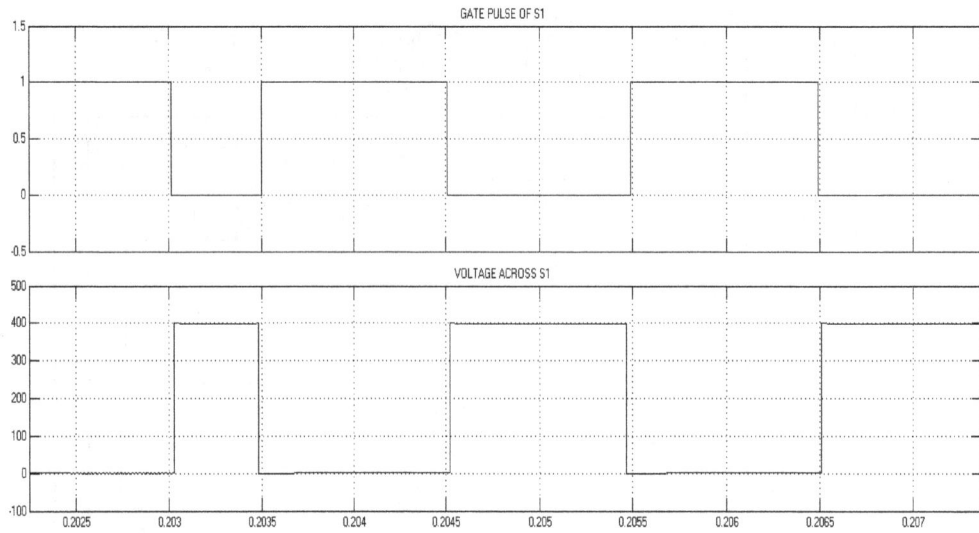

Figure 24: Switch 1 pulse and switch across the capacitor voltage.

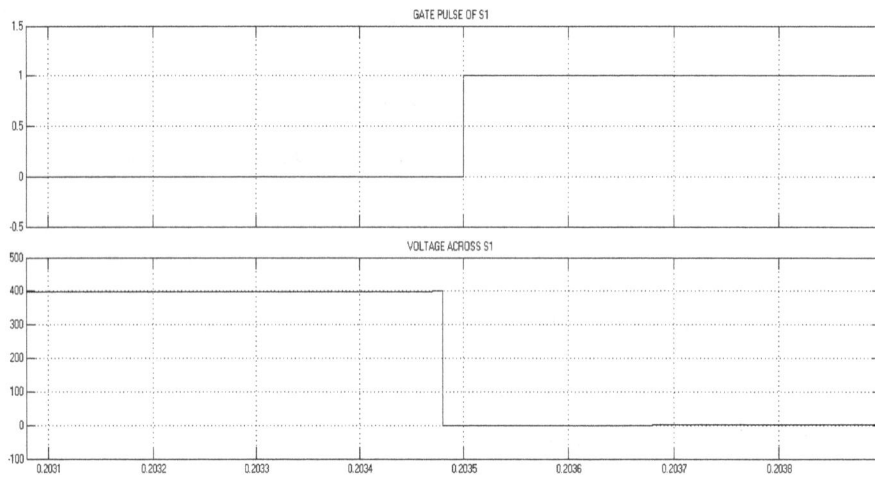

(a)

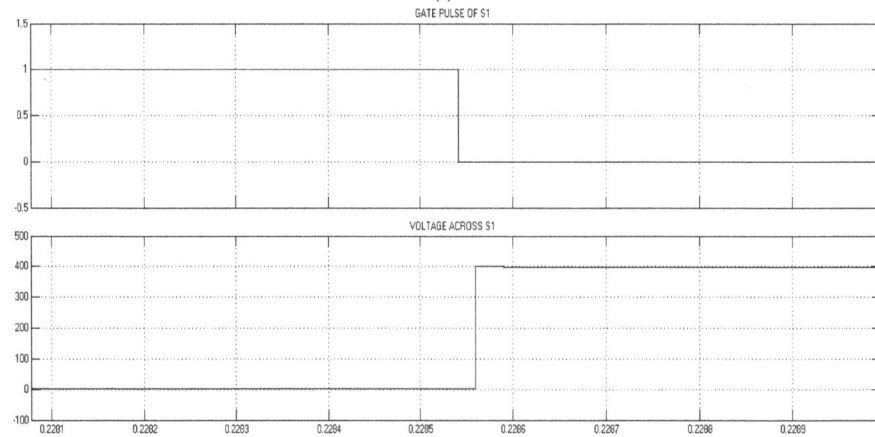

(b)

Figure 25: (a) shows ON state at ZVS condition of the switch 1 pulse and switch across the capacitor voltage and (b) shows OFF state at ZVS condition of the switch 1 pulse and switch across the capacitor voltage.

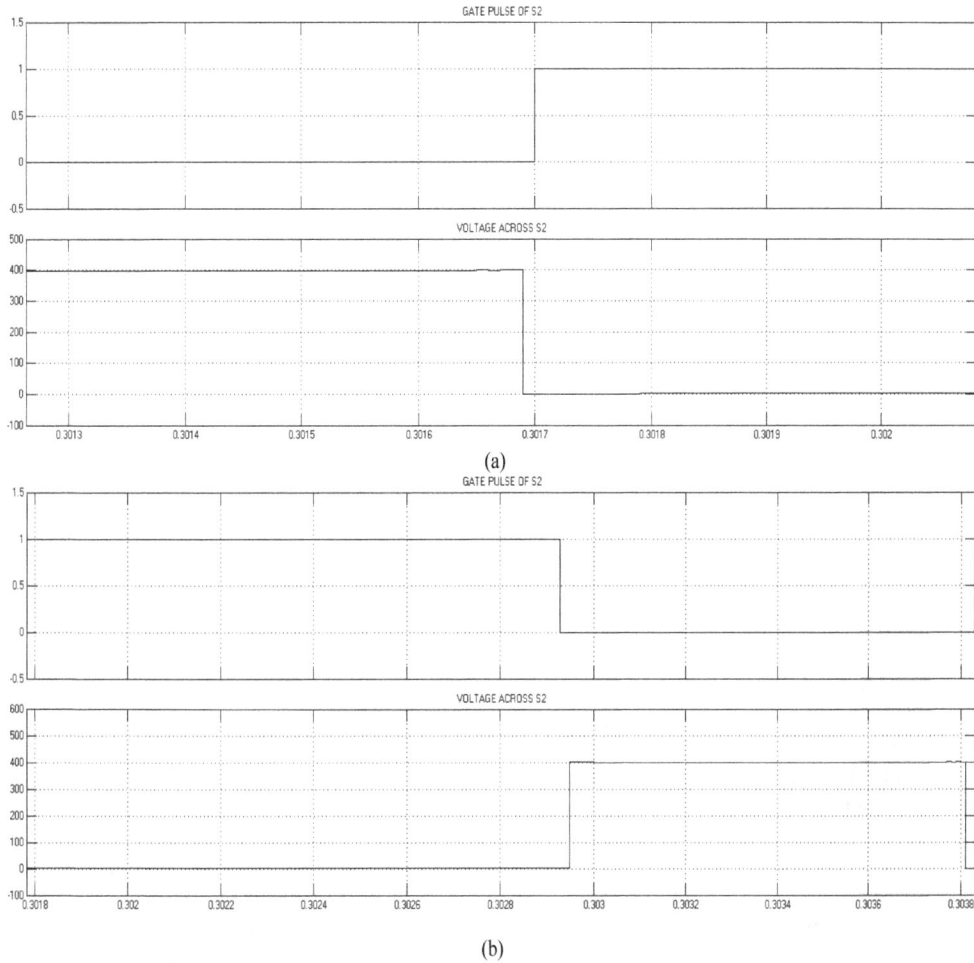

GATE PULSE OF S2

VOLTAGE ACROSS S2

(a)

GATE PULSE OF S2

VOLTAGE ACROSS S2

(b)

Figure 26: (a) shows ON state at ZVS condition of the switch 2 pulse and switch across the capacitor voltage and (b) shows OFF state at ZVS condition of the switch 1 pulse and switch across the capacitor voltage.

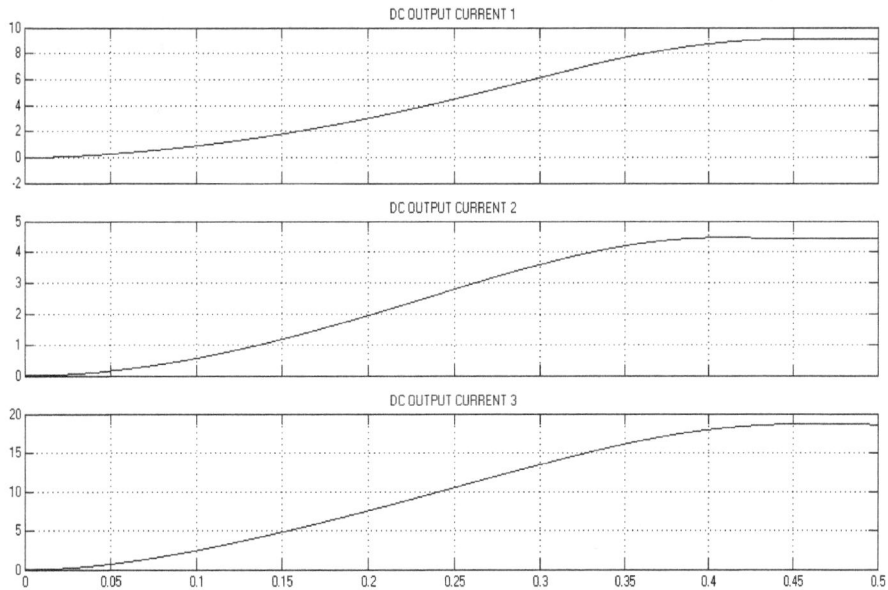

DC OUTPUT CURRENT 1

DC OUTPUT CURRENT 2

DC OUTPUT CURRENT 3

Figure 27: Output current waveforms of closed loop control of the proposed three output converter.

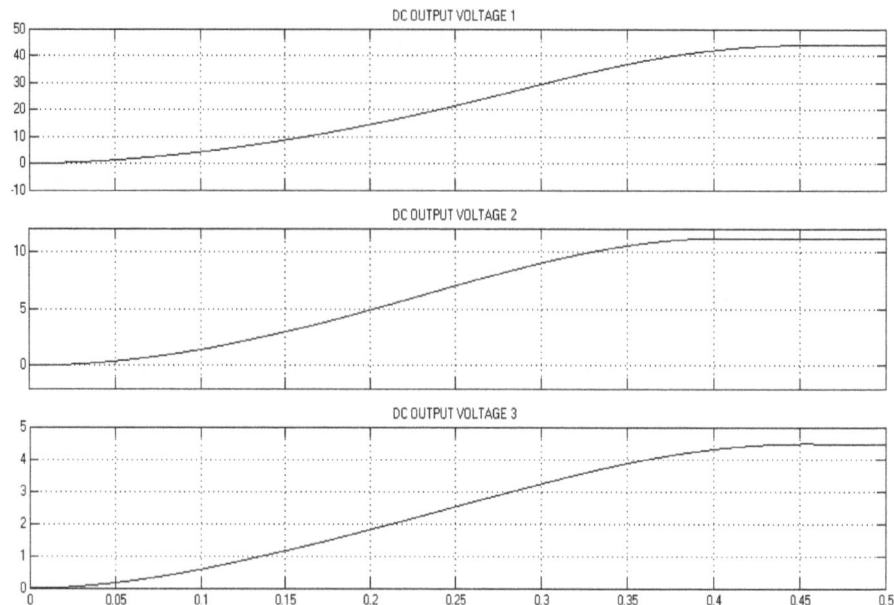

Figure 28: Output voltage waveforms of closed loop control of the proposed three output converter.

efficiency. The operation stages, ZVS condition and control details are also analyzed. Although in the proposed converter there are two and three bridge legs and three and five outputs, it can be extended to number of bridge legs and 2-1 outputs. In the proposed multiple-output ZVS dc–dc converter, there are two and three bridge legs and three and five outputs. However, the number of the bridge legs is not limited. We can add another bridge leg, two transformers and two rectification circuits to produce additional two outputs. A generalized principle is that it can produce 2-1 outputs where the number of bridge legs is. The simulation of three multiple and five multiple output zero-voltage switching DC-DC converter was implemented [3].

References

1. Lin CBR, Chia-Hung C (2013) Analysis of Interleaved Three level ZVS converter with series connected transformers, IEEE Trans on Power Electronics 28: 3088-3099.

2. Mousavi A, Das P, Moschopoulos G (2012) A Comparative study of a new ZCS dc-dc full bridge boost converter with a ZVS active clamp converter, IEEE Trans. on Power Electronics 27: 1347-1357.

3. Alexander JC, Rowden B, Balda JC (2009) A Three level Full bridge Zero voltage Zero current switching converter with a simplified switching scheme, IEEE Trans. on Power Electronics 24: 329-3381.

4. dos Giacomini PGS, Scholtz JS, Mezaroba M (2008) Step-Up/Step-Down DC–DC ZVS PWM Converter With Active Clamping, IEEE Trans on Industrial Electronics 55: 3635-3643.

5. Lin BR, Huang CL, Wan JF (2008) Analysis, Design, and Implementation of a Parallel ZVS Converter, IEEE Trans on Industrial Electronics 55: 1586-1594.

6. Xinke Wu, Junming Zhang, Xin Ye, Qian Z (2008) Analysis and Derivations for a Family ZVS Converter Based on a New Active Clamp ZVS Cell, IEEE Trans on Industrial Electronics 55: 773-781.

7. Wang CM, Su CH, Jiang MC, Lin YC (2008) A ZVS-PWM Single-Phase Inverter Using a Simple ZVS-PWM Commutation Cell, IEEE Trans on Industrial Electronics 55: 758-766.

8. Chen W, Ruan X (2008) Zero-Voltage-Switching PWM Hybrid Full-Bridge Three-Level Converter With Secondary-Voltage Clamping Scheme, IEEE Trans. on Industrial Electronics 55: 644-654.

9. Wu TF, Lai YS, Hung JC, Chen YM (2008) Boost Converter with Coupled Inductors and Buck– Boost Type of Active Clamp, IEEE Trans on Industrial Electronics 55: 154-162.

10. Lee SS, Choi SW, Moon GW (2007) High-Efficiency Active-Clamp Forward Converter with Transient Current Build-Up (TCB) ZVS Technique, IEEE Trans on Industrial Electronics 54: 310-318.

23

Single-Chip Implementation of Level-Crossing ADC for ECG Sampling

Bengtsson L*

Department of Physics, University of Gothenburg, SE-41296 Gothenburg, Sweden

Abstract

This work demonstrates for the first time the implementation of a level-crossing analog-to-digital converter (LC-ADC) in a single, commercially available IC (that costs less than $2). The implementation utilizes adaptive threshold levels in order to prevent overload distortions for fast-changing signals. The entire design is based on a 20-pin PIC16F1769 microcontroller from Microchip and no external components are required. In fact, the only external circuitry required is a single jumper wire. This is due to the fact that the new generation of microcontrollers have integrated core-independent hardware, analog as well as digital. This design takes full advantage of the core-independent logic and analog blocks in a PIC16F17xx circuit to implement the LC-ADC technique that so far has required multiple-circuit designs or ASIC implementation. The design is demonstrated on a standard electrocardiogram (ECG) signal.

Keywords: Asynchronous sampling; Core-independent peripherals; Electrocardiogram; Level-crossing ADC; Microcontroller; Nyquist sampling; Overload distortions; Sparse signals; Synchronous ADC, Threshold levels

Introduction

Background

Traditional ADCs (Analog-to-Digital Converters) acquire samples at regular intervals T_s at a sample rate $f_S=1/T_S$ and needs to adhere to the Nyquist sampling theorem, i.e., $f_S>2 \times f_B$, where f_B is the signal's bandwidth [1]. Due to the periodic nature of traditional ADCs, they are sometimes referred to as *synchronous* ADCs [2,3]. Synchronous ADCs are characterized by a periodicity in time and equidistant quantization levels as illustrated in Figure 1. Due to the fixed equidistant quantization levels, each sample will have an inherent uncertainty U, limited by the ADC resolution:

$$U_{max} = \pm\frac{1}{2} \times \Delta U = \pm\frac{1}{2} \times \frac{U_{ref}}{2^N} \qquad (1)$$

where U_{ref} is the ADC's reference voltage and N represents the ADC's number of resolution bits. This inherent uncertainty in the samples defines the limit of the SNR (Signal-to-Noise Ratio) of synchronous ADCs [4]:

$$SNR=6.02 \times N+1.76 \text{ dB} \qquad (2)$$

The disadvantage of a synchronous ADC is that it generates a great deal of samples that carry no information when "sparse and burst-like" signals are analyzed [5]. Sparse signals are, for example, radar and speech signals and electro cardiograms (ECG).

A sparse signal, like the one in Figure 2, has numerous regions with no, or very-low, activity, resulting in a corollary of identical samples which contain no net information.

In order to compress the data volume in sparse data sets, an *asynchronous* ADC may be used [6]. The asynchronous ADC is also referred to as the *level-crossing* ADC (LC-ADC) and was first suggested by Inose et al. in 1966 [7]. In an LC-ADC, the sampling is triggered by the signal activity rather than by a fixed time interval. Instead of periodically recording the signal's voltage level, the time between predefined level-crossings is recorded. Each sample becomes a 2-tuple:

$$u_n=D,T_n \qquad (3)$$

where T_n is the time elapsed since the last sample and D is the "direction bit"; D indicates whether the upper or lower threshold was crossed. Figure 3 illustrates the same signal as in Figure 1 sampled with an LC-ADC.

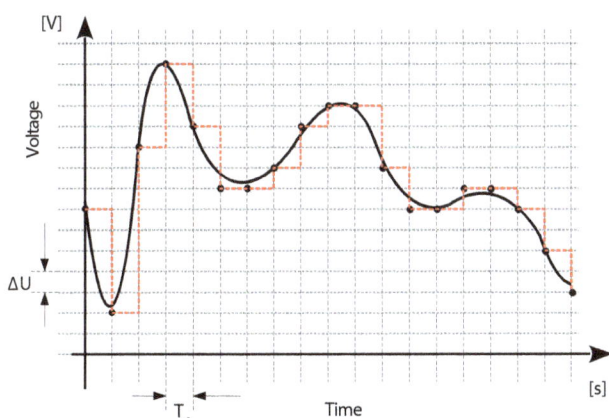

Figure 1: Synchronous ADC. Notice the periodic sampling interval and the inherent error in the sample values.

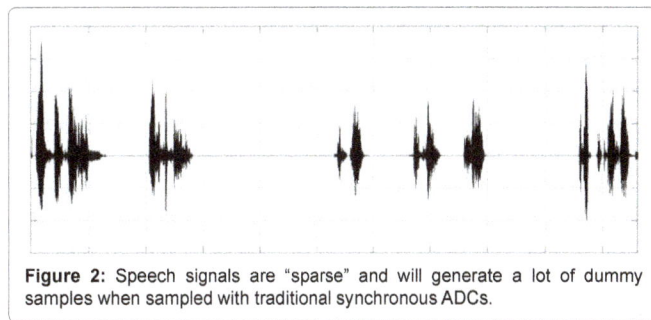

Figure 2: Speech signals are "sparse" and will generate a lot of dummy samples when sampled with traditional synchronous ADCs.

***Corresponding author:** Bengtsson L, Department of Physics, University of Gothenburg, SE-41296 Gothenburg, Sweden
E-mail:lars.bengtsson@physics.gu.se

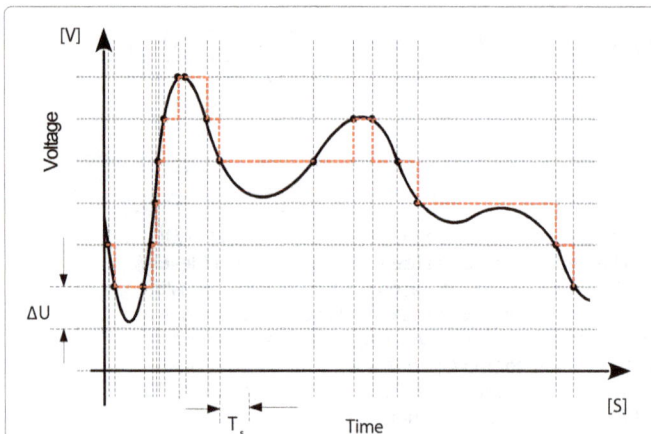

Figure 3: Asynchronous ADC. Notice the irregular sampling intervals and the lack of errors in the amplitude information of the samples.
Notice in Figure 3 how the sample density follows the signal derivative.

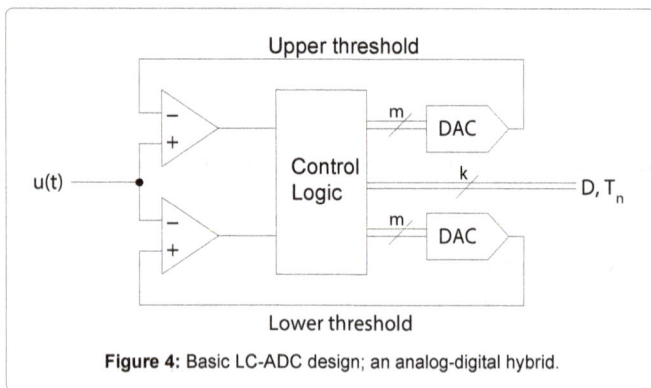

Figure 4: Basic LC-ADC design; an analog-digital hybrid.

Asynchronous sampling has several advantages over synchronous ADCs beyond the obvious inherent compressibility in sparse signal sampling. First of all, the data acquisition problem is transferred into one of quantifying time rather than voltage which is technically less complicated, less expensive and less power consuming. Second, high resolution time samples are easily accomplished; they depend on the reference clock frequency f_{clk} only. Third, the sample number reduction in sparse signals also suggests a reduced power consumption if the ADC host chip is retired to a low-power mode between samples. Finally fourth, the SNR is not limited to Equation (2); in an LC-ADC there is no uncertainty in the voltage levels. Instead, the SNR depends only on the "time resolution rate" R defined as [8]:

$$R = \frac{f_{clk}}{f_B} \qquad (4)$$

(Reference clock's frequency/signal's bandwidth ratio). The SNR of an LC-ADC is [9]:

$$SNR = 20 \times \log R - 11.2 \text{ dB} \qquad (5)$$

Hence for any given signal the SNR depends on the reference clock frequency only.

Basic design idea

Figure 4 illustrates the basic design on which most reported LC-ADCs are based [3,5,10-12]. The analog input signal $u(t)$ is compared to two reference levels (the upper and lower thresholds, respectively) in two analog comparators. If the signal crosses either one of the

threshold levels, the control logic increases/decreases both levels in order to maintain the signal within the boundaries defined by the threshold levels. At every level-crossing a time stamp is generated that represents the data sample.

Related work

The level-crossing ADC technique was first suggested by Inose et al. [7] (but referred to as "asynchronous delta-modulation"). Mark and Todd picked it up in 1981 [6] with the pronounced objective of compressing data in sparse signals. In 2003, Allier et al. [3] implemented an "irregular sampling ADC" based on the LC-ADC technique. The main objective was to reduce power consumption in speech signal analysis and they implemented their design in a micro-pipelined architecture using STMicroelectronics 0.18 μm CMOS technology.

Implementation of "adaptive" LC-ADC algorithms was reported around 2010 [5,10]. In an adaptive LC-ADC the resolution of the threshold levels is reduced with increasing signal slope in order to reduce the overload distortion caused by fast-changing signals. Kózmin et al. [9] suggested a logarithmic distribution of threshold levels for ultrasound applications.

Tang et al. [13] suggested a "fixed window" design of the LC-ADC in 2013; instead of changing the threshold levels for every crossing, the thresholds stay fixed and instead the output from a DAC is subtracted from the signal. If the difference is outside the $\pm$ *Vth* thresholds the DAC input register is adjusted in order to keep the signal within the thresholds. This is illustrated in Figure 5. The objectives in Tangs et al. [13] work were to capture *in vivo* bio-potential signals. A similar design idea was suggested at the same time by Weltin-Wu and Tsividis [14]. The advantage of the fixed threshold window design is that it only requires one DAC and that the threshold levels are fixed. The disadvantage is that extra analog circuitry is required (for subtraction) and a negative reference voltage is required. Because of these adverse design issues, only the implementation scheme in Figure 4 was considered in this work.

The implementation of an LC-ADC is a synthesizing challenge. From Figure 4 it is obvious that the design has to be an analog/digital hybrid. Typical implementations have so far been separated into one digital part consisting of a microcontroller or an FPGA and an analog part consisting of external analog circuits [5,11,15]. Alternatively the entire design is implemented in an ASIC [3,12,13]. These solutions are either expensive or power consuming. This work will for the first time demonstrate a single-chip implementation of an LC-ADC in a low-power, commercially available integrated circuit that costs less than $2.

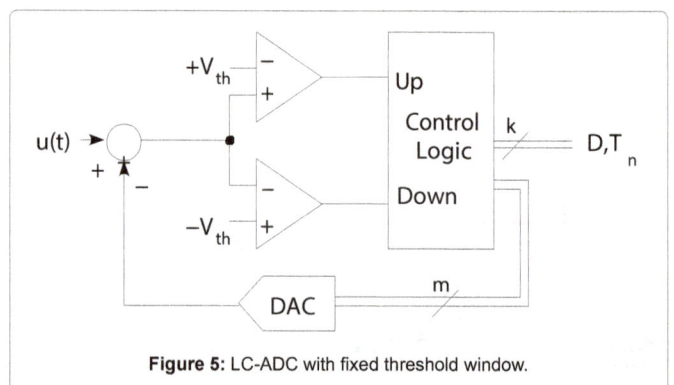

Figure 5: LC-ADC with fixed threshold window.

Method and Material

Hardware

The entire design is based on a PIC16F1769 microcontroller from Microchip [16] (with a list pricing of $1.92, October 2016). This controller has an 8-bit RISC architecture optimized for C programming. Apart from the "usual" microcontroller peripherals found in any commercially available controller (such as Timers, ADCs, UARTs, PWMs etc.) this circuit has "core-independent" peripherals that can run asynchronously and independently of the cpu. Also, these core-independent peripherals are both analog and digital. The digital blocks are 4-input/1-output configurable logic cells (CLCs) that can have one of eight predefined configurations with combinatorial and/or sequential digital electronics. The "intelligent analog peripherals" [16] comprise 10-bit ADCs, operational amplifiers, fast comparators, voltage reference generators and 5- or 10-bit DACs.

In order to implement the LC-ADC, two 10-bit DACs, two comparators and two CLCs are required from the core-independent peripherals. From the "standard" I/Os a 16-bit timer, a UART (Universal Asynchronous Receiver/Transmitter) and a Capture/Compare/PWM module (CCP) is used (for capturing). Figure 6 illustrates the implementation of the LC-ADC in a PIC16F1769 circuit.

If we compare Figure 6 and Figure 4 we can see that almost the entire design is implemented in core-independent hardware (once initiated they run autonomously with no cpu interference). CLC block 1 (CLC1) is configured to detect which threshold was crossed (upper or lower) by using an RS latch, and CLC block 2 (CLC2) is configured as a 2-input OR gate that triggers a capture event of Timer 1 whenever either threshold is crossed. Each CLC block has four inputs and only two are used in both CLC1 and CLC2; the remaining inputs are properly grounded or connected to V_{DD}.

Notice in Figure 6 that the only external hardware is limited to a single wire that connects pin 14 to pin 15; everything else is either configurable by registers or controlled by software. (The micro controller used here does not allow the negative comparator inputs to be connected internally).

Software

The software is written in the C programming language and it is straight forward but depends on several subtle details. The following two major issues need to be considered:

- The available on-chip data memory is limited and only a limited number of samples can be stored on-chip; samples must eventually be transferred to a host computer. In the proposed design, the sample heap is off-loaded during the quiescent signal intervals.

- In order to prevent overload distortion caused by fast-changing signals, the distance between the threshold levels ("steps") must be dynamically adapted depending on the last sample value (i.e., the signal slope).

The work presented here was designed with the distinct objective of capturing transients (such as ECG signals) with an LC-ADC sampler. The hardware does nothing until the transient arrives and then samples are temporarily stored in on-chip RAM memory and transferred to the host computer only when the signal level has subsided below the lowest detectable level for some finite time; the data are transferred to the host computer during the quiescent intervals of the probed signal (the T-P interval in the ECG case).

The first part of the software is the main function which is responsible for off-loading the sample stack and transfers the samples to the host computer (Figure 7). The other part is the data sampler which is hosted in the interrupt service routine (isr) triggered by the capturing event. Samples are stored on the sample heap, the step size is determined depending on the size of the acquired sample and the threshold levels are adjusted accordingly. The ISR is illustrated in Figure 8.

The threshold levels are adapted to the signal's derivative as follows;

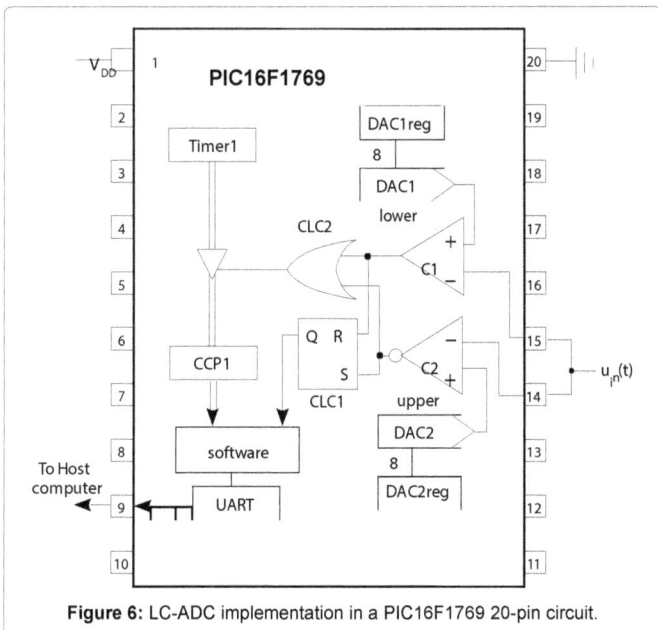

Figure 6: LC-ADC implementation in a PIC16F1769 20-pin circuit.

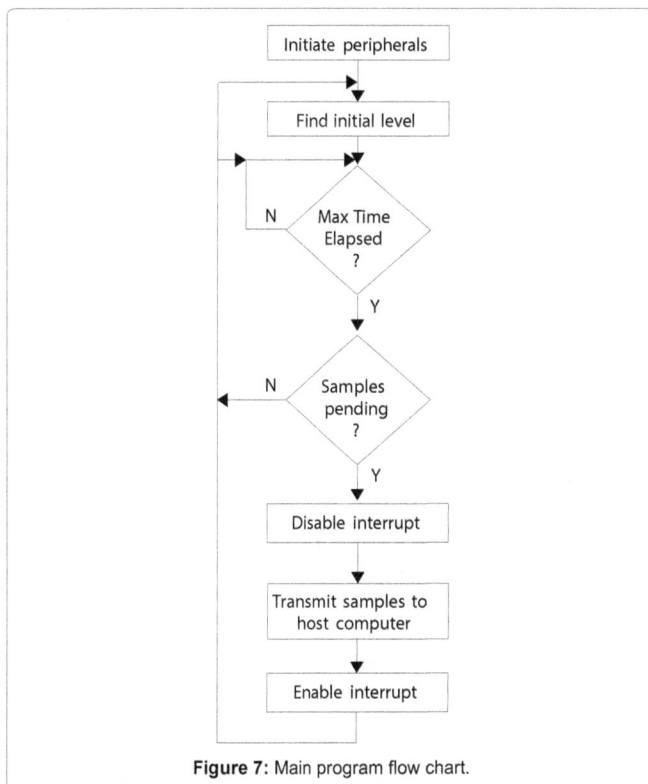

Figure 7: Main program flow chart.

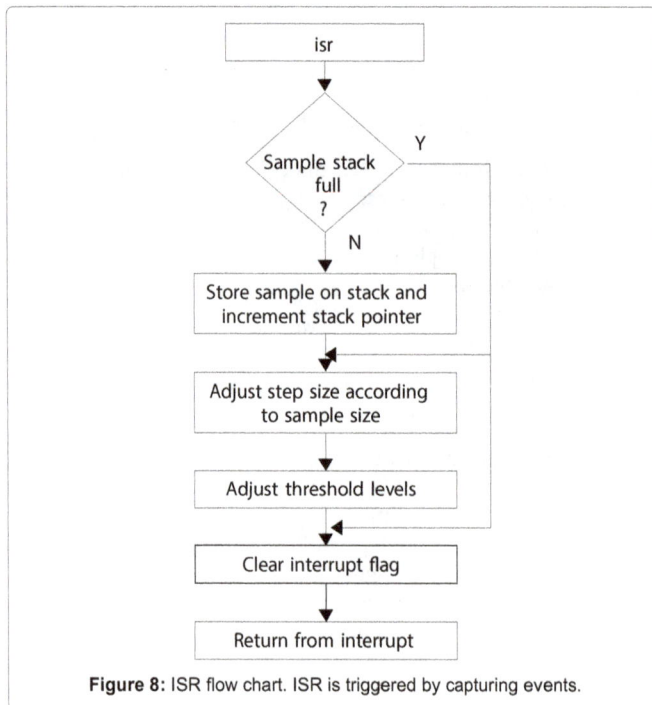

Figure 8: ISR flow chart. ISR is triggered by capturing events.

depending on the sample size (=the captured time value) the magnitude of the threshold levels are adjusted according to a 4-level logarithmic scale [9]. These levels were implemented as generic variables for easy adjustment.

The captured time samples are 16-bit integer numbers and the on-chip sample heap size was set to 100. This maximum heap size will depend on the particular controller used; the sample heap occupied 44% of the available data memory in the PIC16F1769 used in this work. Hence, a heap twice as large is possible to implement, but for this work 100 samples was sufficient.

The main program was designed to start off-loading the sample heap as soon as the signal line was "silent" or when the heap was full. In order to guarantee uncorrupted data transfers to the host computer, interrupt was disabled during data transfers.

Both DACs use 8-bit resolution (i.e., only the eight most significant bits are used). The entire software is hundreds of C code lines distributed over 23 source/header files (including *main.c*). However, by taking advantage of Microchip's Code Configurator (MCC), which is a graphical tool for initialization of peripherals, the designer's concern is the application code only. The application code is less than 100 lines of C code.

Setup/Data acquisition

In order to validate the design a Hewlett-Packard 33120A waveform generator was used to produce transient input signals to the signal input in Figure 6. The generated signal was captured on a digital oscilloscope and transferred via a thumb drive to the host computer and processed in MATLAB for presentation. The sampled data from the LC-ADC sampler was connected to a USB port on the host computer via an RS232-to-USB converter cable [17] and displayed in a standard terminal window (from which data could be easily imported to MATLAB). The baud rate used was 115,200 bits/sec.

Results

In order to adjust design parameters for optimal performance,

some standard transients (with "decent" behavior) were initially generated. This made it possible to adjust threshold levels and step sizes in the software and the timer period in the hardware.

The first signal used was a triangular shaped transient; this is the simplest transient since it has constant slopes. Figure 9 illustrates an example of the generated signal and the signal reproduced by the samples produced by the LC-ADC. In Figure 9, the triangular waveform signal has a "frequency" of 15 Hz.

Next, the design was stressed by a transient with higher frequency and non-constant derivative; a sinusoidal signal with a "frequency" of 30 Hz was generated. Figure 10 illustrates the transient and the samples.

The experiments presented in Figures 9 and 10 made it possible to find the optimal design parameters (by a few trial-and-error adjustments) and finally a "realistic" ECG signal was injected into the system and the result is presented in Figure 11. The ECG signal in Figure 11 represents one heartbeat from a 1 Hz ECG (60 bpm) (from the HP33120A waveform generator) and the reproduced waveform from the LC-ADC samples.

Analysis

Figures 9-11 verifies the potential of the proposed design. Design parameters were optimized to accurately capture the vital information of an ECG signal (P, QRS and T waves [18]) and Figure 11 confirms that this is successfully achieved.

The time resolution in this design is 16 bits, but from Equation (3) it is clear that each sample will be 17 bits long; the time stamps are in the range ± 65,535. Hence, each sample consists of six ASCII characters that need to be transmitted by the UART interface. Also, a CR+LF character couple (Carriage Return, Line Feed) is appended

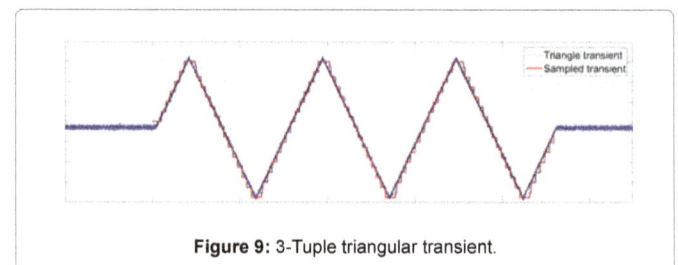

Figure 9: 3-Tuple triangular transient.

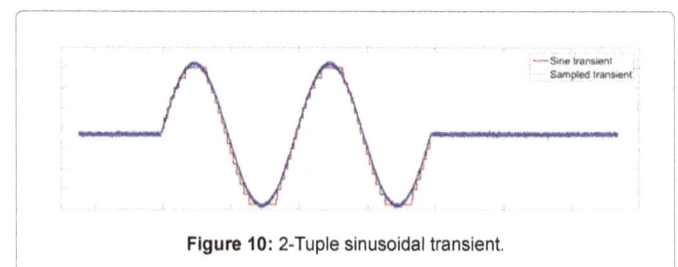

Figure 10: 2-Tuple sinusoidal transient.

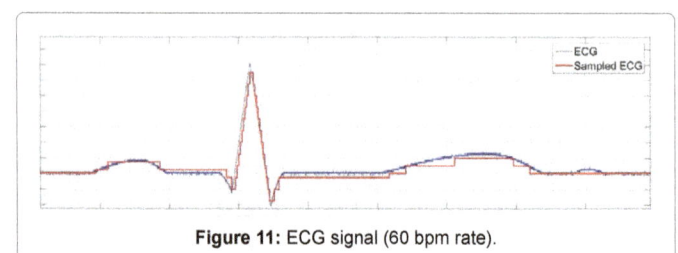

Figure 11: ECG signal (60 bpm rate).

to each sample in order to display the samples on separate rows on the receiving terminal. Every sample then consists of eight ASCII characters. The ECG signal in Figure 11 contains 34 samples that need to be transmitted to the host computer during the quiescent T-P [18] interval of the signal. In the RS-232 asynchronous serial interface, the transmission overhead is a single start bit and (at least) one stop bit (parity is optional). Hence, it will take 10 bits to transmit an ASCII character. The number of bits required to transmit the ECG signal in Figure 11 is therefore

10 bits × 8 ASCII characters × 34 samples=2,720 bits.

Using a bit rate of 115,200 baud means that it takes 2,720/115,200=2.36 ms to off-load the sample heap. At a heartbeat rate of 60 bpm (1 Hz), the quiescent time interval between the T wave and the next P wave is at least 200 ms, which means that there is plenty of time to transfer samples to the host computer; the off-loading of the sample heap occupies approximately 10% of the available "silent" time in a 60 bpm ECG signal.

In Figure 11, the average count number between samples is 3919.7 and the count rate (f_{clk}) was 2 MHz. This indicates an average sample rate of 2×10⁶/3919.7=510 S/s.

Conclusions

Level-crossing ADCs have the potential of saving a great deal of data space in sparse-signal sampling and/or reducing the need for data transfers to a host computer [6]. They are straight-forward designs based on standard components. However, the analog/digital hybrid nature makes them complicated to implement on a printed circuit board and has so far required either a multi-circuit design using a combination of analog and digital circuits or an ASIC design. This work has proved that it can be implemented in a single, low-cost integrated circuit with an absolute minimum of external hardware (a single wire). In this work the objective was to sample an ECG signal of a specific rate but the design is not limited to this particular signal. The main design parameters are the size of the sample heap, the threshold levels, the time resolution and the time range. All of these parameters are configurable in software and can easily be adjusted for other signal types.

The "absolute limits" of this design have not (yet) been investigated since they depend on so many different parameter settings; they need to be determined for each signal to be sampled.

Discussion

Software structure

The flow charts in Figures 7 and 8 indicate some alternative ways to design the software (the dashed arrows). The software used assumes that the base line level (the quiescent signal interval) does not change. The software determines this level at the beginning of the program and then assumes that it stays constant for the rest of the measurement. If it does change, the small algorithm that finds the initial level will have to be included in the super loop of the main function; this is indicated by the dashed arrow in Figure 7). This design has also assumed a maximum number of samples in the transient; if the number of samples suddenly increases above that number, information is lost. If this is critical, the sample heap must be expanded or threshold levels must be adjusted in order to reduce the number of samples. However, if the sample heap is full, the software used here just ignores any additional samples. A compromise could be to keep tracking the signal (i.e., still adjusting

the thresholds) without acquiring new samples. That would guarantee that the sampler is maintained at the right level when the sample heap is off-loaded and sampling resumes. This is indicated by the dashed arrow in Figure 8.

Sampling rate, SNR and reconstruction

Uniformly sampling ADCs must adhere to the sampling theorem [1] in order to be able to accurately reconstruct the sampled signal, i.e., the sample rate must exceed the signal bandwidth by a factor of 2. Since the LC-ADC produces non-uniform samples, the sampling theorem is not valid. However, it has been claimed that for LC-ADCs, the *average* sampling rate must exceed 2 × (signal bandwidth) for correct reconstruction [3]. In this work the average sample was 510 S/s. The main ECG power is concentrated between 0.5 and 30 Hz [19]. The "patient monitoring required bandwidth" for ECG signals is specified to 0.05 to 30 Hz [20] and even if we stretch that to "diagnostic grade monitoring", which is up to 100 Hz [20], the sampling rate still exceeds 2×(signal bandwidth) with a satisfactory marginal.

Assuming a signal bandwidth of 100 Hz, we can also calculate the time resolution rate R according to Equation (4):

$$R = \frac{2 \times 10^6}{100} = 20000$$

Hence we have an SNR of Equation (5):

20 × log 20,000−11.2=74.8 dB

Inserting this SNR into Equation (2) gives as the necessary resolution of 12 bits required for an equivalent synchronous ADC.

Uniformly sampled signals are reconstructed by interpolation of sinc functions [21]. The reconstruction of non-uniformly sampled signals is less straight-forward. Most reconstructions are based on straight line interpolation and smoothing filters [6]. Several other suggestions have been reported [8,22,23]. In 2009, Kozmin et al. [9] suggested the "Adaptive-weight conjugate Toeplitz method" for reconstruction of non-uniformly sampled signals which is based on trigonometric polynomials. So far though, no method for exact reconstruction of non-uniformly sampled signals has been reported.

Future work

The main objective of this work was to demonstrate the capability of the "core-independent" hardware available in low-cost microcontrollers and how they can be adapted to improve data acquisition of standard bio-metrical signals. A possible design flaw in the proposed design is that the sampler is "blind" for a short time during the off-loading of the sample stack (since the interrupt is disabled). The occurrence of asynchronous transients between the expected ECG transients could potentially lead to loss of information, which of course could be a serious problem. However, unless they are shorter than 2.34 ms (=the sample heap off-loading time) and occur exactly during the offloading window, at least some of the transient will be captured and would indicate a need for adjustment of the design parameters.

Future work is aimed at taking care of this permanently; the problem is only a matter of software design. Instead of waiting for the entire transient to pass before the off-loading starts (and disabling sampling during off-loading) the sample heap can be off-loaded continuously. New samples are continuously transferred to the host computer (without disabling the sampling). A pointer variable points to the next sample to be off-loaded and another pointer points to the next available location to store the next sample (on a circular sample heap).

That way no information will be lost and any transients exceeding the threshold levels will be captured since the UART transmission is core independent.

The next generation of this LC-ADC design will also encompass a 32-bit timer option in order to extend the range of the time intervals. Also, a separate 32-bit timer will be implemented to measure the period of the QRS wave.

References

1. Nyquist H (1928) Certain topics in telegraph transmission theory. Trans AIEE 47: 617-644.

2. Petrellis N, Birbas A, Kikidis J, Birbas M (2010) Asynchronous Analog-to-Digital Conversion Techniques, (1st edn.), INTECH Open Access Publisher.

3. Allier E, Sicard G, Festquet L, Renaudin M (2003) A New Class of Asynchronous A/D Converters Based on Time Quantization. Proceedings of the 9th Int Symp on Asynch Circuits and Systems, pp: 196-205.

4. Hauser MW (1991) Principles of Oversampling A/D Conversion. J Audio Engin Soc 39: 3-26.

5. Agrawal R, Trakimas M, Sonkusale S (2009) Adaptive Asynchronous Analog to Digital Conversion for Compressed Biomedical Sensing. IEEE Biomedcial Circuits and Systems Conference, pp: 69-72.

6. Mark JW, Todd TD (1981) A Nonuniform Sampling Approach to Data Compression. IEEE Trans. on Comm 29: 24-32.

7. Inose H, Aoki T, Watanabe K (1966) Asynchronous delta-modulation system. Electron Lett 2: 95-96.

8. Sayiner N, Sorensen HV, Viswanathan TR (1996) A Level-Crossing Sampmling Scheme for A/D Conversion. IEEE Trans Circ Syst–II: Anal Dig Sign Proc 43: 335-339.

9. Kózmin K, Johansson J, Delsing J (2009) Level-Crossing ADC Performance Evaluation Toward Ultrasound Application. IEEE Trans Circ Syst-Reg Pap 56: 1708-1719.

10. Trakimas M and Sonkusale SR (2011) An Adaptive Resolution Asynchronous ADC Architecture for Data Compresssion in Energy Constrained Sensing Applications. IEEE Trans Circ Syst- Reg Pap 58: 921-934.

11. Silva VML, Catunda SYC (2014) Flexible A/D Converter Architecture Targeting Sparse Signals. Proceedings of the IEEE International Instrumentation and Measurement Technology Conference.

12. Ravanshad N, Rezaee-Dehsorkh H, Lotfi R, Lian Y (2014) A Level-Crossing Based QRS-Detection Algorithm for Wearable ECG Sensors. IEEE J Biomed Health Informatics 18: 183-192.

13. Tang W, Osman A, Kim D, Goldstein B, Huang C, et al. (2013) Contiunuous Time Level Crossing Sampling ADC for Bio-potential Recording System. IEEE Trans on Circ Syst- Reg Pap 60: 1407-1418.

14. Weltin-Wu C, Tsividis Y (2013) An Event-driven Clockless Level-Crossing ADC with Signal-Dependent Adaptive Resolution. IEEE J Sol State Circ 48: 2180-2190.

15. Baums A, Grunde U, Greitans M (2008) Level-Crossing sampling using microprocessor based system. ICSES 2008 International Conference on Signals and Electronic Systems.

16. Microchip Tech Inc (2015) PIC16(L)F1764/5/8/9-14/20-Pin, 8-Bit Flash Microcontrollers. Datasheet DS40001775B.

17. Future Technology Devices International (2016) TTL-RS232R to USB Serial Converter Range of Cables. Datasheet FT_000054.

18. Ashley EA, Niebauer J (2004) Cardiology Explained. London: Remedica.

19. Merri M, Farden DC, Mottley JG, Titlebaum EL (1990) Sampling Frequency of the Electrocardiogram for Spectral Analysis of the Heart Rate Variability. IEEE Trans Biomed Engin 37: 99-106.

20. Texas Instruments (2016) ECG and EEG applications. Quick reference guide.

21. Ghate AH, Khanchandani KB (2013) Difference between Reconstruction from Uniform and Non-Uniform Samples using Sinc Interpolation. Int J Research Rev Eng Sci Techn 2: 17-21.

22. Tertinek S, Vogel C (2008) Reconsruction of nonuniformly sampled bandlimited signals using a differentiator-multiplier cascade. IEEE Trans Circ Syst 55: 2273-2286.

23. Feichtinger HG, Gröchenig K, Strohmer T (1995) Efficient numerical methods in non-uniform sampling theory. Num Math 69: 423-440.

24

A New Survey for Optimum Power Flow with Facts Devices

Mahmood Taha Alkhayyat[1]* and Sinan Mahmood Bashi[2]

[1]Electrical Engineering Dept, Northern Technical University, Mosul, Iraq
[2]Electrical Engineering Dept, Mosul University, Mosul, Iraq

Abstract

Optimal Power flow considered the backbone tool in the complex power system. The expanding in demands lead to increasing in generation that requires increase the transmission capacity, for these reasons the problem of optimal power flow OPF still under many studies in order to minimize the cost, losses, emission of harm gases, etc. FACTS is the main articles of this paper include the last nontraditional OPF methods, hybrid methods, multi-objective OPF, and OPF with FACTS devices. Also there are three Tables contain the recent stander and hybrid methods used, mostly, in solving OPF problems with their advantages, disadvantages, and their applications that may help the researchers in this field, eventually some important points have been concluded.

Keywords: Optimal power flow; FACTS; Artificial intelligence

Introduction

Optimal power flow has an important for solving the complex problems in large power systems at many reality constraints. The problems are nonlinear, non-convex, non-differentiable, and non-smooth, so many researches adopted the modification of theories used for solving OPF based on minimizing the consuming time beside achieving the best fitness. For these reasons it might be of interest to represent the historical survey of the researches dealing with this field. In our work, generally, OPF researches are classified in two categories OPF and OPF with FACTS devices. The objectives of OPF without FACTS devices is made to minimize:

1) Total generation cost of thermal units.

2) Total real power loss of transmission system.

3) Gas emission from thermal units.

4) Voltage deviation of system buses.

The first and second objectives could be taken as the main objectives for all optimization problems and could be considered as the own required objective to be solved, but the third and fourth objectives are often taken with the first two main objectives to be multi-objective OPF problem in order to obtain more and more modified state variables.

In the case of OPF with FACTS devices the objectives are the same as listed above plus the additional objectives made to obtain

5) Minimum capacity of the FACTS device(s).

6) Optimal values of controlling variables.

7) Optimal location of FACTS device(s).

The last objective is made based on the sensitivity methods rather than the well-known OPF methods. The last two objectives are done to satisfy one or more of the other objectives, the problem can be extended to contain two or more of the above objectives to be multi-objective OPF. In addition, security constraints optimal power flow SCOPF has the important partition in the OPF researches. However, our review not includes this subject.

OPF Reviews

There are a lot of researches presents a reviews for OPF focusing on the methods used for solving economic dispatch HAPP [1], IEEE group [2], Carpentier [3], Chowdhury and Rahman [4], Momoh et al. [5]. Pandya and Joshi [6] tried to categorize the methods for OPF suitable for the certain objective(s) and compare the traditional OPF methods with AI methods and presents their advantages and drawbacks. When the effect of the large steam turbine valve is included in economic dispatch the characteristic of the cost has local minimum points and the method solving such problem known as the optimal dynamic dispatch ODD. Xia and Elaiw [7] present the literature review of the ODD and categorized their study to conventional methods, artificial intelligence methods, and hybrid methods they concluded that traditional methods like newton methods, lambda iteration method etc. are not effective for solving the non-convex and non-smooth problems but AI can obtain global optimization for this problem successfully with sacrificing more time, the hybrid methods can solve the problem faster. Mary B Cain et al. [8] presented a history of OPF focusing on the power flow, economic dispatch, and OPF and review the most elements for solving PF and OPF, the authors concluded that even 50 years ago for the beginning the problem but at yet there is no theory could solve OPF commercially and showed that an approximation in solving the problem reduces consuming time that important for controlling variables but it cause a huge financial impact. All literature reviews acknowledged that Carpentier [9] covered the most of the bases, used at yet, for solving an optimal power flow. Although our research focusing on OPF with FACTS it is of interest to define the main fields of OPF studies.

Economic dispatch calculation (EDC): is performed to dispatch, or schedule, a set of online generating units to collectively produce electricity at a level that satisfies a specified demand in an economical manner.

Optimal power flow OPF: refers to full AC power problem solved to optimize real and reactive power flow subject to a certain constraints.

*Corresponding author: Mahmood Taha Alkhayyat, Electrical Engineering Dept, Northern Technical University, Mosul, Iraq
E-mail: mtmahmoud@yahoo.com

Multi-objective OPF: refers to optimization (minimize) two or more objective functions simultaneously subject to unignorable constraints.

DC OPF: refers to single objective OPF under assumption that all voltage magnitudes are fixed and all phase angles are close to zero, it is not refers to the solution of direct current power flow.

Dynamic OPF: refers to include the effect of variation in variables in OPF calculation, for example the effect of valves that causes local optimization points.

On-line OPF: refers to the solution of OPF directly online, the OPF controlling system receives the information as sampling data and sends the commands in order to satisfy the optimization for the objective function.

Security constraints SC OPF: refers to satisfy optimization after contingency being happens in the system. A contingency is defined as an event that causes one or more important components such as transmission lines, generators, and transformers to be unexpectedly removed from service.

OPF with Artificial Intelligence AI Methods

Conventional methods such as Linear Programming, Newton-Raphson and Non-linear Programming, quadratic programming, mixed integer programming, and interior point, methods were previously offered to handle the complexity of the OPF. However, the common barriers for these methods that may closet to local optima because the complicated features of the mathematical functions of the objectives and constraints due to the nature of real OPF that may be nonlinear, non-convex, non-differentiable and non-smooth optimization problem with discontinuous solution space. This kind of optimization problem is very hard, if not impossible, to solve using traditionally deterministic optimization methods, moreover, if the effects of valve-point loading of thermal generators and/or nonlinearity of FACTS devices are included that may increase the possibility for trapping in local optima or premature convergence Abido [10]. With the emergence of artificial intelligence, many novel techniques such as Artificial Neural Networks, Genetic Algorithms, Particle Swarm Optimization and other Swarm Intelligence techniques have also received great attention. Sumpavakup et al. [11] solved OPF using Artificial Bee Colony ABC and compared it with GA and PSO using IEEE 14 and 30 bus test system. The results showed that BC converge faster than the rest methods and give more accurate solution. Seyed Reza Moasheri and Masoud Khazraei [12] modified GA for solve OPF with IEEE 30 bus as test system. They compared the results with PSO, DE, and conventional GA, the results showed that the modified GA gives best fitness.

OPF with Hybrid Methods

The practical and industrial control applications required fast controller system that needs, beside a very high frequency microprocessor, an effective method capable of reaching a best fitness with less iteration or time. For this reason many attempts still at yet not for proposed new method only, but for merge the methods to satisfy this target. Worawat Nakawiro and István Erlich [13] combined GA with ANN for solving OPF and obtain 5 time faster than GA alone maintaining the error less than 5%. In 2011 and 2012 Sumpavakup et al. [14], proposed a hybrid cultural bee colony to solve single objective and multi-objectives OPF respectively, authors tried to prove that the proposed method is faster than some other AI methods. It is observed that their results not always best than the compared methods but

depend on the size of the problem. Sivasubramani [15], combined sequential quadratic programming SQP, which is known as traditional method, once with differential evolution DE and then with Harmony Search HS by two ways to overcome the stagnation and premature convergence in both methods and showed that the proposed hybrid method is quite suitable for non-convex problems. Amjady [16] added the DE searching attribute to BF method to enhance its exploration capability. Authors compared the results of the proposed method with about twenty evolution methods showed that it obtains best fitness. If the individuals are conside on the global best and if both the velocity and the weight are not zero, the method suffers from stagnation. The main contribution of Ahmed and Antônio [17] that they added the GA mutation feature to the PSO method (hybrid particle swarm optimizer with mutation HPSOM) to overcome the stagnation problem when solving OPF to minimize active power loss in simple system, the results as compared with PSO gave best fitness and fast convergence. Sai H Ling et al. [18], have the same target for adding the mutation ability to PSO method using wavelet. Authors compared their method with some standard and hybrid PSO methods to show its good performance over them. Karthik and Chandrasekar [19] proposed a hybrid technique for identifying the proper place for fixing the IPFC. The proposed hybrid technique utilizes genetic algorithm and neural network to identify the proper place for fixing the IPFC. The training dataset is generated using the genetic algorithm. For creating training dataset different combinations of buses are taken and their power values are analyzed and the best combination of buses to which the IPFC must be connected to attain maximum system stability is selected. By using this dataset the neural network is trained. This technique identified the proper place for fixing the IPFC, also the losses in the compensated line is decreased.

Multiobjective MO OPF

About 50 years ago, as mentioned in previous section, Carpentier established the main basics for solving OPF adopted at yet and became a traditional way for solution. It is of convenient to use optimal active power flow, optimal reactive power flow, and optimal active and reactive power flow for solving various objective functions. OPF with single objective had been studied by many authors. However, due to the fact that real life problems involve several objectives and the decision maker would like to find solution, which gives compromise between the selected objectives [20]. Firstly, MOOPF solved by weighting the objectives and combined them as a single objective, this approach has the disadvantage of finding only a single solution and it cannot find trade-off between the different objectives in single run [M10], and requires multiple runs as many times as the number of desired Pareto-optimal solutions [21]. Debjani Chakraborti et al. [22] used GA with fuzzy goal programming for solving MOOPF, the objectives are the fuel cost, gas emission, and voltage deviation, these objectives are described fuzzily and incorporate the power loss in an equality constraint equation, the properties of Particle Swarm Optimization PSO technique are fast convergence, less parameters to tune, easier searching in very large problem spaces, and effectively optimizes the multidimensional discontinuous nonlinearities. However, the tuning of its parameters needs some experience. Yong Zhang [23], proposed a bare bone PSO that has the ability for self-tuning parameters and applied the proposed method in IEEE 30 bus for solving environmental/economic dispatch multi-objective function and compare the results with families of GA and PSO methods, the results proved the efficient of this method.

OPF with Facts Devices

With continuous increase in power demand many problems arises due to the need for extended generation units and additional transmission lines. An expanding in transmission system consumes long time and facing the right of way problem. In addition, the transmission system becomes more complex with additional interconnected between buses, for this reasons a powerful controlling system is invented by Hingorani et al. [24]. FACTS devices are used in power system for controlling power, increase transmission capacity near their thermal limits, enhance system stability and hence its security, reduced reactive power flow, thereby permitting greater active power flow, and reduced cost of energy received due to enhanced line capacity. Direct current transmission system is one of the powerful controller device used in power system to transfer a huge power for long distance and improving the stability of the system. However, in this paper we focusing on the new version of controlling devices, so called, FACTS. Many authors investigate the problem of OPF in presence of FACTS devices. Referring to the equation of power transfers, most of the controlling devices depend on variation of phase angle between the two buses of transmission line and its impedance [25], in this research an attention is closest on three categories, OPF methods with FACTS dvices, hybrid methods and multi-objective OPF, incorporating the last version of controlling devices (UPFC and IPFC) an OPF can be achieved without re-dispatching active power of generator [26], the IPFC has an effect on the voltage profile of the buses neighboring to the compensating bus [27], efficient for minimizing TL loss and reducing reactive power flow [28,29], and it is more efficient compared with UPFC for improving available transfer capability ATC of TL [26]. Many efforts had been done for modeling and simulation of FACTS devices with OPF study and their contribution circulating about the common targets for accelerating the convergence using modified models [30], and adaptive and flexible method [31], PSO is more efficient for solving optimization problems [32]. Moreover, OPF control method used to obtain the minimum of total required capacity of FACTS devices inserted in the system [33]. Although long time ago for the first application of Newton-Raphson method in power flow analysis, it is still extensively used at yet for solving power flow and optimum power flow, Marcos Pereira and Luiz Cera Zanetta [34] used a new current injected model for simulating UPFC and solving the problem with Newton-Raphson NR method. Some authors used NR method with AI method for solving OPF with FACTS [35]. Sreejith et al. [36] used TCSC with OPF and showed that differential evolution DE is more efficient than Lagrangian method. E.S. Ali, S.M. Abd-Elazim [37], used bacterial foraging BF method for optimization the simultaneous tuning of TCSC controller, the results give robust damping performance over a wide range of operating conditions in compare to optimized TCSC controller based on GA. With the increment in power flow controlling devices, additional problems may appear and require a suitable coordination between them [20], authors, pointed out some problems associated with increasing the number of FACTS devices. The voltage-sourced converter (VSC)-based Interline Power Flow Controller (IPFC) was first proposed in 1998, as the latest component of the Flexible AC Transmission System (FACTS) device family [26]. Its unique capability of simultaneously compensating multiple transmission lines at a given substation has since aroused great interest of researchers and power industries around the world, especially when now manufacturing of VSCs are becoming more economical. Like Unified Power Flow Controller (UPFC), the IPFC is a kind of combined compensators, in which at least two Static Synchronous Series Compensators (SSSCs) are combined via a common DC voltage link. If there is no energy storage system installed in the apparatus, this DC voltage link is usually modeled as a DC capacitor. It is this link that provides the IPFC with the path through

which different transmission lines can exchange active power. Teerathana et al. [33] proposed the utilization of the IPFC to mitigate overload problem with Optimal Power Flow (OPF) control method. The OPF control method for a satisfied solution of the minimum cost and the entire power flow balance was also discussed. Radhakrishnan and Rathika [38] deals with the development of IPFC using fuzzy technology. In the proposed scheme, series and shunt configuration employing an interline power flow controller using fuzzy technology is designed. Most of the compensation is provided by series controller. In case of excess real power demand shunt controller is used. The authors proved that the offset time required for the oscillations to settle down after compensation has reduced when using FL controller instead of PI controller. Karthik and Chandrasekar [39] presented the Separated IPFC, which eliminates the common DC link of the IPFC and enable the separate installation of the converters. Without location constrain, more power lines can be equipped with the S-IPFC, which gives more control capability of the power flow control. Instead of the common dc link, the exchange active power between the converters is through the same ac transmission line at 3rd harmonic frequency. Every converter has its own dc capacitor to provide the dc voltage. The 'master' converter can adjust the voltage magnitude, transmission angle, and line impedance. The 'slave' converter provides the active power for 'master' converter and at the same time adjusts its own line reactance. However, the active power loss in the line is increased also the authors did not show the practical validity of this method. EL-Sadek, et al. [40], incorporated the IPFC by PQ/PQ/PQ IPFC model and by the injection power flow IPFC model. The two models are incorporated in a MATLAB power flow program based on N-R algorithm. These IPFC power flow models are modified to set control of power flows of multiline. Based on these models it is possible to estimate the IPFC control variables and its ratings. Numerical comparisons between PQ/PQ/PQ IPFC model and the injection power flow IPFC model are presented. Irusapparajan and Rama Reddy [41], modeled and simulated the 30 bus systems with and without IPFC using Matlab/Simulink. The voltage profile is improved by adding IPFC. The authors showed that the IPFC pushes more power to the buses with higher load and the real and reactive power can be easily controlled with the help of IPFC. Sankar and Ramareddy [42], described interline Power flow controller in power system. The different controller's circuits are simulated using PSPICE software package. IPFC is used to improve the power flow and to provide a power balance of a transmission system. Circuit model with phase difference and voltage difference were simulated to study the real and reactive power flows. The circuit model for open loop and closed loop systems are presented. The authors observed that the real and reactive powers are increased by the presence of IPFC. Naresh Babu and Sivanagaraju [43], proposed a new intelligent search evolution algorithm (ISEA) to minimize the generator fuel cost in optimal power flow (OPF) control with IPFC. In the proposed algorithm, the two-steps initialization process had been adopted which eliminates the mutation operation and also it gives optimal solution with less number of generations. The proposed algorithm has been examined and tested on a standard IEEE-30 bus system without and with IPFC. The results reveal that the generation cost is less with IPFC. Hans Glavitsch and Rainer Bacher [44] attempted to review various optimization methods used to solve OPF problems. A new powerful technique to implement FACTS devices is presented in this paper for the congestion management in the open power market. The merits of this method are that there is no requirement to modify the power mismatch equations to implement the FACTS devices. Application of this technique to Optimal Power Flow has been explored and tested. The simulation results show that this simple algorithm can give a good

No.	Theory	Advantages	Drawbacks	Applications
1	Classical methods	Reliable and fast convergence	1-Weak in handling constraints 2-Poor convergence 3-Convergence becomes too slow. if the numbers of variables are large	
2	Stochastic methods			
3	Artificial Neural Network ANN	1-Fast 2-Appropriate for non-linear models	1-Large dimensionality 2-The choice of training methodology	1-Real time control 2-Optimization
4	Fuzzy Logic FL	Represents constraints more accurate		1-Real time control
5	Genetic Algorithm GA	Rarely trapped in local optima	Consumes long time for global convergence	
6	Evolutionary programming EP	1-Adaptability to change, 2-Ability to generate good solutions 3-Rapid convergence		
7	Bacterial Foraging BF			
8	Bee Colony BC			
9	Ant Colony AC	1-Positive feedback for recovery of good solutions, 2-Robust against premature convergence.		1-To find shorter rout in TLs 2-Dispatching Gens. 3-Optimal unit commitments
10	Particle swarm optimization PSO	1-Fast convergence 2-More accurate and robust for global convergence 3-Less parameters need adjustment 4-Able to search in large problem with non-differentiable	1-Need experience for adjusting parameters 2-Sometimes suffers from stagnation	1 large area optimization 2 multi-objective OPF
11	Simulating Annealing SA			
12	Deferential Evolution DE		1-Stagnation 2-Premature convergence	
13	Tabu search TS		1-trapped in local optima 2-converge to global optima in long time	
14	Honey bee mating HBM			
15	Harmony search HS		1-Stagnation 2-Premature convergence	

Table 1: AI theories advantages/drawbacks, and applications.

No.	Hybrid theory		Advanced features	Application
1	ANN	Fuzzy		
2	GA	ANN	Minimize iterations to reaching best fitness	Optimization and optimal location
3	GA	PSO		
4	DE	SQP	Overcome the premature convergence and stagnation	
5	HS	SQP	Overcome the premature convergence	
6	BF	GA	Enhance the exploration capability	

Table 2: Hybrid AI theories features and applications.

result using only simple modifications. Finally, Tables 1 and 2 illustrates some advantages, draw backs, and applications of various AI methods and hybrid methods respectively.

Conclusion

The competition between OPF methods, regarding they impact with the same problem or objective, to reach best fitness with minimal time depends on:

1) Modifications of methods.

2) The proposed model of FACTS device(s) when these devices incorporated in the PS.

3) The Selected methods combined as a hybrid method.

4) The size of the problem.

Factors effect for obtaining advanced results.

1) Transfer the constraint optimization problem into unconstraint.

2) The size of the problem.

At yet there is no method can solve OPF for global problems and

obtain best fitness precisely with minimal time, but the modified and hybrid methods success in achieved more accurate results and less solving time. In many optimization formulas solving multi-objective OPF, the minimization of this objectives cannot precisely have obtained simultaneously, for example if the process intended to minimized cost, the loss may increase, this is depending on the objectives coefficients. It is of important to note that MOOPF approaches not always means the combination of two or more of the known objectives like cost, losses, and gas emission, but in some researches the single objective with transformed constraints is known as MOOPF. When the number of FACTS devices is increased, the new problems may appear and need a suitable coordination between them that lead to an increase the complication in controller system.

References

1. Happ HH (1977) Optimal power dispatch-A comprehensive survey. IEEE Trans Power Apparat Syst 90: 841-854.

2. IEEE working group (1981) Description and bibliography of major economic-security functions part-II and III. IEEE Trans Power Apparat Syst 100: 215-235.

3. Carpentier J (1985) Optimal power flow, uses, methods and development. Planning and operation of electrical energy system Proc. Of IFAC symposium, Brazil.

4. Chowdhury BH, Rahman (1990) Recent advances in economic dispatch. IEEE Trans Power Syst 5: 1248-1259.

5. Momoh JA, El-Harwary ME, Ramababu A (1999) A review of selected optimal power flow literature to 1993 part- I and II. IEEE Tran Power Syst 14: 96-111.

6. Pandya KS, Joshi SK (2008) A Survey of Optimal Power Flow Methods. Journal of Theoretical and Applied Information Technology.

7. Xia X, Elaiw1 AM (2010) Optimal dynamic economic dispatch of generation: A review. Electric Power Systems Research, Elsevier.

8. Cain MB, O'Neill RP, Castillo A (2012) History of Optimal Power Flow and Formulations. Optimal Power Flow.

9. Carpentier J (1962) Contribution á l'étude du dispatching économique. Bulletin de la Société Française des Électriciens (8) 3: 431-447.

10. Abido M (2002) Optimal power flow using particle swarm optimization. International Journal of Electrical Power and Energy Systems 24: 563-571.

11. Sumpavakup C, Srikun I, Chusanapiputt S (2010) A Solution to the Optimal Power Flow Using Artificial Bee Colony Algorithm. IEEE, International Conference on Power System Technology.

12. Moasheri SR, Khazraei M (2011) Optimal Power Flow Based on Modified Genetic Algorithm. IEEE Conference, Asia Pacific.

13. Nakawiro W, Erlich I (2009) Voltage security assessment and control system using a hybrid intelligent method. PowerTech, IEEE conference, Bucharest.

14. Sumpavakup C, Srikun I, Chusanapiputt S (2012) A Solution to Multi-Objective Optimal Power Flow using Hybrid Cultural-based Bees Algorithm. Power and Energy Engineering Conference (APPEEC), Asia-Pacific.

15. Sivasubramani S (2011) Economic Operation of Power Systems Using Hybrid Optimization Techniques. PhD Thesis, department of electrical engineering, Indian institute of technology MADARS.

16. Amjady N, Fatemi H, Zareipour H (2012) Solution of Optimal Power Flow Subject to Security Constraints by a New Improved Bacterial Foraging Method. IEEE Transactions on Power System.

17. Esmin AAA, Lambert-Torres G, Zambroni de Souza AC (2005) A Hybrid Particle Swarm Optimization Applied to Loss Power Minimization. IEEE Transactions on Power Systems.

18. Ling SH, Iu HHC, Chan KY, Ki SK (2007) Economic Load Dispatch: A New Hybrid Particle Swarm Optimization Approach. Australasian Universities Conference on Power Engineering.

19. Karthik B, Chandrasekar S (2011) Modeling of IPFC without Common DC Link for Power Flow Control in 3-Phase Line. European Journal of Scientific Research 61: 282-289.

20. Anantasate S, Bhasaputra P (2011) A Multi-objective Bees Algorithm for Multi-objective Optimal Power Flow Problem. The 8th Electrical Engineering/ Electronics, Computer, Telecommunications and Information Technology (ECTI) Association of Thailand, Conference.

21. Abido MA (2008) Multiobjective Particle Swarm Optimization for Optimal Power Flow Problem. 12th International Middle-East Power System Conference.

22. Chakraborti D, Biswas P, Mukhopadhyay A (2011) Bio-inspired computational technique to multiobjective optimal planning of electric power generation and dispatch. IEEE International Conference on Communication and Industrial Application (ICCIA).

23. Zhang Y, Gong DW, Ding Z (2011) A bare-bones multi-objective particle swarm optimization algorithm for environmental/economic dispatch. Information Sciences: an International Journal 192: 213-227.

24. Hingorani NG, Gyugyi L (1999) Understanding FACTS: Concepts and Technology of Flexible AC Transmission Systems. Wiley-IEEE Press.

25. Hermet DV, Verboomen J, Belmans R, Kling WL (2005) Power flow controlling devices: an overview of their working principles and their application range. IEEE International Conference on Future Power Systems, Belgium.

26. Parhizgar N, Dehghani Z, Roopaei M, Esfandiar P (2011) Comparison between PST-UPFC and IPFC on Power Flow Control and Profile Voltage in Power System. Australian Journal of Basic and Applied Sciences 5: 711-723.

27. Babu AVN, Sivanagaraju S, Padmanabharaju Ch, Ramana T (2010) Power Flow Analysis of a Power System In The Presence Of Interline Power Flow Controller (IPFC). ARPN Journal of Engineering and Applied Sciences.

28. Mohamed KH, Rao KSR, Md. Hasan KNB (2009) Application of particle swarm optimization and its variants to Interline Power Flow Controllers and Optimal Power Flow. International Conference on Intelligent and Advanced Systems.

29. Mohamed KH, Rao KSR (2010) Intelligent Optimization Techniques for Optimal Power Flow using Interline Power Flow Controller. IEEE, international conference on power and energy.

30. Amhriz-PBrez H, Acha E, Fuerte-Esquivel CR (2000) Advanced SVC Models for Newton-Raphson Load Flow and Newton Optimal Power Flow Studies. IEEE Transactions on Power Systems 15: 129-136.

31. Irusapparajan G, Ramareddy S (2011) Digital Simulation of Thirty Bus System with Interline Power Flow Controller. International Journal of Computer and Electrical Engineering.

32. Mohamed KH, Rao KSR, Md. Hasan KNB (2010) Optimal parameters of interline power flow controller using particle swarm optimization. International Symposium on Information Technology.

33. Teerathana S, Yokoyama A, Nakachi Y, Yasumatsu M (2003) An Optimal Power Flow Control Method of Power System by Interline Power Flow Controller (IPFC). The 7th International Power Engineering Conference.

34. Pereira M, Zanetta LC (2013) A current based model for load flow studies with UPFC. IEEE Transactions on Power Systems.

35. Mohamed KH, Rao KSR, Md. Hasan KNB (2009) Optimal Power Flow and Interline Power Flow Controllers using Particle Swarm Optimization Technique. IEEE Region Conference TENCON.

36. Sreejith S, Chandrasekaran K, Simon SP (2009) Application of Touring Ant colony Optimization technique for optimal power flow incorporating thyristor controlled series compensator. IEEE Region 10 Conference TENCON.

37. Ali ES, Abd-Elazim SM (2012) TCSC damping controller design based on bacteria foraging optimization algorithm for a multimachine power system. Elsevier.

38. Radhakrishnan G, Rathika M (2011) Application of IPFC Scheme in Power System Transients and Analysed using Fuzzy Technology. International Journal of Computer Applications 25: 24-29.

39. Karthik B, Chandrasekar S (2011) A Hybrid Technique for Controlling Multi Line Transmission System Using Interline Power Flow Controller. European Journal of Scientific Research 58: 59-76.

40. EL-Sadek MZ, Ahmed A, Mohammed MA (2007) Incorporating of IPFC in Load Flow Studies.

41. Irusapparajan G, RamaReddy S (2010) Simulation Results of Current Fed Interline Power Flow Controller Using Simulink. The Annals of "Dunarea De Jos" University Of Galati, Fascicle Iii.

42. Sankar S, Ramareddy S (2007) Simulation of Closed Loop Controlled IPFC System. International Journal of Computer Science and Network Security 7: 245-249.

43. Babu VN, Sivanagaraju S (2012) A New Approach for Optimal Power Flow Solution Based on Two Step Initialization with Multi-Line FACTS Device. International Journal on Electrical Engineering and Informatics.

44. Glavitsch H, Bacher R (1998) Optimal Power Flow Algorithms. Swiss Federal Institute of Technology Swiss Federal Institute of Technology, Zurich, Switzerland.

25

Denial of Service (DoS) Attacks using PART Rule and Decision Table Rule

Aladesote O Isaiah*

Department of Computer Science, Federal Polytechnic, Ile – Oluji, Ondo State, Nigeria

Abstract

Network Security has become a major and critical issue as a result of the vast growth in the field of Information Technology. This paper adopted the result of an existing extraction or attributes selection of KDD '99 dataset. The dataset was run on data de-duplicated software developed using C# Programming Language and final mining analysis was carried out on Waikato Environment for Knowledge Analysis (WEKA) with the adoption of PART and Decision Table algorithms. The performance evaluation was carried out with some related existing works based on certain intrusion detection metrics. The Classification Rate of Decision Tree Rule, Part Rule and JRIP Rule are 98.14%, 99.4% and 99.1%, respectively. The False Alarm Rate of Decision Tree Rule, Part Rule and JRIP Rule are 0.86%, 0.43% and 0.55% respectively. The Sensitivity of Decision Tree Rule, Part Rule and JRIP Rule is 92.6%, 98.3% and 97.2% respectively while the Specificity of Decision Tree Rule, Part Rule and JRIP Rule is 99.1%, 99.6% and 99.4% respectively.

Keywords: Waikato Environment for Knowledge Analysis (WEKA); PART rule; Detection metrics; Decision table rule; Data deduplication

Introduction

Intrusion detection is an efficient method of dealing with network security related problems [1]. Network Security has become a serious concern due to the development and expansion in the field of Information Technology [2]. This appreciable improvement in network technologies has showed a way for invaders or hackers to devise an unauthorised means into a network system. Therefore, an effective and timely Intrusion Detection System, which helps to enhance the security of a network, is needed when attack(s) is/are noticed [3]. Intrusion detection is a security approach used to protect computer networks from unauthorised access [1].

An intrusion can be defined as any attempt that violates the basic elements of information security: confidentiality, integrity and availability [4]. There is necessity to apply data mining in Intrusion Detection System owing to the huge amount of existing intrusion dataset and also recently emerging network dataset [5]. There is need for effective and efficient intrusion system as conservative intrusion detection approach can no longer match the newly emerging dataset.

Coupled with enormous data available today with lots of record duplications, which to use for optimal data analysis becomes challenging. Data deduplication thus, helps to remove such bottlenecks, thereby leaving a copy of each record in a set of data; this leads to the reduction in the amount of data to be moved into the network [6].

Research Motivation

In the work of ref. [4], Hypothesis Testing was applied on KDD dataset. The significant attributes or features of the dataset were extracted; the records of the thirteen significant attributes were used in the research. The training set was run on an existing Decision Tree algorithm which resulted in some rules. The mean of each rule was determined and later used to form hypothesis. The accuracy of the system was tested using some detection metrics. Meanwhile there is the need to valid the accuracy of the existing result by applying data deduplication with other mining algorithm on the intrusion dataset to help offer more accurate classification.

Research Objective

The objectives of the research work are to develop deduplicated program, classify intrusion dataset using PART and Decision table Rules and also to carry out performance evaluation on the KDD dataset.

Methodology

Review of few existing works was carried out. The NSL-KDD dataset which is an improvement upon KDD '99 data was used. The records of Denial of Service (DoS) attacks and normal traffic based on the thirteen significant attributes were extracted, this contains Eighteen thousand, One hundred and Thirteen (18113) records. The dataset was run on data deduplication program developed using C#.

Decision table and PART Rules were used to classify the Denial of Service (DOS) attacks and normal traffic from WEKA data mining implementation. The performance of the system would be tested on the test data using classification rate, detection rate and false alarm rate, after which the comparative analysis would be carried out against the work of Oladunjoye [7].

Result and Discussion

Data deduplication

Table 1 shows the result obtained when the dataset was run on Data deduplicated program. 9711 records of Normal traffic were reduced to 7761, which amount to 20.1% reduction. 737 records of Apache2 were reduced to 440, which is 40.3% reduction. 359 records of Back were reduced to 65, which is 82% reduction. 7 records of Land were reduced to 3, resulting in 57.1%. 293 records of Mail bomb were reduced to 4, which amount to 98.6% reduction. 4557 records of Neptune were reduced to 295, which is 93.5% reduction. 41 records of Ping of Death (PoD) were reduced to 14, which equate to 65.8% reduction. 685 records of Processstable were reduced to 367, which is 46.4% reduction. 665 records of smurf were reduced to 10, which is equivalent to 98.5%

*Corresponding author: Aladesote O Isaiah, Department of Computer Science, Federal Polytechnic, Ile – Oluji, Ondo State, Nigeri
E-mail: isaaladesote@fedpolel.edu.ng, lomaladesote@yahoo.com

Attacks/Normal Traffic	Before Deduplication	After Deduplication
Normal	9711	7761
Apache2	737	440
Back	359	65
Land	7	3
Mailbomb	293	4
Neptune	4557	295
PoD	41	14
Processtable	685	367
Smurf	665	10
Teardrop	12	2
Udpstorm	2	1
Warezmaster	944	180
Total	18113	9142

Table 1: Result obtained when the dataset was run on Data Deduplicated Program.

Attacks/Normal Traffic	TCC	TWC	TUC	TOTAL
Apache2	117	23	0	140
Back	23	0	0	23
Land	0	2	0	2
Mailbomb	0	1	0	1
Neptune	93	0	0	93
Normal	2303	20	0	2323
Ping of Death (PoD)	3	0	0	3
Processtable	107	0	0	107
Smurf	1	1	0	2
Teardrop	0	0	0	0
Udpstorm	0	0	0	0
Warezmaster	45	4	0	49

TCC: Test Correctly Classified; TWC: Test Incorrectly Classified; TUC: Test Unclassified

Table 2: Performance of Rules Generated on Test Data.

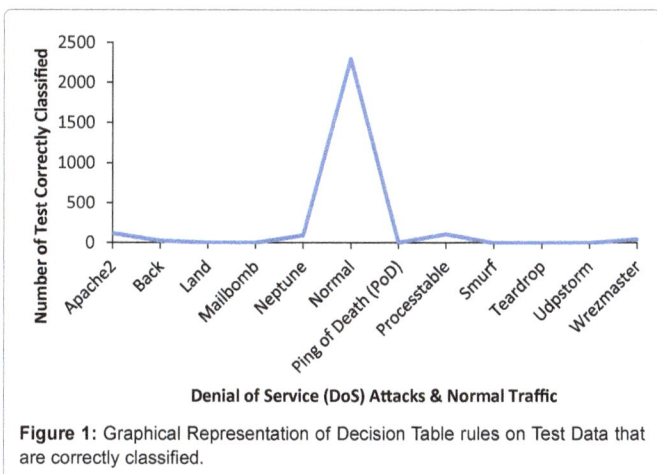

Figure 1: Graphical Representation of Decision Table rules on Test Data that are correctly classified.

reduction.12 records of teardrop were reduced to 2, which corresponds to 83.3% reduction. 2 records of teardrop were reduced to 1, which is 50% reduction while 994 records of warezmaster were reduced to 180, which is 80.9% reduction.

Performance of rules generated using decision table rules

The performance of rules generated on test data using Decision Table Rules from Table 2, Figures 1 and 2 show that out of 2303 records of Normal traffic, 2303 were correctly classified while 20 were wrongly classified. Out of 140 records of Apache2, 117 were correctly classified

while 23 were wrongly classified. All records of Back, Neptune, PoD and processtable were correctly classified. A record of mail bomb was wrongly classified. Out of 2 records of Smurf, 1 was correctly classified while the remaining 1 was wrongly classified. Out of 49 records of warezmaster, 45 were correctly classified while 4 were wrongly classified. Teardrop and udpstorm have no record in the test data.

Performance of rules generated using part rules

The performance of rules generated on test data using PART Rules from Table 3, Figures 3 and 4 show that all records of Apache2, Back, Mail bomb, PoD processtable and Smurf were correctly classified. The 2 records of Land were wrongly classified. Out of 92 records of Neptune, 92 were correctly classified and 1 was wrongly classified. Out of 2323 records of Normal traffic, 2313 were correctly classified while 10 were wrongly classified. Out of 49 records of warezmaster, 45 were correctly classified while 4 were wrongly classified. Teardrop and udpstorm have no record in the test data.

Confusion matrix obtained from denial of service (dos) and normal traffic using decision table rules

Table 4 shows the confusion matrix obtained from the Decision Table Rules Classification when DOS attacks and Normal Traffic test data were used. Out of 140 records of Apache2, 117 were correctly classified, while 21 and 2 were incorrectly classified as Neptune and Normal respectively. All records of Back, Neptune, Ping of Death (POD) and processtable were correctly classified. The 2 records of Land were incorrectly classified as Neptune. A record of Mail bomb was incorrectly classified as Normal. Out of 2323 records of Normal

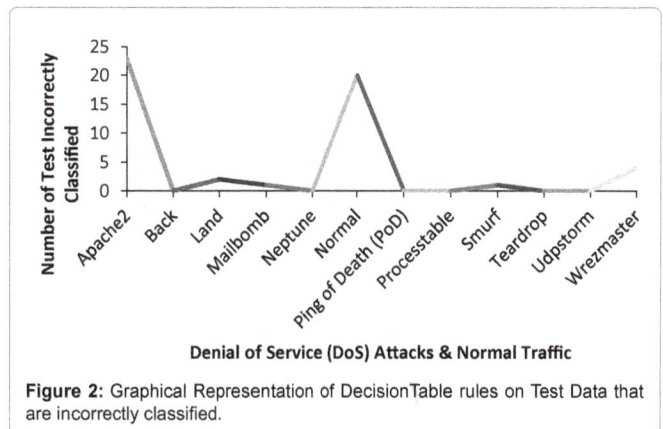

Figure 2: Graphical Representation of DecisionTable rules on Test Data that are incorrectly classified.

Attacks/Normal Traffic	TCC	TWC	TUC	TOTAL
Apache2	140	0	0	140
Back	23	0	0	23
Land	0	2	0	2
Mailbomb	1	0	0	1
Neptune	92	1	0	93
Normal	2313	10	0	2323
Ping of Death (PoD)	3	0	0	3
Processtable	107	0	0	107
Smurf	2	0	0	2
Teardrop	0	0	0	0
Udpstorm	0	0	0	0
Warezmaster	45	4	0	49

TCC: Test Correctly Classified; TWC: Test Incorrectly Classified; TUC: Test Unclassified.

Table 3: Performance of Rules Generated on Test Data.

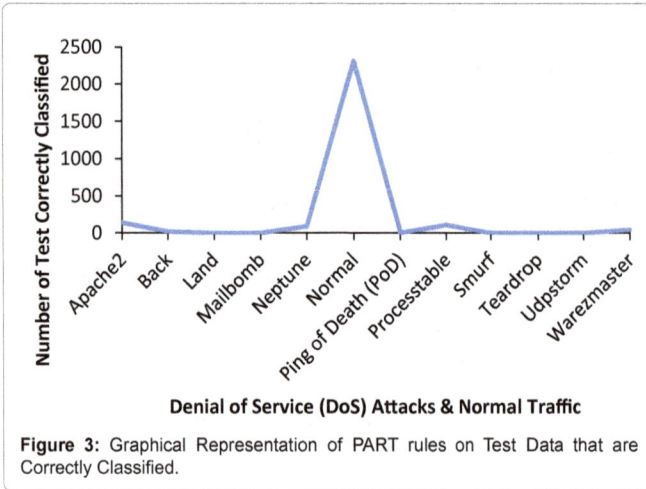

Figure 3: Graphical Representation of PART rules on Test Data that are Correctly Classified.

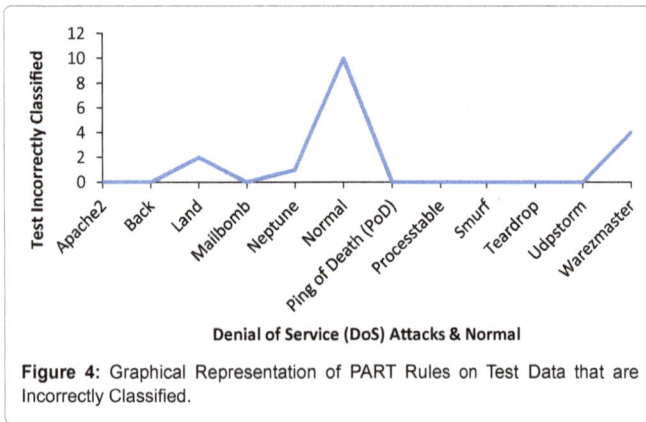

Figure 4: Graphical Representation of PART Rules on Test Data that are Incorrectly Classified.

	Ap	Ba	La	Ma	Nep	Nor	Pod	Pro	Smu	Tea	Udp	Wam
Ap	117	0	0	0	21	2	0	0	0	0	0	0
Ba	0	23	0	0	0	0	0	0	0	0	0	0
La	0	0	0	0	2	0	0	0	0	0	0	0
Ma	0	0	0	0	0	1	0	0	0	0	0	0
Ne	0	0	0	0	93	0	0	0	0	0	0	0
Nor	11	1	0	0	7	2303	1	0	0	0	0	0
Pod	0	0	0	0	0	0	3	0	0	0	0	0
Pro	0	0	0	0	0	0	0	107	0	0	0	0
Sm	0	0	0	0	0	0	1	0	1	0	0	0
Te	0	0	0	0	0	0	0	0	0	0	0	0
Ud	0	0	0	0	0	0	0	0	0	0	0	0
Wa	0	0	0	0	0	4	0	0	0	0	0	45

Table 4: Confusion Matrix obtained from decision table rules system on test data.

Traffic, 2303 were correctly classified while 11, 1, 7 and 1 were incorrectly classified as Apache2, Back, Neptune and Ping of Death (POD) respectively. 1 of the 2 records of Smurf was correctly classified while the other was incorrectly classified as POD. Out of 49 records of warezmaster, 45 were correctly classified while 4 were incorrectly classified as Normal.

TN=2303; FP=20; FN=21; TP=389

$$\text{Classification Rate }(CR) = \frac{TP + TN}{TP + TN + FP + FN} = 98.14\%$$

$$\text{False Alarm Rate }(FAR) = \frac{FP}{TN + FP} = 0.86\%$$

Sensitivity=$(100 \times TP/TP+FN)$

=92.6%

Specificity=$(100 \times TN/TN+FP)$=99.1%

Confusion matrix obtained from denial of service (dos) and normal traffic using part rules

Table 5 shows the confusion matrix obtained from the PART Rules Classification when DOS attacks and Normal Traffic test data were used. All records of Apache2, Back, Mail bomb, Ping of Death (POD), Processtable and Smurf were correctly classified. The 2 records of Land were incorrectly classified as Neptune. Out of 93 records of Neptune, 92 were correctly classified while 1 was incorrectly classified as Apache2. 2313 records of Normal were correctly classified out of 2323 while 1, 1, 1, 5 and 2 were incorrectly classified as Apache2, Back, Mail bomb, Neptune and Warezmaster respectively. Out of 49 records of Warezmaster, 45 were correctly classified while 4 were incorrectly classified as Normal.

NO˙=Normal, WM˙=Warezmaster, US˙=Udpstorm, TD˙=Teardrop, SM˙=Smurf, PR˙=Processtable, PD˙=Pod, NE˙=Neptune, MB˙=Mailbomb, LA˙=Land, BA˙=Back, AP˙=Apache2.

TN = 2313; FP = 10; FN = 7; TP = 413

$$\text{Classification Rate }(CR) = \frac{TP + TN}{TP + TN + FP + FN} = 99.4\%$$

$$\text{False Alarm Rate }(FAR) = \frac{FP}{TN + FP} = 0.43\%$$

Sensitivity=$(100 \times TP / TP + FN)$=98.3%

Specificity=$(100 \times TN / TN + FP)$=99.6%

Performance evaluation with existing system

Table 6 shows the number of records that are correctly classified incorrectly classified and not classified for each denial of services attacks and normal traffic.

Figure 5 reveals that the % of the record correctly classified using decision tree rules 98.14%, 99.43% when PART rules methods are

	Ap	Ba	La	Ma	Ne	No	Po	Pr	Sm	Te	Ud	Wa
AP˙	140	0	0	0	0	0	0	0	0	0	0	0
BA˙	0	23	0	0	0	0	0	0	0	0	0	0
LA˙	0	0	0	0	2	0	0	0	0	0	0	0
MA˙	0	0	0	1	0	0	0	0	0	0	0	0
NE˙	1	0	0	0	92	0	0	0	0	0	0	0
NO˙	1	1	0	1	5	2313	0	0	0	0	0	2
PD˙	0	0	0	0	0	0	3	0	0	0	0	0
PR˙	0	0	0	0	0	0	0	107	0	0	0	0
SM˙	0	0	0	0	0	0	0	0	2	0	0	0
TD˙	0	0	0	0	0	0	0	0	0	0	0	0
US˙	0	0	0	0	0	0	0	0	0	0	0	0
WM˙	0	0	0	0	0	4	0	0	0	0	0	45

Table 5: Confusion Matrix obtained from PART rules system on Test Data.

	Classification Rate (%)	False Alarm Rate (%)	Sensitivity (%)	Specificity (%)
DecisionTree Rules	98.14	0.86	92.6	99.1
PART Rules	99.4	0.43	98.3	99.6
JRIP Rules	99.1	0.55	97.2	99.4

Table 6: Performance evaluation with an existing work.

used and 99.1% for JRIP rules. It can be deduced that PART rules is competitively better with this type of classification than the other two methods.

Figure 6 shows the % of the normal connections that are not correctly classified in the training and testing sets. The result show that FAR is 0.86 when decision tree rules is applied, 0.43 when PART rules is used and 0.55 JRIP rules is used. This indication that the percentage of records that are misclassified is minimal when rules in PART used. Therefore, PART rules are preferably better in term of false Alarm rate for this type of classification.

Figure 7 show the % of the number of attacks connection that is correctly classified. The result indicates that the number of attacks that are correctly classified when decision tree Rules in used is 92.6%, 98.3% when PART rules in used whiles 97.2% when JRIP rules in used. PART rules perform better than the two other methods in term of sensitivity.

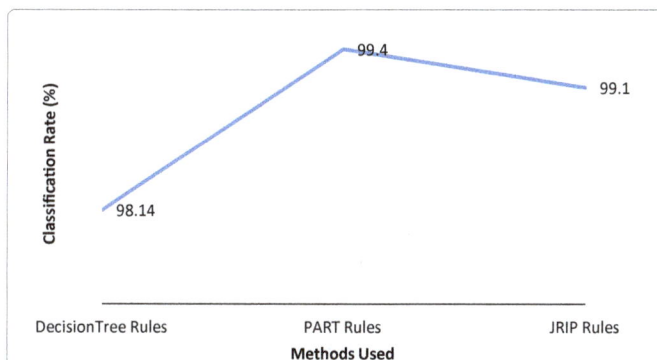

Figure 5: Graphical Representation of Classification Rate of different methods.

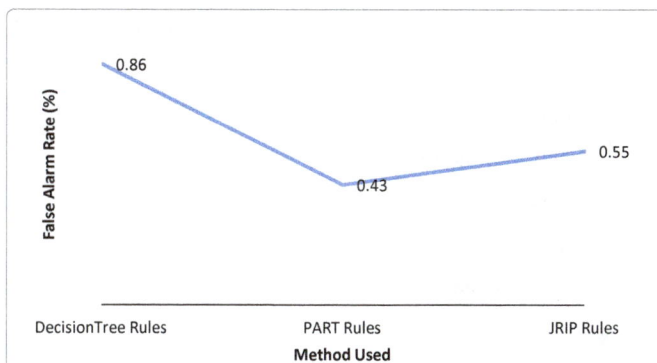

Figure 6: Graphical Representation of False Alarm Rate (FAR) of different methods.

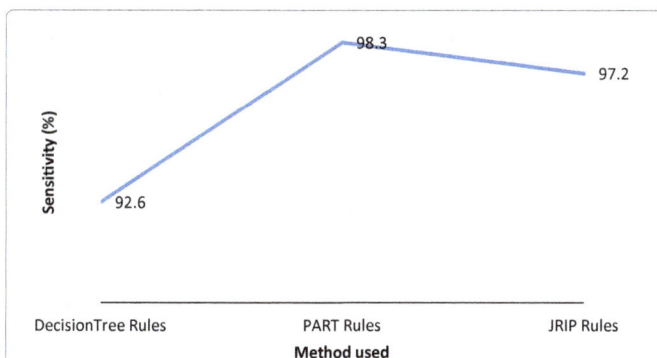

Figure 7: Graphical Representation of Sensitivity of different methods.

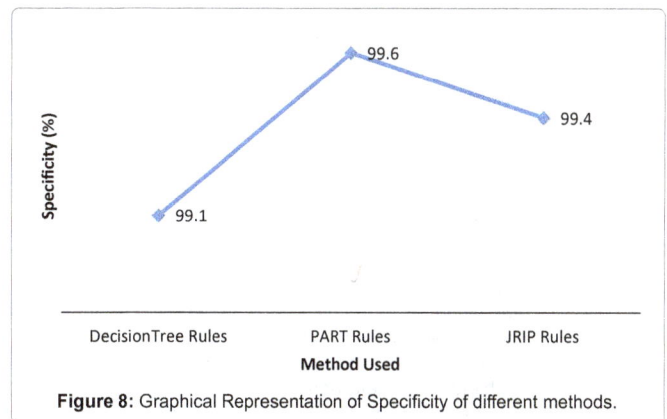

Figure 8: Graphical Representation of Specificity of different methods.

Figure 8 shows the specificity is 99.1% when decision tree is used, 99.6% when PART rules is used and 99.4% when a JRIP rule is used.

Conclusion

The system shows that PART Rules performed better than other methods in terms of Classification Rate, False Alarm Rate, Sensitivity and Specificity.

References

1. Manandhar P (2014) A Practical Approach to Anomaly-based Intrusion Detection System by Outlier Mining in Network Traffic. Masdar Institute of Science and Technology.

2. Amudha P, Karthik S, Sivakumari S (2015) A Hybrid Swarm Intelligence Algorithm for Intrusion Detection Using Significant Features. The Scientific World Journal.

3. Jaiganesh V, Sumathi DP, Mangayarkarasi S (2013) An Analysis of Intrusion Detection System using Back Propagation Neural Network. IEEE Computer Society Publication.

4. Aladesote OI, Boniface KA, Dahunsi F (2014) Intrusion Detection Technique using Hypothesis Testing. Proceedings of the World Congress on Engineering and Computer Science.

5. Shona D, Senthilkumar M (2016) International Journal of Applied Engineering Research 11: 4161-4166.

6. Meister D (2013) Advanced Data Deduplication Techniques and their Application. Johannes Gutenberg University, Mainz.

7. Oladunjoye F (2015) Intrusion Detection System using JRIP. Unpublished manuscript, Rufus Giwa Polytechnic, Owo, Nigeria.

26

Analog Multiplier Based Single Phase Power Measurement

Pandey S[1] and Singh B[2]*

[1]Krishna Engineering College; Ghaziabad, Uttar Pradesh, India
[2]Ajay Kumar Garg Engineering College; Ghaziabad, Uttar Pradesh, India

Abstract

Present paper proposes a power measurement technique of a single phase electrical load. The proposed method is a low cost power measurement technique. The load which is taken into consideration is either resistive load such as bulb or inductive load like single phase induction motor. This method is based on a low cost analogue IC (AD633) which does the analogue multiplication of the two input signals. In fact, AD633 is well suited for applications such as power measurement, modulation and demodulation, automatic gain control, voltage-controlled amplifiers, and frequency doublers. In present work AD633 is employed for power measurement. The AD633 output voltage give the multiplication of corresponding current and voltage signals of the load and produces pulsating signal/voltage at the double of the supply frequency. The higher frequency output of AD633 is attenuated by a low pass filer of an appropriate cut-off frequency (90 Hz). The corresponding DC component of the multiplier IC output obtained at the output of the active RC low- pass filter is found to be proportional to the average load power consumed. The output of RC low pass filter was plotted against different values of load using MATLAB code. The linearity of these plots was checked through the linear curve fit and the original plots for the validation purposes.

Keywords: Power measurement; Analogue multiplication; Frequency doublers; Curve fitting

Introduction

The requirement of the load power measurements is routine in the electrical engineering labs and installations. There are equipments available that can be used to measurement these quantities. Accurate measurement of power and other AC quantities is extremely important at all levels of the electrical power system, and is of value for both for power distributors and power consumers.

The objective of this paper is to design and fabricated power measurement system of an electrical load. The loads considered in the present study are resistive load (bulb) and inductive load (single phase induction motor, 220 V, 1 HP). The AD633 is a low cost multiplier comprising of a trans linear core, a buried Zener reference, and a unity gain connected output amplifier with an accessible summing node. AD633 is a complete four-quadrant multiplier offered in low cost 8-lead SOIC and PDIP packages. The result is a product that is cost effective and easy to apply. No external components or expensive user calibration are required to apply this IC. Monolithic construction and laser calibration make the device stable and reliable. High (10 MΩ) input resistances make signal source loading negligible. Power supply voltages can range from ±8 V to ±18 V. The internal scaling voltage is generated by a stable Zener diode; which gives multiplier accuracy supply insensitive [1-3].

Analog Multiplier Based Single Phase Power Measurement System Block Diagram and Schematic Diagram

This paper presents a power measurement technique of an electrical load. The proposed method is a low cost power measurement technique. The load taken into consideration is resistive load (such as bulb) and inductive load (like single phase induction motor, 220 V, 1 hp). This method employs analogue circuit (AD633 IC) which does the analog multiplication of the two signals: one signal transducted via current transformer and the other one transducted through the voltage transformer. The AD633 finds various applications, such as power measurement, modulation and demodulation, automatic gain control,

voltage-controlled amplifiers, and frequency doublers. In present application AD 633 is used for power measurement. The multiplication of signals corresponding current and voltage of the load results in the pulsating power at the double of the supply frequency. The higher frequency output of AD 633 is attenuated by filtering it through a low pass filer of appropriate cut-off frequency (90 Hz). The DC average component of the multiplier output obtained at the output of the active RC low- pass filter is proportional to the average load power consumed.

CT: Current Transformer

VT: Voltage Transformer

I-V CONVERTOR: Current to Voltage Convertor

PS: hase Shifter

AM: Analog Multiplier

LPF: Low Pass Filter

CRO: Cathode Ray Oscilloscope

The block diagram of the single phase power measurement circuit is depicted in the Figure 1. The CT and the VT employed for bringing the levels of the current and voltage signals to the accepted range of the AD 633. The acceptable maximum input supply voltages of analogue multiplier AD 633 IC is ±18 volts. The current transformer and potential transformer are so chosen that the maximum input to AD 633 IC is <18 volts. $Y_1 Y_2$ is the voltage signal from the VT. A resistor of 10 kΩ with power rating of 10 W is connected at the output of CT (Figure 2) which is then fed to terminals of multiplier. Analog multiplier AD633 multiplies these signals and produces output voltage proportional to

---

*Corresponding author:** Singh B, Ajay Kumar Garg Engineering College Ghaziabad, Uttar Pradesh, India, E-mail: bhupals_21@yahoo.co.in

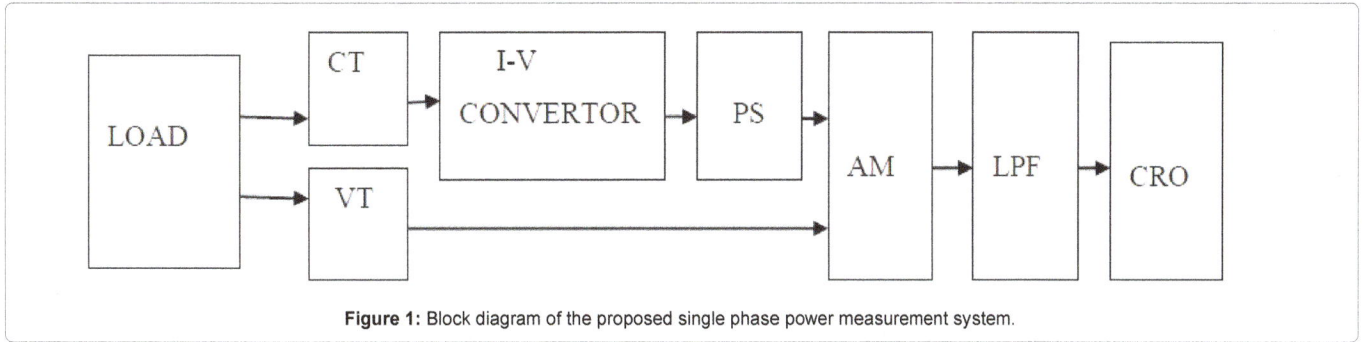

Figure 1: Block diagram of the proposed single phase power measurement system.

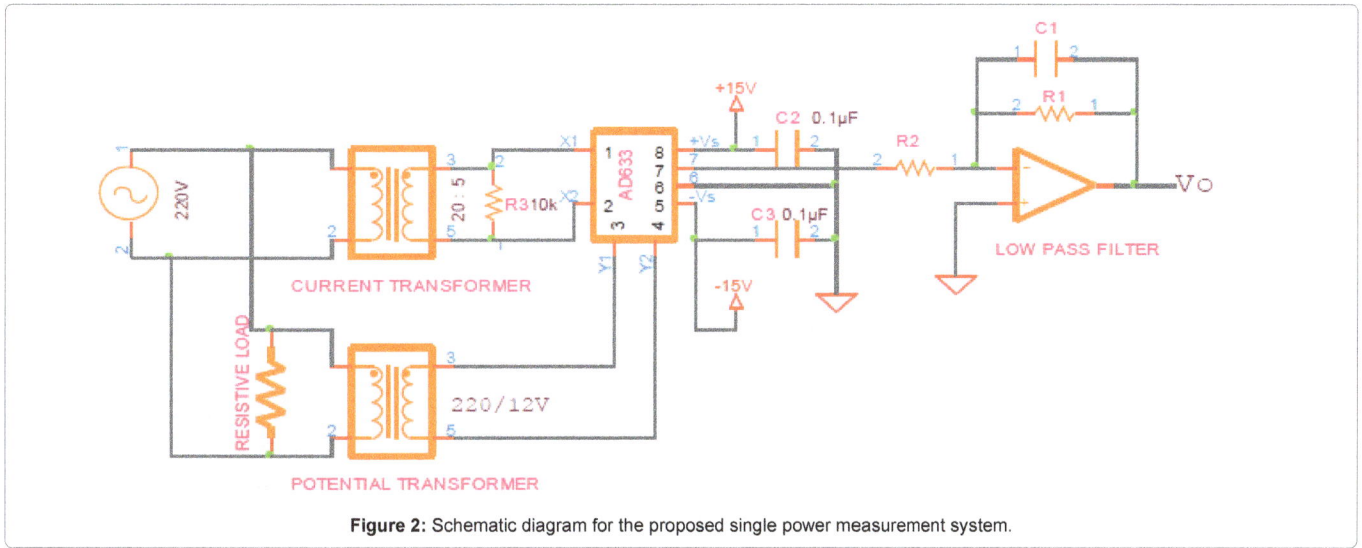

Figure 2: Schematic diagram for the proposed single power measurement system.

the input quantities which in turn is fed to active RC low pass filter to suppress pulsating component. The output voltage V_o of the low pass filter is proportional to the average load power consumed and displayed through cathode ray oscilloscope [4-6].

Experimental Results

First Case- Power measurement of resistive load

The schematic diagram with the components ratings and electrical connection is depicted in the Figure 2. This diagram is drawn using the or CAD (Table 1).

R_1=0.18K, R_2=0.18K and C_1=10 µF

Second case-Power measurement of inductive load using phase shifter

In power measurement, it is necessary that the phase of secondary winding current shall be displaced by exactly 180^0 from that of the primary winding current. It is seen that the phase difference is different from 180^0 by an angle β (phase error). Thus in power measurement, owing to use of CT, secondary winding current not being 180^0 out of phase with the primary winding current. Thus, the angle β is compensated to make current out of phase by using a variable phase shifter prior to AD633 IC. A resistor of 10 kΩ is connected to convert current into voltage at the output of CT. With the help of a CRO the voltage signal across the secondary of the VT and across the secondary of the CT are compared. By changing the potentiometric probe of the

phase shifter the phase difference between the signals is made zero Figures 3and 4.

Subsequently, this set up is used for measurement of inductive load and following parameter are recorded as given below in Table 2.

Curve fitting with mathlab

Curve fitting is a useful exercise for representing a data set in a linear or polynomial term. We are performing curve fitting to establish relation between the output of filter and load. There are two such function available in MATLAB which can be used for this purpose Polyfit (Polynomial curve fitting) and Polyval (Polynomial evaluation). The Polyfit (input data, output data and order) is a function that approximates the inputs/outputs data sets in terms of polynomial of chosen order in the sense of minimum mean square error. "Polyval" evaluates a polynomial for a given set of x values. So, polyval actually generates a curve to fit the data based on the coefficients found using polyfit.

Here x are the different load values and y is their corresponding output at low pass filter.

The blue line is original plot and the green line is the polyfit plot

The equation of a line is y=mx + c

$V_{0=}\alpha W + C$

α=0.0094 and c=- 0.4034

Resistive load (W)	CT Output voltage, V_x	PT Output voltage, V_y	Output of AD633 at pin 7	Output of low pass filter, V_o (DC)
200 W	1.9 V	12 V	2.3 V	1.5 V
300 W	3.17 V	12 V	3.8 V	2.5 V
400 W	4.10 V	12 V	4.9 V	3.3 V
460 W	4.84 V	12 V	5.8 V	3.8 V
500 W	5.38 V	12 V	6.5 V	4.3 V
560 W	6.14 V	12 V	7.4 V	4.9 V
600 W	6.41 V	12 V	7.5 V	5.2 V
625 W	6.96 V	12 V	8.5 V	5.6 V
660 W	7.25 V	12 V	8.5 V	5.8 V

Table 1: First case- power measurement of resistive load.

Inductive load (W)	CT Output voltage, V_x	PT Output voltage, V_y	Output of AD633 at pin 7	Output of low pass filter, V_o (DC)
150.0 W	7.20 V	12 V	8.6 V	5.6 V
297.8 W	7.45 V	12 V	9 V	6 V
444.0 W	8.20 V	12 V	10 V	6.4 V
587.5 W	8.62 V	12 V	10.6 V	6.8 V
678.0 W	9.16 V	12 V	11 V	7.2 V
744.5 W	10.10 V	12 V	12.2 V	7.6 V

Table 2: Second case-Power measurement of inductive load using phase shifter.

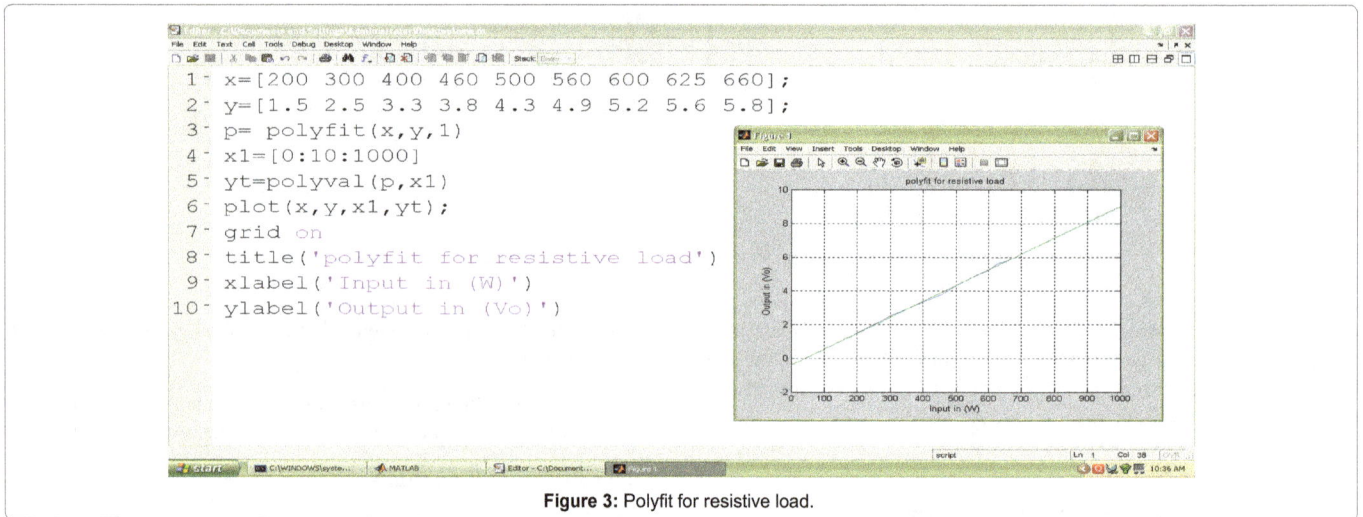

Figure 3: Polyfit for resistive load.

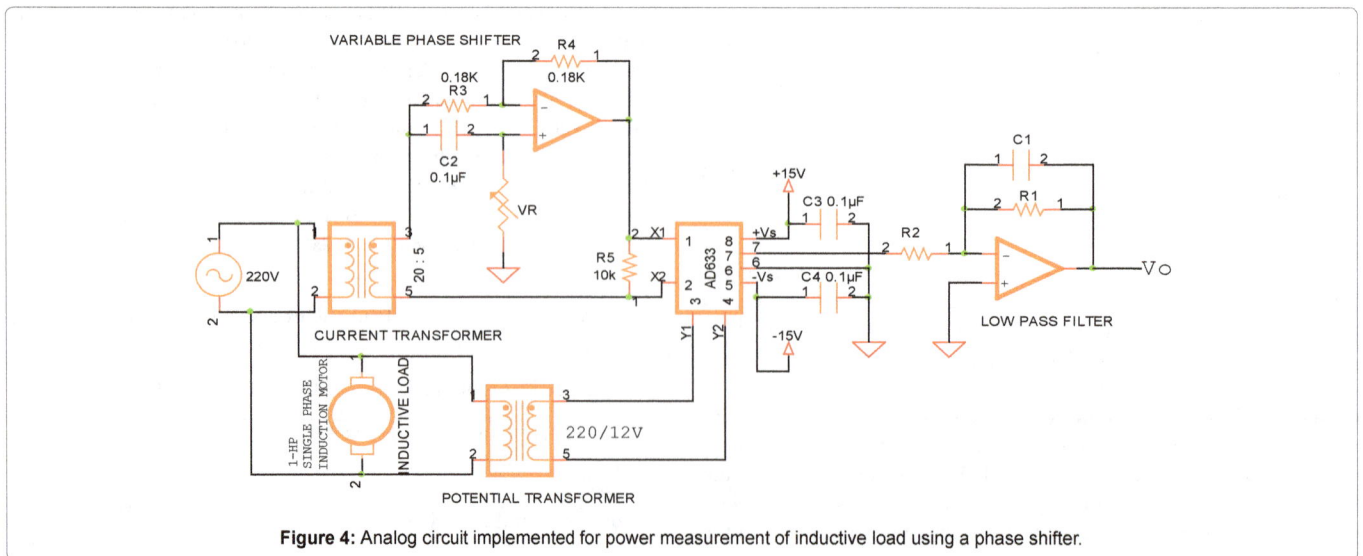

Figure 4: Analog circuit implemented for power measurement of inductive load using a phase shifter.

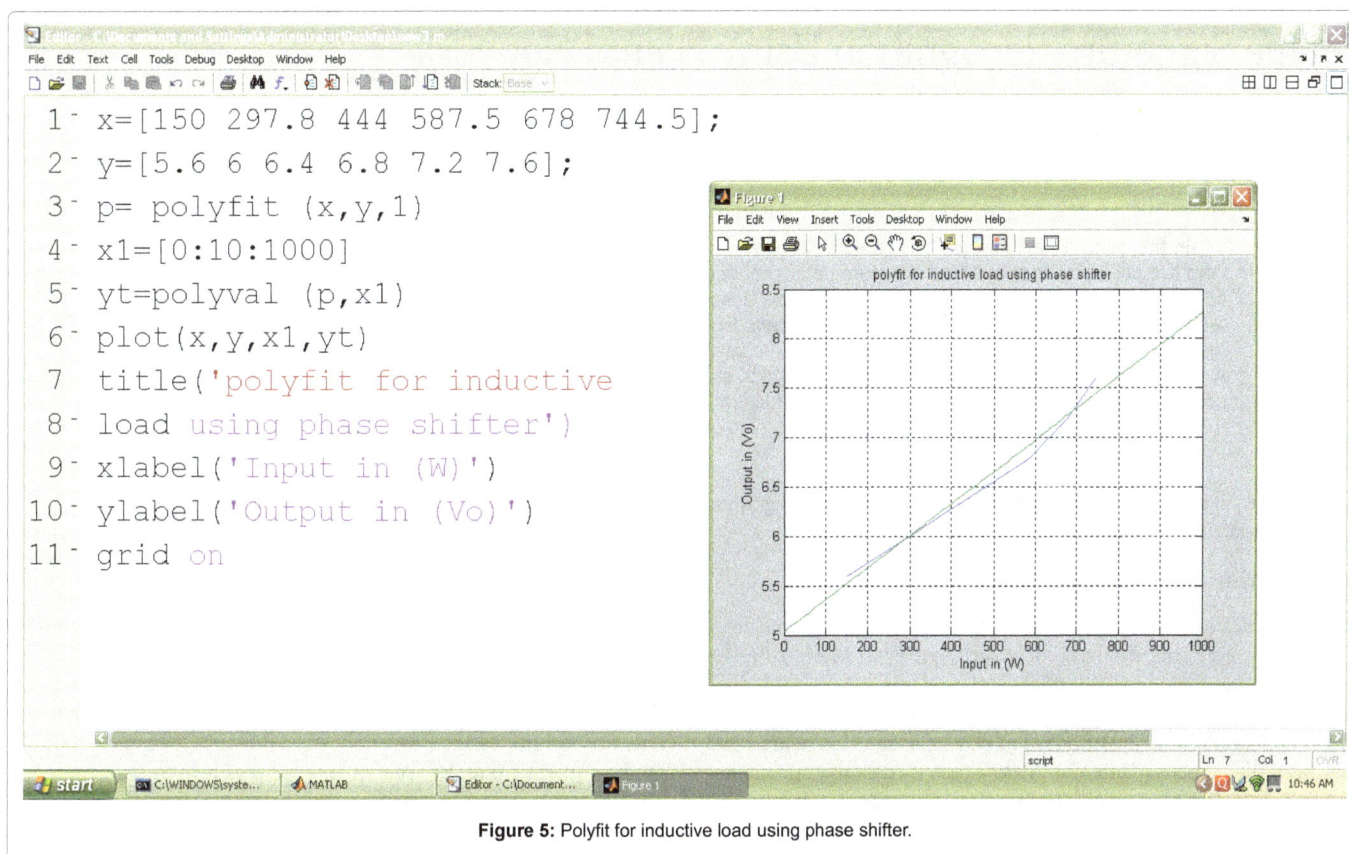

Figure 5: Polyfit for inductive load using phase shifter.

$$W = \frac{V_0 + 0.4034}{0.0094}$$

$$W = 106.38 V_0 + 42.914$$

By putting the value of V_0 the value of W (Load power) can be known.

The blue line is original plot and the green line is the polyfit plot.

$$V_0 = \alpha W + C$$

$\alpha = 0.032$ and $c = 5.0407$

$$W = 31.5 V_0 - 157.52$$

Thus for the inductive load we obtain the relationship between load and output of filter which can be used to compute the wattage. Still it seems that there is some nonlinearity. It may be due to indirect measurement of power consumed by load, instead it should have been using the wattmeter Figure 5.

Conclusion and Future Scope

In this project, it has been established the relationship of dc resistive load and ac inductive load through a linear approximation. However, the setup is not tested for capacitive load. The linearity as seen from the plot in the range of interest shows that set up can be employed for power measurement. Further same set up can be use in conjunction with the DSP or Microcontroller and can be used for other supply parameter estimation specially the supply frequency, energy consumed and power factor that too using a single measurement, that may be the part of future work and scope.

References

1. Salivahnanan S (2008) Linear integrated circuits, analog multipliers, Tata McGraw Hill.

2. Google, Low Cost Analog Multiplier AD633 IC.

3. Wijngaards DDL, Wolffenbuttel RF (2001) IC fabrication compatible processing for instrumentation and measurement applications. IEEE Transactions on Instrumentation and Measurement 50: 1475-1484.

4. Sawhney AK, Sawhney P (2003) Electrical and electronics measurements and measuring instruments. Measurement and measurement Systems, Dhanpat Rai.

5. José IA, Urriza I, Acero J, Barragán LA, Navarro D, Burdío JM (2009) Power Measurement by Output-Current Integration in Series Resonant Inverters. IEEE Transactions on Instrumentation and Measurement 56: 559-567.

6. Simpson RH (1998) Instrumentation, measurement techniques, and analytical tools in power quality Studies. IEEE Transactions on Industry Applications 34: 534-548.

27

Online Signature Recognition Using Neural Network

Babita P*

Department of Electrical Engineering, Jorhat Engineering College, Jorhat, Assam, India

Abstract

In this work a new method of signature recognition with neural network is proposed. The features are extracted from two raw databases of ATVS signature database: one consisting of 25 signature samples of 350 persons and other 46 signatures of 25 persons. The features include 9 features computed by DS Guru and HN Prakash and proposed features are no of pen downs, magnitude of average velocity, magnitude of average acceleration and length to height ratio. Signature features are pre-processed with a scaling method and brought to a value having same decimal point. A feed forward neural network is trained using back propagation learning method. With each features removed, an accuracy rate is calculated to check the feature which will be better for signature verification. Accuracy of recognition up to 98% and 89% are obtained using signature samples of 10 persons from each database respectively.

Keywords: Signature recognition; Neural network; Back propagation; Confusion matrix

Introduction

Authentication of user is becoming very important to do business transactions, accessing data and for security purpose. Many different techniques are applying for authentication purpose. User IDs and passwords, PIN codes, ATM card, PAN card are many different ways which are common today, but the problem of such systems are that, they need remembering different PINs or passwords, carry such items and they need to be kept secret from others. Signature is a behavioral biometric. Automatic signature authentication is now becoming popular in research areas because of its acceptance in legal and social areas and its widespread use for authentication purpose.

Since signature for different individuals vary with the variation of individuals, so it is a very robust biometric to authenticate a user. Signature verification is a very difficult pattern recognition problem. Since intra class variations occur, even experts get difficulty to recognize the forgery signature. And also it is not very difficult to forge a signature. Signature is believed to be a reflex action which produces its dynamic properties unconsciously.

Biometrics can be broadly divided into physiological and behavioral biometrics. Some examples of physiological biometrics are fingerprint, iris and face; and behavioral biometrics is signature, voice and hand writing. Authentication of signature is done by detecting forgeries. Forgeries can be divided into random forgery, simple simulated forgery and simulated skilled forgery. Figure 1 shows an example of forgery signatures. Random forgeries are produced randomly without any information about the name and person for whom signature is produced. Random forgery are generated when forger do not have any available access to the signature. They may have different shape and size from the authentic signature. Simple simulated forgery may have same semantic meaning like authentic one but overall shape and size may differ. Skilled simulated forgery signatures are the signatures which are given by a large number of practices.

For online signature verification, signature data are generally taken using capacitive tablet or personal data assistance (PDA) which gives the x-y coordinate, pressure readings etc. From these raw data different features can be computed. Signature authentication problem can be solved by two ways: dynamic and static, which needs computation of dynamic and static features respectively. Dynamic measurement of signature features can be obtained by electronic tablet or PDA; static features can be computed from images obtained either by camera or scanning the photo of the signatures. Dynamic features [1] are functions of time and static features are time independent. Even if a skilled forger produces the same looking signature like the authentic one, they cannot easily learn to produce the same pressure produced by the authentic one. Hence dynamic features help to detect forgery. The use of pen dynamics over shape of signature would be more useful in forgery detection because dynamic features of a signature are not readily available to forger as in the shape of offline signature.

Handwritten signatures can be represented by multiple modals i.e. global and local, shape based and time based. Local shape based signature and their advantages are discussed in Automatic On-Line Signature Verification by Nalwa [2]. Three signature databases are collected and analyzed for a different period of time. Different reasons for inaccuracies are also discussed. Global feature based technique is applied in Signature Verification method [3]. Three global features i.e. projection moment, upper envelope based characteristics and lower envelope based characteristic are used and then multiple neural network is applied for the classification purpose. To remove noise they have applied median filter. A fusion of local and regional features is discussed in Fusion of Local and Regional Approaches for On-Line Signature Verification by Aguilar [4]. The local function based features are classified with dynamic time warping (DTW) and regional features are classified with hidden markov model (HMM). In Pattern recognition [5], signature verification based on logarithmic spectrum is done. Principle components of the logarithmic spectrum are compared with the reference signature and similarity value is calculated between the enrolled and reference signature. A stroke based method for shape and dynamics of signature is discussed in Multiexpert Verification of Hand-Written Signature by Bovino [6]. Two level strategies using soft and hard rule are implemented in it.

***Corresponding author:** Babita P, Department of Electrical Engineering, Jorhat Engineering College, Jorhat, Assam, India, E-mail: 825babita@gmail.com*

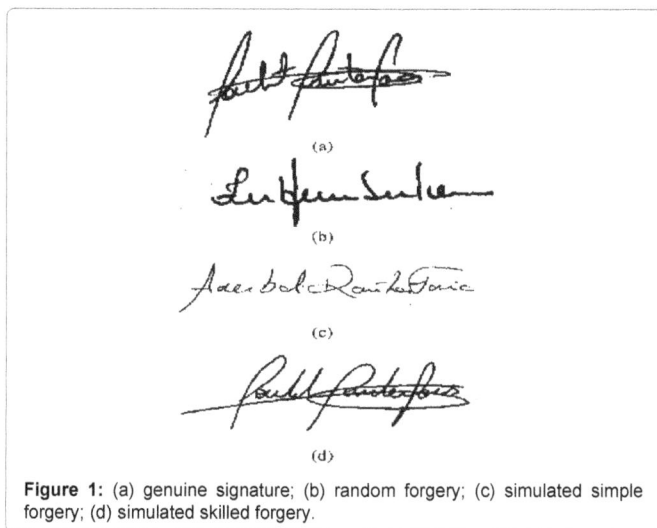

Figure 1: (a) genuine signature; (b) random forgery; (c) simulated simple forgery; (d) simulated skilled forgery.

The signature authentication is done in two types of problems: signature verification and signature recognition. In verification, the features of the test signature are compared with a few number of stored features of the signature of the claimed person, and need to verify if the signature belongs to that particular person or not. But in recognition the features of the test signature are compared with a few no of stored features of a no of persons and we have to recognize whether the test signature belongs to the one of the enrolled persons and identify the person.

Online signature verification can be broadly classified into two groups based on their feature extraction method: parametric approach and function based approach. In parametric approach a set of parameters (e.g. Speed, displacement, position, pen up pen down, wavelet transform etc.) extracted can be used as a feature to form a signature pattern, and those feature patterns can be used as reference and test signature to examine the authentication of the signature. In function based approach the features are the function of time (e.g. velocity, acceleration, pressure, direction of pen movement etc.). Online signatures are characterized as a time function.

In any verification task there are two types of error involved i.e. false rejection and false acceptance. False rejection occurs when the authentic signature is rejected and false acceptance occurs when a forged signature is accepted to be authentic. When percentage of false rejection rate (FRR) is equal to the percentage of false acceptance rate (FAR) we call it equal error rate. Equal error rate is the measure of the performance of a biometric system. Average error rate is the average of FAR and FRR. The authenticity of test signature is evaluated by matching it with that of reference signature. There are many techniques available for matching e.g. dynamic time warping (DTW) [7], Hidden Markov model (HMM), support vector machine (SVM) and neural network (NN). When functions are considered, the matching technique must take into account the variation of duration of signature. A method of similarity measure for signature verification and recognition using symbolic representation is done [8]. Here the following are discussed:

1. Dynamic time warping technique is generally used for function based parameter. But the time complexity of DTW is more of the order of (O_2).

2. HMM performs stochastic matching using probability distribution of the features. They can compute both similarity and variability of the pattern. But they require a large dataset to train and are complex.

3. Support vector machine classifies one class of data from the other by finding the hyper plane that maximizes the separation between classes. SVM have algorithmic complexity, and requires large storage for large scale task

4. Neural network have ability of generalization. They can be used to detect nonlinear equations for dependent and independent variables. They can train large amount of database. Easily implemented in parallel architecture.

In [9] a method of string matching or dynamic time warping is done. Here local features and stroke based global features are extracted and the results are compared, by varying the values of absolute and relative speed of the signature. Then signature feature vectors are formed in symbolic interval value and test data are inserted to check if it lies within that interval or not. After that interval value is set by calculating mean, variance and standard deviation. Finally writer dependent and feature dependent threshold are set in the database.

Many applications of artificial neural network are there. A work using autoregressive artificial neural network is done [10]. Here a monthly inflow forecasting is done using static and dynamic autoregressive artificial neural network and both radial and log-sigmoid activation function is carried in this work. A 47 years of discharge is used, with 42 years of discharge is used for training and remaining 5 years of discharge is used for forecasting. In the neural network architecture, weights are initialized by Nguyen and Widrow's weight initialization method and it is giving a good result with a hidden layer having 17 neurons. From this work it is found that, the dynamic autoregressive artificial neural network is better than the static autoregressive artificial neural network. When results are compared with radial and log- sigmoid function, log-sigmoid function has given a superior result. In the successive work [11], the monthly inflow of Taleh Zang station is carried on. A comparison using both static and dynamic model of autoregressive moving average (ARMA), autoregressive integrated moving average (ARIMA) and autoregressive artificial neural network is done with the same database. And the results are compared by comparing the root mean square error (RMSE) and mean square error (MSE). From this work, it was found that, the ARIMA model is better than the ARMA model and dynamic autoregressive model is better than the static autoregressive model. From the result it can be concluded that, the ARIMA model can be used for forecasting of inflow for next 12 months and the autoregressive artificial neural network model can be used to forecast the inflow for next 5 years. In their next work [12] a comparison of performance ability of autoregressive static and dynamic artificial neural network is done for monthly inflow forecasting with the same network parameters. From this, it is found that, neural network architecture is very advantageous for data analysis of large database.

Feature Extraction

From ATVS signature sub corpus, two sets of database are collected. First database contains 25 genuine signature samples for each user. In the second database each users have 46 samples each. The databases contains raw data values of the signature i.e. x-coordinate, y-coordinate, time stamp, pen up pen down and pressure signal. From these data the features are extracted. Initially from these raw data 33 features were computed. For training in neural network feature selection is done manually where some features, which give better results were kept. Features are introduced as:

Total duration of signature

It is the time taken to complete a signature. It can be calculated as the difference between the last time stamp and the first time stamp.

Number of pen ups

The number of times pen is removed from the pad/paper.

Sign changes of dx/dt and dy/dt

dx/dt and dy/dt may be positive or negative value. So when it changes the value from positive to negative or negative to positive it is counted.

Average jerk

Jerk is change in acceleration with respect to time. Average jerk is the mean of the jerk.

Standard deviation of velocity in y-direction:

Standard deviation of v_y;

Where, velocity in y-direction: $v_y = dy/dt$ $\qquad$ (1)

Standard deviation of acceleration in y-direction

Standard deviation of a_y

Where, acceleration in y-direction: $a_y = dv_y/dt$ $\qquad$ (2)

Number of local maxima in x direction

Local maxima can be calculated from change in x with respect to time.

Standard deviation of acceleration in x-direction: standard deviation of a_x; where

Velocity in y-direction: $v_x = dx/dt$

Acceleration in y-direction: $a_x = dv_x/dt$ $\qquad$ (3)

Standard deviation of velocity in x-direction

Standard deviation of v_x; where

Velocity in y-direction: $v_x = dx/dt$ $\qquad$ (4)

Length to width ratio

It is the ratio of length of the signature to the width of the signature i.e. ratio of number of sample point covered by the signature in x-coordinate to number of sample point covered by the signature in y-coordinate.

Number of pen downs

Number of times pen touches the pad/paper to complete a signature.

Average magnitude of velocity

Velocity at every sample points changes. Average velocity is the mean of the velocities at every sample point.

Average magnitude of acceleration

Acceleration also changes at every sample points. Therefore average acceleration is the mean of the acceleration at every sample points.

Methodology

The main motive of this work is to find some features suitable for signature verification and to check the performance of the neural network with those features. Here we have used neural network back-propagation algorithm for training a network. In neural network approach, the main procedure to implement is: first of all the features need to be extracted and then the network is to be trained to learn the relationship between the pattern and its class. After training, validation of the network is to be checked by few features to see if the network is giving a satisfactory result or not. After validation, the network is to be tested using features which are completely unknown for the network, to see the performance of the network. Data bases are collected from ATVS signature sub corpus [13]. The reason of using two dataset is to test the generalization capability of the network. Generally it is easy for the network to generalize more with more sample data. In second database sample data is more so it should give more accurate result. Now, first database i.e. dataset I consists of 25 signature data for each individual. Second dataset i.e. dataset II consist of 46 signature sample for each individual. From the raw data set consisting of information of x coordinate, y coordinate, time stamps, pressure and pen up and pen down, features were extracted. Then the databases are divided into three parts for training, validation and testing. From dataset I, 15 signatures were extracted for training, for validation next 5 signatures were extracted and for testing also remaining 5 signatures were extracted. From dataset II, 30 signatures were extracted for training, for validation next 8 signatures were extracted and for testing remaining 8 signatures were extracted. After division of data, they are randomized. The matching technique used is neural network approach. At first the extracted data are brought to 10th decimal point. Then the data is normalized so that the pattern values are between 0 and 1. During training weights are updated to minimize the difference between the desired output and the actual output i.e. error. The fixed weights after training can be used for the task in pattern recognition and classification. The neural network structure used have three layers: one input layer, one hidden layer and one output layer. Two models are designed for the comparison purpose. The first model with 9 features have number of input nodes 9, hidden layer nodes 60 and the output layers have 10 nodes. The second model with 13 features have 13 input nodes, 60 hidden layer nodes and 10 output layer nodes. For the second dataset the number of hidden layer nodes taken is 80. Activation provides the measure of confidence of corresponding decision of the classifier. Commonly used activation functions are sigmoid functions. The activation function used here is log sigmoid function (logsig). It gives the activation label between 0 and 1. If activation is 1 it means confidence is high and if it is 0 means confidence is zero.

A. Vector matrix form of back propagation algorithm

1. An input pattern is presented and calculated the outputs of the network at all the internal layers.

2. For each of the layers, the sensitivity vector is calculated according to

$D^{(s)} = G(v^{(s)})(d_q - x_{out}^{(s)})$ for output layer $\qquad$ (5)

$D^{(s-1)} = G(v^{(s-1)})W^{(s)T}D^{(s)}$ for all hidden layers $\qquad$ (6)

The synaptic weights are updated for the network according to

$W^{(s)}(k+1) = W^{(s)}(k) + \alpha^{(s)}D^{(s)}x_{out}^{(s-1)T}$ $\qquad$ (7)

3. Continue steps 1 through 3 until the network reaches the desired mapping accuracy.

Where,

$D^{(s)}$ = sensitivity vector of layer of particular layer s

$G(v^{(s)}) = \text{diag } [g(v_1^{(s)}), g(v_2^{(s)}), \ldots g(v_{ns}^{(s)})]$

$g(v)$ = derivative of activation function v

d_q = desired output vector

x_{out} = actual output vector of the neural network

$W^{(s)}$ = weight vector for layer s

k = iteration number

$\alpha^{(s)}$ = learning rate parameter associated with the particular layer s

Result

Neural network is a generalization tool. The reason why the neural network approach is chosen among the number of other classification method is that, neural network is easy to use and can solve complex problems with ease. From this work it is realized that, when variation of data is more, neural network finds it difficult to generalize. That is why normalization of database is important. When introduced pre-processing of data by converting them to 10th decimal point or same decimal point value the generalization becomes more and classification error decreases. Confusion matrix is a table which helps the visualization of the performance of supervised machine learning. The diagonal boxes in the table from left (up) to right (down) gives the true positive classification. And other boxes show the true negative classification. Figure 2 is showing the confusion matrix of genuine signatures of first dataset. The data consist of 15 genuine signatures of 10 users. From this confusion matrix, the true positive rate found is 98% i.e. false rejection rate (FRR) is 2%. Figure 3 shows the confusion matrix of forgery signatures of first dataset. The data consist of 5 forgery signatures of 10 users. From this confusion matrix, when forgery signatures were taken, the false acceptance rate is 8%.

From the confusion plot the accuracy of the system can be calculated by:

$$\text{Accuracy} = (100 - (FAR + FRR)/2) \% \qquad (8)$$

The accuracy rate of the system is 95 %

Table 1 show the result obtained from dataset I using all the features extracted. And Table 2 shows the result obtained from dataset I and dataset II using all the features extracted and also from removal of one or more features. Table 2 gives the comparison of the accuracy of result with 9 and 13 features respectively for different number of classes (users).

From the Table 3 it is clear that features i.e. Average jerk, Standard deviation of acceleration in y direction, Standard deviation of acceleration in x direction, Standard deviation of velocity in x direction and average acceleration are not very suitable in case of dataset I. And in case of dataset II Time duration of signature and Average acceleration are not suitable. From the final accuracy result it is found that, result of dataset with more sample number gives more accurate result, which reflects the generalization capability of neural network. From Table 3 it is observed that when features i.e. length to width ratio, number of pen downs, average magnitude of velocity and average magnitude of accelerations are added with previous 9 features, the network gives a better accuracy. It can also be observed that as the number of classes increases the accuracy decreases in both the cases. It implies that the features mentioned are not very good when database with more number of classes are taken.

Conclusion

The main objective of this work is to construct a signature recognition system using some feature values so that to get a maximum accuracy label. To do this, some features were extracted and formed pattern from them. The features acts as an input pattern to the neural network and corresponding targets are constructed. In neural network, the patterns are trained according to the target, where weights are updated to get a minimum error. When a stopping condition is reached the iteration stops. In neural network training, many times trial and error is done to get satisfactory neural network architecture. Parameters like hidden nodes in a neural network, initial learning rate parameter, learning rate schedule are needed to be adjusted again and again. Neural network architecture is dependent on these parameters. In this work the nodes in the hidden layers for the first dataset is 60 and second.

Dataset is 80. Initial learning rate parameter is 1 and learning rate scheduled at 300. These parameters give a false acceptance rate of 8% and a false rejection rate of 2%. And accuracy rate when calculated gives an accuracy rate of 95% for the dataset I. And for the dataset II an accuracy of 89% is obtained using the same parameter. When experiment is performed taking different number of classes, it is found that with increase in the number of classes the accuracy is decreasing. It

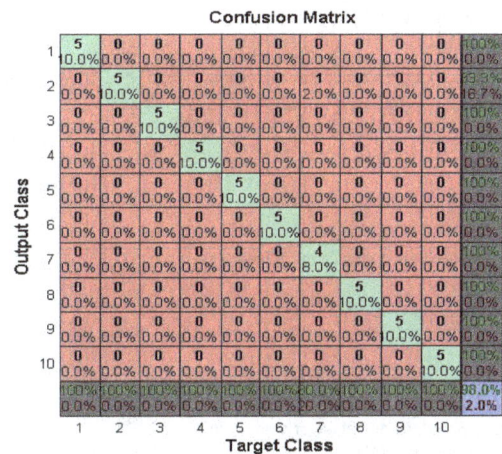

Figure 2: confusion matrix for genuine signature data of 10 users.

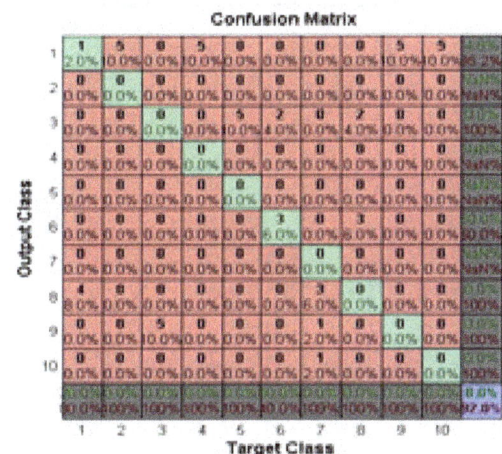

Figure 3: confusion matrix for forgery signature data of 10 users.

Features	True acceptance rate (%)	False rejection rate (%)	False acceptance rate (%)	Accuracy (%)
All the features mentioned	98	2	8	95

Table 1: Result showing the accuracy of network using dataset I using 10 persons.

No of persons taken	Accuracy with 9 features	Accuracy with 13 features
10	85	98
25	80	89
50	50	49.2
100	36	40.4
200	25.5	27.2
350	16.5714	19.02

Table 2: Comparison of accuracy of neural network with 9 features and 13 features.

Feature/features removed	True acceptance rate (%) Using dataset I	True acceptance rate (%) Using dataset II
(using all features)	74	
Time duration of signature	72.5	75.2
Number of pen ups	70.5	70.4
Sign changes of dx/dt and dy/dt	73.5	72.8
Average jerk	79	63.2
Standard deviation of acceleration in y direction	76.5	69.6
Standard deviation of velocity in y direction	65.5	72
No of local maxima	62	66.4
Standard deviation of acceleration in x direction	76	63.2
Standard deviation of velocity in x direction	74	69.6
Length to height ratio	65	77.6
No of pen down	70	67.2
Average velocity	70	78.2
Average acceleration	74	- - - -
Average jerk, Standard deviation of acceleration in y direction	86	- - - -
Average jerk, Standard deviation of acceleration in y direction, Standard deviation of acceleration in x direction	87.5	- - - -
Average jerk, Standard deviation of acceleration in y direction, Standard deviation of acceleration in x direction, Standard deviation of velocity in x direction, average acceleration	89	- - - -
Time duration of signature, Average acceleration	- - - -	73.6
Time duration of signature, Average acceleration	- - - -	80

Table 3: Result showing the accuracy of network for dataset I and dataset II.

is because of the fact that, the features considered in the experiment are not very suitable for database with large number of classes.

Since neural network has a generalization capability, once trained its weight need not be changed again. In testing it gives the result from the trained architecture itself. The main drawback of neural network training is that, for larger dataset it is very difficult to adjust the parameters by trial and error method. And the time consumption is more.

Reference

1. Nelson W, Kishon E (1991) Use of dynamic features for signature verification, Proc. IEEE Int'l Conf. Systems, Man and Cybernetics 1: 201-205.

2. Nalwa VS (1997) Automatic on-line signature verification, Proc. Third Asian Conf. Computer Vision 1: 10-15.

3. Bajaj R, Chaudhary S (1997) Signature verification using multiple neural classifiers, Pattern Recognition 30: 1-87.

4. Aguilar JF, Krawczyk S, Garcia JO, Jain AK (2005) Fusion of Local and Regional Approaches for On-Line Signature Verification, Proc. Int'l Workshop Biometric Recognition System.

5. Wu QZ, Lee SY, Jou IC (1998) On-Line Signature Verification Based on Logarithmic Spectrum, Pattern Recognition 31: 1865-1871.

6. Bovino L, Impdevo S, Pirlo G, Sarcinella L (2003) Multiexpert verification of hand-written signature, Proc. Int'l Conf. Document Analysis and Recognition 932-936.

7. Marcos FZ (2007) On-line signature recognition based on VQDTW, Pattern Recognition 40: 981-992.

8. Guru DS, Prakash HN (2009) Online signature verification and recognition: An Approach Based on Symbolic Representation, IEEE transactions on pattern analysis and machine intelligence 31.

9. Jain AK, Griess F, Colonnel S (2002) On-line signature verification, pattern recognition 35: 2963-2972.

10. Valipour M, Banihabib ME, Behbahani SMR (2012) Monthly Inflow Forcasting using Autoregressive Artificial Neural Network. J Appl Sci 12: 2139-2147.

11. Valipour M, Banihabib ME, Behbahani SMR (2013) Comparison of the ARMA, and the autoregressive artificial neural network models in forecasting the monthly inflow of dez dam reservoir, Journal of hydrology 476: 433-441.

12. Valipour M, Banihabib ME, Behbahani SMR (2012) Comparison of Autoregressive Static and Artificial Dynamic Neural Network for the Forecasting of Monthly Inflow of Dez Reservoir.

13. Description of Atvs-SSig DB, National Laboratory of Pattern Recognition (NLPR), Institute of Automation, Chinese Academy of Sciences (CASIA).

28

Performance Evaluation of Energy Detection in Spectrum Sensing on the Cognitive Radio Networks

Rayan Abdelazeem Habboub Suliman[1]*, Khalid Hamid Bilal[2] and Ibrahim Elemam[3]

[1]Department of Electrical Engineering, Engineering College, Qassim University, Kingdom of Saudi Arabia
[2]Department of Communication, Faculty of engineering/university of science and technology, Khartoum, Sudan
[3]Department of Communication, El NeelainUniversity, Khartoum, Sudan

Abstract

Cognitive radio technology is a modernistic technology by which the idle licensed spectrum can be used by an unlicensed user which are called cognitive radio CR or secondary users. The CR technology was found to overcome the spectrum incompetence and inefficiency usage troubles. The major motivation of CR technology is current heavily underutilized of spectrum. Surly the primary user PU had all rights to use his spectrum band so CR mustn't interfere it on that band. Here we have to discuss about energy detection which is one of spectrum sensing techniques in CR functions and how to obtain the best available spectrum, to detect the spectrum hole, to estimate the optimum threshold voltage that produce the minimum Probability of false alarm and finally to enhance energy detection sensing algorithm by using MATLAB simulation.

Keywords: Cognitive radio; Energy detection; Threshold; Probability of false alarm

Introduction

Cognitive radio is an a enhanced to software-defined radio (SDR) that its automatically detects the surrounding RF, catalysts and smartly accommodates its operating parameters to the infrastructure of network according to meet user demand [1], if this band is further used by a licensed user, the cognitive radio stirs to other spectrum band or remains in the same band with altering its level of the transmission power or modulation scheme all of that avert interference, calibrations the congestion due to spectrum participating [2].

The unemployed spectrum was called white space or spectrum holes. The CRN is development of Dynamic Spectrum Access DSA and SDR by enabling the spectrum smartly and highly configuring radio transmitter/receiver.

There are four functions of CRN spectrum sensing, spectrum decision, spectrum management and spectrum sharing.

- **Spectrum sensing:** A CR user can occupy only the vacant spectrum band. Therefore a CR user should superintend the available bands spectrum, pick up their information, and therefore it can check the white space [1].

- **Spectrum decision:** Depend on the availability of spectrum on internal (and possibly external) policies [3], CR users can allocate a spectrum band.

- **Spectrum sharing:** Regard to there may be multiple cognitive radio users all of them are trying to access the spectrum, access on CRN should be arranged to prevent multiple users clash in interference bands of the spectrum [4].

- **Spectrum mobility:** Cognitive radio users are considered as guests on the spectrum. So if the particular servant spectrum band is wished for the primary user, the communication has to be continued in another idle spectrum band [5].

The spectrum sensing is the important one cause the CR technology are depend on it by which the spectrum holes are sensed [6], and it enables CR to configure to the environment, according to this point one of spectrum sensing techniques had taken to discuss and simulate on this paper.

Spectrum Sensing Review

It is a necessary issue of CR is to sense the spectrum holes, it's based on to be aware of and sensitive to changes on it surrounded environment [1], spectrum sensing algorithms enable the CR to adjust and adapt to the environment by detecting PUs which receiving data across the communication scope of CR users [7].

The sensing on CRN has two-main points:

a) To ensure that the secondary user doesn't caused interference to a primary user

b) To aid secondary user to recognize and utilize the white space for the required quality of service [2].

The spectrum sensing is to decide and determined between two hypotheses which are:

$$x(t) = w(t), \text{ H0 (PU absent)} \tag{1}$$

$$x(t) = h\, n(t) + w(t), \text{ H1 (PU present)} \tag{2}$$

Where x(t) is the received signal by the unlicensed user, n(t) is the transmitted signal of the licensed user, w(t) is the Additive White Gaussian Noise (AWGN), h is the amplitude gain of the channel. H0 is a null hypothesis, means that there is no signal from licensed user [2], according to the primary user these hypotheses means:

H0: The spectrum band is idle, H1: The spectrum band is occupied [2].

The used sensing cognitive radio algorithm should be sensitive enough to characterize between the noise power and the signal power [7]. The main objective of this paper is to explicate and simulate specific

*Corresponding author: Rayan Abdelazeem Habboub Suliman, Department of Electrical Engineering, Engineering College, Qassim University, Kingdom of Saudi Arabia, E-mail: rere.abdoo2014@gmail.com

spectrum sensing algorithm and improve the performance of it for a low SNR environment.

As spectrum sensing is the key enabling in cognitive radio to adapt it is environment by detecting spectrum holes, so it could have categorized to three main types transmitter detection or non-cooperative sensing, cooperative sensing and interference based sensing. Transmitter detection technique is further classified into matched filter detection, cyclostationary feature detection and energy detection which it discussed in this paper.

Energy Detection

It unveils the basic signal based on the sensed energy. According to its artlessness and no need to a prior knowledge of the signal of the primary user, energy detection (ED) is the most popular sensing technique [6]. Figure 1 shows the block diagram of ED technique.

In this method, signal is passed through band pass filter of the bandwidth W and is integrated over time interval, the output of the block of an integrator is then compared to a predefined threshold. The goal of this comparison is to discover the existence or absence of the licensed user. The value of threshold can set to be variable or fixed based on the conditions of the channel [6]. The Blind signal detector is the another name of ED [2], because it ignores the signal structure i.e., it estimates the presence of the signal by comparing the energy received with a known threshold v derived from the statistics of the noise [8], analytically signal can be formalized as a hypothesis test

$$Y(k)=n(k)....H0 \tag{3}$$

$$Y(k)=h*s(k)+n(k)....H1 \tag{4}$$

In sample the y(k) is analyzed at each instant k and n(k) is the noise of variance σ^2, say y(k) be as sequence of received samples k $\in\{1,2....N\}$ at the signal detector, therefore a decisionrule can be stated as

$$H0....if\ \varepsilon<v \tag{3.1}$$

$$H1....if\ \varepsilon> \tag{4.1}$$

Where $\varepsilon = |\ Ey(k)|^2$ the estimated energy of the received signal and v is is chosen to be the noise variance.

Descriptive Analysis of ED

The spectrum sensing algorithms are care about sensing the spectrum holes, these holes are immigrating with time and frequency so the spectrum holes detection should be real time. Spectrum sensing have a high computational and it needs to implement it a special hardware, furthermore the detection of these holes are affected by low SNR, so the used spectrum sensing algorithm should be able to distinguish the signal power from the noise power [2]. The ED method calculate the input signal energy and compares it with a threshold energy value ,if the signal energy exceeds these threshold value the signal is presented in this particular frequency otherwise the frequency is empty [6].

The main objective is to determine the presence or the absence of a signal in a period of time. The spectrum of the input signal is computed

using eqn. 1 where X(ω) is the Fourier transform of the input signal and t2-t1 is the time period over which the input time samples are observed.

The threshold voltage is given by the expression in eqn. 3.1. The threshold voltage is related to the probability of false alarm and the noise power. The threshold energy spectrum given in the eqn. 5 can be computed by the following method.

The threshold value is defined by the classical Radar detection theory which limitation a threshold voltage [9] .For the input signal the energy spectrum of the signal and the energy spectrum of the threshold had calculated firstly to calculate the energy spectrum of the signal using energy spectrum computation mentioned in eqn. 5, secondly the energy spectrum of the threshold had to calculate by using calculation of threshold and threshold energy spectrum in eqns. 5-7. So the comparison had done between energy of the threshold and energy of the signal [9], if the energy of the threshold is greater than the energy of the signal that implies to presence of the frequency holes and the cognitive radio user can use it, otherwise there is no frequency holes at all [10].

All of that done based on Principal component analysis (PCA) which is a mathematical technique to minimize dimensionality of data.

Mathematical Model

$$\text{Energy spectrum} = |\ x(\omega)|^2\ (t_2-t_1) \tag{5}$$

$$V_{T=}\sqrt{2\sigma_n^2\ \log\left(\frac{1}{P_{fa}}\right)} \tag{6}$$

$$V_T(\omega) = \int_{t_1}^{t_2}V_T e^{-j\omega t}dt$$

$$=V_T\left[\frac{e^{-j\omega t}}{-j\omega}\right]_{t_1}^{t_2}$$

$$= j\frac{V_T}{\omega}\left[e^{-j\omega t}\right]_{t_1}^{t_2}$$

$$= \frac{V_T}{\omega}e^{j[\frac{\pi}{2}-\omega(t_2-t_1)]}\ ....$$

$$|\ V_T(\omega)|^2 = \frac{V^2}{\omega^2}(t_2-t_1)$$

$$P_{fa} = 10^{-\frac{V_T^2}{2\sigma_n^2}} \tag{7}$$

Where:

σ_n^2 =Noise Power

V_T=Threshold Voltage

ω=Frequency

P_{fa}=Probability of False Alarm

$$X(n)=[x_1...x_M\ x_{M+1}...x_{2M}...\ x_{Mn}] \tag{8}$$

$$L(n)=[x_1\ x_2...x_M]^T$$

$$L(n)=[x_{M+1}\ x_{M+2}...x_{2M}]^T$$

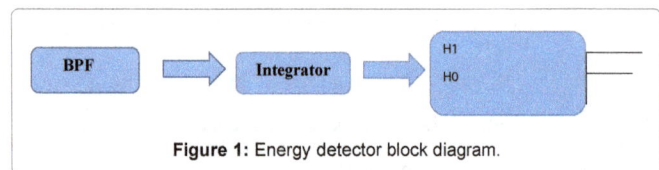

Figure 1: Energy detector block diagram.

$$LL_i = L(i) \times L^T(i) \tag{9}$$

$$W_i = \sum_{i=1}^{n} LL_i \tag{10}$$

$$W_{x.N} E_j = Y_j E_j \; ; \; j=1 \text{ to } M \tag{11}$$

$$\text{Signal power} = \frac{1}{P} \sum_{i=1}^{P} Y_i \tag{12}$$

$$\text{Noise power} = \frac{1}{M-P} \sum_{i=P+1}^{M} Y_i \tag{13}$$

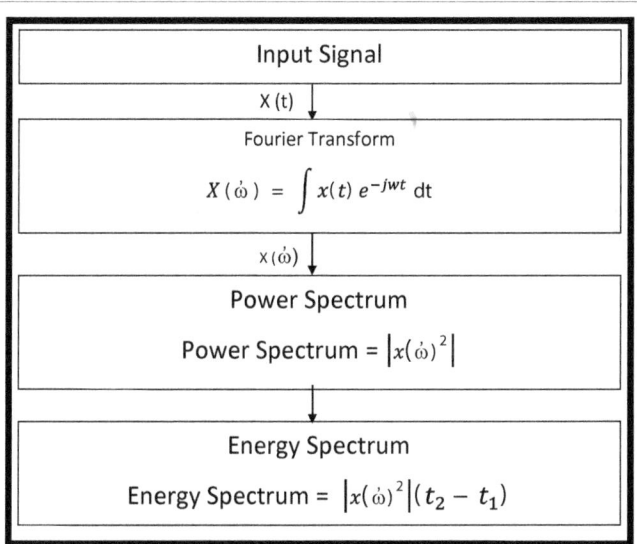

Figure 2: Flow for energy spectrum computation.

Computer Model

- Flow for energy spectrum computation (Figure 2).

- Conventional energy detection method (Figure 3).

Simulation Environment

Simulation environment is set as given in Table 1.

The simulation code had implemented Using MATLAB to find the relationship between Signals to Noise Ratio and Probability of False Alarm PFA for energy detection algorithm. Also the relationship among power of signal in Watt and signal to noise ratio in dB had implemented by using PCA technique.

Results

After execution of simulation code the results explained in form of graphs as follows:

Plot of Probability of False Alarm versus Signal to Noise Ratio for Energy Detection (Sensing Algorithm) at Vth=4 (Figure 4).

Plot of Probability of False Alarm versus Signal to Noise Ratio for Energy Detection (Sensing Algorithm) at Vth=8 (Figure 5).

Parameters	Values
Noise power (np)	0.1 , 0.2 , 0.3 , 0.4 , 0.5 , 0.6 , 0.7 , , 0.8 , 0.9 , 1
Energy of threshold (evt)	-
Input signal (xt)	Random
Energy of the signal (ext)	-
Threshold (VTh)	4,8,12,16,18

Table 1: Simulation environment.

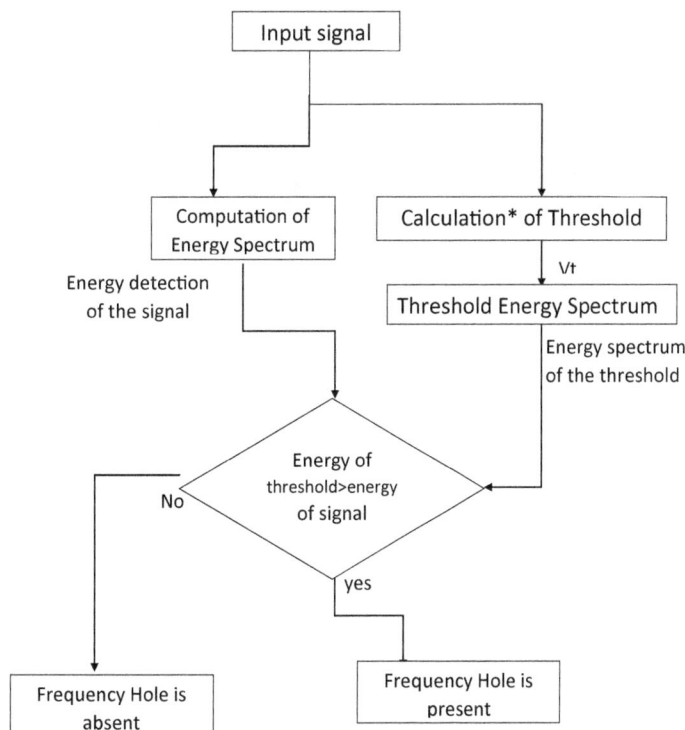

Figure 3: Conventional energy detection method.

Plot of Probability of False Alarm versus Signal to Noise Ratio for Energy Detection (Sensing Algorithm) at Vth=12 (Figure 6).

Plot of Probability of False Alarm versus Signal to Noise Ratio for Energy Detection (Sensing Algorithm) at Vth=18 (Figure 7).

Plot of Power in Watts vs. SNR in dB for Energy Detection using PCA Technique (Figures 8 and 9).

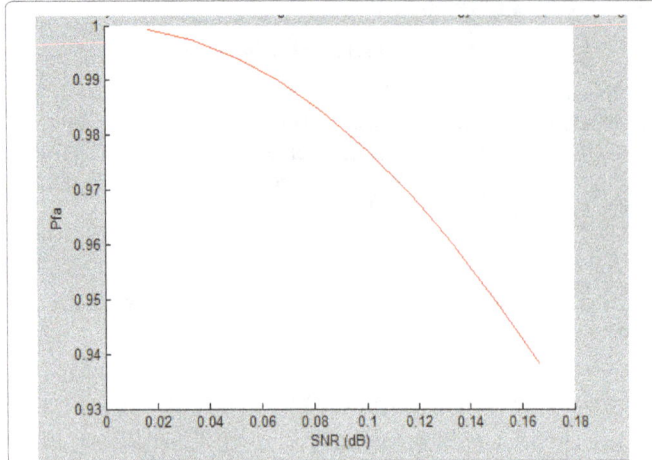

Figure 4: Plot of Probability of False Alarm versus Signal to Noise Ratio for Energy Detection (Sensing Algorithm).

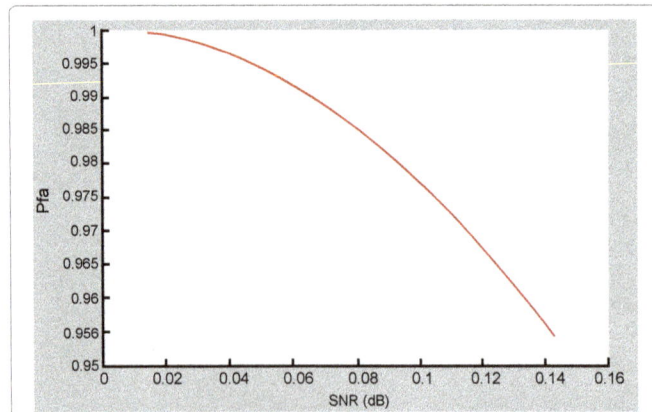

Figure 7: Plot of Probability of False Alarm versus Signal to Noise Ratio for Energy Detection (Sensing Algorithm).

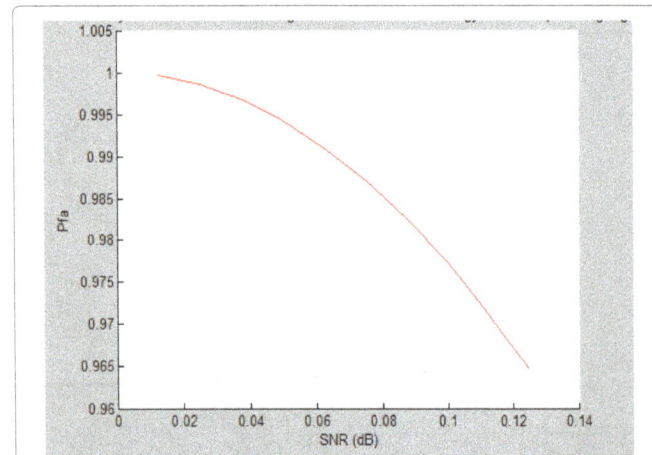

Figure 5: Plot of Probability of False Alarm versus Signal to Noise Ratio for Energy Detection (Sensing Algorithm).

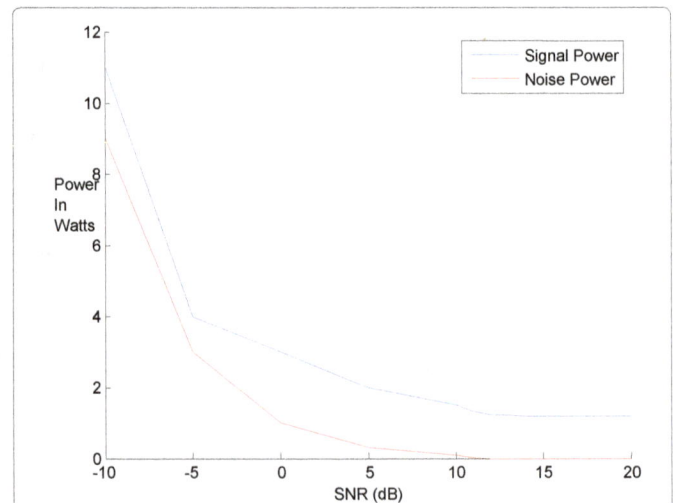

Figure 8: Plot of Power in Watts Vs SNR in dB for Energy Detection Using PCA Technique.

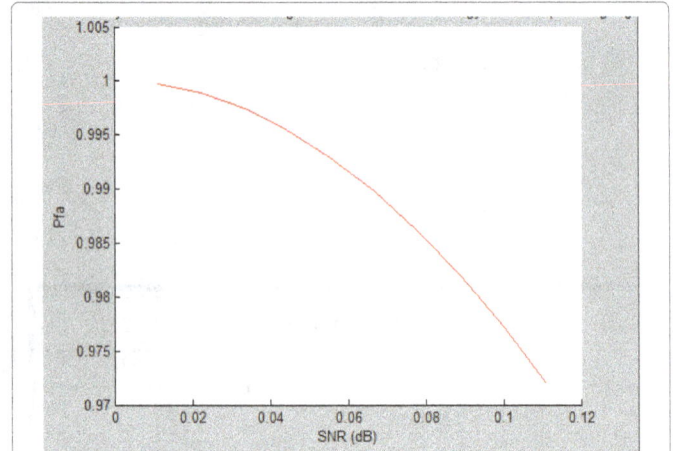

Figure 6: Plot of Probability of False Alarm versus Signal to Noise Ratio for Energy Detection (Sensing Algorithm).

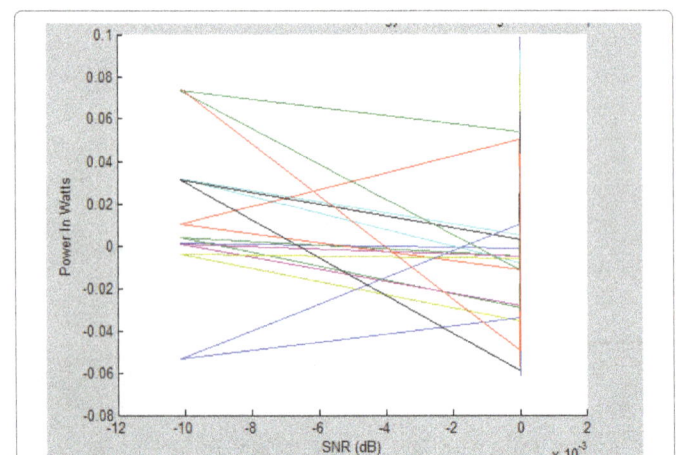

Figure 9: Plot of power in Watts Vs SNR in dB for Energy Detection Using PCA Technique by random data entry ranges.

Results Discussion

From eqns. 5-7, the signal to noise ratio and probability of false alarm are calculated and the relationship between them is plotted in Figures 4-7. From graphs the relationship between the signal to noise ratio (SNR) and probability of false alarm (Pfa) is found to be as follows:

SNR is reversely proportional to pfa, i.e., as SNR increases Pfa decreases when threshold (VTh) is constant for the varying values of noise Power from 0.1 up to 1 (step 0.1). The predefined range of threshold is between 4 and 18, the smallest values of (SNR) is found to be when using (VTh=18 which is max of Vth), then the (Pfa) is proportional to (VTh).

Conclusion

Cognitive radio technology is a modernistic technology by which the idle licensed spectrum can be used by an unlicensed user which is called cognitive radio CR. The CR technology was found to overcome the spectrum incompetence and inefficiency usage troubles. The major motivation of CR technology is current heavily underutilized of spectrum. Surly the PU have all rights to use his spectrum band so CR mustn't interfere it on that band. This paper discussed about energy detection which is one of spectrum sensing techniques, explained how to obtain the best available spectrum, to detect the spectrum hole, to estimate the optimum Threshold voltage that produce the minimum Probability of false alarm, calculate the power of signal in special range and random one and finally to enhance Energy Detection Sensing Algorithm.

References

1. Yucek T, Arslan H (2009) A survey of spectrum sensing algorithms for cognitive radio applications. IEEE Communications Surveys and Tutorials 11: 116-130.

2. Lassila P, Penttinen A (2009) Survey on performance analysis of cognitive radio networks. COMNET Department, Helsinki University of Technology, P.O. Box 3000, FIN 02015 TKK, Finland.

3. Kaur MJ, Uddin M, Verma HK (2010) Analysis of Decision Making Operation in Cognitive radio using Fuzzy Logic System. India International Journal of Computer Applications 4: 35-39.

4. Hou YT, Shi Y, Sherali HD (2008) Spectrum sharing for multi-hop networking with cognitive radios. IEEE Journal on Selected Areas in Communications 26: 146-155.

5. Verma S, Chawla M (2015) A Survey on Spectrum Mobility in Cognitive Radio Network. International Journal of Computer Applications 119: 33-36.

6. Haykin S (2005) Cognitive Radio: Brain-Empowered Wireless Communications. IEEE Journal on Selected Areas in Communications 23: 201-220.

7. Ghasemi A, Sousa ES (2005) Collaborative Spectrum Sensing for Opportunistic Access in Fading Environment. IEEE DySPAN, pp: 131-136.

8. Ghasemi A, Sousa ES (2008) Spectrum sensing in cognitive radio networks: requirements, challenges and design trade-offs. IEEE Communications Magazine 46: 32-39.

9. Matinmikko M, Mustonen M, Sarvanko H, Höyhtyä M, Hekkala A, et al. (2008) A Motivating Overview of Cognitive Radio: Foundations, Regulatory Issues and Key Concepts. First International Workshop CogART, Aalborg.

10. Chincholkar AA, Thakare CH (2014) Matlab implementation of spectrum sensing methods in cognitive radio. Global Journal of Engineering Science and Research Management, pp: 19-28.

29

Straightening Uniformly Folded Document Image

Patel HN and Modi PR*

Electronics and Communication, Engineering Department, A. D. Patel Institute of technology New Valla Bhai Vidyanagar, Gujarat, India

Abstract

Image processing is very much important area of research nowadays and is being taken interest deeply as the mobile phones and smart cameras are demands of the emerging global markets. Document image processing is the sub branch of it which includes processing the images of the paper may or may not be containing any write up or say textual contents, graphics, tables or any informative thing in form of written or printed stuff. Many a time's we people are habituated to fold the document papers in our hands and put them in pocket or somewhere. If then we want them to be scanned or apply OCR (OPTICAL CHARACTER RECOGNITION) on them then we need to capture an image of such folded documents. For image acquisition there are many sources like digital cameras, mobile phone cameras, printers etc. are available around us. When we need to scan them to convert them into electronic form (to perform OCR on them), then the problems occur because the folded documents cannot be easily scanned as they have folds and due to such folds there are huge possibilities of shadows in fold parts which can distort the quality of the document as well as this may make the readability of the document poorer. In this paper, the authors have proposed an algorithm to straighten such folded document image to get accurate OCR results in accordance with least distortion in quality of the same. The derived method helps to straighten the uniform folds digitally. So that the processed document can be applied to get improved OCR results with improved in terms of accuracy.

Keywords: Image processing; Corner points; Sub-images; Projective transform

Introduction

When images of folded documents are to be straightened, they need to be pre-processed in order to restore them otherwise scanning such folded documents is not much accurate and the quality of original data in document images get affected due to it, which is not desirable. Also content in document images may distort causing their quality degradation. Folded documents may need to be restored to perform OCR on them or to scan them. It is a very real time application because nowadays all government policies as well as private firms operate digitally. In case such documents were folded by we then that algorithm become helpful when we need to upload the scanned images to the websites suggested by them. For folded document images issues are shadow in fold parts and warping in horizontal as well as vertical direction.

Character and line segmentation techniques were used to know which part in image is warped and then the restored images after straightening them were obtained using Thin Plate Splines (TPS) [1]. Segmentation part provides the warping direction. Line segmentation involve connected component analysis and projection which gives baseline, while character segmentation involve the same along with envelop analysis to provide key points in order to get the destination image from the original warped image. TPS performs global point to point mapping to have restored image finally. Resultant images were obtained by using fine dewarping to straighten the warped images followed by coarse dewarping [2,3]. Coarse dewarping uses projective transform to map curved surface to rectangular area. After that fine dewarping was performed which is based on word detection by its pose normalization. Employs difference between horizontal and vertical projection profile and ends up with concluding that noise was greatly removed with horizontal projection profile [4]. The histogram based technique was used to obtain both the projection profiles. Vertical projection was obtained on binarized image with white background and black data, by summing the black pixels of each column and computing the energy at number of angels and based on angle at which the maximum energy was found, at that angle, image was

rotated. Energy was computed as the sum of square of black pixels for each column in document images. The horizontal projection is same but works for rows instead of columns. They have used projective transform to estimate the 3-D shape of an object when a thick bound book was scanned by scanner [5,6]. They also performs shade removal and x, y-axis corrections while scanning. In proposed work, different uniformly folded images were taken as input images [7-11] (Figure 1).

Images of folded documents were captured at different illumination conditions to justify efficiency of the proposed algorithm.

Material and Methods

The objective of the method is to straighten the document image in the way that the warping is removed and the content in that document image is not distorted after straightening it and to have better OCR

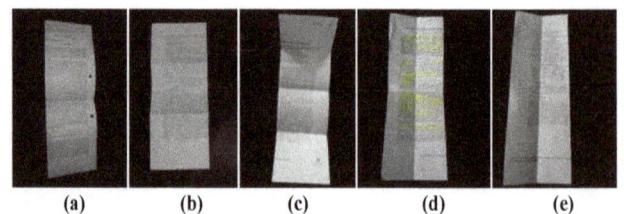

Figure 1: Examples of folded document images (a) single horizontal fold, (b) two horizontal folds, (c) three horizontal folds, (d) horizontal and vertical fold, (e) vertical fold only.

*Corresponding author: Modi PR, Electronics and Communication, Engineering Department, A. D. Patel Institute of Technology New Valla bhai Vidyanagar, Gujarat, India, E-mail: modipayal98@gmail.com

accuracy [12-16]. The conditions to work the designed method in good way were that the images should be captured with high quality (resolution) camera so that the images were not blurred because if the input image, already is blurred then the resultant straightened image will not produce good OCR accuracy and the corners of the document should not be torn because the proposed method of straightening is based on processing the corners of the document [16-20] (Figures 2-4).

Images of uniformly folded documents were captured. Below are specifications of database prepared for algorithm. Images are in RGB plane and captured by high resolution (5MP to 16 MP) and dpi (72 to

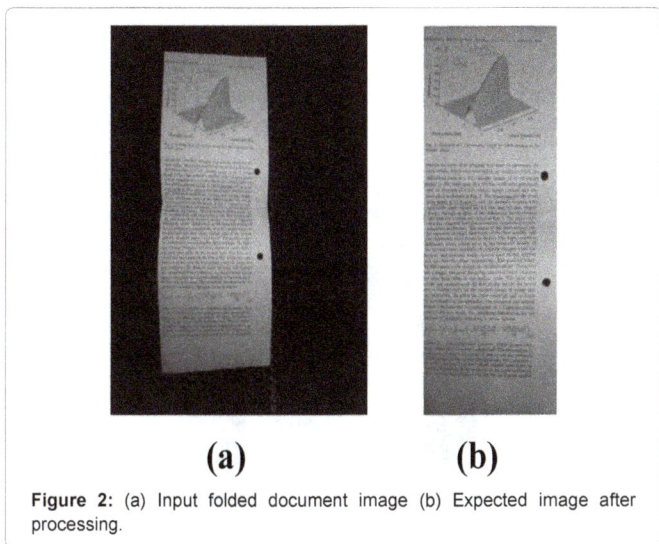

(a) **(b)**

Figure 2: (a) Input folded document image (b) Expected image after processing.

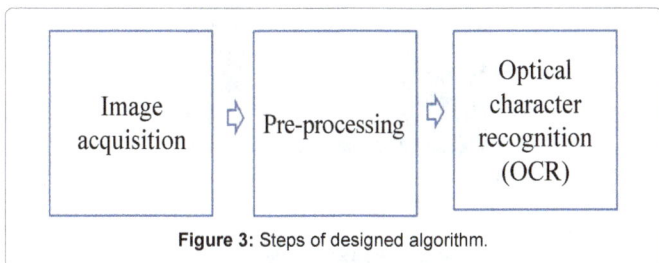

Figure 3: Steps of designed algorithm.

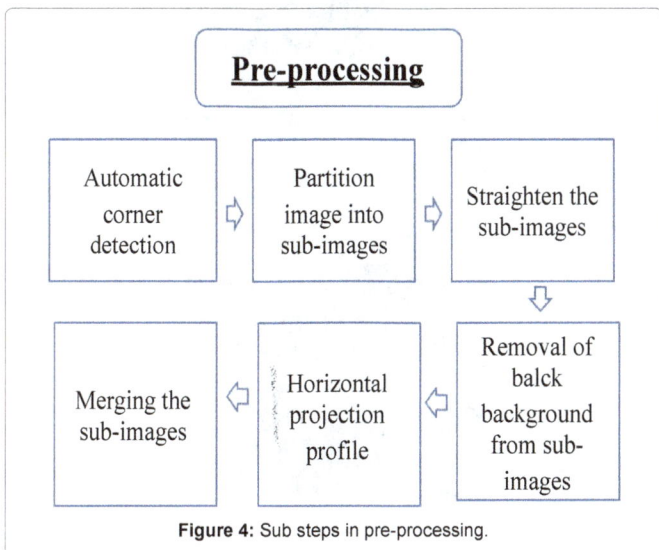

Figure 4: Sub steps in pre-processing.

96 dpi) cameras. Horizontal folded images, vertical folded images and images with both types of folds were captured. Images were taken in different lighting conditions as well as at different times to justify the algorithm. Images should be non-blurred. Corners should not be torn. Pre-processing is a second step in which entire straightening procedure were carried out to make the straightened image which is ready to apply optical character recognition on it. First of all the corners were obtained where there are folds, and based on that input images are divided into sub-images. All these sub-images were processed individually and straightened by applying projective transformation on them. The next pre-processing step was to make sure whether there were characters on fold line by horizontal projection profile for every single sub-image. If yes then the fold line was shifted downward where there were no characters. Sub-images were then resized and merged. All the sub steps in pre-processing are explained in detail as follows.

The folded input images have number of corner points according to type of folds like left up, right up, right down and left down corner locations of folded parts.

For example Figure 5 has horizontal three folds and ten corner points for four sections of the same image. These sections are based on folds. If it is horizontal one fold image it has two such sections, if horizontal two fold image, then three such sections. Further every section has four corners as shown which are upper left corner (LU), upper right corner (RU), lower right corner (RD), lower left corner (LD). Top most sections LD; RD is LU and RU for second section exactly below it respectively. For corner detection edge of folded image needs to be extracted. For that first of all images were converted to binary. When it is done there may be noise or generation of holes due to variation in lightening conditions, which is shown in Figure 6. These holes or noises are harmful because it too would be considered while finding corner points as it's explained in following part and must be removed. Holes in images are filled but still noise won't be totally removed. There is pepper noise in background which will disturb edge detector in its work. So image is filtered by two dimensional order statistic filter which replaces each element in image by the minimum of its north, east, south, and west neighbors resulting in noise-free image in most cases. The resultant images after all these steps are as below.

$$\text{Where, Filter} = \begin{bmatrix} 0 & 1 & 0 \\ 1 & 0 & 1 \\ 0 & 1 & 0 \end{bmatrix}$$

Filtered image is eroded twice in order to remove text or any white

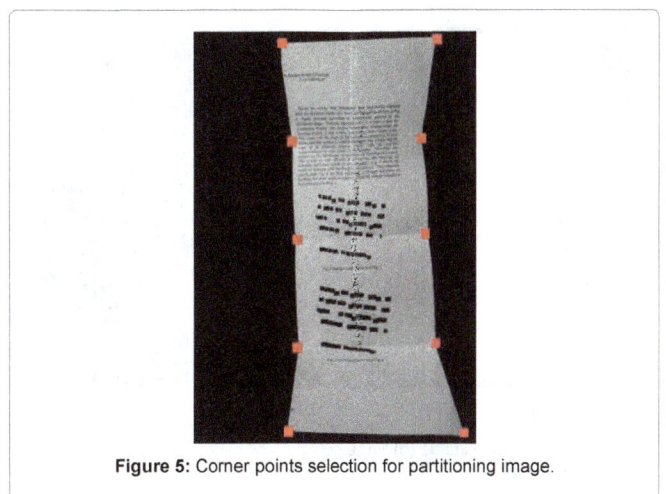

Figure 5: Corner points selection for partitioning image.

dots still remaining outside image in black background because these letters too will have edges which are not required and provide wrong edge information for corner detection [21,22]. For erosion square structuring element is used with all one and size 3*3. Erosion contracts an object in image based on structuring element.

$$im\Theta se = \left\{ x \in z^2 \mid (se)_x \subseteq im \right\}....\qquad(1)$$

Where,

Se=Structuring element=$\begin{bmatrix} 1 & 1 & 1 \\ 1 & 1 & 1 \\ 1 & 1 & 1 \end{bmatrix}$

im=image.

Finally for image is ready for edge detection. Canny edge detector is used for that as it provides best results than any other methods. The Canny method finds edges by looking for local maxima of computed gradient of image which is calculated by taking the derivative of a Gaussian filter. The canny method uses two thresholds for detection of strong as well as weak edges, and includes the weak edges in the output only if they are connected to strong edges. So this method is not "fooled" by noise, and more likely to detect true weak edges. Thresholds are selected in such a way that ratio of low to high threshold in 0.4 depending on intensity of image. The twice eroded and edge detected images are shown below in Figure 7.

Then edge is detected and corners of edge of an image are to be found. It's having horizontal three fold document image, it is not possible to find all the ten corners at the same time. So first the

concentration is on finding prime corners i.e. top most left, right and fourth section's bottom left, right corners. For that the steps as below are followed: Construct an array to store all points of edge. Euclidean distance between (1,1) and array; (1,n) and array; (m,n) and array; (m,1) and array is found. For every points where there is a minimum Euclidean distances are the prime corners LU, RU, RD and LD. Size (im)=m*n. Where,

m=total number of rows, n=total number of columns (Figure 8).

Still six corners are undetected. For finding them column difference is useful. Left side: columns of LU up to LD are subtracted from 1 (original image's first column). Right side: column of RU up to RD are subtracted from n (original image's last column) (Figure 9).

Where there are sharp changes in the plot of distance versus row number those points are considered as sub-corners. Peaks suggesting sharp changes are so many as image is not smooth in the plot. If polynomial fitting of order 1(i.e. line y=mx+b: m-slope, b-intercept) is selected then it provides 2 points: first slope and second intercept. The first parameter is useful as where there is a sudden sign change

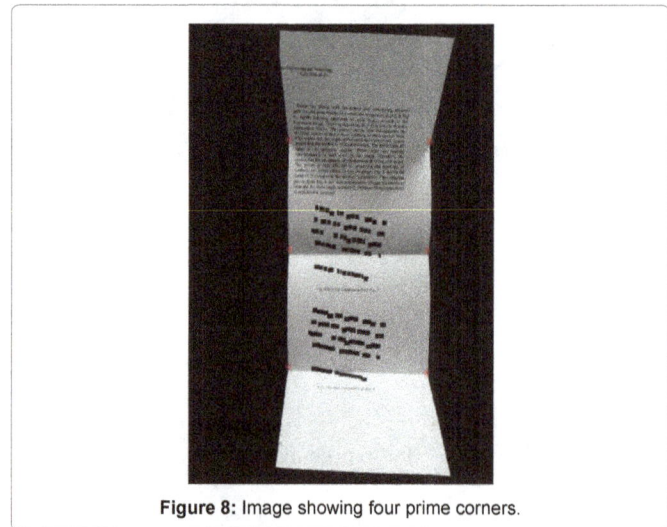

Figure 8: Image showing four prime corners.

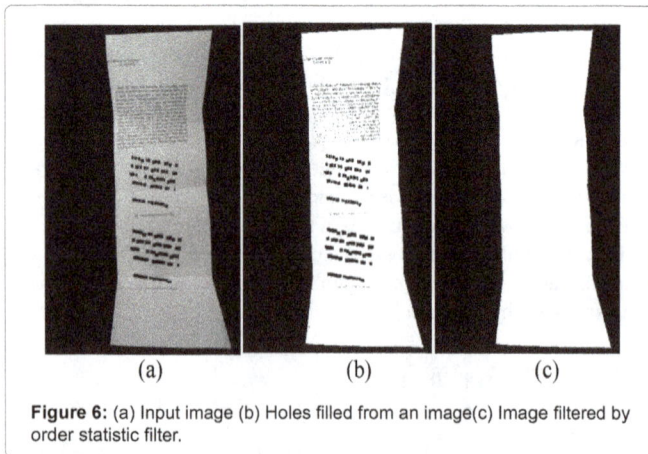

Figure 6: (a) Input image (b) Holes filled from an image(c) Image filtered by order statistic filter.

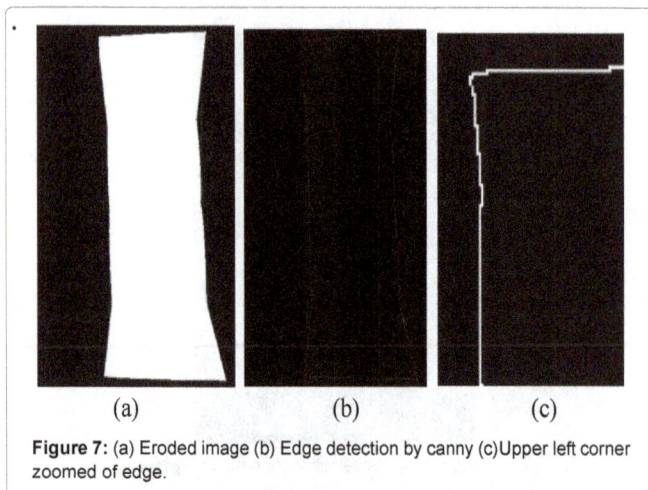

Figure 7: (a) Eroded image (b) Edge detection by canny (c)Upper left corner zoomed of edge.

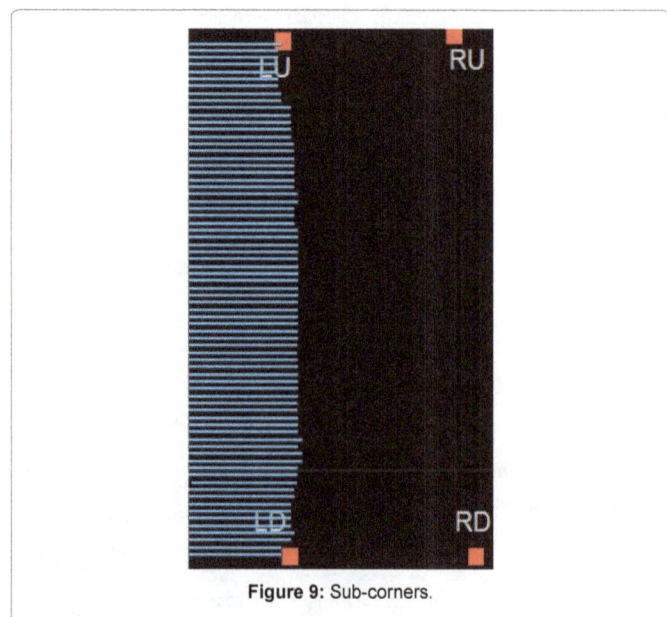

Figure 9: Sub-corners.

after many regular signed values those respective rows can be taken as peak. Plots are not smooth so Gaussian filtering is applied with frame size [800,800] and standard deviation 80. Window size for polynomial fitting is selected 50. This method provides true sub-corners rows and with the help of same, respective column too can be found and is accurate in 72.63% folded images of every type when tested on 365 images with different type of folds. Error rate in corners is maximum + or –100 (Figure 10).

In this algorithm, we have separated the folded parts in a document image by using corner points for e.g. if it is horizontal one fold image, then it is separated into two parts, upper part above the fold and lower part below the fold. Two horizontal folded images produces three sub-images, three horizontal fold images can be separated into four sub-images (Figure 11). The same way it is applicable to all the uniform folds described above. Same way vertical one fold image can be separated into two sub-images, one is left of the fold and other is right of the fold. Image with vertical and horizontal fold can be divided into four sub-images. Initially these corner points are obtained manually as upper left, upper right, down right and down left corners. The conditions for validity of this algorithm are that the images should be captured

the way so they are not blurred and the corners are not torn. Here the entire algorithm is explained for horizontal three folds image.

Then we have used the projective transform in order to straighten these individual sub-images obtained in this. Image transform there are parts of folded image or four sub-images and they are needed to straighten individually. For that projective transform is used. In such transformation technique, the spatial transformation structure is required. The logical thing in this entire algorithm is to design the general spatial transformation structure that can straighten all images for our case. Creating spatial transformation structure requires input points and base points which work as control points. For input points (x,y) co-ordinates are set such that all the corner points of horizontal lines are having same y locations and vertical lines are having same x locations. While base points which work as control points are obtained by setting the left, up, down and right parameters according to the mean of them. For e.g. Left is the mean of upper left and lower left's x co-ordinate. Right is the mean of upper right and lower right's x co-ordinate. Up is the mean of upper left and upper right's y co-ordinate. Down is the mean of lower left and lower right's y co-ordinate as shown in algorithm described below. Now the inter-section of Left and Up gives the final upper left point of resultant image, the inter-section of Up and Right gives the final upper right point of the resultant image and same way the other two points of resultant image are obtained. Finally applying the spatial transformation structure on every individual sub-images of horizontal three fold image we will get the projective transformed straightened images in Figure 12.

All the sub-images are straightened but they are containing black background which is not required. So, the sub-images are cropped such way that the black background is removed and we are left with the sub-images without background (Figure 13).

When all the sub-images of original folded documents are straightened and background is removed too, then the final step is to merge them to get the entire image which is straightened version of the original image. Before merging images, horizontal projection profile is used to check whether there are characters on fold line on gray scale image. Because if there are characters on fold line then while resizing

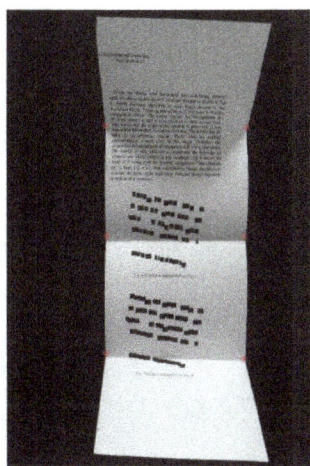

Figure 10: Sub-corners detected automatically.

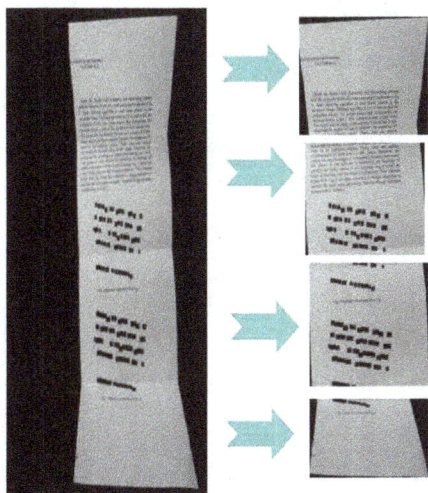

Figure 11: Horizontal three folds image and sub-images after partitioning it.

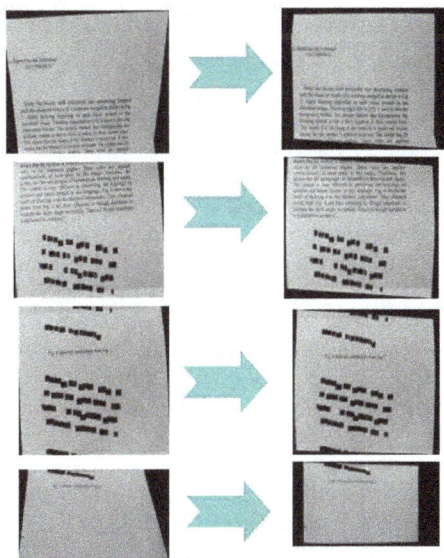

Figure 12: Sub-images and the images projectively transformed.

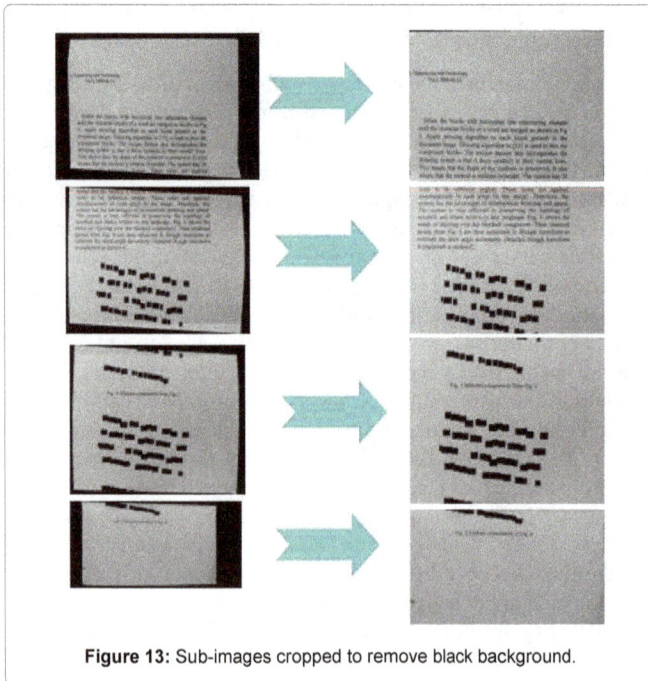

Figure 13: Sub-images cropped to remove black background.

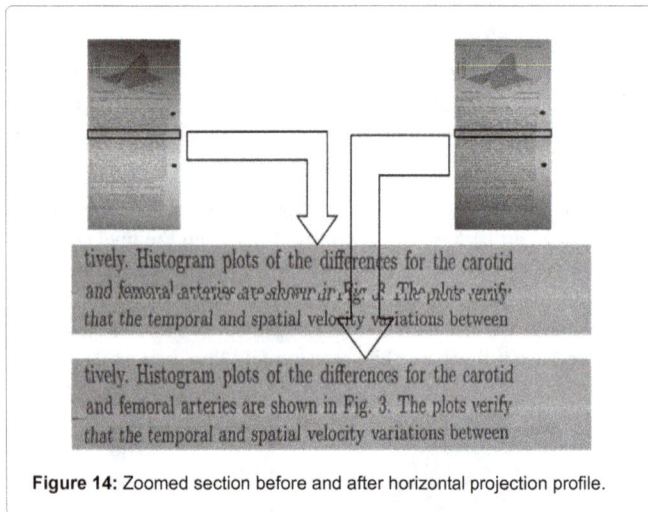

Figure 14: Zoomed section before and after horizontal projection profile.

before merging these characters will be shifted and not well-aligned resulting in reduced OCR accuracy at that line (fold line) also making it not readable. In projection profile every section of images except top section are processed. Sum of pixels in columns are carried out for each row after converting image to grey scale. If the value of sum is maximum then it means there is white part meaning no character on fold line. So no need to shift fold line. If not then initial 25 lines are checked where value of sum is maximum i.e. where the white line with no characters and first row is shifted upward making white line found by projection profile the first row. Then images are resized. Resizing procedure is done by providing scale in form of number of columns and number of rows of one of the sub-image which computes the number of rows or columns automatically in order to preserve the image aspect ratio. The sub-images are resized first the way that all of them are having same number of columns. Then they can be easily merged which is the straighten version of the input horizontal three folds images in Figures 14 and 15.

The same method can be applied for straightening any uniform folded images as suggested in Figure 1.

OCR helps to convert physical document in electrical or say soft form. When OCR is applied on the resultant images which are straightened after procedure explained above the results were quite promising. For OCR, ABBYY Fine Reader 12 software is used which is trustworthy and reliable. It includes an analysis of OCR results on original folded image, output straightened image without projection profile and with projection profile. When OCR performed on original image accuracy of readability on data in it was 81.25% which is not so bad but time consumed while performing the same was quite high. Whereas the second case is of OCR on straightened image but with characters on fold line and no horizontal projection profile, found resultant accuracy has fallen to 12.50 which is very poor. When the OCR performed on output straightened images with shifted fold line by horizontal projection profile, OCR results rise to 93.75%, which to in less consumption of time compared to both the cases before for same number of images (1.39 sec) (Table 1).

Algorithm

This chapter involves all the steps explained above in summarized form and more discriminative study of the same. The algorithm will give clear picture of what is the goal of the system and how to achieve it.

1) Partition the image into sub-images according to folds by using corner points of original folded image based on concept of Euclidean distance for prime corners and slope by polynomial fitting of column difference for sub-corners as explained above.

2) All the sub-images are needed to be processed individually. Each sub Image has 4 corners: upper left LU, down left LD, upper right RU and downright RD.

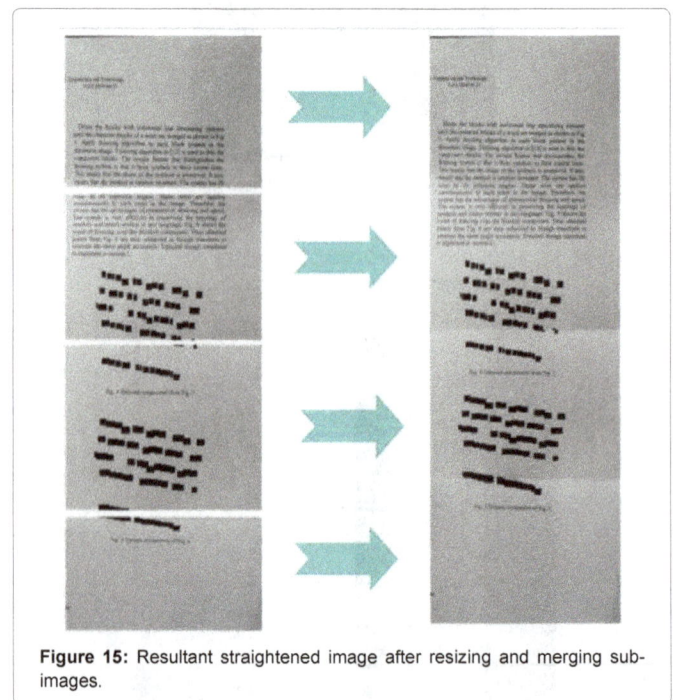

Figure 15: Resultant straightened image after resizing and merging sub-images.

Images	Character on fold line(%)	Character on fold line(%)
Folded document images	12.50	93.75

Table 1: Percentages on character on Fold line.

3) The sub-images are straightened using projective Transform. Projective transform is explained briefly in section below. For projective transform, spatial transformation structure needs to be designed using input points and base points. Let the corner points of sub-image be C1=[LU(1) LU(2), RU(1) RU(2), RD(1) RD(2), LD(1) LD(1)]. To get input points C2=[LU1, LD1, RU1, RD1], which are nothing but the documents corner points by not considering black background, for designing spatial transformation structure following mechanism is applied:

LU(1)=upper left corner's column number,

LU(2)=upper left corner's row number,

LD(1)=down left corner's column number,

LD(2)=down left corner's row number,

RU(1)=upper right corner's column number,

RU(2)=upper right corner's row number,

RD(1)=down right corner's column number,

RD(2)=down right corner's row number.

LU1's column value depends on whether LU's column value is bigger or smaller than LD's column value. If it is smaller, then LU1(1) is having 1 making it first column otherwise it is equals to 1 more than the difference between LU and LD's column number to make it smaller. And LU1's row can be obtained the same way but the comparison is made between LU and RU's row. Such type of designing will help the sub-images to make the straightening easier. Do the same for RU1, RD1 and LD1 to get the column and row co-ordinates of them.

4) Corner points detected are now:

C2=[LU1(1) LU1(2), RU1(1) RU1(2), RD1(1) RD1(2), LD1(1) LD1(1)].

5) Base points, which are to work as control points in constructing spatial transformation structure, are CR=[L U, R U, R D, L D]. They are obtained by taking mean of respective components of C2. For example L is the mean of LU1 and LD1's column value, U is the mean of LU1 and RU1's row value, R is the mean of RU1 and RD1's column value and D is the mean of RD1 and LD1's row value. Finally C2 and CR work as input and control points respectively to construct spatial transformation structure. Then this projective transformation structure is applied on sub-images obtained. That way inter-section of L and U points, U and R points, R and D points, D and L points will construct the straightened image. It will straighten the sub-images.

6) Resultant straightened sub-images are cropped to remove black background which is not of the interest to get the straightened sub-images.

7) Our ultimate goal is to get the image which is straightened version of folded image, not the sub-images. So we need to merge the straightened sub-images obtained in the previous step. But before that the lower sub-images rows are applied horizontal projection profile to check whether there are characters on fold line. If so, then fold line is shifted down wards and at the same value upper sub-images shifted down ward. Also those sub-images are of different sizes. So they need to be resized the way that all the sub-images are having same columns. We have nothing to do with the rows. Finally they are merged at the end to get image which is straightened.

Experimental Results

The same method explained above can straighten any uniformly folded images. But if the images captured are blurred then it will not provide good OCR results. The experimental results for horizontal one fold image, horizontal two folds image, horizontal three folds image, horizontal and vertical fold image, vertical one fold image are shown below. All the images were captured by good quality cameras and all images are JPEG images with colour representation RGB (Figure 16).

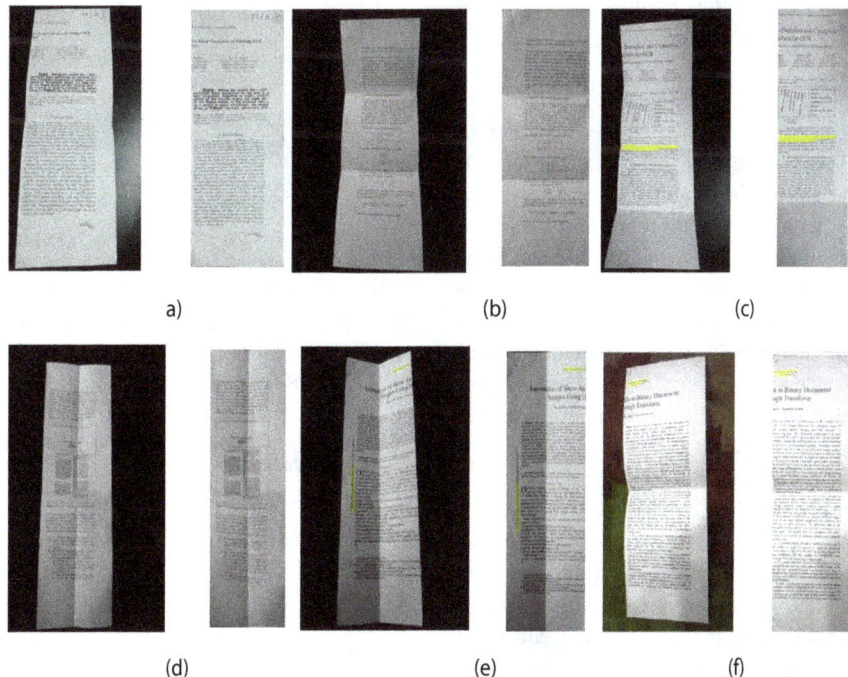

Figure 16: Input images and resultant images for (a) single horizontal fold, (b) two horizontal folds, (c) horizontal three folds, (d) horizontal and vertical fold, (e) vertical fold only, (f) single horizontal fold image with colored background.

In Figure 16(a) shows the document image with the one horizontal fold at the centre of the document. The image is captured in brighter lighting conditions and on black surface. After straightening the resultant image is shown.

In Figure 16(b) is an input image with two horizontal folds so the warping in middle portion is seen in image. The image is captured at darker lighting conditions and on black surface. Above the fold-lines shadow would be there. But resultant image shows that it does not affect the resultant straightened image.

In Figure 16(c) is a horizontal three folds image of document taken at brighter lighting conditions than in (b). It is much curvy than above two examples because it has much folds and therefore much warping too. The resultant image is the straightened form of input image. For straightening it needs to be divided into four sub-images by using corner points. Then they are straightened by using projective transformation. So it requires processing four sub-images individually and then merging them as explained in algorithm.

In Figure 16(d) shows the document image with horizontal one fold and vertical one fold. Here to the four sub-images are obtained but the corner points differ then the previous case of horizontal three folds image. Because horizontal fold is partitioned to provide two sub-images: one above the fold and one below the fold. Then after upper and lower both sub-images are again partitioned as left of the vertical fold and right of the vertical fold. Then all sub-images are processed the same way and resultant straightened image is obtained.

In Figure 16(e) is an input image with vertical one fold which is processed in similar way as the horizontal one fold image. The image is partitioned into two sub-images; the only difference is the corner point selection. It has so much warping in horizontal direction which is hard to read the write up in it. But after straightening the warping is removed and it is good to read then.

In Figure 16(f) horizontal one fold document image with coloured background is there with shadow at lower fold portion. The resultant is an output straightened image.

Conclusion

The derived algorithm gives very good quality of content in resultant straightened document image even in the folded parts. It can be proved from the image above. The proposed algorithm has less complexity and good speed. The algorithm has robustness because it can work well with images in all illumination conditions and shadow conditions background independently. OCR results rise to 93.75% for content containing fold line which would be shifted a little when required while straightening than that 12.5% at fold line of input folded images when there were characters on fold line and fold line was not shifted by horizontal projection profile, while folded input image had OCR accuracy of readability 81.25%. So all in all, just by giving the path we can have straight image from the folded document image.

Future Scope

The algorithm can be optimized for straightening of document images with non-uniform folds. Moreover, the algorithm can be optimized so that when straightened the fold is equalized and document image is perfectly straight and fold is not visible in resultant image. If background is white then automatic corner detection is challenging and can be considered as future scope.

References

1. Zhang Y, Liua C, Dinga X, Zoub Y (2008) Arbitrary Warped Document Image Restoration Based on Segmentation and Thin-Plate Splines. 19th international conference on pattern recognition, ICPR, Tampa, FL.

2. Stamatopoulos N, Gatos B, Pratikakis I, Perantonis S (2008) A two-step dewarping of camera document images. Eighth IAPR workshop on document analysis systems, DAS 08: 209-216.

3. Stamatopoulos N, Gatos B, Pratikakis I, Perantonis S (2011) Goal-Oriented Rectification of Camera-Based Document Images. IEEE Transactions on image processing 20: 910-920.

4. Jain B, Borah MA (2014) AComparison paper on skew detection of scanned document images based on horizontal and vertical projection profile analysis. International journal of scientific and research publications 4: 1-4.

5. Tiwari L, Kumar B, Patnaik T (2014) Robust Camera Captured Image Mosaicking for Document Digitization and OCR Processing. International Conference on Information: 100-105.

6. Zhang Z, Tan C, Fan L (2004) Estimation of 3D Shape of Warped Document Surface for Image Restoration. 17th international conference on Pattern Recognition 1 Date of Conference: 486-489.

7. Negi A (2001) Document Image Processing and Optical Character Recognition Systems for Indic Scripts. School of Computer and Information Sciences University of Hyderabad.

8. Zhang L, Tan L (2006) Restoring warped document images using shape-from-shading and surface interpolation. In Proc 18th International Conference on Pattern Recognition 1: 642-645.

9. Wada T, Ukida H, Matsuyama T (1997) Shape from Shading with Inter reflections Under a Proximal Light Source: Distortion- Free Copying of an unfolded Book. International Journal of Computer Vision 24: 125-135.

10. Brown M, Seales W (2004) Image restoration of arbitrarily warped documents. IEEE Trans On Pattern Analysis and Machine Intelligence 26: 1295-1306.

11. Zhang Z, Tan C (2003) Correcting document image warping based on regression of curved text lines. International Conference on Document Analysis and Recognition, Edinburgh, Scotland: 589-593.

12. Wahl F, Wong K, Casey R (1982) Block segmentation and text extraction in mixed text/image documents. Comput Graph. Image Process 20: 375-390.

13. Okun O, Pietikäinen M, Sauvola J (1999) Document skew estimation without angle range restriction. International Journal on Document Analysis and Recognition 2: 132-144.

14. Tang Y, Suen C (1993) Image Transformation Approach to Nonlinear Shape Restoration. IEEE Trans Systems Man and Cybernetics 23: 155-171.

15. Lavaille O, Molines X, Angella F, Baylou P (2001) Active Contours Network to Straighten Distorted Text Lines. Proc Int'l Conf Image Processing: 1074-1077.

16. Weng Y, Zhu Q (1996) Nonlinear Shape Restoration for Document Images. Proc IEEE CS Conf Computer Vision and Pattern Recognition: 568-573.

17. Tsoi YC, Brown MS (2004) Geometric and Shading Correction for Images of Printed Materials-A Unified Approach Using Boundary. Proc IEEE CS Conf Computer Vision and Pattern Recognition 1: 240-246.

18. Burger W, Burge M (2008) Digital Image Processing. 1st (edn) An Algorithmic introduction using Java. Springer.

19. Gonzalez R, Woods R (2012) Digital Image Processing. 3rd (edn) Pearson Education.

20. Jayaraman S, Esakkirajan S, Veerakumar T (2011) Digital Image Processing. 1st (ed) Tata McGraw Hill.

21. http://www.img.cs.titech.ac.jp/~akbari/pmwiki/uploads/Site/Felipe-rep.pdf

22. http://math.stackexchange.com/questions/96662/augmented-reality-transformation-matrix-optimization

30

A New Energy Efficient Clustering-based Protocol for Heterogeneous Wireless Sensor Networks

Nuray AT* and Daraghma SM

Department of Electrical and Electronics Engineering, Anadolu University, Turkey

Abstract

Energy efficiency is one of the most important design goals for wireless sensor networks (WSNs). To this effect, clustering is mostly used to prolong the lifetime of WSNs. In this study, we propose a new energy efficient (EE) clustering-based protocol for single-hop, heterogeneous WSNs. The proposed protocol uses channel state information (CSI) in the selection process of Cluster Heads (CHs). It is shown through simulations in MATLAB that the proposed protocol has 1.62 to 1.89 times better stability period than that of a well-known protocols including LEACH, DEEC, and SEP.

Keywords: Wireless sensor network, Sensor node, Clustering, Routing protocols, Residual energy, Energy efficient routing, Network lifetime

Introduction

Sensor networks can contain hundreds to thousands sensing nodes. It is desirable to make these nodes as cheap and energy-efficient as possible and rely on their large numbers to obtain high quality results. Wireless Sensor Networks (WSNs) are used for a variety of tasks including detection, localization and tracking objects of interest. In [1], an accurate estimate of a source location is obtained by using energy readings of sensors.

One challenge faced by WSNs is the power management issue. As sensors have limited energy, it needs to be used wisely so that a WSN has a long lifetime. Clustering [2-4] is one of the successful techniques to improve the lifetime of a WSN. In this approach, a network field is divided into sub regions, called clusters. Each cluster has a Cluster Head (CH) which is responsible for collecting data from member nodes within its cluster and transmitting the data to the Base Station (BS). However, this technique comes with its own problems such as deciding the number of clusters and CH to member ratio, selection and rotation process of CH(s).

Most of the analytical results in the literature assume that all sensor nodes in the network have equal energy. These networks are called homogeneous networks. In some cases on the other hand, some nodes in the network have different energies. For example, as the lifetime of a sensor network is limited there is a need to re-energize the network by adding more nodes. These nodes will be equipped with more energy than nodes that are already in use which will create heterogeneity in the network resulting heterogeneous networks [5].

The communications between entities on a network are governed by protocols. Network protocols must be designed to achieve fault tolerance in the presence of individual node failure(s) and to minimize energy consumption. Moreover, since the limited wireless channel bandwidth is shared among all sensors in the network, routing protocols should be able to perform a local collaboration to reduce bandwidth requirements.

A protocol called LEACH (Low-Energy Adaptive Clustering Hierarchy) was developed in [6,7]. It is a clustering-based protocol that minimizes energy dissipation in a sensor network. It is shown that LEACH outperforms classical clustering algorithms since it uses adaptive clusters, rotates CHs, and allows energy requirements of the system to be distributed among all sensors in the network. But then, when aCH dies in LEACH that cluster will become useless since data gathered at the CH will never reach to the base station.

In this study, we assume that the BS is not energy limited, the dimensions of the field and the coordinates of the BS are known. We propose a new clustering-based energy-efficient (EE) protocol for single-hop, heterogeneous WSNs. In EE-Heterogeneous LEACH, CHs are selected by using weighted probabilities. These weighted probabilities are evaluated based on the ratio between residual energy and the best channel of each node and average energy of the network. The rotating epoch (time interval) for each node is different according to its initial and residual energy. Nodes with high initial and residual energy will be more likely to become CHs per round per epoch. CHs collect data from member nodes in their respective clusters, aggregate the received data and send it to the BS using single-hop communication. Simulation results show that the proposed protocol extends network lifetime and improves energy consumption compared to other well-known protocols including LEACH, DEEC, and SEP.

The outline of the paper is as follows. In Section 2, the previous work on the subject is summarized. The proposed protocol is presented in Section 3 and its performance is evaluated through simulations via MATLAB in Section 4. Some concluding remarks are given in Section 5.

Previous Work

Single-hop dynamic clustering protocols like LEACH [8] suffer from the problem of cluster head rotation overhead in each round. LEACH elects cluster heads based on randomly generated value between 0 and 1. If this randomly generated value is less than the threshold value T(s) is given in (10) then the node become cluster head for the current round. Advanced LEACH [9] tries to improve the performance of LEACH by selecting best suited node for cluster

---

*Corresponding author: Nuray AT, Department of Electrical and Electronics Engineering, Anadolu University, Turkey, E-mail: nat@anadolu.edu.tr

head and improves the threshold equation of LEACH by introducing two terms: General Probability (GP) and Current State Probability (CSP). In ALEACH, nodes make autonomous decision without any central intervention taking into account residual energy. In reference [10], an energy-efficient hierarchical clustering algorithm (EEHCA) is proposed for WSNs. The EEHCA adopts a new method for CH selection along with the concept of backup CHs to improve the performance of the WSNs. Furthermore, when the CHs have finished the data aggregation, the head clusters transmit aggregated data to the BS node by a multi-hop communication approach. It is shown that the EEHCA achieves a good performance in terms of network life time by minimizing energy consumption for communication and balancing the energy load among all the nodes. Re-cluster-LEACH [11] protocol is based on nodes density, which considers the density of nodes inside the cluster for Cluster Head formation. LEACH-F [12] is an algorithm in which the number of clusters will be fixed throughout the network life time and the cluster heads are rotated within its clusters. Steady state phase of LEACH-F is identical to that of LEACH.LEACH-B [13] is a decentralized algorithm in which a sensor node only knows about its own position and position of final receiver and not the position of all sensor nodes. E-LEACH [14] provides improvement in selection of cluster heads of LEACH protocol. It uses residual energy of the nodes as the main factor whether these sensor nodes turn into the cluster head or not in the next round. LEACH-HPR [15] is an energy efficient cluster head election method and using the improved Prim algorithm to construct an inter-cluster routing in the heterogeneous WSN.

System Model

In this study, we use a radio energy dissipation model given in Figure 1. Here, L bit data packets are transmitted to a receiver (Rx) located at a distance d from the transmitter (Tx). Eelec is the amount of energy needed in Tx or Rx hardware to send or receive data. Due to path loss and multipath fading phenomena that occur in wireless channels, Tx is equipped with an amplifier. The amplifier has a gain of where denotes the path loss exponent. Note that the value of the path loss exponent is between 2 and 4 in general.

To transmit L-bit message to a distance d:

$$E_{Tx}(L,d) = \begin{cases} LE_{elec} + L\varepsilon_{fs}d^2, & d \leq d_0 \\ LE_{elec} + L\varepsilon_{mp}d^4, & d \geq d_0 \end{cases} \quad (1)$$

where is the amplifier energy per bit per square meter (m²) when free space model is used for the channel and is the amplifier energy per bit per m4 when multipath propagation model is used. The threshold distance d0 in (1) is given by

$$d_0 = \sqrt{\frac{\varepsilon_{fs}}{\varepsilon_{mp}}}$$

and is set to 87 m in this study. Similarly, at the Rx side:

$$E_{Rx} = LE_{elec} \quad (2)$$

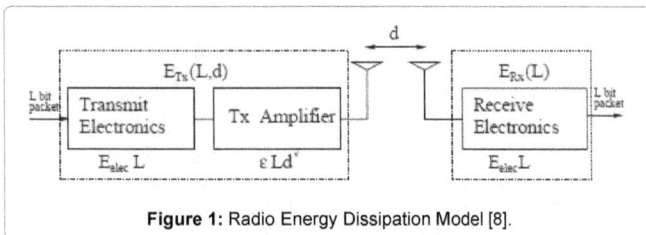

Figure 1: Radio Energy Dissipation Model [8].

Since transmitting a message is a costly operation in wireless channels, protocols used in WSNs should try to minimize not only the transmit distances but also the number of transmit and receive operations for each message.

Optimal clustering

We assume that N nodes are uniformly distributed over a square field (D x D). The square field has an area of D2 square meters and the BS is located at the center of the field for simplicity. The field is divided into Ksub regions, clusters. For each cluster, one node is assigned as the cluster head. During transmission, each non-cluster head (non-CH) node sends L bit data to the CH node within its cluster. Thus, the energy used by a non-CH node is [16]:

$$E_{non-CH} = LE_{elec} + L\varepsilon_{fs}d_{CH}^2 \quad (3)$$

where is the average distance between a cluster member and its corresponding CH [17]

$$d_{CH}^2 = \frac{D^2}{2\pi K}$$

Similarly, the energy dissipated in aCH is given by

$$E_{CH} = \left(\frac{N}{K} - 1\right)LE_{elec} + \frac{N}{K}LE_{DA} + LE_{elec} + L\varepsilon_{fs}d_{BS}^2 \quad (4)$$

where is the data aggregation processing cost and is the average distance between a CH and the BS given by [16]:

$$d_{BS} = 0.765\frac{D}{2}$$

The total energy dissipated during one round is

$$E_{round} = L(2NE_{elec} + NE_{DA} + \varepsilon_{fs}(Kd_{BS}^2 + Nd_{CH}^2)) \quad (5)$$

By differentiating with respect to and equating to zero, the optimal number of clusters is found to be [16,17]:

$$K_{opt} = \sqrt{\frac{N}{2\pi}}\frac{D}{d_{BS}} = \sqrt{\frac{N}{2\pi}}\frac{2}{0.765} \quad (6)$$

If a significant percentage of nodes are farther away from the BS (a distance greater than d0), then the optimal number of clusters is given by [16]

$$K_{opt} = \sqrt{\frac{N}{2\pi}}\sqrt{\frac{\varepsilon_{fs}D}{\varepsilon_{mp}d_{BS}^2}} \quad (7)$$

The optimal probability of a node being selected as a CH is computed as [16]

$$p_{opt} = \frac{K_{opt}}{N} = \frac{1}{0.765}\sqrt{\frac{2}{N\pi}}\sqrt{\frac{\varepsilon_{fs}}{\varepsilon_{mp}}} \quad (8)$$

The quantity plays an important role in the operations of WSNs. If the clusters are not constructed in an optimal way, the total energy consumed during a round will increase in a nonlinear fashion [17,18].

Proposed protocol

Like LEACH, our proposed protocol also consists of two phases (1) Setup Phase and (2) Steady Phase.

During the Setup Phase, the BS broadcasts a message at a certain power containing its identification information. This message makes each node aware of the BS. The BS decides the number of and optimal size of clusters depending upon the size of a network area and density of nodes. The BS then sends control packets to each node informing about the protocol to be used. These control packets consist of all necessary information required for steady state working of the protocol including threshold energy value for a CH change, TDMA slots for intra-cluster communication, CDMA code for communication with the BS along with node identities. This makes each node aware of the other members of its cluster, current round CH, CH rotation sequence, sleep and wake-up patterns to reduce collision and to save energy. Threshold energy is the least amount of energy to have for a CH node [19].

During the Steady Phase, non-CH nodes collect data and send to its CH (inter-cluster communication). Similarly, each CH sends data to the BS (intra-cluster communication). After each round, a CH checks its remaining energy to see if it is reached to the threshold value (Eth) for a CH change. If so, it generates a beep signal and lets its cluster members know a CH is about to change. Thereafter, it behaves like a non-CH node, i.e., it collects data and sends to the new CH. The process of CH rotation continues till each node within the cluster gets a chance to be a CH. After the last node is reached to the threshold value, the algorithm starts over and runs the same CH schedule.

In short,

1. The BS is fixed (not mobile) and located far from sensor nodes.

2. All sensor nodes in the network are uniformly distributed over a square field.

3. Sensor nodes and the BS are left unattended after their deployment.

4. A network consists of heterogeneous nodes in terms of energy.

5. All sensor nodes are of equal significances.

6. The BS has a rechargeable battery.

LEACH is an iterative algorithm and each iteration in the algorithm is called a 'round'. The round r is referred to a time interval where all cluster members transmit to the CH. In LEACH, during the setup phase, each node generates a random number between 0 and 1. If this random number is less than a specific value, the threshold value T(s), then the node becomes a CH for that round. The threshold value is chosen as:

$$T(s) = \begin{cases} \dfrac{p_{opt}}{1 - p_{opt}\left(r \bmod\left(\dfrac{1}{p_{opt}}\right)\right)}, & \text{if } s \in G \\ 0, & \text{otherwise} \end{cases} \qquad (10)$$

where is the desired percentage of CHs among all nodes, r is the current round number, G is a set of sensor nodes that have not been selected as CHs in the last $\left(\dfrac{1}{p_{opt}}\right)$ rounds, and s is the current CH node.

Most energy-efficient schemes in the literature aim at minimizing the average wasted energy. To this affect, they only take into account the residual energy. On the other hand, other energy-efficient schemes focus on minimizing the average transmission energy and take into account the CSI. It is a well-known fact that network lifetime depends on both the average wasted energy and the average transmission energy Since the ultimate goal is to maximize the network lifetime, one needs to balance the both schemes. In this study, we consider the case where the CSI is available [20].

The received signal at the BS due to the CH is:

$$y_{iBS}(n) = \sqrt{P_i} h_{iBS}(n) x_i(n) + \eta_{iBS}(n) \qquad (11)$$

where is the transmission power of the CH node i, is the channel gain between the CH node i and the BS (circularly-symmetric Gaussian random variable with zero mean and variance), and is the M-PSK modulated transmitted signal (M=2k with k positive integer) with unit average power, and is an additive noise term, circularly-symmetric Gaussian random variable with zero mean. The variance of is given by

$$\sigma_{iBS}^2 = \beta D_{iBS}^{-\alpha} \qquad (12)$$

where is the distance between CH node i and the BS, is the path loss exponent, and is a constant whose value depends on the propagation environment [21].

We propose a new threshold value T(s), an improvement to (10), as follows:

$$T(s) = \begin{cases} \dfrac{p_{opt}}{1 - p_{opt}\left(r \bmod\left(\dfrac{1}{p_{opt}}\right)\right)} \left(\dfrac{E_{MRE} E_{bestchannel}}{E_{av}}\right) \end{cases} \qquad (13)$$

where denotes the maximum of minimum residual energies, is the largest variance (or, energy) of channel gains , and is the average energy of the network.

The average energy of the network at the rth round is given by

$$E_{av}(r) = \dfrac{1}{N} E\left(1 - \dfrac{r}{R}\right) \qquad (14)$$

where R is the network lifetime, which assumes that every node consumes the same amount of energy in each round. Suppose also that all nodes die at the same time; hence, R is the total number of rounds in which the network is alive. Let denote the energy consumed by the WSN in each round. Then R is given by

$$R = \dfrac{E_{total}}{E_{round}} \qquad (15)$$

In this study, we consider a WSN that uses a single-hop communication. An illustrative example of a single-hop communication with four clusters is depicted in Figure 2 where each non-CH node communicates directly (single-hop) with its CH and each CH communicates directly with the BS.

The data retrieval operation is initiated by the BS which broadcasts the request to send (RTS) message to activate sensors. CHs are enabled

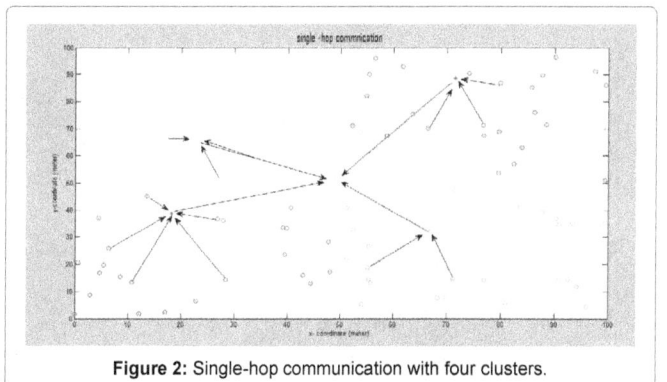

Figure 2: Single-hop communication with four clusters.

and data packets are transmitted to the BS through the wireless channel. All sensor nodes that have received RTS will contend for the data channel according to the non-cooperative schemes presented in [20]. At any time instant, only one CH occupies the channel to transmit a data packet to the BS. When one transmission is being held, other CH nodes in the network are in a sleeping state.

The average BER performance for a non-cooperative node with M-PSK modulation is upper-bounded by [22]:

$$BER \leq \frac{AN_0}{bP_i\sigma_{iBS}^2 \log_2 M} \quad (16)$$

where N0 is the power spectral density of the additive noise term , A and b are defined as

$$A = \frac{(M-1)}{2M} + \sin\left(\frac{2\pi}{\frac{M}{4\pi}}\right), b = \sin^2\left(\frac{\pi}{M}\right) \quad (17)$$

If the average BER is required to be less than or equal to a given value ζ, that is, $BER_{iBS} \leq \zeta$, the minimum transmission power is

$$P_{min} = \frac{AN_0}{b\zeta\sigma_{iBS}^2 \log_2 M}$$

Thus $P_i \in [P_{min}, P_{max}]$, where P_{max} denotes the maximum transmission power.

Simulation Results

The performance of the proposed clustering-based protocol is evaluated using MATLAB both for homogeneous and heterogeneous networks. In the network, 100 nodes are randomly deployed in a 100 m × 100 m region where the BS is located at the center as illustrated in Figure 3.

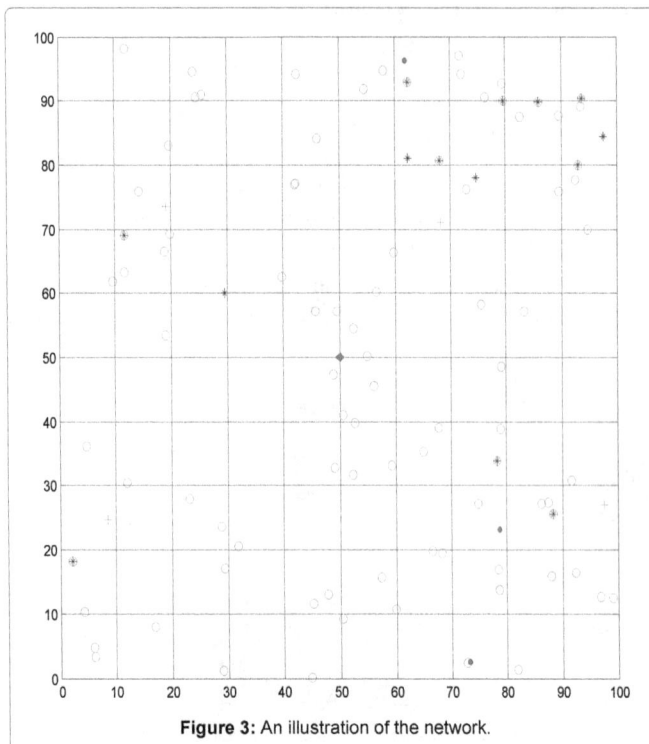

Figure 3: An illustration of the network.

Parameter name	Value
E_0	0.5 J
Packet size	4000 bits
E_{elec}	50 nJ/bit
$E_{tx} = E_{rx}$	50 nJ/bit
E_{fs}	10 pJ/bits/m²
E_{mp}	0.0013 pJ/bits/m²
E_{DA}	5 nJ
F_c	1 GH
N_0	-40 dBm
BER	10^{-3}
η	1
α	2
P_{max}	0.2
Data rate R	10^4
Modulation type	BPSK
E_c	2×10^{-4}
E_{cs}	2×10^{-4}
E_{as}	10^{-4}

Table 1: Network model parameters.

The performance evaluation of the proposed network is done with respect to the following parameters:

Stability period: The time interval between the start of the network operation and the death of the first sensor node also known as stable region.

Instability period: The time interval between the death of the first sensor node and of the last sensor node.

Network lifetime: The time interval between the start of the network operation and the death of the last sensor node.

Number of alive nodes: The number of sensor nodes that have not yet depleted their energy.

Number of dead nodes: The number of sensor nodes that have consumed all of their energy and are not able to do any kind of functionality [22-24].

Throughput: The rate of data sent from cluster heads to the base station.

The proposed algorithm is compared with LEACH, DEEC, and SEP in terms of dead and alive nodes per round, energy consumption of the network, and overall through put. The total number of rounds used in our experiments is 8000.

Network model parameters are summarized in Table 1.

Figure 4 shows the number of dead nodes per round indicating stability time of the networks. The death of the first node occurs at the round 1270 in the proposed protocol whereas the death of the first node occurs at rounds 984, 1140, and 912 in LEACH, DEEC, and SEP, respectively. The death of the last node occurs at the round 2529 in the proposed protocol whereas the death of the last node occurs at rounds 1450, 1554, and 2115 in LEACH, DEEC and SEP, respectively. Hence, the proposed protocol has better stability time and network lifetime as compared to the other networks.

The proposed protocol has also better energy consumption and higher throughput than the other protocols considered which can

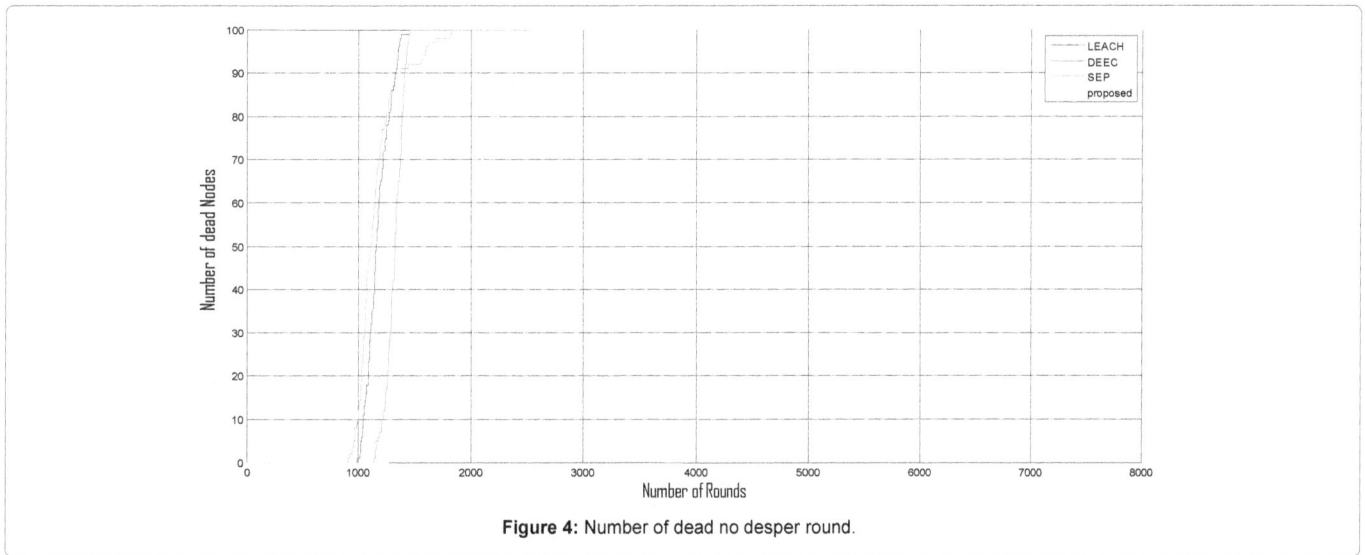

Figure 4: Number of dead no desper round.

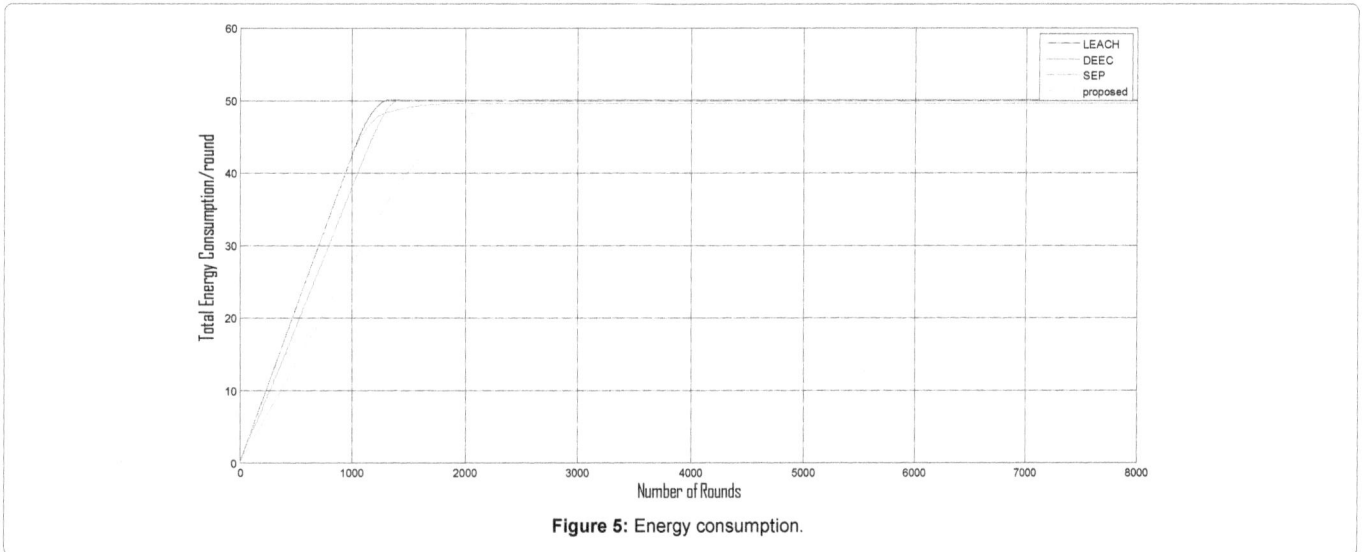

Figure 5: Energy consumption.

be seen from the Figures 5 and 6. Initial energy E0 of the network is consumed at the round 2100 in the proposed protocol whereas the initial energy E0 of the network is consumed at rounds 1100, 1200 and 1300 in LEACH, DEEC and SEP, respectively (Figure 5). Similarly, Figure 6 shows the superiority of the proposed algorithm in terms of the throughput thanks to the wiser selection of CHs [25].

Here all nodes have different amount of initial energies. The initial energies are uniformly distributed on [0.5, 1] resulting the average initial energy of 0.75 J. Several experiments are conducted, the average stability periods are calculated, and the results are shown in Figures 7-9. The results show that on the average the proposed algorithm has 1.62 to 1.89 times better stability period compared to other protocols considered. Figure 7 shows the number of dead nodes per round indicating stability time of the networks. The death of the first node occurs at the round 1894 in the proposed protocol whereas the death of the first node occurs at rounds 1036, 1322, and 1243 in LEACH, DEEC, and SEP, respectively. Hence, the proposed protocol has better stability time and network lifetime as compared to the other networks. The proposed protocol has also better energy consumption and higher

throughput than the other protocols considered which can be seen from the Figures 8 and 9.

Conclusions

This paper presents a new clustering-based protocol, EE-Heterogeneous LEACH, for WSNs. In the proposed protocol, nodes with different energy levels are considered causing heterogeneity in the network. Moreover, a single-hop transmission approach is adopted for intra-cluster and inter-cluster communication. We proposed an optimized routing scheme where the main focus is to enhance cluster head selection process. CHs are selected in each cluster on the basis of residual node energy and the best channel.

From the conducted experiments, it is seen that:

Stability period of the network is enhanced compared to the other protocols LEACH, DEEC, and SEP. Superior network lifetime is obtained for different scenarios. Last but not least, the throughput of the proposed protocol is significantly better than the other protocols

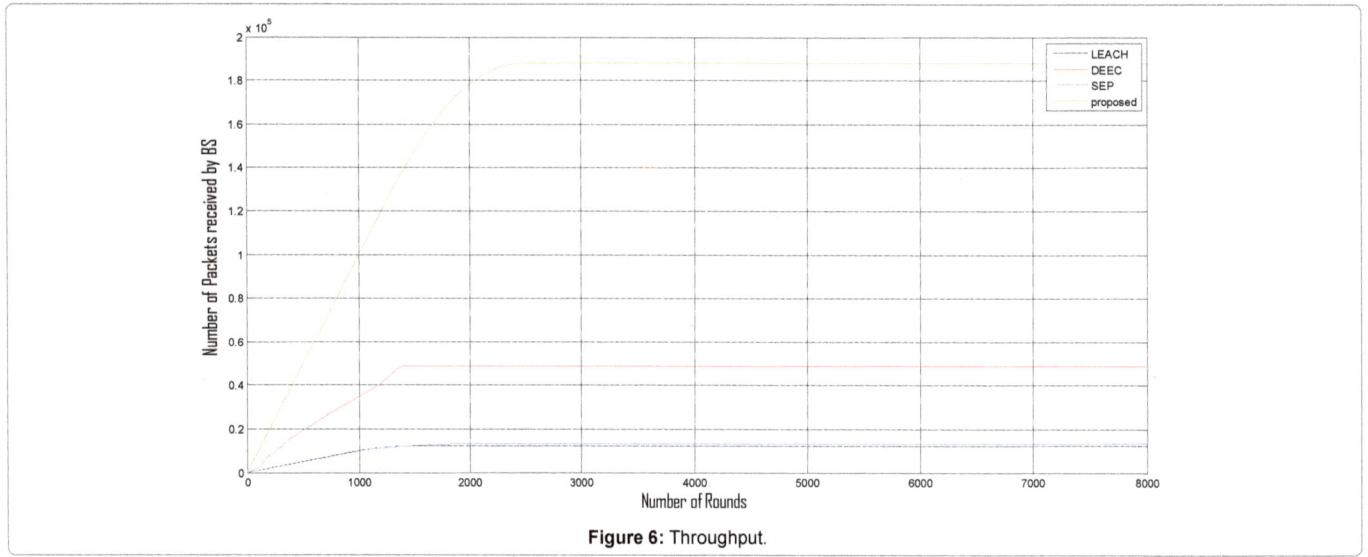

Figure 6: Throughput.

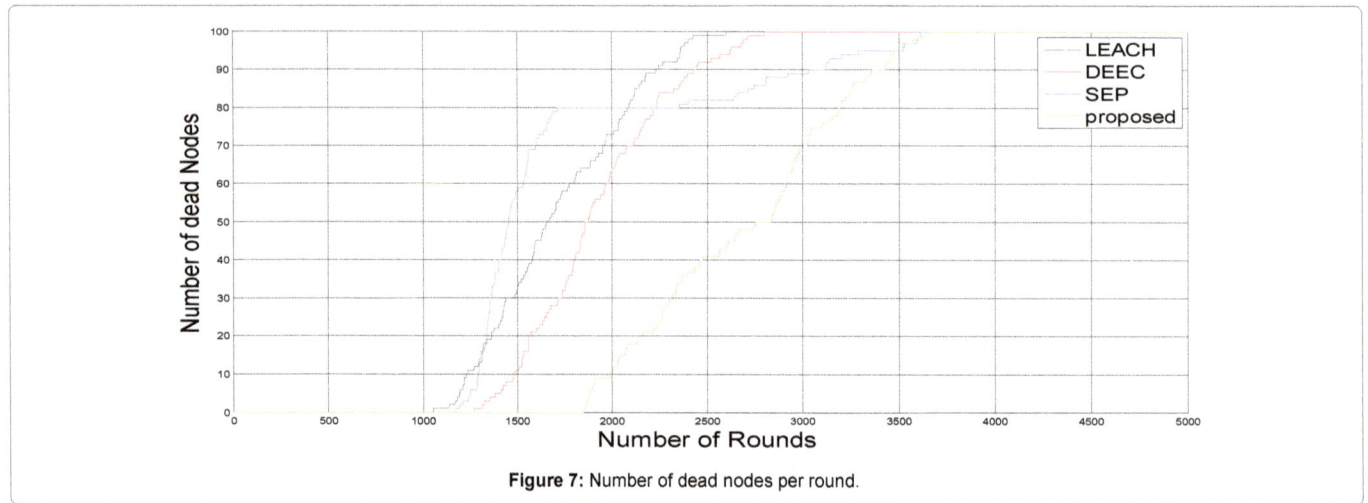

Figure 7: Number of dead nodes per round.

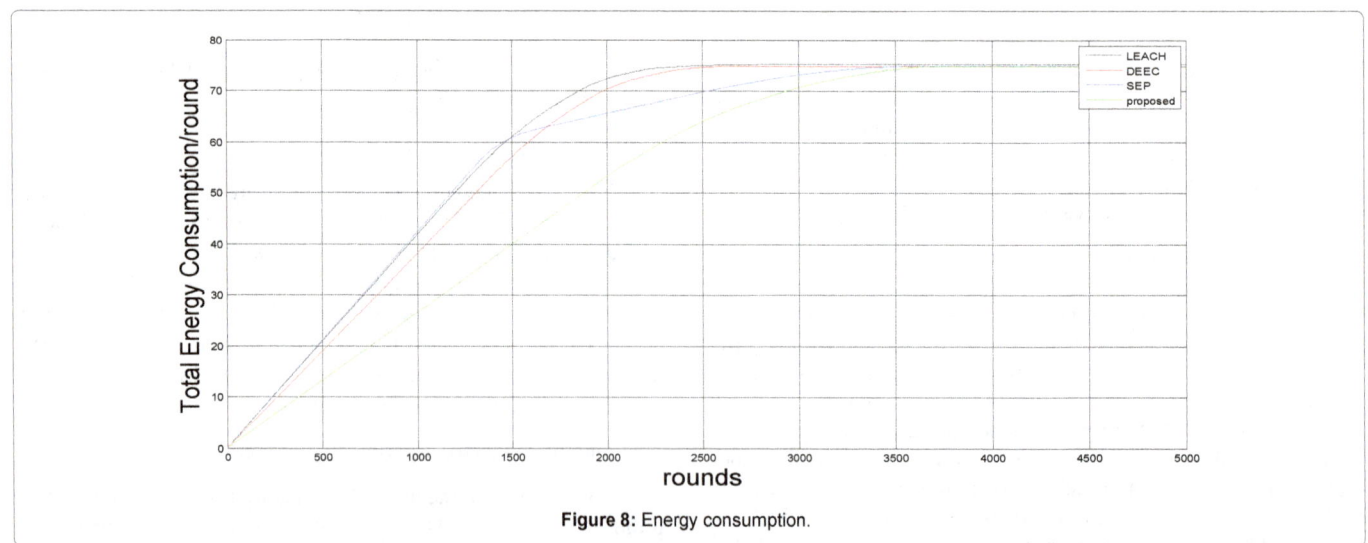

Figure 8: Energy consumption.

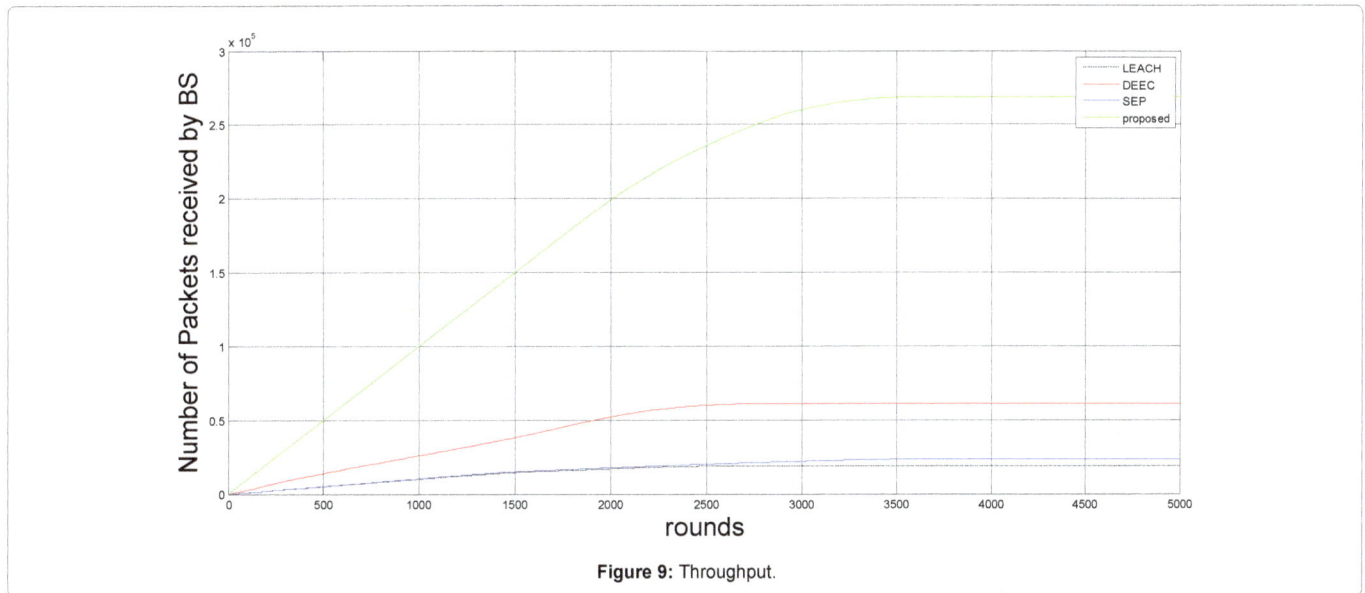

Figure 9: Throughput.

considered. Thus, we claim that the proposed protocol has improved energy efficiency and effective in prolonging the network lifetime.

References

1. Kumar V, Jain S, Tiwari S (2011) Energy Efficient Clustering Algorithm in Wireless Sensor Networks A survey, IJCSI International Journal of Computer Science 8.

2. Kumar P, Singh MP (2010) A New Clustering Protocol Based on Energy Band for Wireless Sensor Network, International Journal of Information Technology and Knowledge Management 3.

3. Lee D, Kaliappan VK, Duckwon C, Dugki M (2008) An energy efficient dynamic routing scheme for clustered sensor network using a ubiquitous robot Research, Innovation and Vision for the Future, IEEE International Conference 198-203.

4. Chang YC, Lin ZS, Chen JL (2006) Cluster based self-organization management protocols for wireless sensor networks, In Proceeding of the IEEE transaction on consumer electronic 52: 75-80.

5. Mhatre V, Rosenberg C (2004) Homogenous vs. heterogenous clustered sensor networks: A comparative study, in Preceedings of 2004 IEEE, International Conference on Communications.

6. Heinelman W, Chandrakasan A, Balakrishnan H (2002) Application specific protocol architecture for wireless sensor networks, IEEE Transaction on wireless Networking 660-670.

7. Amanjeet P, Nitin M, Singh DP, Chauhan RS (2010) Improved Leach communication protocol for WSN, National Conference on Computational Instrumentation CSIO Chandigarh 177-181.

8. Heinzelman W, Chandrakasan A, Balakrishnan H (2002) Energy-efficient communication protocol for wireless sensor networks, in the proceeding of the Hawaii International Conference System Science, Hawaii.

9. Ali MS, Dey T, Biswas R (2008) ALEACH: Advanced LEACH Routing Protocol for Wireless Microsensor Networks, 5th International Conference on Electrical and Computer Engineering Conference on Electrical and Computer Engineering ICECE 909-914.

10. Xin G, Yang WH, Gang DD (2008) EEHCA: an energy-efficient clustering algorithm for wireless sensor network, Inf Technol 7: 245-252.

11. Yi G, Guiling S, Weixiang L, Yong P (2009) Recluster-LEACH: A recluster control algorithm based on density for wireless sensor network, 2nd International Conference on Power Electronics and Intelligent Transportation System 3: 198-202.

12. Heinzelman WB (2000) Application-Specific Protocol Architectures for Wireless Networks, PhD thesis, Massachusetts Institute of Technology.

13. Depedri A, Zanella A, Verdone R (2003) An Energy Efficient Protocol for Wireless Sensor Networks In Proc. AINS 1-6.

14. Fan X, Song Y (2007) Improvement on leach protocol of wireless sensor network, In Proceeding of the International Conference on Sensor Technologies and Applications 260-264.

15. Han L (2010) LEACH-HIR: An energy efficient routing algorithm for Heterogenous WSN, IEEE International Conference on Intelligent Computing and Intelligent Systems (ICIS) 2: 507-511.

16. Kumar D, Aseri TC, Patel RB (2009) EEHC: energy efficient heterogeneous clustered scheme for wireless sensor networks, Elsevier Comput Commun 32: 662-667.

17. Smaragdakis G, Matta I, Bestavros A (2004) SEP: a stable election protocol for clustered heterogeneous wireless sensor networks, Proc.Int Workshop Sensor and Actor Network Protocols and Applications, Boston, MA 251-261.

18. Xianging F, Yulin S (2007) Improvement on LEACH protocol of wireless sensor networks, Proc.Int.Conf. Sensor Technologies and Applications 260-264.

19. Sharma N, Verma V (2013) Heterogeneous LEACH Protocol for Wireless Sensor Networks, Int J Advanced Networking and Applications 1825-1829.

20. Chen Y, Zhao Q (2005) Maximizing the lifetime of sensor network using local information on channel state and residual energy, in Proc. IEEE CISS'05.

21. Himsoon T, Siriwongpairat WP, Han Z, Liu KJR (2007) Lifetime maximization via cooperative nodes and relay deployment in wireless networks, IEEE J Select. Areas Commun 25: 306-317.

22. Su W, Sadek AK, Liu KJR (2005) SER performance analysis and optimum power allocation for deode-and-forward cooperation protocol in wireless networks, in Proc. IEEE WCN'05 2: 984-989.

23. Henielman W, Chandrakasan A, Balakrishman H (2000) Energy-Efficient Routing Protocols for Wireless Microsensor Networks, Proceeding of 33rd Hawaii International Conference System Sciences, Maui 4-7.

24. Qing L, Zhu Q, Wang M (2006) Design of a distributed energy-efficient clustering algorithm for heterogenous wireless sensor networks, Computer. Comput, Comm 29: 2230-2237.

25. Smaragdakis G, Matta IA (2004) SEP: A Stable Election Protocol for Clustered Heterogenous Wireless Sensor Networks, Proceeding of SANPA.

31

A Concept for Molecular Addressing by Means of Far-reaching Electromagnetic Interactions in the Visible

Langhals H*

LMU University of Munich, Department of Chemistry, Butenandt Str-13, D-81377 Munich, Germany

Abstract

A pronounced concentration dependence of the time constant of fluorescence decay was found, disproving Strickler-Berg's equation restricting molecular light emission to a local process. Thus, interactions were found significantly extending to more than 100 nm and make such systems promising for interfaces between molecular dots or more complex molecular arrangements and conventional macroscopic electronics.

Keywords: Molecular electronics; Strickler-Berg-equation; Interfaces; Fluorescence

Introduction

The increasing demands of integration in electronic devices [1] makes the addressing of molecular structures attractive [2] where the operating frequency should be extended into the petahertz region [3]. Basic structures of molecular operating functional materials are promising candidates therefore where fluorescent chromophores exhibit the advantage of intermediate storage of the energy of excitation for further processing [4]. However, the addressing of local molecular structures starting by means of macroscopic electronics has still to be developed; the scanning tunnelling [5] and atomic force microscope are useful for such applications, however, hardly applicable for routine devices. Optical addressing is more attractive, however, limited by the optical resolution given by about half the wavelengths of about 250 nm in the visible known as Abbe's diffraction limit. Resolution beyond this limit is possible, however, hardly practicable for routine applications in electronics. An interface between the well-accessible conducting technology nodes of 25 nm to molecular structures of about 1 nm remains a challenge [6].

Materials and Methods

Dye **1** (CAS RN 110590-84-6) was prepared according to literature [7] Nile Blue A (CAS RN 3625-57-8) was purchased from Acros Organics, product code: 415690100. All solvents were used in spectroscopical grade, chloroform was purchased from Sigma Aldrich (≥ 99.8%, contains 0.5–1.0% ethanol as stabilizer, product code: 1001744040), ethanol from Merck (≥99.9%, water content ≤0.05%, index-No: 603-002-00-5). UV/V is spectra were obtained with a Varian Cary 5000 spectrometer. Fluorescence spectra were obtained with a Varian Cary Eclipse spectrometer, slit width 5 nm unless otherwise specified. Fluorescence lifetimes were measured with a NKT-Laser SuperK Extreme EXB-4 from NKT Photonics A/S as light source, an Edinburgh Instruments Ltd. monochromator and detected with a PicoHarp 300 from PicoQuant GmbH in combination with a PMA-C 192-N-M photomultiplier. Fluorescence lifetimes were obtained by exponential fitting of the deconvoluted fluorescence decays with a resolution of 8 ps. The fluorescence lifetime equals the negative reciprocal exponent. Excitation and detection wavelengths were steadily the corresponding relative spectral maxima, except the local maximum at 576 nm was applied in case of the detection of **1**. All spectroscopic measurements were carried out in 1.000 cm quartz cuvettes.

Results and Discussion

Molecular addressing in the visible region (about 0.6 PHz) with operating chromophores with dimension of about 1 nm can be basically performed by means of electric dipole-dipole interactions where a theoretical concept was developed by Perrin [8] and further extended by Förster [9-12] (Förster resonance energy transfer, FRET) having established equation (1) for the quantitative description of the distance function **R**. For more recent discussion see [13].

$$k_{FRET} = \frac{1000 \cdot (\ln 10) \cdot \kappa^2 \cdot J_{DA} \cdot \Phi_D}{128 \cdot \pi^5 \cdot N_A \cdot \tau_D \cdot |R_{DA}|^6} \tag{1}$$

There are some mathematical constants in equation (1), the factor 1000 for the adaption to SI units and Avogadro's number N_A, some spectroscopic data of the involved chromophores where J_{DA} means the integral of the spectral overlap between the fluorescence spectrum of the energy donor and the absorption spectrum of the acceptor, Φ_D the fluorescence quantum yield and τ_D the fluorescence lifetime of the donor and the orientation factor κ. Important for molecular addressing is the dependence of the rate constant k_{FRET} on the inverse 6th power of the distance **R** causing a very fast damping. As a consequence, addressing by energy transfer extends over several nm where technology nodes of 25 nm or even larger and more distant seem to be hardly reachable. Influencing the emission of light from electronically excited molecules seems to be even more restricted where the basic theory for light emission was developed Perrin [14], Förster [15-16], Lewis and Kasha [17] and Strickler and Berg [18]. The latter established the generally accepted equation (2) known as the Strickler-Berg-Equation where τ_0 means the natural fluorescence lifetime and $A_{u \to l}$ Einstein's transition probability coefficient concerning to the lower (*l*) and upper (*u*) energetic level.

$$\frac{1}{\tau_0} = A_{u \to l} = 1000 \frac{8\pi \ln(10) c \tilde{v}_{u \to l}^2 n^2}{N_A} \frac{g_l}{g_u} \int \varepsilon \, d\tilde{v} \tag{2}$$

---

*Corresponding author: Langhals H, LMU University of Munich, Department of Chemistry, Butenandt Str-13, D 81377 Munich, Germany
E-mail: Langhals@lrz.uni-muenchen.de

$$\frac{1}{\tau_0} = \frac{2.880 \cdot 10^{-9}}{cm \cdot s} n^2 \tilde{v}_0^2 \frac{g_l}{g_u} \int \varepsilon \, d\tilde{v} \tag{3}$$

$$\tau = \phi \cdot \tau_0 \tag{4}$$

N_A means Avogadro's constant $\tilde{v}_{l \to u}$ the wavenumber of the electronic transition, c the velocity of light, n the index of refraction and g_l and g_u the degeneracies of the lower and the upper energetic state, respectively. There are some mathematical constants, the factor 1000 for the adaption to SI units and the integral extending over the absorption band, thus representing the oscillator strengths. The ratio of the degeneracies becomes unity for most complex organic chromophores because of low symmetry. The factors can be combined to the constant in equation (3) with $\tilde{v}_0$ being the mean wavenumber of the absorption band. The molar absorptivities of the maxima are proportional to the oscillator strengths for identical shapes of UV/Vis spectra. Equation (1) was established for strongly light-absorbing dyes ($\varepsilon \geq 8000$) with small geometrical distortion upon light absorption indicated by small Stokes' shifts [8]. The natural lifetime τ_0 can be interrelated with the apparent lifetime τ and the fluorescence quantum yield φ by means of equation (4).

Equation (2) essentially contains molecular properties of the fluorescent molecule and the index of refraction n of the molecular surrounding medium; a macroscopic influencing for data processing can be hardly expected on this basis.

The constancy of the fluorescence lifetime [19] implied by equation (2) was tested [20-22] by means of the fluorescent dye Nile blue chloride (CAS-RN 2381-85-3) in ethanol where a constant fluorescent lifetime of 1.42 ± 0.15 ns was reported for diluted solutions within 10^{-8} molar and 10^{-3} molar; for comparison: A fluorescence quantum yield [23] of 0.27 was reported for Nile blue perchlorate (CAS-RN 53340-16-2).

$$\tau = a \cdot \ln\left(\frac{c}{c^*} + 1\right) + b \tag{5}$$

We re-examined the concentration dependence of the fluorescence lifetime by means of the more stable Nile blue sulphate (CAS-RN 3625-57-8; the fluorescence lifetime slightly depends on the counter ion) and found the surprisingly strong dependence on the concentration of the dye shown in Figure 1. The concentration dependence of τ can be described by means of the previously developed equation (5) for solvent

effects where c^* is a characteristic concentration and a and b adjustable parameters; b means τ at infinite dilution and a the sensitivity.

There are various interactions of the chromophore of Nile blue with the molecular surrounding, where Coulomb interactions dominate because of the positive charge of the chromophore and the negatively charged counter ion. Thus, we applied the uncharged point symmetrical chromophore 1 (S-13; CAS- RN 110590-84-6, Figure 2) for extended investigation and the exclusion of any interference. The high fluorescence quantum yield [24,25] of 1 close to unity renders τ close to τ_0 and excludes influences by varying φ. The extraordinarily high light fastness [26] of 1 allows measurements even with long acquisition times. Starting with a highly diluted solution of $2.26 \cdot 10^{-5}$ molar of 1 in chloroform a time constant of the fluorescence decay of 5.04 ns was found where the dye molecules are well-separated by a mean intermolecular distance of 23 nm; for comparison the seize of individual dye molecules extend to about 1 nm. A stepwise further dilution [27] until $1.08 \cdot 10^{-7}$ molar caused a decrease of the lifetime to 3.79 ns where the mean intermolecular distance became as high as 138 nm reaching macroscopic dimensions. Again, the concentration dependence of τ could be described by means of equation (5); see curve in Figure 3 and the linear correlation shown in the insert there. Interference by any molecular interactions was excluded by the application of Lambert-Beer's law (Figure 4, upper diagram) where a perfect linear correlation was obtained within the limits given by the precision of the spectrometer; a perfect linear correlation was also found for the concentration dependence of the fluorescence intensity; Figure 4, lower diagram. Moreover, the spectral band types both of absorption and fluorescence spectra were investigated and found to be congruently independent from the concentration. This can be interpreted as additional proofs for isolated, non interacting dye molecules in such highly diluted solutions.

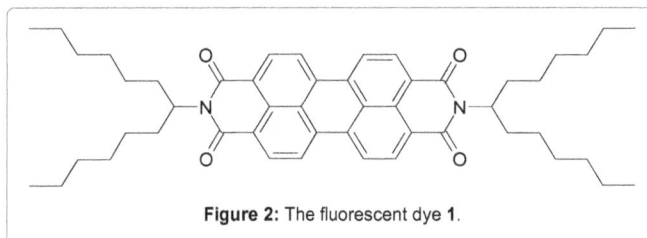

Figure 2: The fluorescent dye 1.

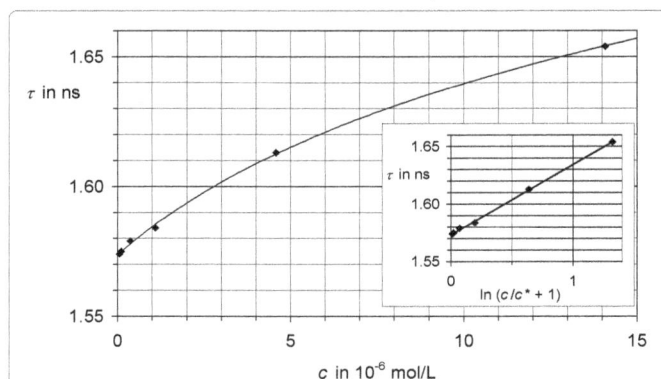

Figure 1: The time constant τ of fluorescence decay of Nile blue sulphate (CAS-RN 3625-57-8) in ethanol (25°C) as a function of the molar concentration c; curve: Application of equation (5). Insert: Linear correlation for the application of equation (5); $c^* = 5.1 \cdot 10^{-6}$ mol·L^{-1} slope 0.061 ns, intercept 1.574 ns, standard deviation 0.0009 ns, correlation number 0.9997 (6 points), $R^2 = 0.9993$).

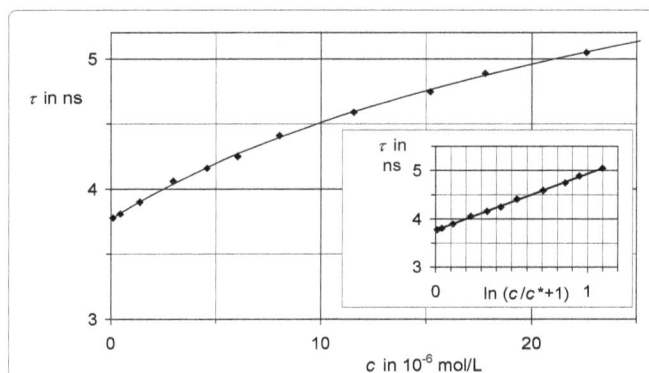

Figure 3: Dependence of the fluorescence lifetime τ of 1 in chloroform on the concentration c (diamonds; see Table 1). Curve: Function $a \cdot \ln(c/c^* + 1)$ $+b$ according to equation (5) with a=1.17 ns, c^*=1.17·10^{-5} mol·L^{-1} and b=3.77 ns. Insert: Linear correlation according to equation (5) (standard deviation 0.015, correlation number 0.9992, coefficient of determination 0.9984, 11 measurements).

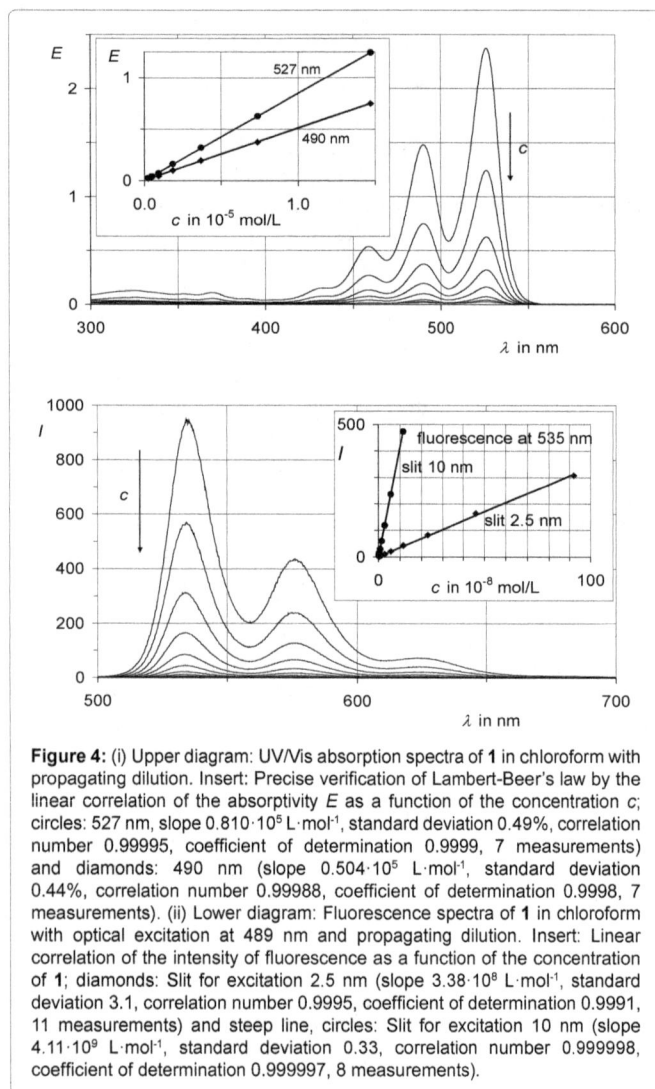

c	τ	l	d
10^{-6} mol·L^{-1} [a]	ns[b]	nm[c]	nm[d]
22.6	5.04	42	23
17.8	4.91	45	25
15.2	4.77	48	26
11.6	4.59	52	29
8	4.41	59	33
6.05	4.29	65	36
4.56	4.13	71	40
2.94	4.05	83	46
1.37	3.9	106	59
0.456	3.82	154	85
0.108	3.79	249	138

Table 1: Fluorescence lifetimes τ of 1 in chloroform depending on the concentration c and the mean molecular distance d, respectively. Fluorescence excitation at 490 nm and detection at 535 nm.

[a] Molar concentration of dye 1 in chloroform. [b] Time constant τ for the exponential decay of fluorescence. [c] Calculated lengths of the cubic volume for one dye molecule [d] Calculated mean inter molecular distance [24] for 1.

Figure 4: (i) Upper diagram: UV/Vis absorption spectra of **1** in chloroform with propagating dilution. Insert: Precise verification of Lambert-Beer's law by the linear correlation of the absorptivity *E* as a function of the concentration *c*; circles: 527 nm, slope 0.810·10⁵ L·mol⁻¹, standard deviation 0.49%, correlation number 0.99995, coefficient of determination 0.9999, 7 measurements) and diamonds: 490 nm (slope 0.504·10⁵ L·mol⁻¹, standard deviation 0.44%, correlation number 0.99988, coefficient of determination 0.9998, 7 measurements). (ii) Lower diagram: Fluorescence spectra of **1** in chloroform with optical excitation at 489 nm and propagating dilution. Insert: Linear correlation of the intensity of fluorescence as a function of the concentration of **1**; diamonds: Slit for excitation 2.5 nm (slope 3.38·10⁸ L·mol⁻¹, standard deviation 3.1, correlation number 0.9995, coefficient of determination 0.9991, 11 measurements) and steep line, circles: Slit for excitation 10 nm (slope 4.11·10⁹ L·mol⁻¹, standard deviation 0.33, correlation number 0.999998, coefficient of determination 0.999997, 8 measurements).

The comparably strong concentration dependence of the fluorescence lifetime τ disproves the validity of equation (1) for real systems and is an indicator for long-reaching interactions up to macroscopic dimensions [28]. There are not only consequences for topics such as fluorescence lifetime spectroscopy (FLIM) including Förster resonance energy transfer processes (FRET), but offer completely novel possibilities such as for molecular electronics [6]. Thus, light emission resembles more the near field of a radio transmitter [29-30] than a molecular processes and is further treated with this concept. Firstly, the relative permittivity ε_r corresponding to the square of the refraction n of the medium has to be considered concerning the geometry. For the high frequency of light around 0.6 Petahertz the n_D at 589 nm is a good choice because this wavelengths is far away from absorption bands with interference by anomalous dispersion in the UV and the NIR of the commonly colorless solvents or polymeric optical media. Thus, resonating conducting structure become important with the dimensions of about $\lambda/(2n^2)$; this means about 125 nm in media such as chloroform ($n_D = 1.45$) or acrylic glass (PMMA, $n_D = 1.49$) and about 50 nm for structuring with $\lambda/(10n^2)$ for arrangements of passive resonators for the generation of travelling waves; compare, for example, the geometry of Yagi-Uda antennae [31-33] for TV frequencies. Such dimensions correspond to the strong

influence found on τ at intermolecular distance of 30 nm remaining still appreciable for more than 60 nm; Table 1.

On the other hand, the influencing of τ of a molecular step of light emission by means of structures extending to macroscopic dimensions offer novel possibilities in technology. There are many concepts for molecular electronics [6], however, the interface to the macroscopic electronics finally necessary for the data processing remains a challenge. The novel findings could help solving these problems by means of the connecting of the meanwhile in technical scale producible conducting technology nodes of 25 nm with operating molecular dots or more complex molecular arrangements.

Conclusion

The limiting Stickler-Berg equation (1) for molecular addressing implies isolated molecular processes of light emission, however, was disproved for real systems where long-reaching interactions until macroscopic dimensions were verified by means of the measurement of the concentration dependence of the fluorescence decay time constants τ and was quantitatively described with equation (5). The dimensions of such interactions resemble more near fields of radio transmitters than molecular processes and the reached significant distances between 30 and more than 60 nm make such systems promising for interfaces between molecular electronics and conventional macroscopic electronics.

Acknowledgements

This work was supported by the Fonds der Chemischen Industrie.

References

1. Hilbert M, López P (2011) The World's Technological Capacity to Store, Communicate, and Compute Information. Science 332: 60-65.

2. Carroll RL, Gorman CB (2002) The Genesis of Molecular Electronics. Angew Chem 41: 4378-4400.

3. Langhals H (2014) Handling Electromagnetic Radiation beyond Terahertz using Chromophores to Transition from Visible Light to Petahertz Technology. J Electr Electron Systems 3: 125.

4. Langhals H (2013) Chromophores for picoscale optical computers. In: Sattler K (ed.), Fundamentals of picoscience. Taylor & Francis Inc CRC Press Inc, Bosa Roca/US, pp: 705-727.

5. Binnig G, Quate CF, Gerber Ch (1986) Atomic Force Microscope. Phys Rev Let 56: 930-933.

6. Cuevas JC, Scheer E (2010) Molecular electronics. World scientific series in nanoscience and nanotechnology: Volume 1, World Scientific, London.

7. Demmig S, Langhals H (1988) Highly soluble and lightfast perylene fluorescent dyes. Chem Ber 121: 225-230.

8. Perrin JB (1925) Fluorescence et radiochimie conseil de chemie. (2nd edn) Solvay, Paris.

9. Foerster T (1946) Energy transport and fluorescence. Naturwiss 33: 166-175.

10. Foerster T (1948) Intermolecular energy transfer and fluorescence. Ann Phys 6 Folge 2: 55-75.

11. Foerster T (1949) Experiments on intermolecular transition of electron excitation energy. Z Elektrochem 53: 93-99.

12. Foerster T (1949) Zeitschr Naturforsch. Chem Abstr 1950, 44: 43074.

13. Langhals H, Esterbauer AJ, Walter A, Riedle E, Pugliesi I (2010) Förster resonant energy transfer in orthogonally arranged chromophores. J Am Chem Soc 132: 16777-16782.

14. Perrin F (1926) La polarisation de la fluorescence des molécules gazeuses diatomiques. J Phys Radium 7: 390-401.

15. Förster T (1951) Fluoreszenz organischer Verbindungen. Göttingen, Vandenhoeck & Ruprecht; Chem Abstr 45: 26285.

16. Förster T (1966) Umwandlung der Anregungsenergie. In Foerst W (eds), 2. Internationales Farbensymposium: Optische Anregung organischer Systeme. Aufnahme und Umwandlung von Lichtenergie durch Farbstoffe und die Einflüsse des Mediums, Verlag Chemie GmbH, Weinheim.

17. Lewis GN, Kasha M (1947) Phosphorescence in Fluid Media and the Reverse Process of Singlet-Triplet Absorption. J Am Chem Soc 67: 994-1003.

18. Strickler SJ, Berg RA (1962) Relationship between Absorption Intensity and Fluorescence Lifetime of Molecules. J Chem Phys 37: 814-822.

19. Lukeš V, Danko M, Andicsová A, Hrdlovič P, Végh D (2013) The synthesis and examination of spectral properties of some 2,2'-bithienyl derivatives with carbonyl-containing substituents. Synth Met 165: 17-26.

20. Grofcsik A, Kubinyi M, Jones WJ (1995) Fluorescence decay dynamics of organic dye molecules in solution. J Molec Structure 348: 197-200.

21. Jose J, Burgess K (2006) Benzophenoxazine-based fluorescent dyes for labeling biomolecules. Tetrahedron 62: 11021-11037.

22. https://en.wikipedia.org/wiki/Nile_blue.

23. Sens R, Drexhage KH (1981) Fluorescence quantum yield of oxazine and carbazine laser dyes. J Lumin 24: 709-712.

24. Chandrasekhar S (1943) Stochastic problems in physics and astronomy. Rev Mod Phys 15: 1-89.

25. Langhals H, Karolin J, Johansson LB-Å (1998) Spectroscopic properties of new and convenient standards for measuring fluorescence quantum yields. J Chem Soc, Faraday Trans 94: 2919-2922.

26. Langhals H (2005) Control of the interactions in multichromophores: Novel concepts. Perylene bisimides as components for larger functional units. Helv Chim Acta 88: 1309-1343.

27. Langhals H, Schlücker T (2015) The use of dilute chromophore fluorescence decay time for concentration determination. Chem Abstr 163: 714018.

28. Langhals H (2017) New possibilities for molecular addressing. Ger Offen DE 102017000953.8

29. Silver HW, Ford SR, Wilson MJ, Applegate A, Bloom A (2015) The ARRL Antenna Book for Radio Communications, (23rd edn.), The American Radio Relay League, Inc, Reynoldsburg (OH 43068).

30. Rothammel K (1976) Antennenbuch, (5th edn.), Telekosmos Verlag, Stuttgart.

31. Yagi H (1928) Beam Transmission of Ultra Short Waves. Proc of the IRE 16: 715-740.

32. Li J, Salandrino A, Engheta N (2007) Shaping light beams in the nanometer scale: A Yagi-Uda nanoantenna in the optical domain. Phys Rev B 76: 245403.

33. Li J, Salandrino A, Engheta N (2009) Optical spectrometer at the nanoscale using optical Yagi-Uda nanoantennas. Phys Rev B 79: 195104.

32

Open-end Winding Induction Motor Drive Using Decoupled Algorithm

Ranjit M*, Sumanjali M and Ganesh B

VNR Vignana Jyothi Institute of Engineering and Technology, Hyderabad, India

Abstract

In now a days modern multi level inverters have emerged to overcome the drawbacks due to the conventional inverters. In various industries inverters with different PWM techniques have been employed to achieve good performance in the context of variable speed drives. But due to multilevel inverters the switching losses are more and the cost of the equipment is also more because of the number of devices is increased in multilevel inverters. These are some drawbacks due to the usage of conventional multilevel inverters in industries. In this proposed work a decoupled algorithm is proposed to overcome the drawbacks due to conventional inverter have been presented. The cascaded connection of asynchronous motor and two 2-level inverters at both ends of motor constitutes to open end winding induction motor drive. The characteristics of dual inverter fed open end winding induction motor drive resembles to those of conventional three level inverter. In this proposed work the performance characteristics of Induction motor with different PWM techniques like CSVPWM, DPWMMAX, DPWMMIN have been analysed using MATLAB/SIMULINK environment and to validate the results the harmonic spectra values has been listed out.

Keywords: Bearing currents; CSVPWM; Decoupled algorithm; Open end winding induction motor drive; Modulation index

Introduction

Conventional two level inverters are extensively used in medium voltage and high power variable speed drive systems because of their inherent switching operation but however have some limitations in operating at high frequency mainly due to switching losses and constraints of device rating. These switching converters can also provokes high dv/dt caused due to the switching transients [1-2]. These zero sequence voltages results into various adverse effects on motors named as bearing currents, conducted electromagnetic interference, ground currents through stray capacitors. In consequence to this premature motor bearing failures will occur [3]. The clear indication of flowing of hazardous bearing currents in the context of motors inside the motors can be shown in the Figure 1.

So concerning to this the hazardous common mode voltages in the context of variable speed motors has to be mitigated [4,10].

The numerous methods for mitigating common mode voltage in inverters can be classified as [4]:

a. Using isolation transformers, Common mode choke, using hybrid active and passive filters, Using dual inverter fed open end winding induction motor drive, Using four phase inverter.

b. Using some advanced modulation techniques like carrier based SVPWM scheme for dual inverter fed open end winding IM drive.

The methods proposed in [a] above increases the system cost as it employs some extra hardware circuitary and complexity in control. So this is mainly focused on the implementation of SVPWM technique for dual inverter fed open end winding induction motor. A schematic of dual inverter fed open end winding induction motor can be represented as shown in the Figure 2.

As dual inverter fed open end winding induction motor drive resembles the performance of three level inverter thus we can achieve multilevel inverter operation using this configuration. Hence the harmonic content of the output voltage waveform decreases significantly, dv/dt stresses are reduced, produces smaller zero sequence voltages therefore stress in the bearings of motor can be reduced, Provides low switching losses and higher efficiency [5-8].

Multi Level Inverter Operation with Open End Winding Induction Motor

A remedy for the production of flowing of bearing currents inside the motors is open end winding induction motor. In this configuration Induction motor is faded by two inverters from either side which are operated be isolated power supplies. A schematic diagram of dual inverter fed induction motor is represented as shown in Figure 3.

Here S_1, S_2, S_3, S_4, S_5, S_6 are the switches of inverter 1 and S_1', S_2', S_3', S_4', S_5', S_6' are the switches of inverter 2. The two inverters are supplied with isolated DC links, If the isolated DC link voltages are equal (i.e., $Vs_1 = V_{DC}/2$ and $Vs_2 = V_{DC}/2$) then the configuration resembles to that of three level inverter drive. If the Isolated DC link voltages are unequal (i.e., $Vs_1 = 2V_{DC}/3$ and $Vs_2 = V_{DC}/3$) then the configuration resembles to that of four level inverter. In this 1 and 0 represents on states of switches of inverter 1, 1' and 0' represents on states of switches of inverter2. For the three level inverter the levels of voltage in the output voltage are $+V_{dc}/2$, 0, $-V_{dc}/2$. For the four level inverter the levels of voltages in the output voltage are $+V_{dc}/2$, $+V_{dc}/6$, $-V_{dc}/6$, $-V_{dc}/2$ (Figure 4).

Here first space vector corresponds to Inverter-1 and the reference vector is switched in sector-1 between first two active vectors V_1 and V_2, at an angle α w.r.t first active vector V_1. And the second space vector corresponds to Inverter-2 and the reference vector is switched in sector-4 between V_4 and V_5 at an angle α i.e., $180^0 + α$ w.r.t V_1, the reference vector of first space vector and reference vector of second space vector are 180^0 apart. But the supply voltages for the two inverters are same i.e., $V_{dc}/2$ for dual inverter (three level operation).

---

*Corresponding author: Ranjit M, VNR Vignana Jyothi Institute of Engineering and Technology, Hyderabad, India, E-mail: ranjit_m@vnrvjiet.in

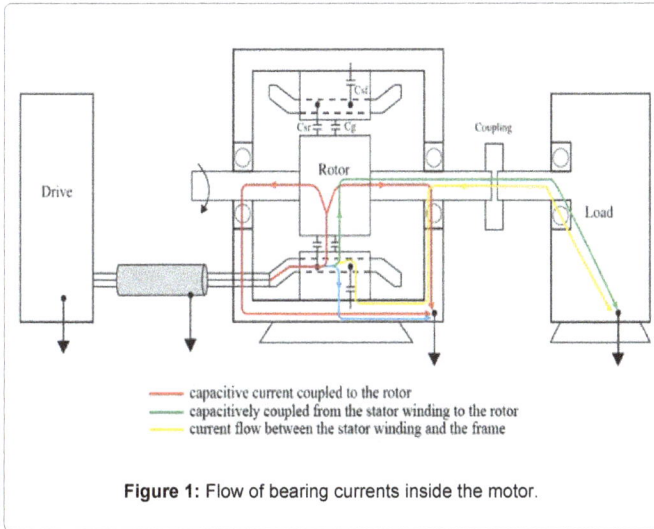

Figure 1: Flow of bearing currents inside the motor.

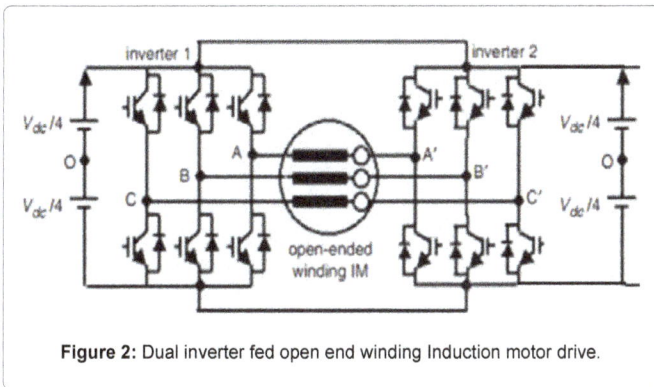

Figure 2: Dual inverter fed open end winding Induction motor drive.

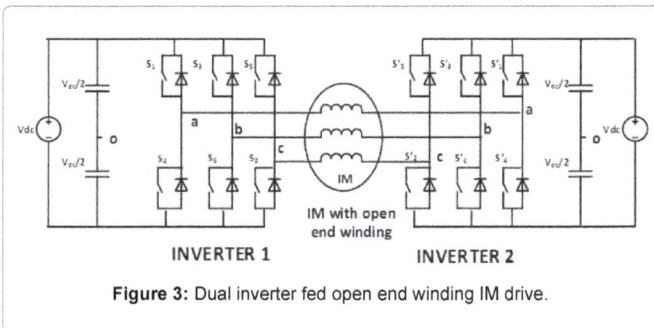

Figure 3: Dual inverter fed open end winding IM drive.

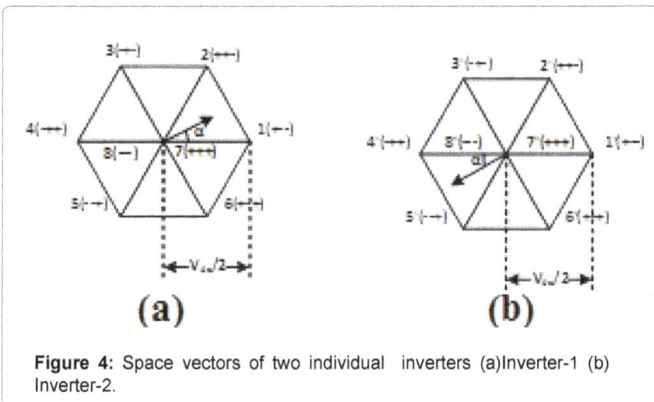

Figure 4: Space vectors of two individual inverters (a)Inverter-1 (b) Inverter-2.

Here V_{A0}, V_{B0}, V_{C0} are the pole voltages of inverter 1, V_{A0}', V_{B0}', V_{C0}' are the pole voltages of inverter 2. V_{AA}', V_{BB}', V_{CC}' are the Phase voltages of the inverter which are supplied to the three phase induction motor but here the sum of all these phase voltages is not equal to zero, which results as zero sequence component in motor due this the bearing currents will flow inside the motor. But here carrier based SVPWM algorithm is proposed to mitigate this Common mode voltage. The three phase voltages of dual inverter fed induction motor drive is given by

$$V_{AA}' = V_{A0} - V_{A0}'$$
$$V_{BB}' = V_{B0} - V_{B0}' \qquad (1)$$
$$V_{CC}' = V_{C0} - V_{C0}'$$

Where V_{A0}, V_{B0}, V_{C0} are the pole voltages of inverter 1,

V_{A0}', V_{B0}', V_{C0}' are the pole voltages of inverter 2

and V_{AA}', V_{BB}', V_{CC}' are the Phase voltages of the inverter

The common mode voltage or Zero sequence voltage is given by

$$CMV \; or \; V_{ZS} = \frac{V_{AA}' + V_{BB}' + V_{CC}'}{2} \qquad (2)$$

The reference voltage in SVPWM modulation technique will be obtained as represented in equation

$$V_{ref} = V_{AA}' + V_{BB}' e^{j2\pi/3} + V_{CC}' e^{j4\pi/3} \qquad (3)$$

Hence by employing this open end winding configuration multilevel inverter operation can be achieved and the problems due to conventional inverters like common mode voltages can be overcome.

Decoupled PWM technique

The procedure described in [6] uses the instantaneous reference voltages and is based on the concept of 'effective Time'. The effective time is expressed as the time "For which the motor is supplied by the inverter voltage and is designated as Teff. The sampling time period is designated as *Ts*. The instantaneous phase reference voltages are acquired by projecting the tip of the reference vector Vsr on to the corresponding phase axes and these projections have to be multiplied with a factor of (2/3). The factor (2/3) is multiplied to the projections because of the 'Two to Three phase' transformation. These instantaneous phase reference voltages are indicated as Va*, Vb* and Vc*. The symbols Tga, Tgb and Tgc respectively signifies the time spell for which a given motor phase is connected to the positive rail of the input DC power supply of the inverter in the given sampling time period Ts. The timings Tga, Tgb and Tgc are labelled as the phase switching times. The procedure to generate the gating pulses for the individual devices using this algorithm is elaborately explained in [10].

For a dual inverter system, there would be two sets of phase switching times, one for each inverter. The phase switching

timings of inverter-1 are represented by the symbols Tga, Tgb and Tgc. while the symbols T'ga, T'gb and T'gc denote the same for inverter-2.

There are two distinct PWM strategies:

(1) Decoupled PWM strategy

(2) The alternative inverter switching strategy.

For achieving three level inverter operation with dual inverter configuration the space vector corresponds to the inverter-2 is overlaid on the space vector of inverter 1. After superimposing space vector of inverter-2 on space vector of inverter-1 there exists a centre hexagon (ABCDEF) with a centre named as O, covered by six sub hexagons signified as OBHGSF, OCJIHA, ODLKJB, OENMLC, OFQPND and OASRQE having centres at A, B, C, D, E, F respectively as shown in the Figure 5.

This Decoupled PWM strategy mainly focusses on the concept that the reference voltage space vector Vsr can be formulated with two opposite components Vsr/2 and -Vsr/2. Subtraction of the second component from the first component achieves the anticipated reconstruction of the reference vector. In other words, it is based on the observation that the effect of applying a vector with inverter-1 while inverter-2 assumes a null state is twice as that of applying the opposite vector with inverter-2 while inverter-1 assumes a null state [11,12]. The decoupled PWM strategy is as shown in the Figures 5 and 6.

It is worth noting that the phase axes of the motor viewed with reference to individual inverters are in phase opposition. In Figure 5 and 6, the vector OT signifies the actual reference voltage space vector, and it has to be synthesized from the dual-inverter system and is specified by |Vsr |∠α. This vector is resolved into two opposite components OT1 is (|Vsr/2 |∠α) and O'T2 is (|Vsr/2 | ∠1800 + α). The vector OT1 is synthesized by inverter-1 by switching among the states (8 1 2 7) while the vector O'T2 is reconstructed by inverter-2 by switching among the states (8' 5' 4' 7'). The advantage with the recommended decoupled control is that the inverter switching timings of both the inverters need not be computed. However, in this strategy, both the inverters have to be switched (Figure 8).

Simulation Results

For single inverter Fed IM drive with SVPWM control

A two level inverter fed induction motor drive is modelled and is simulated by employing space vector pulse width modulation(SVPWM) control technique with Modulation index=0.8(Under modulation) and the Modulated waveform, output pole voltage, Line voltage, phase voltages and the common mode voltage of the inverter are shown in Figure 7.

The performance characteristics of Induction motor drive i.e., Stator currents, Torque response, Speed response at no load are as shown in Figure 7. The motor achieves steady state at 0.3 sec. The Total Harmonic Distortion (THD) for the stator currents is 7.26% for this model.

For dual inverter Fed IM drive (three level inverter operation) with CSVPWM, DPWMMAX, DPWMMIN control

A dual inverter fed induction motor drive is modelled and is simulated by employing space vector pulse width modulation (CSVPWM) control technique and the Modulated waveform, output pole voltage, Line voltage, phase voltage and the common mode voltage of the inverter which resembles the characteristics of three level inverter characteristics are shown in Figure 9.

The performance characteristics of Induction motor drive i.e., Stator currents, Torque response, Speed response at no load are as shown in Figure 10. The motor achieves steady state at 0.25 sec. The Total Harmonic Distortion (THD) for the stator currents is 4.77% for this model and the common mode voltage is mitigated compared to single inverter fed IM drive.

A dual inverter fed induction motor drive is modelled and is simulated by employing space vector pulse width modulation (DPWMMAX) control technique and the Modulated waveform,

output pole voltage, Line voltage, phase voltages and the common mode voltage of the inverter which resembles the characteristics of three level inverter characteristics are shown in Figure 11.

The performance characteristics of of dual inverter fed Induction motor drive(three level inverter operation)with DPWMMAX control i.e., Stator currents, Torque response, Speed response at no load is as shown in Figure 12. The motor achieves steady state at 0.25 sec and the Total Harmonic Distorsion (THD) for the stator currents is 3.97% for this model and the common mode voltage is mitigated compared to single inverter fed IM drive.

A dual inverter fed induction motor drive is modelled and is simulated by employing space vector pulse width modulation (DPWMMIN) control technique and the Modulated waveform, output

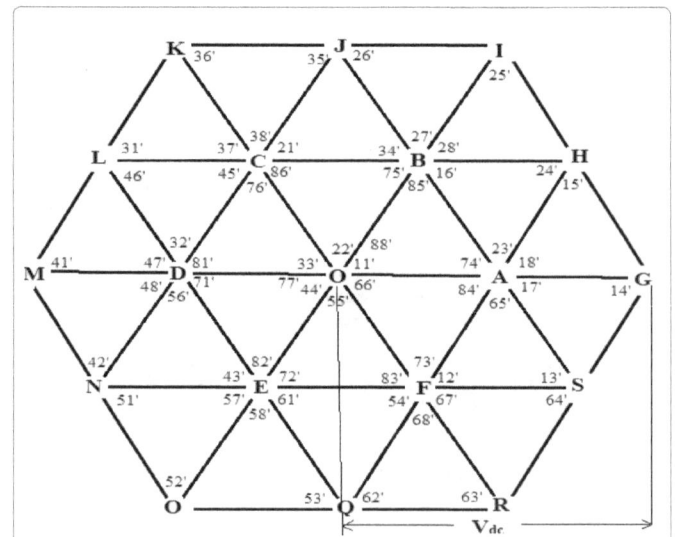

Figure 5: Resultant space vector combinations in the dual inverter scheme

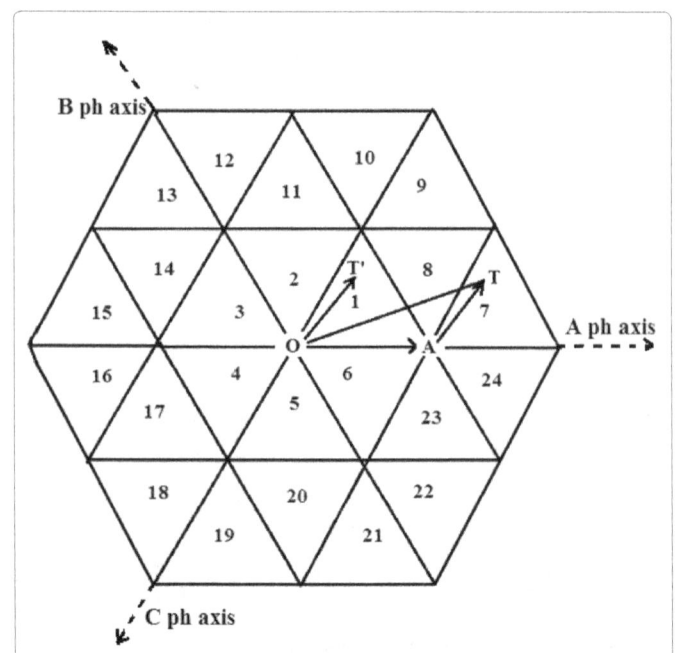

Figure 6: Reference vector for three level voltages.

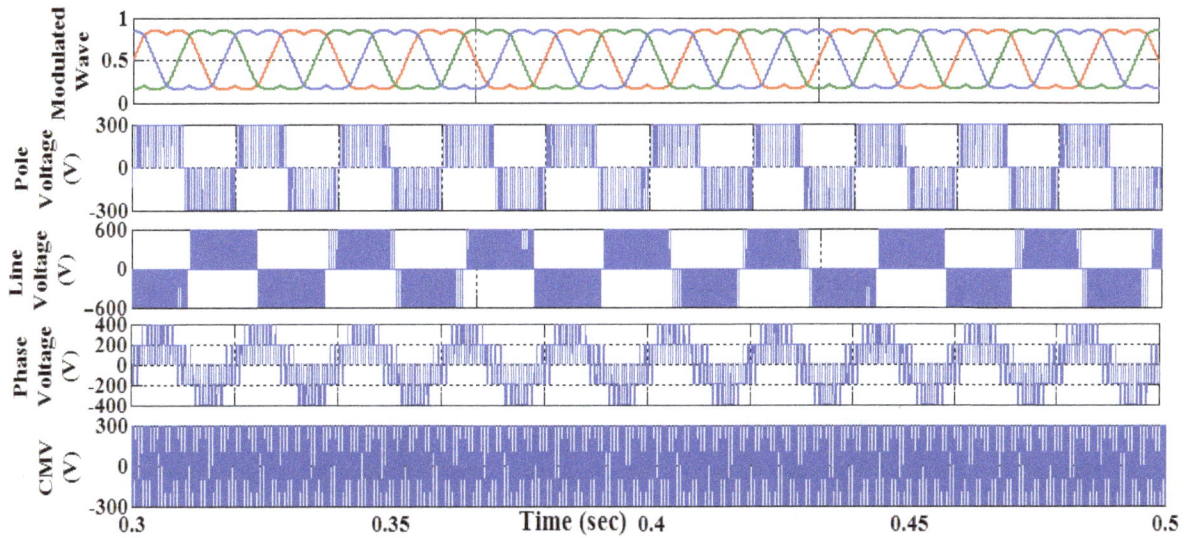

Figure 7: Modulated wave, Pole voltage, Line Voltage, Phase Voltage, CMV for Single inverter with SVPWM.

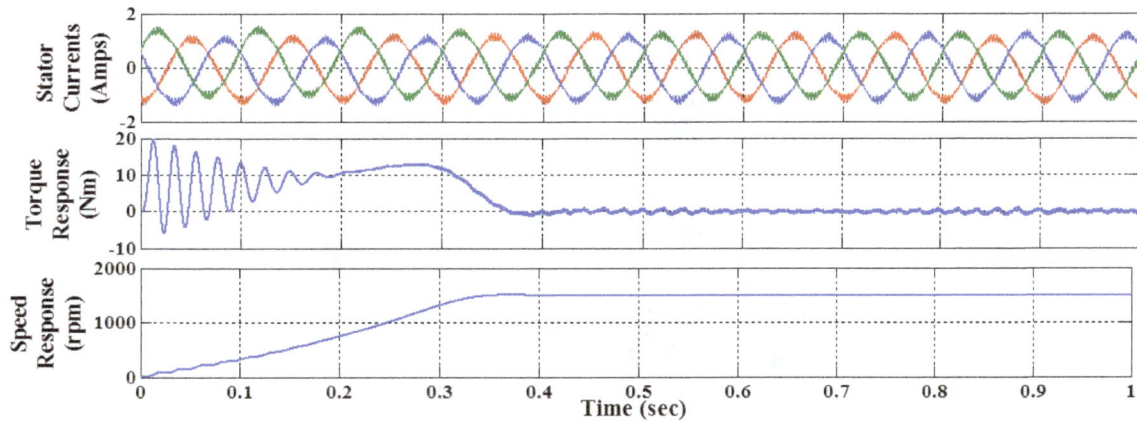

Figure 8: Performance characteristics of IM drive with single inverter (SVPWM) at steady state.

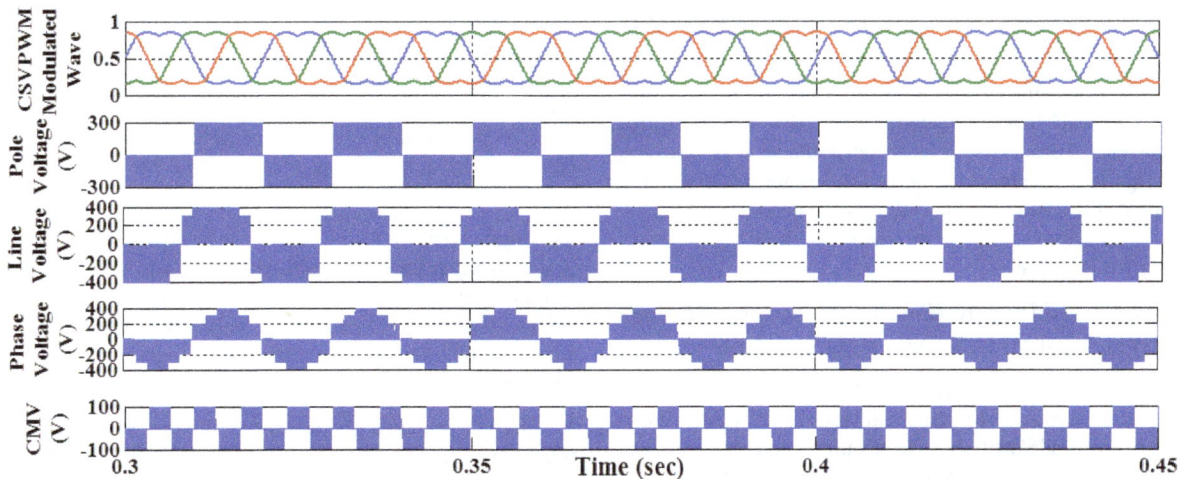

Figure 9: Modulating Wave, Pole voltage, Line Voltage, Phase Voltage, Common mode voltage for dual inverter (Three level inverter operation)with CSVPWM control technique.

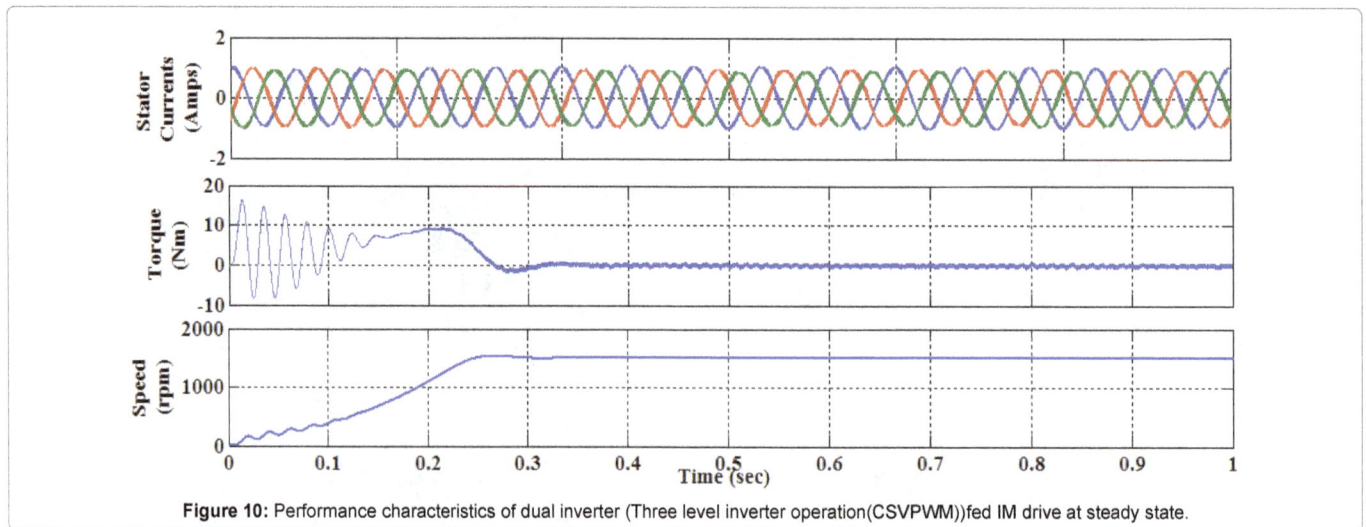

Figure 10: Performance characteristics of dual inverter (Three level inverter operation(CSVPWM))fed IM drive at steady state.

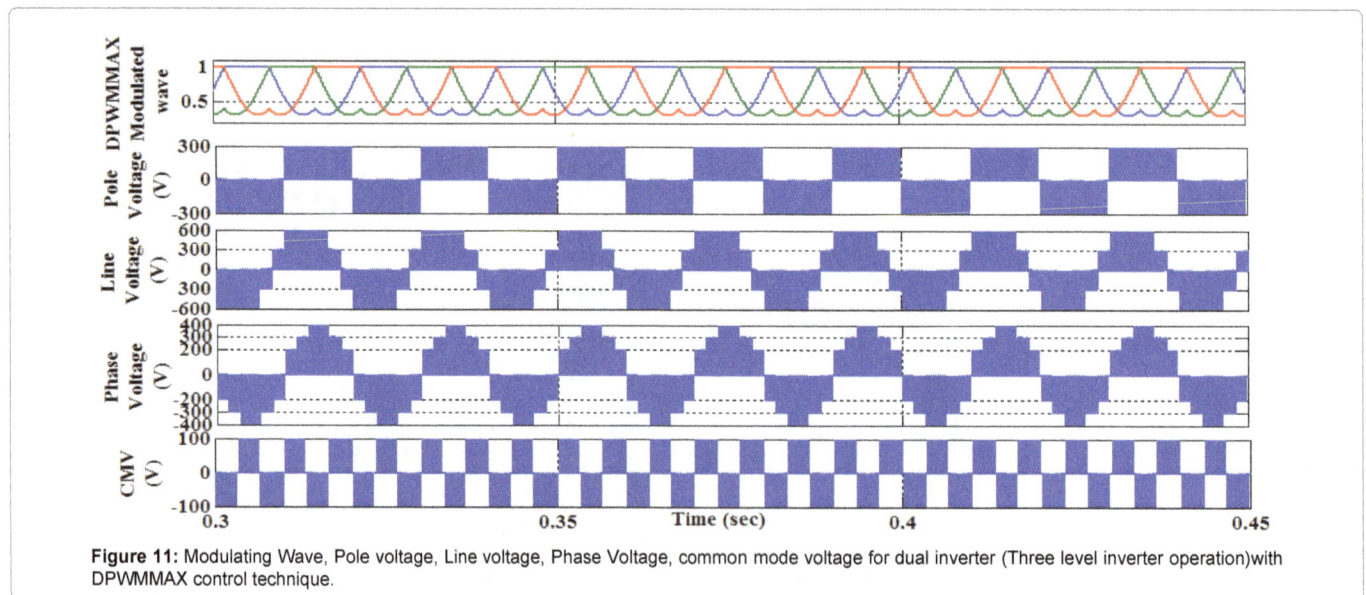

Figure 11: Modulating Wave, Pole voltage, Line voltage, Phase Voltage, common mode voltage for dual inverter (Three level inverter operation)with DPWMMAX control technique.

pole, Line voltage, phase voltages and the common mode voltage of the inverter which resembles the characteristics of three level inverter characteristics are shown in Figure 13.

The performance characteristics of of dual inverter fed Induction motor drive(three level inverter operation)with DPWMMIN control i.e., Stator currents, Torque response, Speed response at no load are as shown in Figure 14. The motor achieves steady state at 0.25 sec. The Total Harmonic Distortion (THD) for the stator currents is 3.85% for this model and the common mode voltage is mitigated compared to single inverter fed IM drive.

For dual inverter Fed IM drive(Four level inverter operation) with CSVPWM, DPWMMAX, DPWMMIN control

A dual inverter fed induction motor drive is modelled and is simulated by employing space vector pulse width modulation(CSVPWM) control technique and the Modulated waveform, output pole voltage, Line voltage, phase voltage and the common mode voltage of the inverter which resembles the characteristics of four level inverter characteristics are as shown in Figure 15.

The performance characteristics of of dual inverter fed Induction motor drive(four level inverter operation)with CSVPWM control i.e., Stator currents, Torque response, Speed response at no load are as shown in Figure 16. The motor achieves steady state at 0.25 sec. The Total Harmonic Distortion (THD) for the stator currents is 4.77% for this model and the common mode voltage is mitigated compared to dual inverter fed IM drive with three level inverter operation.

A dual inverter fed induction motor drive is modelled and is simulated by employing space vector pulse width modulation(DPWMAX) control technique and the Modulated waveform, output pole voltage, Line voltage, phase voltage and the common mode voltage of the inverter which resembles the characteristics of four level inverter characteristics are as shown in Figure 17.

The performance characteristics of dual inverter fed Induction motor drive (four level inverter operation) with DPWMMAX control i.e., Stator currents, Torque response, Speed response at no load are as shown in Figure 18. The motor achieves steady state at 0.25 sec. The Total Harmonic Distorsion (THD) for the stator currents is 3.85% for

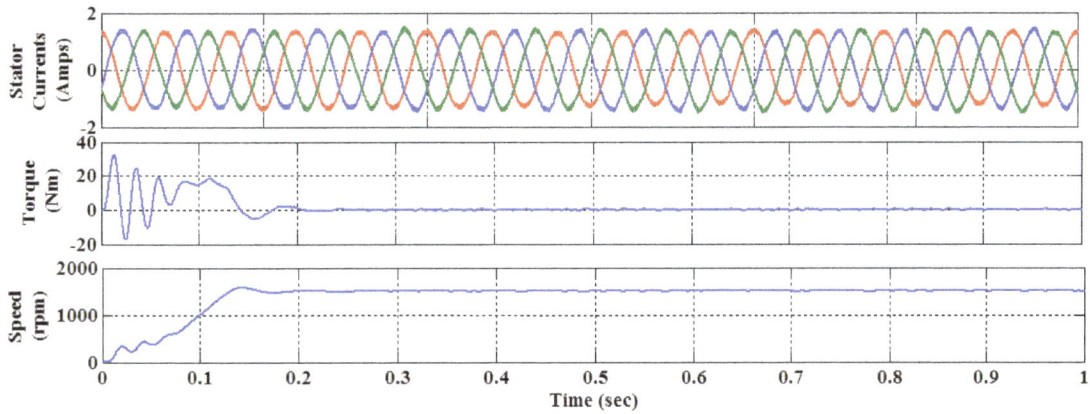

Figure 12: Performance characteristics of dual inverter (Three level inverter operation with DPWMMAX)fed IM drive at no load.

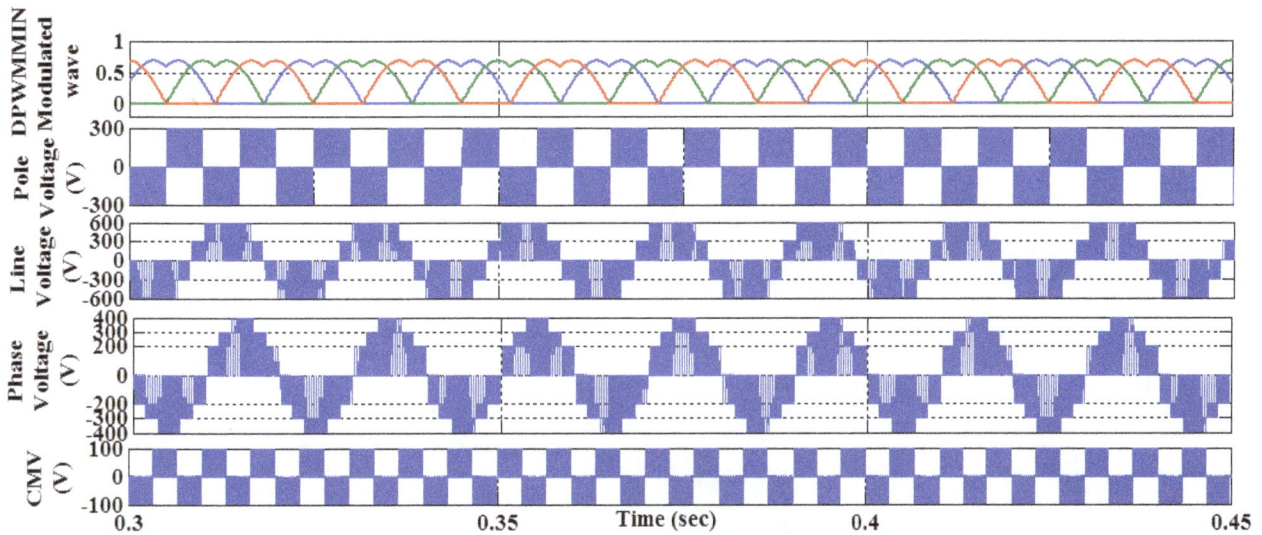

Figure 13: Modulating Wave, Pole voltage, Phase Voltage, common mode voltage for dual inverter (Three level inverter operations) with DPWMMIN control technique.

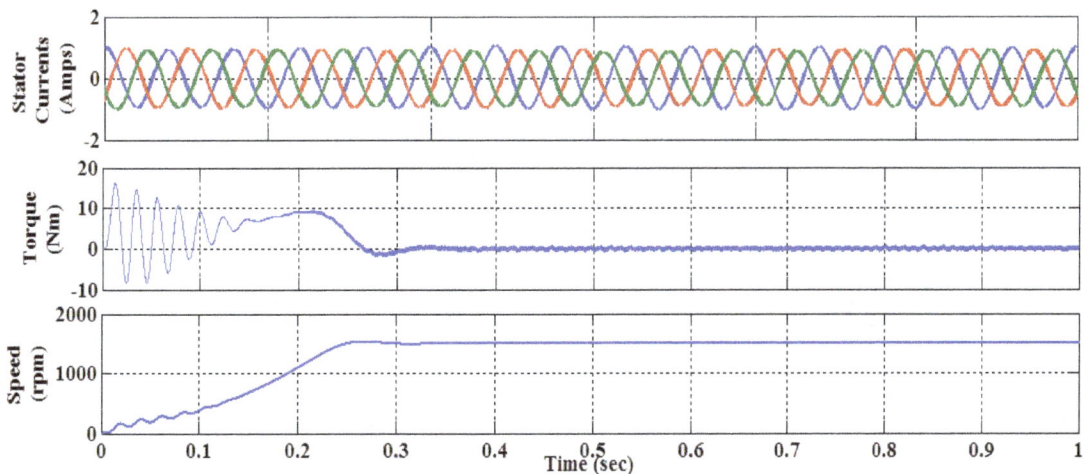

Figure 14: Performance characteristics of dual inverter (Three level inverter operation with DPWMMIN control) fed IM drive at steady state.

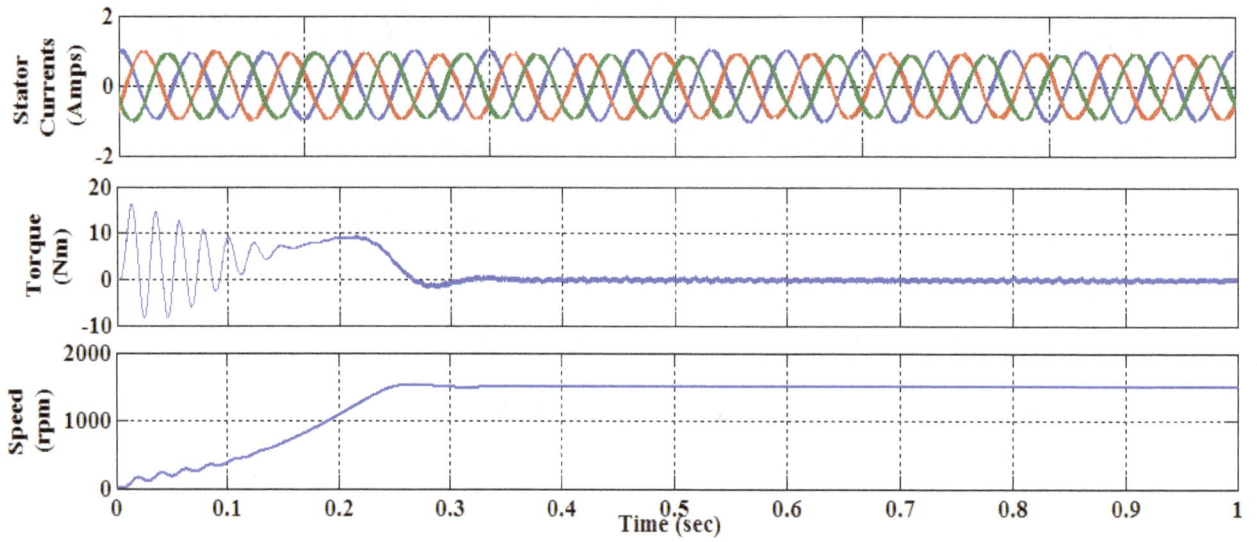

Figure 15: Modulated wave, Pole voltage, Line voltage, Phase voltage, Common mode voltage for dual inverter(Four level inverter operation with CSVPWM control).

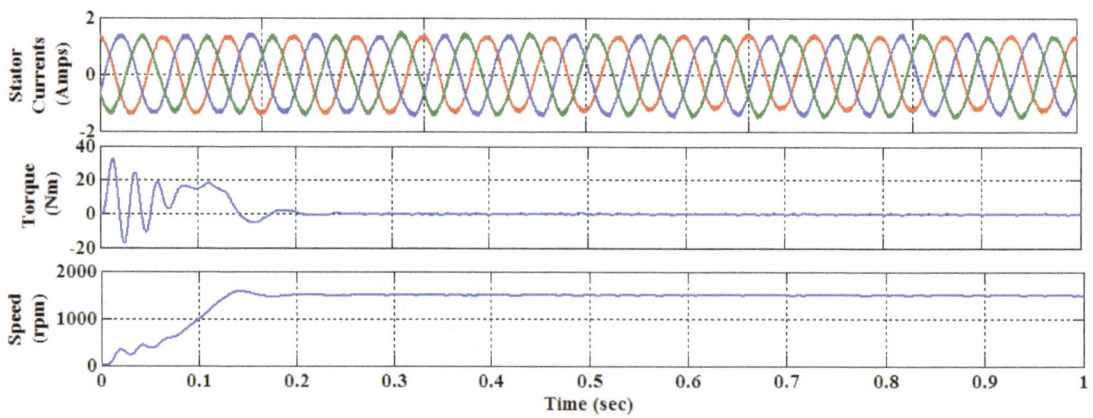

Figure 16: Performance characteristics of dual inverter (Four level inverter operation(CSVPWM))Fed IM drive at steady state.

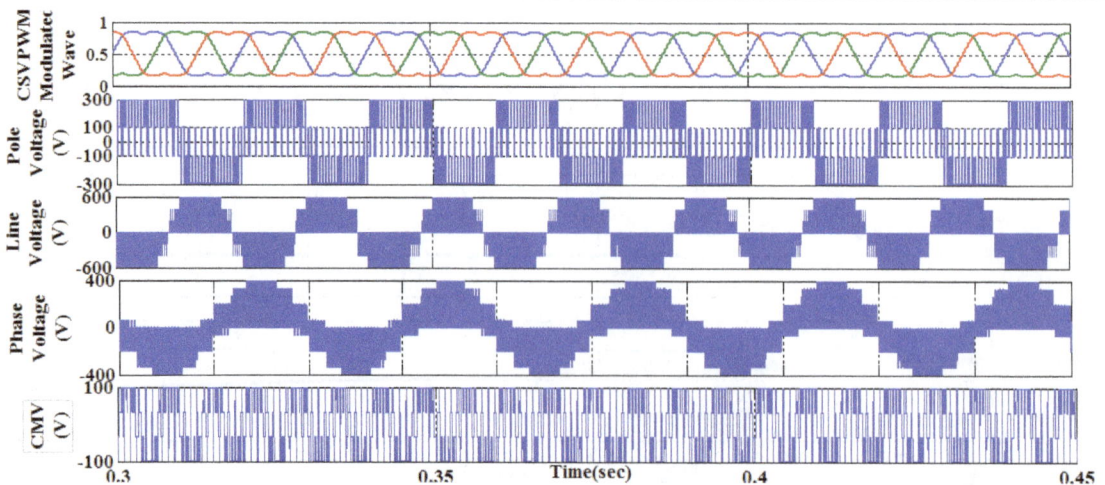

Figure 17: Modulated wave, Pole voltage, Line voltage, Phase voltage, Common mode voltage for dual inverter (Four level inverter operation with DPWMMAX control).

this model and the common mode voltage is mitigated compared to dual inverter fed IM drive with three level inverter operation.

A dual inverter fed induction motor drive is modelled and is simulated by employing space vector pulse width modulation(DPWMMIN) control technique and the Modulated waveform, output pole voltage, Line voltage, phase voltages and the common mode voltage of the inverter which resembles the characteristics of four level inverter characteristics are as shown in Figure 19.

The performance characteristics of of dual inverter fed Induction motor drive(four level inverter operation)with DPWMMIN control i.e., Stator currents, Torque response, Speed response at no load are as shown in Figure 20. The motor achieves steady state at 0.25 sec. The Total Harmonic Distortion (THD) for the stator currents is 3.43% for this model and the common mode voltage is mitigated

compared to dual inverter fed IM drive with three level inverter operation.

THD Comparison

The THDs for stator currents of IM drive is listed out as shown in the Table 1.

Conclusion

In this paper the implementation of dual inverter fed induction motor drive has been done. With the implementation of triangular based SVPWM the machine performance will be improved in the context of harmonic spectra and effective DC bus utilization over the conventional sinusoidal pulse width modulation technique. And the zero sequence voltage problems is also mitigated at a greater level compared to other mitigating techniques. This work can be

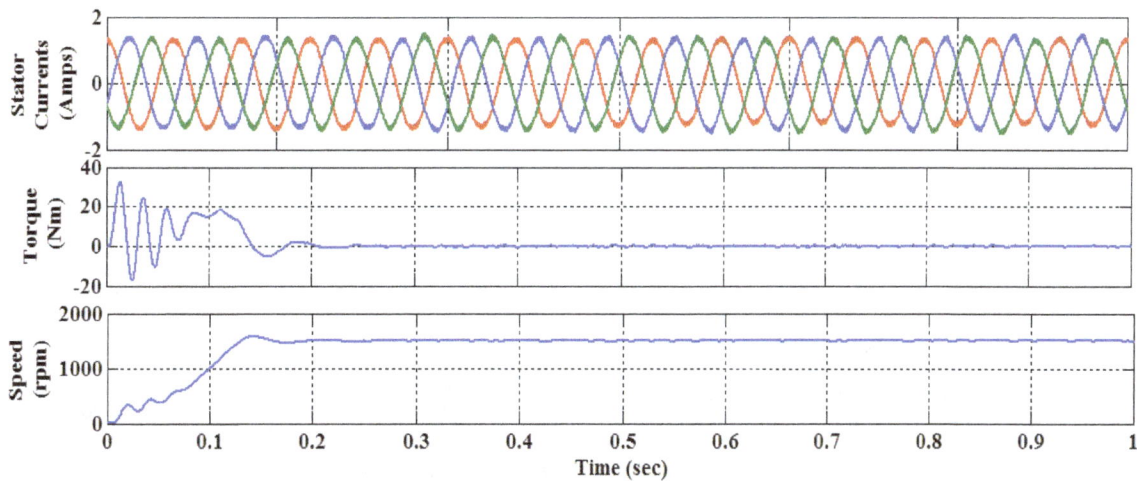

Figure 18: Performance characteristics of dual inverter (Four level inverter operation(DPWMMAX control)) fed IM drive at steady state.

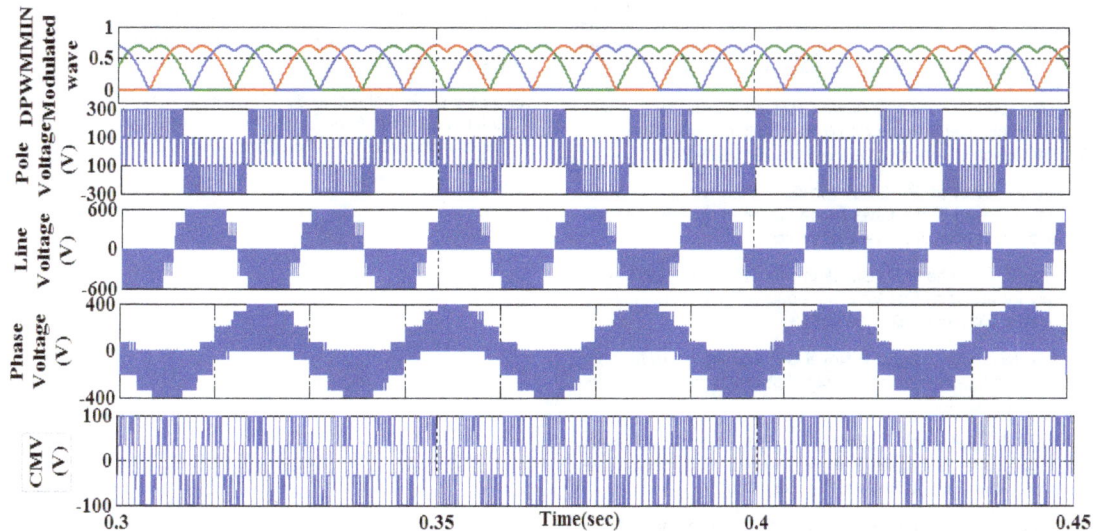

Figure 19: Modulated wave, Pole voltage, Line voltage Phase voltage, Common mode voltage for dual inverter (Four level inverter operation with DPWMMIN control).

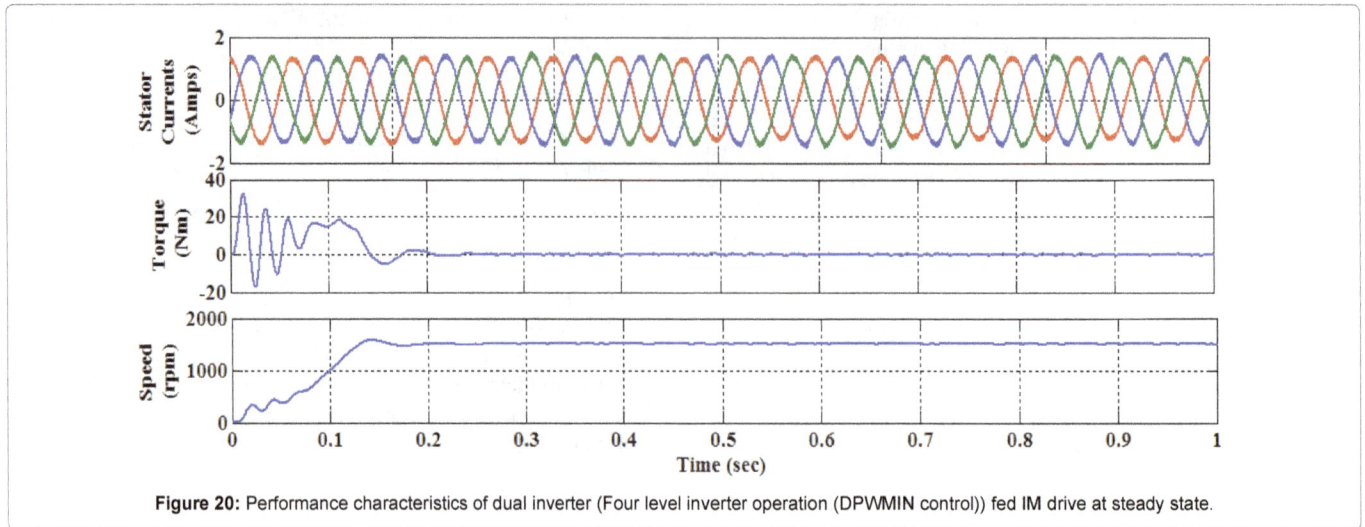

Figure 20: Performance characteristics of dual inverter (Four level inverter operation (DPWMIN control)) fed IM drive at steady state.

Invert-er Type	Control technique	THD of stator currents of the Motor (I_{THD})	THD of inverter voltage (V_{line} THD) In under modulation(M.I=0.8) region	THD of inverter voltage (V_{line}THD) In over modulation (M.I=1.154) region
2 level	SVPWM	7.26%	77.27%	44.11%
3 level	CSVPWM	4.77%	64.97%	22.37%
	DPWMMAX	3.97%	70.09%	24.57%
	DPWMMIN	3.85%	69.99%	23.75%
4 level	CSVPWM	4.51%	63.65%	25.16%
	DPWMMAX	3.85%	72.15%	26.99%
	DPWMMIN	3.43%	71.48%	25.55%

Table 1: Comparison of stator currents THDs for various control techniques.

extended with the implementation of SVPWM for higher level (5-7 levels) for dual inverter fed open end winding induction motor.

References

1. Gopakumar K, Baiju MR, Mohapatra KK, Kanchan RS (2004) A Dual Two Level Inverter Scheme With Common Mode Voltage Elimination for an Induction Motor Drive. IEEE Transactions On Power Electronics 19: 794-805.

2. Tekwani PN, Kanchan RS, Gopakumar K (2007) Dual Five-Level Inverter-Fed Induction Motor Drive With Common-Mode Voltage Elimination and DC-Link Capacitor Voltage Balancing Using Only the Switching-State Redundancy. IEEE transactions on industrial electronics 54: 2609-2617.

3. Mohan MR, Hiralal MS (2008) Five-Level Diode Clamped Inverter to Eliminate Common Mode Voltage and Reduce dv/dt in Medium Voltage Rating Induction Motor Drives. IEEE transactions on power electronics 23: 1598-1607.

4. Chaturvedi P, Jain S, Agarwal P (2012) Carrier-Based Common Mode Voltage Control Techniques in Three-Level Diode-Clamped Inverter. Hindawi Publishing Corporation Advances in Power Electronics.

5. Mondal G, Sivakumar K, Ramchand R, Gopakumar K (2009) A Dual Seven-Level Inverter Supply for an Open-End Winding Induction Motor Drives. IEEE transactions on industrial electronics 56: 1665-1673.

6. Satheesh G, Reddy TB, Babu CS (2011) Novel SVPWM algorithm for open end winding induction motor drive using the concept of imaginary switching times. Int J Advance Sci Tech 2.

7. Rosaiah N, Chalasani HK, Satheesh G, Reddy TB (2012) Decoupled Space Vector PWM for dual inverter fed Open End winding Induction motor drive. International Journal of scientific and Engineering Research 3.

8. Dr. Gowri SK, Venkata SM, Hakeem H (2012) Current Harmonic Analysis Of a Dual Two Level Inverter Fed Open-End Winding Induction Motor Drive Based On SVPWM Switching Strategies, International journal of Modern Engineering Research (IJMER) 2: 2561-2567.

9. Prasad NV, Reddy HV (2012) Comparative Study Of Various Multi Level Inverter Topologies for Vector Controlled Induction Motor Drive. International Journal Of Advanced Scientific And Technical Research 4.

10. Bharatiraja C, Raghu S, Rao P, Paliniamy KP (2013) Comparative Analysis of Different PWM Techniques to Reduce the Common Mode Voltage in Three-Level Neutral-Point-Clamped Inverters for Variable Speed Induction Drives, International Journal of Power Electronics and Drive System (IJPEDS) 3: 105-116.

11. Sajan C, Kumar P, Shiva T, Srilatha K (2014) A New Hybrid Cascaded Multilevel inverter Fed Induction Motor Drive With Mitigation of Bearing Currents and Low Common Mode Voltage, International Journal of Engineering Research and Development 10: 25-29.

12. Satheesh G, Brahmananda RT, Babu S (2013) SVPWM Based DTC of OEWIM Drive Fed With Four Level Inverter with Asymmetrical DC Link Voltages, International Journal of Soft Computing and Engineering (IJSCE) 3.

33

Time Synchronization Mechanism for Radio Interferometer Array

El Houssain Ait Mansour[1]*, Bruno Da Silva[2] and Karl-Ludwig Klein[3]

[1]Laboratoire d''etudes spatiales, et d'instrumentation en, astrophysique – LESIA, 18330 Nanc¸ay, France
[2]St'ephane Bosse, Station de radioastronomie de Nanc¸ay, Route de souesmes, 18330 Nanc¸ay, France
[3]Observatoire de Paris, LESIA, LESIA - Bat 14, 92195 Meudon, France

Abstract

In both wired and wireless networks, synchronization is an important service for a wide range of applications in distributed systems. This includes radio interferometry. However, literature researches show that many distributed protocols cannot satisfy sub-ns time synchronization accuracy due to asymmetric delay errors, accumulated jitters and because of the strategy used to adjust offsets. This paper proposes a global time synchronization algorithm in distributed networks. The algorithm allows us to minimize asymmetric delays with sub-ns accuracy better than 10ps and achieved global time synchronization. The theory of the algorithm used is presented and analysed to prove that it achieves global time synchronization. Simulation results are given to show the performance and limitations of the proposed algorithm.

Keywords: Time synchronization; Algorithm; Asymmetric delay; Adjustment; Sub-ns; Accuracy

Introduction

Time synchronization is a critical problem for infrastructure in any distributed networks including radio interferometer [1]. Many global time synchronization algorithms were developed previously, but few of them can reach sub-ns accuracy in the distributed networks [2-5]. Additionally, several system issues limit the accuracy, such as asymmetric delays, long propagation delays, number of nodes and other sources of delay causes by measurement method.

The Nanc¸ay Heliograph (RH) Radio [6] is an analog radio interferometer array that consists of 48 distributed antennas, It allows one to construct radio images of the solar corona in the 150-450 MHz frequency band [7]. The Radio Frequency (RF) signals are transported in coaxial cables with different lengths (50-3.2 km) from the antenna to the receiver. The main limitations of the current instrument are offset delays errors introduced by manual calibration between each antenna and receiver. Furthermore, the switching time between each frequency analysed introduce a latency in solar images, also reduces the signal-to-noise ratio. In order to increase radio image resolution and data processing flexibility, a new study has started on a digital radio interferometer system. It requires a 1 GHz Analog-to-Digital Converter (ADC) frequency clock for each antenna (full band sampling). The new system enables one to acquire simultaneously radio images [8] at different frequencies with adjustable resolution. The main problems of this system are: the distribution of the high-frequency clock required for each antenna, the different lengths of cables between individual antennas and receiver, which need global time synchronization with sub-ns accuracy. The main objective of this paper is to reach a sub-ns global time synchronization of a radio interferometer array as the Nanc¸ay Radioheliograph. This work is presented with these following steps:

• We propose an algorithm based on the computation of global delay offsets and the theory of global time synchronization adjustment in a distributed network (section III.A).

• We present iterative algorithm with generic implementation to compensate random asymmetric delays between master and slaves based on errors estimation and minimization (section III.B).

• We test the algorithm using measurements from the previous work on a distributed network with cables of lengths up to 3.2 km with different asymmetric delays. The performance and limitations of the algorithm are discussed (section IV).

Synchronization Protocols and Limitations

In distributed networks, physical clock drifts, temperature changing and measurements errors may reduce synchronization accuracy [9]. Identically, networks topology affects slightly synchronization accuracy. However, mean errors increase rapidly with number of nodes [5]. The biggest source of offset and jitter errors in synchronization algorithm systems from random delays between master AND slaves [10]. Offset delays estimation accuracy is important issue in time synchronization system [3]. The most important sources are the following:

• **Send time:** The time it takes the Master to construct a message

• **Access time:** Time delay to access to channel

• **Propagation delay:** Time delay between Master and slave, on the length of the connection

• **Receiver time:** Time it takes the slave to receive messages.

Most time synchronization protocols proved global time synchronization over Ethernet. They estimated and calibrated non deterministic sources of delays, mean offsets and jitters estimation with manual adjustment can increase PTP accuracy to sub-µs (delay ≃ 0, jitter ≃ ±70ns) [3]. Using precision PHTER, PTP protocol may achieved nanosecond software accuracy and sub-ns hardware accuracy (PPS) between master and slave clock [11]. Table 1 illustrates existing time synchronization protocols accuracy and implementation complexity.

---

*Corresponding author: El Houssain Ait Mansour, Department of Electronics, Observatoire de Paris/Station de Rasdioastronomie de Nançay, France
E-mail: eaitmansour@obspm.fr

Protocols	Accuracy	Layer	Implementation
NTP [12]	us	3	Complex
PTP [4,9,17]	ns	2	Complex
PPS [11]	ns	2	Complex
DTP [10]	ns	2	Complex
GPS [10]	ns	2	Complex
WR [15,16]	sub-ns	1-2	Complex

Table 1: Existing time synchronization protocols accuracy and implementation complexity.

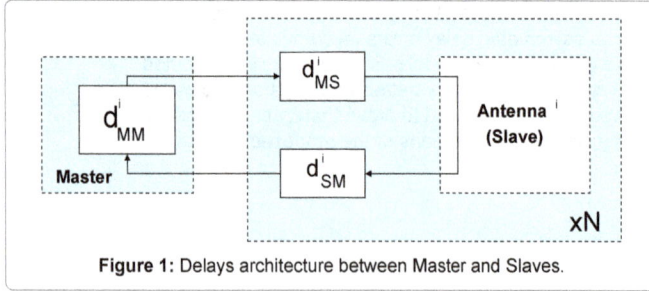

Figure 1: Delays architecture between Master and Slaves.

Full Synchronization Algorithm

Offsets compensation

In this section we present the main contributions of this paper. We describe the proposed algorithm for global time synchronization based on the previous researches in distributed networks. Consider an interferometer array with N antennas (i=1, 2,..., N), where each antenna (i.e., each slave) has round trip delay d_{MM}^i (Figure 1) [12].

$$d_{MM}^i = d_{MS}^i + d_{SM}^i \tag{1}$$

Eq. (1) presents basic model delay between Master and Slave, d_{MS}^i includes transmission delay, channel delay and reception delay, and so does d_{SM}^i. Consider two slaves (i, i+1). Here we make assumptions; we neglected asymmetric delays between master and slave ($d_{MS}^i = d_{SM}^i$). Additionally, Master to Master delays assumed different for each slave ($d_{MM}^i \neq d_{MM}^{i+1}$). Let T_i be the delays between master and slave before offset delays compensation and:

$$\begin{cases} d_{MM}^i = \delta_i \\ d_{MM}^{i+1} = \delta_i + \delta_0^i \\ T_i = d_{MM}^i \end{cases} \tag{2}$$

Let T_i' be the delays between master and slave after offsets compensation. The average between each delay pair T_i' and T_{i+1}' is given by Eq. (3).

$$\Delta T_{i,i+1}' = T_i' - T_{i+1}' = \frac{1}{2}[T_i' - T_{i+1}'] = \Delta T_{i,i+1}' \tag{3}$$

For global synchronization, it is necessary to set to zero time difference between each slave pair, the synchronization condition is $\Delta T_{i,i+1}' = 0$ Eq. (4)-(8) shows how to synchronize between antennas pairs (1,2).

$$\Delta T_{1,2}' = T_1' - T_2' \Rightarrow T_1 = T_2 \Rightarrow d_{MM}^1 = d_{MM}^2 \tag{4}$$

$$\Rightarrow 2\delta_1 + \delta_0^1 = 2\delta_2 + \delta_0^2 \Rightarrow \delta_1 + \frac{1}{2}\delta_0^1 = \delta_2 + \frac{1}{2}\delta_0^2 \tag{5}$$

Let's insert two unknown offsets δx_1 and δx_2 between Master and pair antennas (1,2), then :

$$\delta_1 + \frac{1}{2}[\delta_0^1 + \delta x_1] = \delta_2 + \frac{1}{2}[\delta_0^2 + \delta x_2] \tag{6}$$

$$\delta x_1 = T_2 - T_1 + \delta x_2 \Rightarrow \delta x_1 = \Delta T_{1,2} + \delta x_2 \tag{7}$$

$$\delta x_1 = \Delta T_{i,i+1} + \delta x_{i+1} \tag{8}$$

Offsets compensation vector for N antennas is given by Eq. (9):

$$\begin{bmatrix} \delta x_1 \\ \delta x_2 \\ .... \\ \delta x_{N-1} \end{bmatrix} = \begin{bmatrix} \Delta T_{1,2} \\ \Delta T_{2,3} \\ .... \\ \Delta T_{N-1,N} \end{bmatrix} + \begin{bmatrix} \delta x_2 \\ \delta x_3 \\ .... \\ \delta x_N \end{bmatrix} \tag{9}$$

In the offset delays compensation process, we introduced $\frac{\delta x_N}{2}$ in the path of the slave clocks. According to the assumption (symmetric delays), all clocks arrive simultaneously at the slave side ($\delta x_i + d_{MM}^i = \delta x_{i+1} + d_{MM}^{i+1}, 1 \leq i \leq N$) (Figure 2).

Asymmetric delay error calibration

In this section we present an iterative algorithm to compute offsets vector (9). This algorithm enables us to reach global time synchronization in distributed networks. Here we make assumptions; we considered asymmetric delays between master and slave Eq. (10).

$$\begin{cases} d_{MS}^i \neq d_{SM}^i \\ d_{MM}^i \neq d_{MM}^{i+1} \end{cases} \tag{10}$$

Let μ and M_{in} be the adjustable parameters. Let d_{MS}^A and d_{MM}^A defines master to slave delay and round trip delay after calibration. Where d_{MM}^S present ascending sort delays vector (Figure 1) allows to get a positive difference delay between each slave pairs (i,i+1) ($\Delta T(i)>0$) (Figure 3). In the offsets calibration step in the algorithm, the last slave set as reference (slave with maximum round trip delay) ($\delta x(N)$ set to 0) (Figure 3).

Let M_{in} define synchronization accuracy and μ define algorithm step-size. Values of the two factors affect directly the convergence or divergence and synchronization period (T_{Sync}) to achieve the required accuracy (time convergence). For each iteration, the proposed algorithm estimate global mean error for all slaves after offsets calibration (μe) (Eq. (11)). This error is due to asymmetric delays between master and slaves. When the required accuracy is not reached, the algorithm updates d_{MM} according to Eq. (12).

$$\mu_e = \frac{1}{N}\sum_{i=1}^{N} d_{MS}^{A(i)} - d_{MS}^i \tag{11}$$

$$d_{MM}^{i+1} = d_{MM}^i - sign(\mu_e) \times \mu \times d_{MS}^A \tag{12}$$

Results and Discussions

In the previous sections we proposed global time synchronization theory and a possible implementation. We now proceed to evaluate

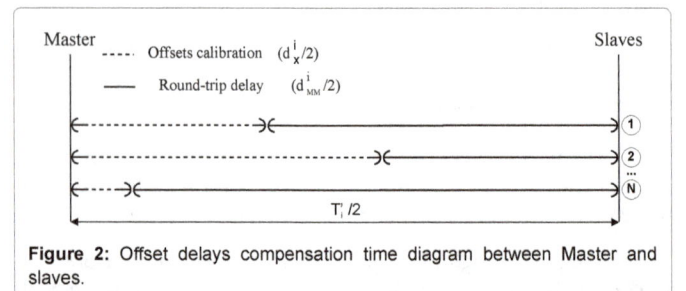

Figure 2: Offset delays compensation time diagram between Master and slaves.

```
variable  μ, Min, T_sync ;
Get  d_MM   for each antennas;
Repeat
         /*  Offset delays computation      */
         d_MM^s   ← sort (d_MM);
         for i=1 to N − 1 do
                   Δ T (i) ← d_MM^s  (i + 1) − d_MM^s  (i);
         end for;
         δ_x (N) ← 0;
         for i=N − 1 to 1 do
                   δ_x (i) ← Δ T (i) + δ_x (i + 1);
         end for;
         /* Offsets calibration      */
         d_MS^A   ← ½ [d_MM^s  + δ_x ];
         /* Error estimation     */
         e ← d_MS^A  − d_MS;
         μ_e ← Mean (e);
         /* Delays updating     */
         if not (|μ_e| < Min)
                   d_MM   ← d_MM  − sign (μ_e) * μ * d_MS^A;
                   T_sync   ← T_sync  + 1;
         end if;
Until (|μ_e| < Min)
```

Figure 3: Pseudo-code for global synchronization algorithm include asymmetric delay calibration between master and slave.

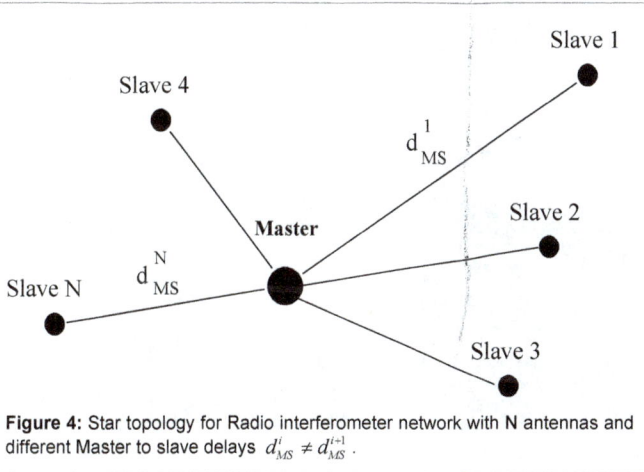

Figure 4: Star topology for Radio interferometer network with N antennas and different Master to slave delays $d_{MS}^i \neq d_{MS}^{i+1}$.

the performance of proposed algorithm and limitations in radio interferometer array. Algorithm delay parameters had chosen [13].

Consider the interferometer array with N antennas with star topology sketched in Figure 4. The 1550 ns and 1390 ns fiber optics link are used between master and each slave. Let L_i be the length of fiber link between master and slave (i).

Asymmetric delay error between master and slave decrease slightly increasing fiber link length (Table 2) [13]. The linear approximation of the average is

$$Average(ns)^i \simeq -7.8 \times 10^{-5} \times L^i + 16 \qquad (13)$$

Let e_{Ad}^i define asymmetric delay error between Master and slave (i) $e_{Ad}^i = d_{MS}^i - d_{SM}^i$ and mean $e_{jit} \sim N(\mu_{jit} = 50ps, \sigma_{jit} = 10ps)$. Consider random jitter due to repeatability of master to slave delay measurement with 50ps and 30ps (Eq. (14) of standard deviation (Table 2) [13]. Then the global error is the accumulation of jitter error and asymmetric delay error.

μ	μ_0 (ps), N=10	μ_0 (ps), N=100	μ_0 (ps), N=1000
10^{-3}	14.11	79.74	98.64
10^{-4}	-7.42	-9.4	-10.23
10^{-5}	1.17	1.15	0.65
10^{-6}	0.10	4.54×10^{-2}	4.32×10^{-2}
10^{-7}	1.112×10^{-2}	1.19×10^{-2}	1.01×10^{-2}
10^{-8}	-7.73×10^{-4}	1.13×10^{-3}	6.97×10^{-3}

Table 2: Offset mean error μ_e after calibration versus μ and number of slaves N.

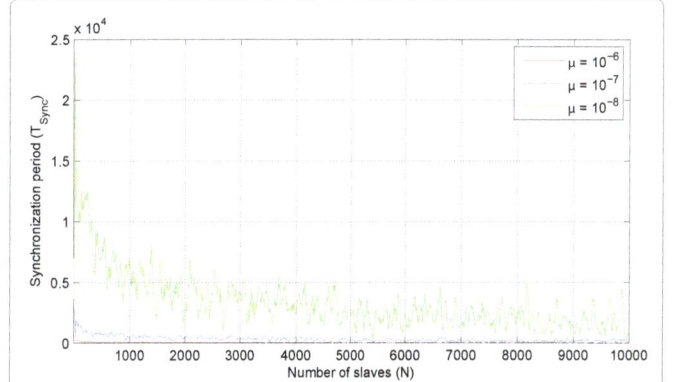

Figure 5: Synchronization period (T_{Sync}) vs. number of slaves (N) for different values of algorithm parameter (μ).

$$e_{jit} \sim N(\mu_{jit} = 50ps, \sigma_{jit} = 10ps) \qquad (14)$$

Assuming the linear approximation for the link delay, one gets.

$$d_{MS}^i (ns) \simeq 9.8 \times L^i + 470 \qquad (15)$$

In order to simulate the overall performance of the synchronization algorithm and its limitations in a distributed radio array, we consider star topology sketched in Figure 4 a cable of length L. Delays and jitters are generated according to Eq. (14)-(15). We set adjustable algorithm parameter M_{in} to 10^{-12}. The number of slaves (N) varies from 10 to 10^4. Step-size μ may take 10^{-4}, 10^{-7} or 10^{-8}. Figure 5 illustrate simulation results.

Figure 5 shows the synchronization period (T_{Sync}) (Figure 6) versus number of slaves (N) for each value of μ. We observe a fast decrease of T_{Sync} for increasing N when N ≤ 1000, it stays slightly steady over 2000 slaves for each value of μ [T_{Sync} (μ=10^{-7}, N=10) = 2397, T_{Sync} (μ=10^{-7}, N=10^4) =191]. However, it is clear that the synchronization period increase significantly with decreasing μ [T_{Sync} (μ=10^{-7}, N=10)=2397, T_{Sync} (μ=10^{-6}, N=10)=209].

We consider number of slave N is set to 100, step-size parameter μ set to 10^{-5}. M_{in} is set to 0 (infinity loop in Figure 3). Figure 6 shows global mean error μ_e (ps) after calibration between master and slaves versus number of iteration (l) (Figure 3). The mean error after calibration decreased linearly versus (l) in the synchronization period T_{Sync}. Thereafter the error oscillated around global algorithm accuracy μ_0 with peak-to-peak jitter e_{pp} (Figure 5). After the simulation analysis the mean error μ_e it can be presented as fallows at the oscillation zone (Finite error in Figure 6):

$$\mu_e = \mu_0 \pm \frac{e_{pp}}{2}, e_{pp} \sim \mu_0 \qquad (16)$$

To evaluate the accuracy of the algorithm and compare it with algorithm in reports [13-15], we analyzed an algorithm mean error (Eq. (16)) for each number of slaves (N=10, 100 or 1000) and varying

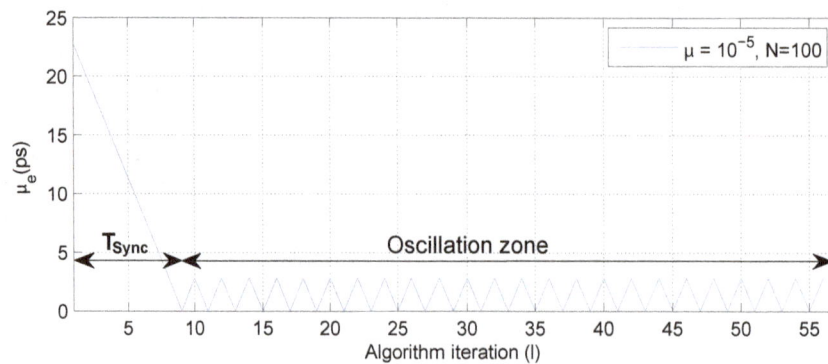

Figure 6: Global mean error μ_e (ps) after calibration between master and slaves versus number of iteration (l) for N=100 and $\mu=10^{-5}$.

μ. M_{in} is set to 0. Table 2 illustrates the accuracy of the global time synchronization μ_0 in picoseconds versus step-size parameter μ. Algorithm maintained sub-ns accuracy for several values of μ and N (Table 2). It is clear that μ_0 decreased rapidly with decreasing μ. It changed from 14.11ps for (N=10, $\mu=10^{-3}$) to -7.73×10^{-4} for (N=10, $\mu=10^{-8}$). On the other hand, we can see that the important is N (N $\geq$ 100), the better accuracy was well when $\mu \leq 10^{-5}$. We also observed that the accuracy was much better when N=10 and $\mu \geq 10^{-4}$ compared to others (Table 2).

Quite surprisingly, the synchronization period (T_{Sync}) decreased rapidly with increasing number of slaves (Figure 5). This may be induced by the varying length of the error estimation vector (N) (Figure 3). The more bigger N, the better is the mean error estimation. This means that the time synchronization period is minimal in widely distributed arrays.

The proposed algorithm may achieve a prescribed accuracy (M_{in}). It depends on step-size parameter (μ) and number of slaves (N) (Table 2). However, in order to obtain a converging algorithm, M_{in} have to be chosen greater than $|\mu e|$ (Eq. (16)), because the limiting accuracy of the algorithm is fixed by μ_e. In the perfect case, M_{in} has to be chosen close to $|\mu e|$. Furthermore, from results shown in Figure 5, we can see that the synchronization period increase rapidly with μ. According to these results, we concluded that we may control global algorithm accuracy by adjusting the parameters M_{in} and μ.

The above evaluation shows that global time synchronization can affectively achieved with sub-ns accuracy in distributed networks. The proposed algorithm can compensate offset asymmetric delays with adjustable accuracy bellow 10 ps. (Table 2). The proposed algorithm performed better then [13,16]. Therefore, accuracy may reduce in real hardware implementation [17,18].

Conclusions and Future Work

Overall, in this paper we suggested global time synchronization algorithm in distributed network such as a radio interferometer array with sub-ns accuracy. The main result of this research is an iterative algorithm for asymmetric offset delays error compensation and minimization. We illustrate that the synchronization period and accuracy can be controlled with algorithm parameters. The proposed algorithm can achieved accuracy better than 10ps. Besides, algorithm hardware implementation may be possible and computational complexity is reduced comparing with algorithms developed previously (Table 2).

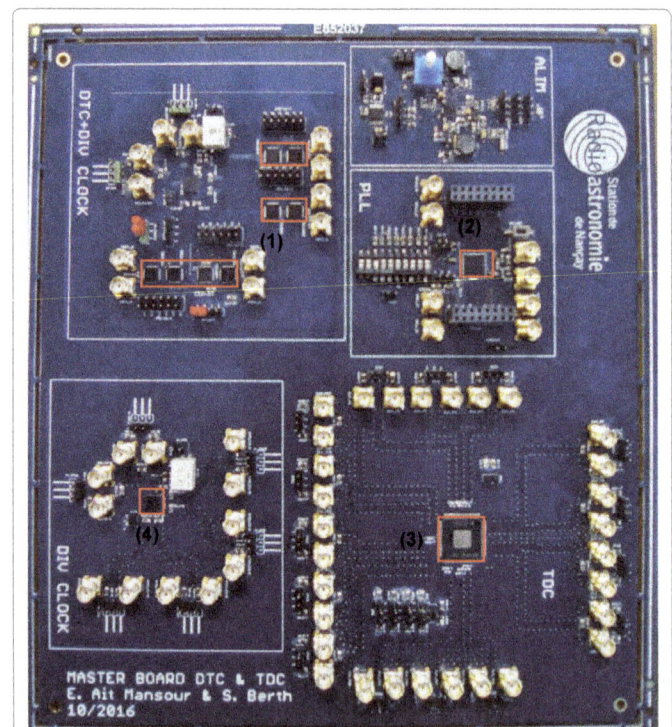

Figure 7: Global synchronization Master hardware architecture basing on TDC and DTC devices.

Master Hardware architecture (Figure 7) for global time synchronization algorithm implementation and testing in radio interferometer networks of 4 slaves is on-going. This board is based on:

i. Programmable delay chips (MC100EP195) for offsets delay adjustment with 10ps accuracy.

ii. Ultra low jitter (100fs) cleaner PLL for clocks generation (LMK03328).

iii. Quad channel time measurement device TDC (THS788) with 8ps accuracy and 13ps jitter

iv. Low skew (8ps) clocks distribution (1:4).

Finally, Virtex-6 FPGA board (ML605) will be used for offset delays computation and adjustment.

Acknowledgment

The authors would like to thank Exploration Spatiale des Environnements Plantaires (ESEP), for thesis funding.

References

1. Sutinjo AT, Colegate TM, Wayth RB (2015) Characterization of a Low-Frequency Radio Astronomy Prototype Array in Western Australia. IEEE Transactions on Antennas and Propagation 63: 5433-5442.

2. Weibel H (2005) High Precision Clock Synchronization According to IEEE 1588- Implementation and Performance Issues. Embedded World, pp: 22-24.

3. Freire I, Sousa I, Klautau A, Almeida I, Lu C, et al. (2016) Analysis and Evaluation of End-to-End PTP Synchronization for Ethernet-based Fronthaul. IEEE Global Communications Conference (GLOBECOM).

4. Mahmood A, Exel R, Sauter T (2014) Delay and Jitter Characterization for Software-Based Clock Synchronization over WLAN Using PTP. IEEE Transactions on Industrial Informatics 10: 1198-1206.

5. Steup C, Zug S, Kaiser J, Breuhan A (2014) Uncertainly aware Hybrid Clock Synchronization in Wireless sensor Networks. UBICOMM 2014: The Eighth International Conference on Mobile Ubiquitous Computing, Systems, Services and Technologies, pp: 246-251.

6. Klein KL, Kerdraon A (2011) Solar Physics at Nancay Radio Observatory: Recent Developments. XXXth URSI General Assembly and Scientific Symposium.

7. Kerdraon A, Delouis JM (1997) Coronal Physics from Radio and Space Observations. Springer LNP 483: 192-201.

8. Kerdraon A, Pick M, Hoang S, Wang Y, Haggerty D (2010) The Coronal and Heliospheric 2007 May 19 event: coronal mass ejection, Extreme Ultraviolet Imager wave, radio bursts, and energetic electrons. The American Astronomical Society.

9. Elsts A, Duquennoy S, Fafoutis X, Oikonomou G, Piechocki R, et al. (2016) Microsecond Accuracy Time Synchronization Using the IEEE 802.15.4 TSCH Protocol. IEEE 41st Conference on Local Computer Networks Workshops (LCN Workshops).

10. Lee KS, Wang H, Shrivastav V, Weatherspoon H (2016) Globally Synchronized Time via Datacenter Networks. SIGCOMM conference, Florianopolis, Brazil, pp: 454-467.

11. Texas Instrument (2013) AN-1728 IEEE 1588 Precision Time Protocol Time Synchronization Performance. Application Report SNLA098A, 10.

12. Novick AN, Lombardi MA (2015) Practical Limitations of NTP Time Transfer. Joint Conference of the IEEE International on Frequency Control Symposium & the European Frequency and Time Forum (FCS).

13. Gong G, Chen S, Du Q, Li J, Liu Y (2011) Sub-nanosecond Timing System Design and Development for LHAASO Project. Proceedings of ICALEPCS2011, Grenoble, France.

14. Lpez MJ, Gutierrez Rivas JL, Alonso JD (2014) A White-Rabbit Network Interface Card for synchronized sensor networks. IEEE SENSORS.

15. Serrano J, Cattin M, Gousiou E (2013) The White Rabbit Project. Proceedings of IBIC2013, Oxford, UK.

16. Dierikx EF, Wallin AE, Fordell T, Myyry J, Koponen P, et al. (2016) White Rabbit Precision Time Protocol on Long-Distance Fiber Links. IEEE Trans Ultrason Ferroelectr Freq Control 63: 945-952.

17. Braun M, Juranek M, Szll A, Sznt P, Marn C (2016) Nanosecond Synchronous Analog Data Acquisition over Precision Time Protocol. European Telemetry and Test Conference, Germany.

18. Serrano J, Alvarez P, Cattin M, Cota EG, Lewis J, et al. (2009) The White Rabbit Project. Proceedings of ICALEPCS2009, Kobe, Japan.

34

Starting Time Calculation for Induction Motor

Abhishek Garg* and Arun Tomar

L&T Construction, Manapakkam, Chennai, Tamilnadu, India

Abstract

This paper presents the starting time calculation for a squirrel cage induction motor. The importance of starting time lies in determining the duration of large current, which flows during the starting of an induction motor. Normally, the starting current of an induction motor is six to eight time of full load current. Plenty of methods have been discovered to start motor in a quick time, but due to un-economic nature, use are limited. Hence, for large motors direct online starting is most popular amongst all due to its economic and feasible nature. But large current with DOL starting results in a heavy potential drop in the power system. Thus, Special care and attention is required in order to design the healthy system.

A very simple method to calculate the starting time of motor is proposed in this paper. Respective simulation study has been carried out using MATLAB 7.8.0 environment, which demonstrates the effectiveness of the starting time calculation.

Keywords: Induction motor; Starting time; Direct online starting

Introduction

Motors in modern industrial systems are becoming larger due the heavy applications requirement. Some are considered large even in comparison to the total capacity of large industrial power systems. Starting of such large motors can cause adverse effects to any locally connected load, other motors and also to buses, which are electrically remote from the point of motor starting.

Ideally, a motor-starting study should be done prior to purchase of a large motor. The motor manufacturer shall provide the value of starting voltage requirement and preferred locked-rotor current. A motor-starting study should be done if the motor horsepower exceeds approximately 30% of the supply transformer(s) base kVA rating, if no generators are present. Whereas, If generator is present, and no other sources are involved, an analysis should be done whenever the motor horsepower exceeds 10-15% of the generator kVA rating, depending on actual generator characteristics [1].

Squirrel cage induction motor is most commonly used motor in the world due to its simple design, less maintenance and simple operation. The rating of induction motor is available from fraction of watts to Mega-Watts. It can be used for different type of applications based on torque speed characteristics requirements such as constant power, constant torque, torque increases in proportion to speed, torque increases with the square of speed, torque decreases in inverse proportion to speed [2].

Many conference papers have been published on motor starting with various starting methods and starting time calculations, but none of them compares the outcome with the simulation software.

This paper presents the comparative study between mathematical calculation and MATLAB result of motor starting time [3].

Basics of Induction Motor

The basic operation of induction motor is similar to transformer, where the stator acts as the primary side of transformer and the rotor as secondary of transformer as shown in Figure 1. At time of starting, the voltage induced in the induction motor rotor is maximum because slip will be maximum (S=1). Since the rotor impedance is low, the rotor current is excessively large. Due to transformer action this large rotor

current is reflected in the stator [4]. This results in large starting current (nearly 6 to 8 times the full-load current) in the stator at low power factor and consequently the value of starting torque is low. This large current does not harm the motor due to short duration.

However, this large starting current will produce large drop in line-voltage [1]. Table 1 represents the minimum allowable voltage levels required, when motor starting is taken into consideration.

The starting torque and starting current also depends on motor class as shown in Table 2. This change is due to change in value of X1 and X2 with different class of motors as shown in Figure 1.

So if we are changing X1 and X2, then starting time will also change. If this time is more than limit value, this large current will adversely affect the other electrical equipment connected to the same bus. Figure 2 shows the equivalent circuit of induction motor.

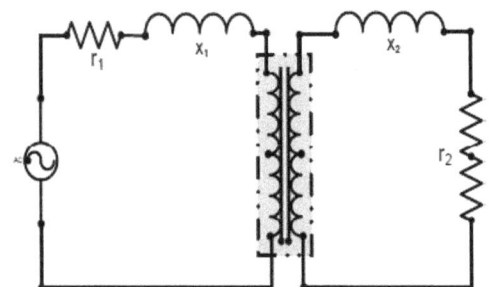

Figure 1: Basic circuit of induction motor.

*Corresponding author: Abhishek Garg, Sr. Design Engineer, L&T Construction, Manapakkam, Chennai, Tamilnadu, India
E-mail: abhishekgarg201@gmail.com

Table 1: Summary of representative critical system voltage levels when starting motors (IEEE 399-1997).

Voltage drop location or problem	Minimum allowable voltage (% rated)
At terminals of starting motor	80%
All terminals of other motors that must reaccelerate	71%
AC contactor pick-up (by standard) (see 9.8, NEMA standards)	85%
DC contactor pick-up (by standard) (see 9.8, NEMA standards)	80%
Contactor hold-in (average of those in use)	60%ª-70%
Solid-state control devices	90%
Noticeable light flicker	3% change

Table 2: Empirical distribution of leakage reactance in induction motors (IEEE 112).

Motor Class	Description	Fraction of X 1 + X 2	
		X 1	X 2
A	Normal starting torque, normal starting current	0.5	0.5
B	Normal starting torque, low starting current	0.4	0.6
C	High starting torque, low starting current	0.3	0.7
D	High starting torque, high slip	0.5	0.5
Wound rotor	Performance varies with rotor resistance	0.5	0.5

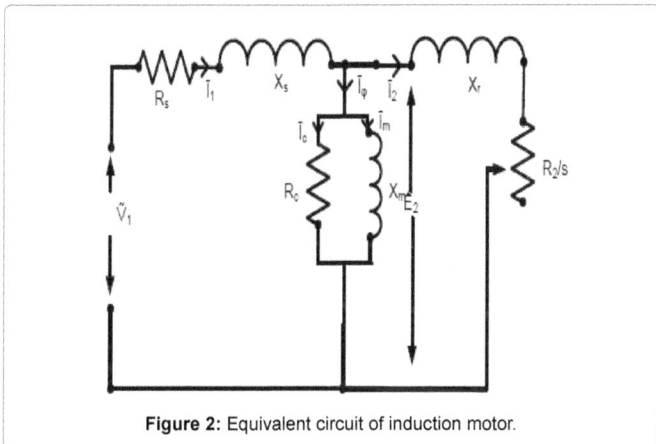

Figure 2: Equivalent circuit of induction motor.

Therefore, it is desirable and necessary to reduce the magnitude of stator current during starting. Several methods have been discovered to reduce the starting current. Some of popular starting methods amongst all are

1) Direct-on-line starting

2) Stator resistance starting

3) Autotransformer starting

4) Star-delta starting

5) Rotor resistance

In practice, any one of the first four methods is used for starting squirrel cage motors, depending upon the size of the motor. But slip ring motors are mostly started by rotor resistance starting.

Stator resistance starting

In stator resistance starting, external resistances are connected in series with each phase of stator winding during starting. This causes voltage drop across the resistances so that voltage available across

motor terminals is reduced and hence the starting current. The starting resistances are gradually cut out in steps (two or more steps) from the stator circuit as the motor picks up speed. When the motor attains rated speed, the resistances are completely cut out and full line voltage is applied to the rotor [5].

This method suffers from two drawbacks. First, the reduced voltage applied to the motor during the starting period lowers the starting torque and hence increases the accelerating time. Secondly, a lot of power is wasted in the starting resistances.

Autotransformer starting

This method also aims at connecting the induction motor to reduced supply at starting and then connecting it to the full voltage as the motor picks up sufficient speed. The tapping on the autotransformer is so set that when it is in the circuit, 65% to 80% of line voltage is applied to the motor. At the instant of starting, the change-over switch is thrown to "start" position. This puts the autotransformer in the circuit and thus reduced voltage is applied to the circuit. Consequently, starting current is limited to safe value. When the motor attains about 80% of normal speed, the changeover switch is thrown to "run" position. This takes out the autotransformer from the circuit and puts the motor to full line voltage [6]. Autotransformer starting has several advantages few of them are low power loss, low starting current and less radiated heat. For large machines (over 25 H.P.), this method of starting is often used. This method can be used for both star and delta connected motors.

Star-delta starting

The stator winding of the motor is designed for delta operation and is connected in star during the starting period. When the machine is up to speed, the connections are changed to delta.

At the instant of starting, the changeover switch is thrown to "Start" position which connects the stator windings in star. Therefore, each stator phase gets volts where V is the line voltage. This reduces the starting current. When the motor picks up speed, the changeover switch is thrown to "Run" position which connects the stator windings in delta. Now each stator phase gets full line voltage V.

The disadvantage of this method is: when there is Star-connection during starting, stator phase voltage is $\sqrt{}$ times of line voltage. Consequently, starting torque is times the value it would have with D-connection. This method becomes rare for a large motor due to large reduction in starting torque [7].

Importance of Starting Time

Large motor over current protection is normally set to trip prior to the locked-rotor withstand time (LRWT) provided by the motor manufacturer, after the calculated motor start time. The locked-rotor withstand time is determined by the motor designer based on the heating of the rotor parts for locked-rotor condition, where the motor continuously requires a large value of inrush current.

At the time of starting, an induction motor draws high values of current (motor is a constant impedance device during the starting condition), that are very close to the motor's locked rotor value and remains at this value for the time required to start the motor. This is the reason why the locked-rotor withstand time is used as an allowable time limit for starting the motor across the line, full voltage [1].

The capability to calculate motor starting time for large induction motor is important in order to evaluate the relative strength of the power system. Typically the motor designer may calculate the motor starting time at a couple of selected values such as 90% or 85% and then plot the results on the time versus current curves for the supplied motor.

System designers can use these starting times, but it is sometimes necessary to calculate the motor starting times using the results of a power system studies for the maximum voltage dip at critical power system conditions. This method can then be used, using the maximum voltage dip, to avoid application problems, set motor protective devices and perform coordination studies with other protective devices on the system.

Calculation using Acceleration Torque and Acceleration Time

Figure 3 shows the torque speed curve of induction motor. By this we can see that when we applied voltage on an induction motor it speed increase from zero with the torque. This is clear that at starting slip of induction motor is 1 and with the increase speed there is reduction in slip. Motor come to rated speed near its synchronous speed and the time taken to reach this point is called starting time. By this Figure 3 it is also clear that if gap between the motor torque and load torque is more motors will start very fast and the gap between these two torques is called acceleration torque.

Calculation for acceleration torque and acceleration time is given below with explanation.

Acceleration torque

A load can only be accelerated when the driving motor provides a greater torque than the load requires at the time. The difference of both of these is called the acceleration torque $\tau\alpha$. This acceleration torque will be equal to the multiplication of moment of inertia of motor and angular acceleration of motor. Mathematical formula to calculate this is given in equation 1. Further this angular acceleration can change in term of angular speed and time of starting. In this case the simplified assumption is made that during acceleration period load torque is constant. By calculating average load torques this assumption can be fulfill and replace the variable motor torque by a constant mean acceleration (Figure 4).

Acceleration torque

$$s T_{st} = \frac{\sum j * \Delta n}{9.55 * M_{\alpha}} \quad \tau_{\alpha} = \tau_{motor} - \tau_{load} \tag{1}$$

$$\tau_{load} = j * \alpha = \frac{W * j}{T_{st}} = \frac{j * 2\pi * n}{60 * T_{st}} = \frac{j * n}{9.55 * T_{st}} \tag{2}$$

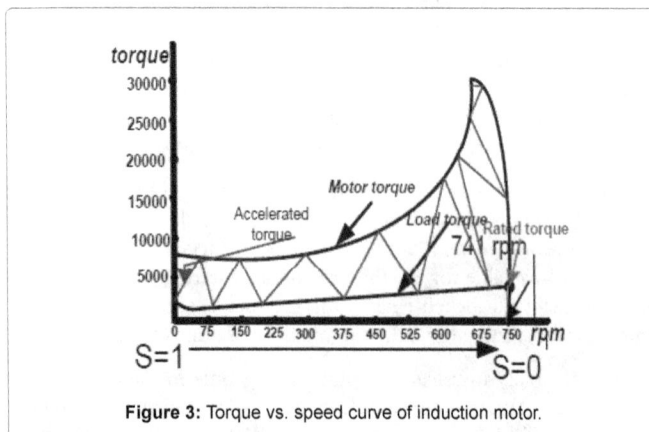

Figure 3: Torque vs. speed curve of induction motor.

Where

τ_{motor} = motor torque in Nm

τ_{load} = load torque in Nm

T_{st} = starting time in s

α = angular acceleration /s2

n= motor speed/min

ω = angular speed/s

τ_{α} = mean acceleration torque in Nm

j = moment of inertia in kg-m² reduced to the motor shaft

Acceleration Time

The acceleration time (starting time) tst of induction motor can be determined from equation 2 by doing some manipulations. As told above that time is time taken by induction motor to reach at rated speed and rated torque at rated voltage. Now if the mean acceleration torque Ta and moment of inertia is known we can easily calculate the value of starting time.

Starting time $T_{st} = \frac{j * n}{9.55 * \tau_{\alpha}}$ (3)

One example has been shown below to understand the concept carefully.

Example: Let an eight-pole SCIM motor with Nr=741 rpm, P=400 kW have an inertia of J=30 kgm² at no-load and have an average acceleration torque τ_{α} =1.5 τ_{motor} (at no load),

the maximum time specified by the manufacturer is 15 sec.

Calculate

a) The starting time at no-load?

b) The starting time together with driven equipment torque with reference to motor is 19 Kgm².

Solution:

a)

Rated torque of the motor

$\tau_{motor} = \frac{9.81 * Rated\,kW * 974}{Rated\,rpm}$

So

$\tau_{motor} = \frac{9.81 * 400\,kW * 974}{741} = 5157.86$ (4)

But at no load

$\tau_{\alpha} = 1.5 * \tau_{motor} - 0$

As $\tau_{load} = 0$

So $\tau_{\alpha} = 1.5 * 5157.86 = 7736.79 Nm$

Now by Equation (2)

$T_{st} = \frac{30 * 741}{9.55 * 7736.79} = 0.3008\,sec$

(b) Acceleration (starting) time with load Rated torque of the motor

So $\tau_{\alpha} = 1.5 * \tau_{motor} - \tau_{load}$

But $\tau_{load} = \tau_{motor}$ at load $\hspace{3cm}$ (5)

$\tau_\alpha = 1.5 * \tau_{motor} - \tau_{motor}$

So $\tau_\alpha = 0.5 * \tau_{motor}$

$\tau_\alpha = 0.5 * 5157.86 = 2578.93 Nm$

At load total moment of inertia in kg-m^2=30+19=49kg-m^2

Now by Equation (2)

$$T_{st} = \frac{49 * 741}{9.55 * 257.93} = 1.474 \text{ sec}$$

In this motor the acceleration time T_{st} is lesser than the maximum time specified by the manufacturer. Unloaded motors and motors with only little additional centrifugal masses reach their idle speed very quickly. But when large centrifugal masses are to be accelerated, starting times are generally quite high. This is called heavy starting, which is the case, for example, in centrifuges, ball mills, transport systems and large fans. These applications often require special motors and corresponding switch gears.

If the curve of the load torque τ_{load} is complex and the motor torque τ_{motor} is not constant, it is advantageous to divide the computation into individual zones as shown in Figure 5.

Now the acceleration times for the individual zones plus the average acceleration torques which take effect in the segment are computed and added up for each and every individual speed segments.

Now to calculate the starting time of non-constant acceleration torque is given by equation 6.

Acceleration time for non-constant torques

$$\text{in } s\, T_{st} = \frac{\sum j * \Delta n}{9.55 * M_\alpha} \hspace{2cm} (6)$$

Where the meaning of each is same as in equation 2, except $\sum j * \Delta n$.

This shows sum all zones with multiplication of moment of inertia and change in speed.

Simulation Result

Simulation study is carried out by using MATLAB 7.8.0 software with the same eight-pole SCIM motor with Nr=741 rpm, P=400 kW as considered in the previous example. Figure 5 shows that there is nearly six time of rated current during motor starting which is

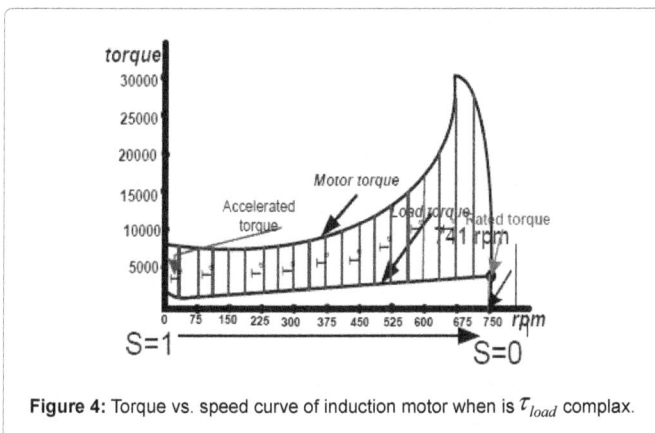

Figure 4: Torque vs. speed curve of induction motor when is τ_{load} complax.

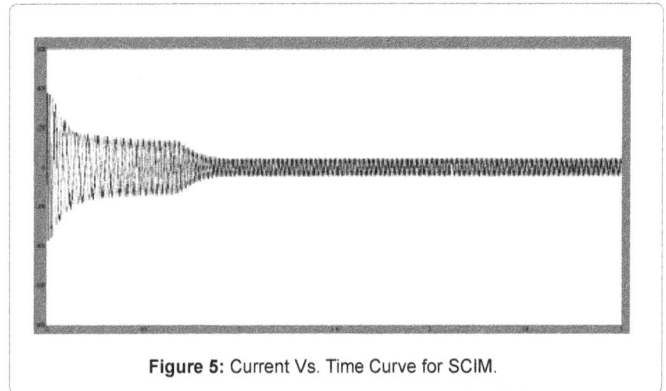

Figure 5: Current Vs. Time Curve for SCIM.

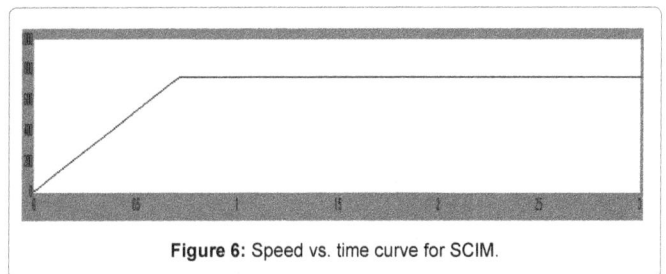

Figure 6: Speed vs. time curve for SCIM.

gradually decreasing and reaches to the rated current within 0.7 sec. This time is nearly equal to the time calculated by equation 5. Figure 6 shows the speed vs. time curve obtained by simulation it is clear by the graph that speed is increasing with time and coming to rated value in approximately 0.7 seconds, which is in line with the result calculated by the mathematical equation.

Conclusion

In this paper the importance of induction motor starting time has been described along-with different type of starting methods. A mathematical formula to calculate the starting time using acceleration torque, motor speed and moment of inertia of an induction motor is presented and the derived results from the calculation has been verified against the simulation study carried on MATLAB environment. Results of both the approaches are found to be in line with each other.

References

1. (1990) IEEE Recommended Practice for Industrial and Commercial Power System Analysis, IEEE Std 399-1990.

2. Venkataraman B, Godsey B, Premerlani W, Shulman E, Thakur M, et al. (2005) "Fundamentals of a Motor Thermal Model and its Applications in Motor Protection."

3. Abbas M, Majeed MA, Kassas M, Ahmad F (2011) "Motor starting study for a urea manufacturing plant," Power Engineering, Energy and Electrical Drives (POWERENG).

4. Hu H, Mao CX, Ji ML, Yu YX (2008) "The torque oscillation study in the motor soft starting process with discrete variable frequency method," Electrical Machines and Systems,ICEMS. International Conference 1686-1690.

5. Grewal GS, Pocsai S, Hakim M (1997)"Transient motor re-acceleration study in an integrated petrochemical facility," Industrial and Commercial Power Systems Technical Conference, 1997. Conference Record, Papers Presented at the 1997 Annual Meeting, IEEE 102-106.

6. "IEEE Houston Section Continuing Education on Demand Seminar", 2007.

7. IEEE Guide for Construction and Interpretation of Thermal Limit Curves for Squirrel-Cage Motors Over 500 Hp, (1987) IEEE.

35

Design and Implementation of High Gain, High Unity Gain Bandwidth, High Slew Rate and Low Power Dissipation CMOS Folded Cascode OTA for Wide Band Applications

Vadodaria MU*, Patel R and Popat J

Department of Electrical Engineering, Marwadi Education Foundation's, Rajkot-Morbi Road, Gujrat, India

Abstract

A novel differential input pair and single output OTA is designed in this paper. This FCOTA is designed by using current mirror and its enhanced gain parameter and also is reduce the power dissipation and increase the gain bandwidth. Simulation results are performed using Mentor Graphics software model for CMOS TSMC 180 nm process technology. The supply voltage is given 1.8 v. The designed FCOTA has different capacitive load and according its Gain, Unity Gain Bandwidth, Power Dissipation, Phase Margin and Slew Rate are measured. Gain, Unity Gain Bandwidth and Slew Rate are optimizing up to 97.37 db, 20.83 GHz and 3.5075 KV/µs respectively. The power dissipation reduces up to 8.31 µW. This FCOTA is designed for wide band application because of high gain and high bandwidth and low power dissipation.

Keywords: Folded cascode structure; Current mirror; Gain; Unity gain bandwidth; Power dissipation; Slew rate

Introduction

Amplifier is one of the basic and important circuits which have a wide range of applications in sever. This is accomplished by the continual integration of complex analog building blocks on a single chip [1]. Gain and Speed both are most important parameter in amplifier. Power dissipation is most important factor in any analog circuit. Designing high-performance base band analog circuits is still a hard task to reduced power consumption and increased frequency and gain [2]. Current tendency focus on some radio-software receivers which suppose a RF signal conversion after the antenna [3]. Thus, a very higher sampling frequency and resolution analog-to-digital converter design is required. Folded Cascode OTA is a solution of Telescopic Cascode OTA. There have some limitations of the voltage swing. To remove the drawback of telescopic OTA i.e. limited output swing and difficulty in shorting the input and output a Folded Cascode OTA is used. This design follows three stages such as (i) Input Pair, (II) Current mirror Cascode Stage and (III) Biasing Stage. Current mirror is a one type of approach to copy current at output side [4].

Design Approach

One way of increasing the impedance is to add some MOSFETs at the output side or in second stage to include for using an active load. MOSFETs are stacked on top of each other. The MOSFETs are called "cascode", and will increase the output impedance and thereby increase the gain [5]. Here in this cascode pair, there is given self biasing using current mirror. The current mirror is one of the main parts of the most analogue and mixed-signal integrated circuits to copy current such as OTAs [5,6].

Here the Figure 1 is consisting of M1 and M2 is called Current Mirror. In the current mirror channel length modulation is neglecting.

$$I_{ref} = \mu_n C_{ox} \left(\frac{W}{L} \right)_1 \left(V_{gs} - V_{th} \right)$$

$$I_{ref} = \mu_n C_{ox} \left(\frac{W}{L} \right)_2 \left(V_{gs} - V_{th} \right)$$

So, take ratio of the I_{out} and I_{ref} $\dfrac{g_{m1} = G_{m1}}{r_{08}}$

$$I_{out} = \frac{(W/L)2}{(W/L)1} I_{ref}$$

Constructing the bias circuit of an amplifier to provide the required bias current [7]. Transistor trans-conductance is the most important parameters in amplifier that must be stabilized. In general biasing of an amplifier is to ensure the proper operation of the circuit.

Design Analysis

This schematic design is consisting of three stages. First stage is NMOS input pair, second stage is cascode stage and third stage is biasing circuit. The input stage is designed by N-MOS input pair. Due to the greater mobility of NMOS device, PMOS input differential pair has a lower transconductance than carrier a NMOS pair. Thus, NMOS MOSFET has been chosen to ensure the largest gain required. For a folded OTA bandwidth performance is high. So here N-MOS input pair is applied for high transconductance. So as shown in Figure 2 M1 and M2 are NMOS input pair MOSFETs.

MOSFETs M1 and M2 are input pair of the OTA. M11 transistor work as a load register of NMOS pair. Second Stage is the Cascode OTA with current mirror or it's also called current mirror biasing.

Cascode stage design for higher isolation at input and output side, higher input impedance, higher output impedance, higher gain and higher bandwidth. So (M3, M4), (M5,M6) are NMOS pair and (M7,M8), (M9,M10) are PMOS pair are working as a current mirror self biased. Third stage is biasing circuit. The width of the M12 is low because if the width of M13 is higher the internal resistor value will

---

*Corresponding author: Vadodaria MU, Department of Electrical Engineering, Marwadi Education Foundation's, Group of Institutions, Rajkot-Morbi Road, Gujrat, India, E-mail: vadodaria.maulik91@gmail.com

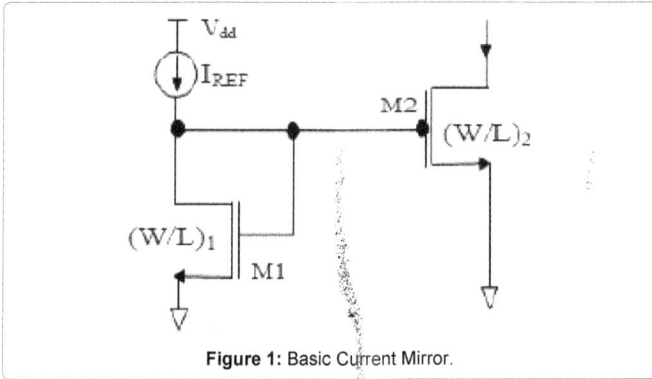

Figure 1: Basic Current Mirror.

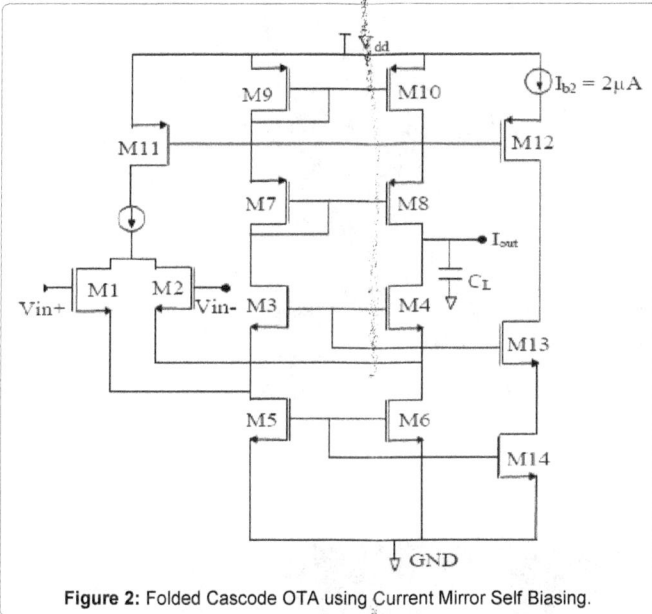

Figure 2: Folded Cascode OTA using Current Mirror Self Biasing.

be low and its gate terminal is directly connected to the M11, which is work as a PMOS load resistor.

$$g_{ml} = \mu_n C_{ox} \left(\frac{W}{L} \right)_1 \left(V_{gs} - V_{th} \right)$$

And internal resistance is defined by R_{out}

$$R_{out} = (g_{m8} r_{08} r_{10}) \| (g_{m4} r_{04} r_{08})$$

So, Voltage gain A_v

$$A_v = G_{ml} R_{out}$$

$$A_v = g_{ml} R_{out}, \text{ where, } g_{ml} = G_{ml}$$

Internal resistance of M8 is represent as r_{08} and it's defined as

$$r_{08} = \frac{1}{\lambda I_{d8}}$$

Where, λ is channel length modulation and I_{d8} is the drain current.

$$I_{d8} = \frac{1}{2} \mu_n C_{ox} \left(\frac{W}{L} \right)_8 \left(V_{gs} - V_{th} \right)^2$$

There are two poles exits and it is denoted by P1 and P2

$$P_1 = \frac{1}{R_{out} C_L}$$

$$P_2 = \frac{1}{R_{08} C_L}$$

Where $R_{08} = g_{m8} r_{08} \left(2r_{08} \| r_{10} \right)$

According to mathematical calculation, there are minor changes in mathematical calculation and theoretical calculation. Now, there is present waveform of FCOTA's schematic.

Performance Parameter

Table 1 presents different sizes of MOSFETs according to its operation.

Table 2 represents the summary of proposed performance parameter using 180 nm technology. There are different capacitive load such as 1 pF. 2 pF and 5.6 pF.

Simulation Result

All simulation results are measured by using 180 nm or 0.18_m technology using Mentor Grphics-Pyxis (Figures 3-6).

Conclusion

A novel design of Folded Cascode Operational Transconductance Amplifier (FCOTA) using current mirror as self biasing and biasing of the driving stage has been presented in this research work. The proposed design of FCOTA has been simulated and analyzed using Mentor Graphics tools. Improvement in performance parameters such as Gain up to 97.36 db, UGB up to 20.83 GHz , Reduction in Power Dissipation is 8.31 W and Slew Rate enhanced upto 3.5075 KV/µs has been achieved with respect to reference papers designs. So this FCOTA is designed for low power, high gain and high UGB. And it's used for wide band applications.

MOSFETs	Width (µm)	MOSFETs	Width (µm)
M1	14	M8	20
M2	14	M9	21
M3	5	M10	21
M4	5	M11	21
M5	5	M12	7
M6	5	M13	5
M7	14	M14	5

Note : All MOSFETs's length is fixed L = 0.18 µm.

Table 1: Size of MOSFETs.

Parameters	CL=1pF [1]		CL=2pF [2]		CL=5.6pF [3]	
	CFC	PFC	CFC	PFC	CFC	PFC
Gain (db)	58.37	97.37	59	97.37	68.48	97.36
Gain Margin(db)	N.A	14.89	N.A	20.86	N.A	29.77
Phase Margin (°)	62.5	38.32	86	55.66	26.3	79.51
UGB	315 MHz	20.83 GHz	86.1 MHz	12.7 GHz	247.1 MHz	5.25 GHz
Slew Rate	N.A	3.5075 KV/µs	N.A	3.5075 KV/µs	92.8V/S	3.5075 KV/µs
Power Dissipation	540 µW	8.31 µW	750 µW	8.31 µW	2.943 mW	8.31 µW

Note: CFC = Conventional Folded Cascode

Table 2: Performance parameter.

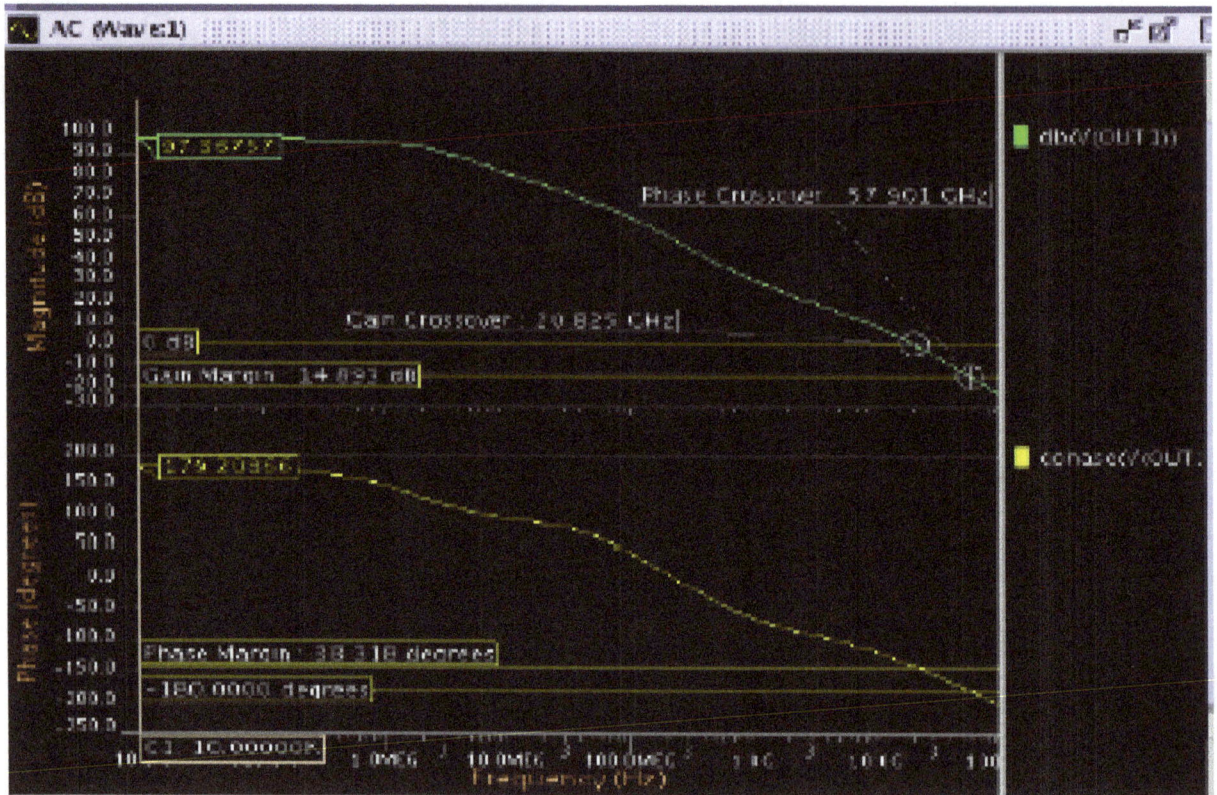

Figure 3: 1 pF Capacitive Load.

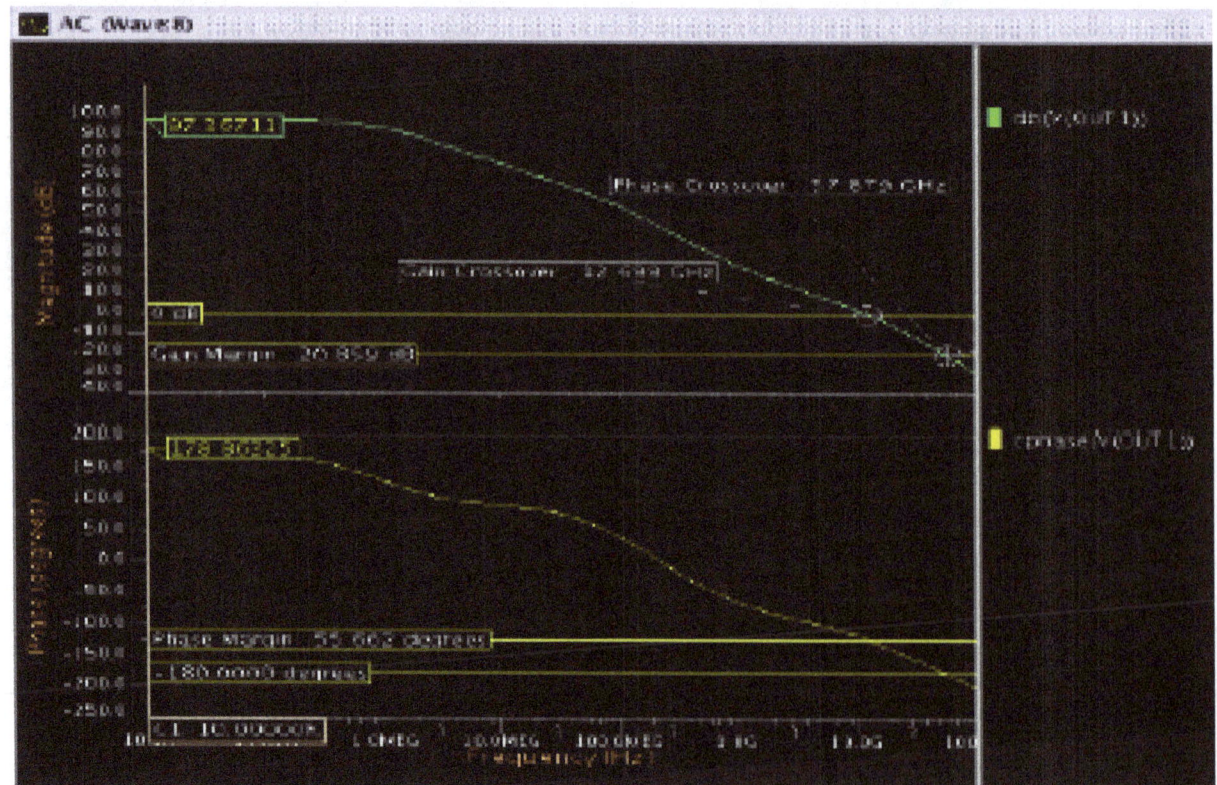

Figure 4: 2 pF Capacitive Load.

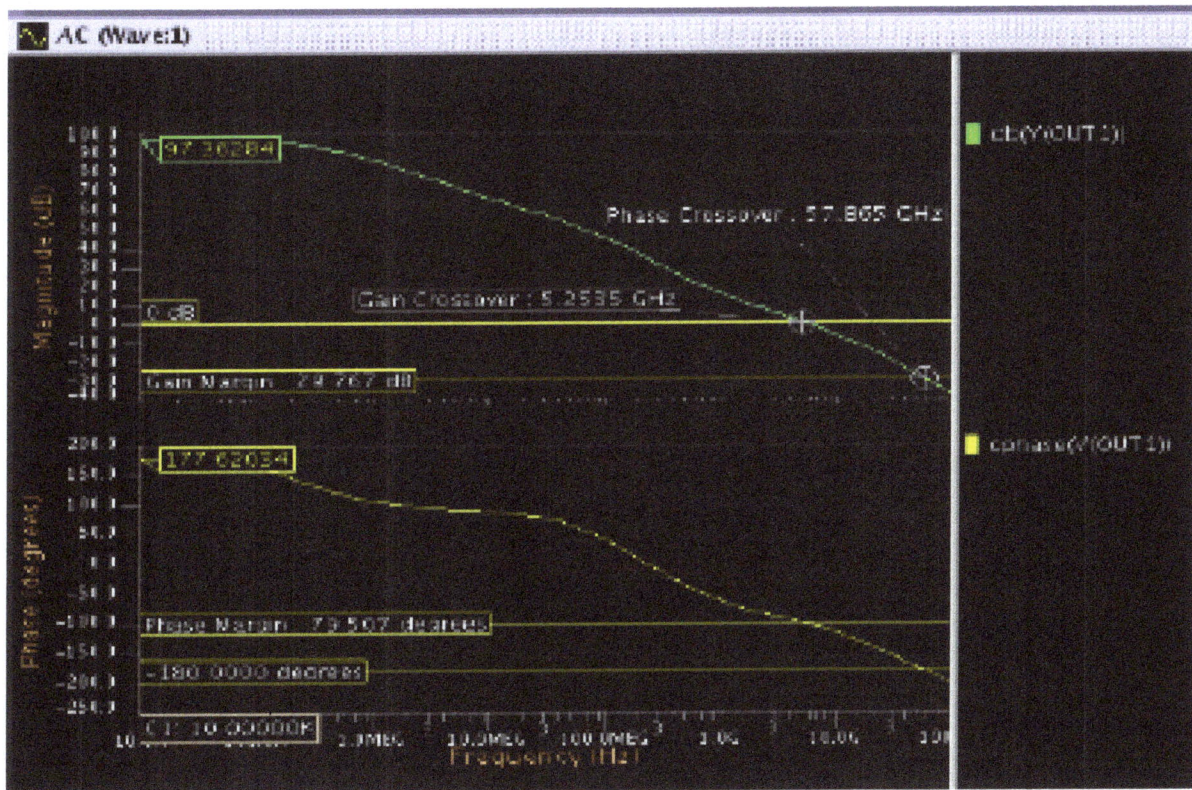

Figure 5: 5.6 pF Capacitive Load.

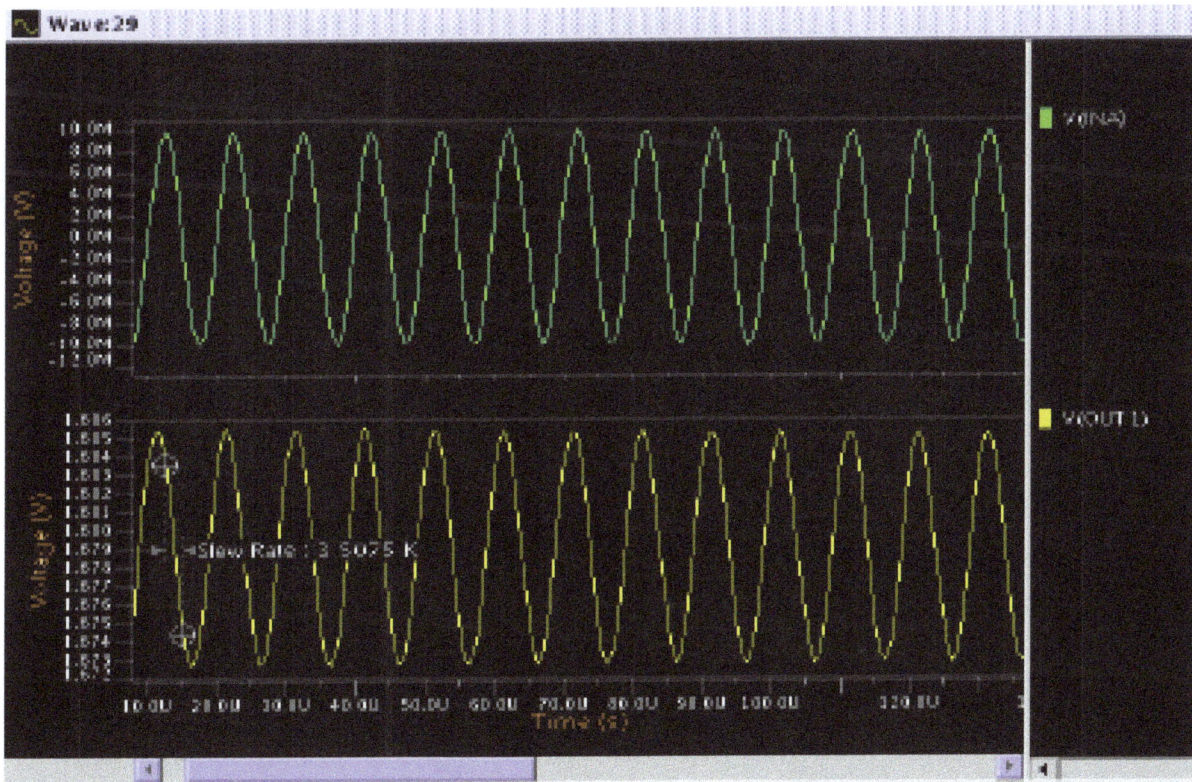

Figure 6: Slew Rate.

References

1. Biabanifard S, Akbari M, Asadi SM, Yagoub CE (2014) High performance folded cascode OTA using positive feedback and recycling structure, School of Electrical Engineering and Computer Science, University of Ottawa, Canada.

2. Patra P, Jha PK, Dutta A (2013) An Enhanced Recycling Folded Cascode OTA with a Positive Feedback, IEEE.

3. Daoud H, Salem BS, Zouari S, Loulou M (2006) Folded Cascode OTA Design for Wide Band Applications, Information technologies and electronic laboratory, LETI National engineers school of Sfax, IEEE, Tunisia.

4. Akbari M, Nazari M, Javid A (2014) Enhancing Phase Margin of OTA using Self Biasing Cascode, Laboratory, Shahid Beheshti University, GC, Iran.

5. Patra P, Jha PK, Dutta A (2013) An Enhanced Recycling Folded Cascode OTA with a Positive Feedback, IEEE.

6. Razavi B (2000) CMOS Circuit design, layout and simulation, professor of electrical engineering University of California, McGraw-Hill Publication, Los Angeles.

7. Daoud H, Salem BS, Zouari S, Loulou M (2006) Folded cascode OTA design for wide band applications, information technologies and electronic laboratory, LETI National engineers school of Sfax, IEEE, Tunisia.

Permissions

List of Contributors

Saifullah Khalid
Department of Electrical Engineering, IET Lucknow, India

Ranjit M
Department of Electrical and Electronics Engineering, VNR Vignana Jyothi Institute of Engineering and Technology, Hyderabad, India

Barakat S
Electrical Engineering Department, Beni Suief University, Beni Suief, Egypt

Eteiba MB and Wahba WI
Electrical Engineering Department, Fayoum University, Fayoum, Egypt

Alidou Koutou and Mamadou Lamine Doumbia
Department of Electrical Engineering, Université du Québec à Trois-Rivières, Trois-Rivières, Québec, G9A 5H7, Canada

Michael Chen, Chyong-Hua Chen, Yin-Chieh Lai, and Yi-Hsin Lin
Department of Photonics, National Chiao Tung University, Hsinchu, 30010, R O C, Taiwan

James Bruiners
Robert Gordon University, London, London UK

Mohammed M Shabat and Muin F Ubeid
Department of Physics, Faculty of Science, Islamic University of Gaza, Palestine

Yuriy Shckorbatov
Institute of Biology, Kharkiv National University, Ukraine

Heera K
Electrical and Electronics Engineering, Francis Xavier Engineering College, Tirunelveli, India

Ankit Rana
B. Tech. (ECE), Bharati Vidyapeeth's College of Engineering, A-4, Paschim Vihar, New Delhi, India

Abounada A, Brahmi A, Chbirik G and Ramzi M
Electrical Engineering Department, Faculty of Sciences and Technology, University of Soultan Moulay Sliman, Laboratory of Automatic, Energy Conversion and Microelectronic, P.B: 523 Mghila, Beni Mellal 2300, Morocco

Jayabharata S and Marimuthu CN
Department of ECE, Nandha Engineering College, Erode, Tamil Nadu, India

Saifullah Khalid
Department of Electrical Engineering, IET Lucknow, India

Langhals H
LMU University of Munich, Department of Chemistry, Butenandtstr-13, D-81377 Munich, Germany

Aiswarya S and Aravinth S
Embedded System and Technologies Velalar College of Engineering and Technology Erode, Tamilnadu, India

Manikandan SK
Department of Electrical and Electronics Engineering Velalar College of Engineering and Technology, Erode, Tamilnadu, India

Hugo J. Niggli
BioFoton AG, Rte. D'Essert 27, CH-1733 Treyvaux, Switzerland

Anand BB and Ramesh V
Department of Electrical Engineering, VIT University, Vellore, Tamil Nadu, India

Ahmed J Jameel
Department of Telecommunication Engineering, Ahlia University, Manama, Bahrain

Maryam M Shafiei
Department of Information Technology, Ahlia University, Manama, Bahrain

Ndombou GB
Laboratory of Electronics and Signal Processing, Faculty of Science, University of Dschang, Dschang, Cameroon

Marquié P
Laboratory of Electronic, Informatics and Image (LE2I), University of Burgundy, Dijon Cedex, France

Fomethe A, Yemélé D and Jeutho MG
Laboratoire de Mécanique et de Modélisation des Systèmes Physiques L2MSP, Faculté des Sciences, Université de Dschang, Cameroon

Kenmogne F
Laboratory of Modelling and Simulation in Engineering, Biomimetics and Prototype, Faculty of Science, University of Yaoundé I, Yaoundé, Cameroon

Hideki Omori, Shinya Ohara, Noriyuki Kimura and Toshimitsu Morizane
Department of Electrical and Electronics Systems Engineering, Osaka Institute of Technology, Japan

Masahito Tsuno
Nichicon Co.Ltd., Kyoto Japan

Mutuo Nakaoka
University of Malaya, Kuala Lumpur, Malaysia

Ibraheem Mohammed Khaleel
Department of Computer Communications Engineering, Al-Rafidain University College, Baghdad, Iraq

Prasad JS
LBR College of Engineering, Mylavaram, Andhra Pradesh, India

Obulesh YP
KL University, Vijayawada, Andhra Pradesh, India

Babu CS
JN Tech. University, Kakinada, Andhra Pradesh, India

Bengtsson L
Department of Physics, University of Gothenburg, SE-41296 Gothenburg, Sweden

Mahmood Taha Alkhayyat
Electrical Engineering Dept, Northern Technical University, Mosul, Iraq

Sinan Mahmood Bashi
Electrical Engineering Dept, Mosul University, Mosul, Iraq

Aladesote O Isaiah
Department of Computer Science, Federal Polytechnic, Ile – Oluji, Ondo State, Nigeria

Pandey S
Krishna Engineering College; Ghaziabad, Uttar Pradesh, India

Singh B
Ajay Kumar Garg Engineering College; Ghaziabad, Uttar Pradesh, India

Babita P
Department of Electrical Engineering, Jorhat Engineering College, Jorhat, Assam, India

Rayan Abdelazeem Habboub Suliman
Department of Electrical Engineering, Engineering College, Qassim University, Kingdom of Saudi Arabia

Khalid Hamid Bilal
Department of Communication, Faculty of engineering/university of science and technology, Khartoum, Sudan

Ibrahim Elemam
Department of Communication, El Neelain University, Khartoum, Sudan

Patel HN and Modi PR
Electronics and Communication, Engineering Department, A. D. Patel Institute of technology New Valla Bhai Vidyanagar, Gujarat, India

Nuray AT and Daraghma SM
Department of Electrical and Electronics Engineering, Anadolu University, Turkey

Langhals H
LMU University of Munich, Department of Chemistry, Butenandt Str-13, D-81377 Munich, Germany

Ranjit M, Sumanjali M and Ganesh B
VNR Vignana Jyothi Institute of Engineering and Technology, Hyderabad, India

El Houssain Ait Mansour
Laboratoire d'´etudes spatiales, et d'instrumentation en, astrophysique – LESIA, 18330 Nanc¸ay, France

Bruno Da Silva
St´ephane Bosse, Station de radioastronomie de Nanc¸ay, Route de souesmes, 18330 Nanc¸ay, France

Karl-Ludwig Klein
Observatoire de Paris, LESIA, LESIA - Bat 14, 92195 Meudon, France

Abhishek Garg and Arun Tomar
L&T Construction, Manapakkam, Chennai, Tamilnadu, India

Vadodaria MU, Patel R and Popat J
Department of Electrical Engineering, Marwadi Education Foundation's, Rajkot-Morbi Road, Gujrat, India

Index

www.ingramcontent.com/pod-product-compliance
Lightning Source LLC
Chambersburg PA
CBHW080536200326

41458CB00012B/4451